Rethinking Power Sector Reform in the Developing World

SUSTAINABLE INFRASTRUCTURE SERIES

Rethinking Power Sector Reform in the Developing World

Vivien Foster and Anshul Rana

1818 H Street NW, Washington, DC 20433
Telephone: 202-473-1000; Internet: www.worldbank.org

1 2 3 4 22 21 20 19

ISBN (paper): 978-1-4648-1442-6
ISBN (electronic): 978-1-4648-1443-3
DOI: 10/1596/978-1-4648-1442-6

Cover design: Bill Pragluski, Critical Stages, LLC.

Library of Congress Control Number: 2019915533.

Contents

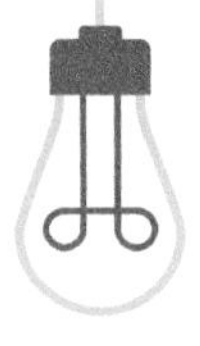

Maps

Tables

Foreword

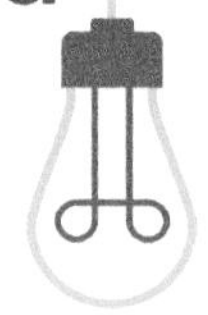

Thirty years ago, a new paradigm emerged to fundamentally alter power sector organization. It aimed to improve the financial and operational performance of utilities, ensure reliable power supply, and attract private sector participation and fair market forces while setting up the public sector to take on a regulatory role.

Yet, after almost three decades, only about a dozen developing countries have been able to fully implement the 1990s model. For many countries, the model simply did not fit the economic preconditions of their power sectors; for many others, the approach encountered political challenges in implementation. Many of those who have adopted the reforms have done so selectively, leading to a situation in which elements of market orientation coexist with a strong state presence—something the designers of the 1990s model did not anticipate.

Moreover, since the turn of the twenty-first century, the power sector has been overtaken by important policy shifts and momentous technological changes. In recent years, the world has embraced the Sustainable Development Goal on Energy (SDG7), which aims to achieve universal access to sustainable, affordable, and modern energy by 2030. We are also witnessing a swift global transition to low-carbon and renewable energy sources in line with the Paris Accord's commitment to fight climate change. Technological disruption is ushering new, decentralized actors into the sector and reshaping business models.

However, the various reform approaches based on the 1990s model alone will not be sufficient to deliver on global energy objectives. We also need complementary, targeted policies to reach the 840 million people who live without access to electricity today and to rapidly increase the share of clean energy in the global energy mix.

Rethinking Power Sector Reform in the Developing World comes at a crucial time. The world is changing—and so must the power sector. The principles that guided policy makers and stakeholders in the 1990s remain strong today. Financial sustainability and good institutional governance in the power sector are still just as critical, even as the scope of private sector participation is increasing and technological disruptions and the benefits of competition energize the sector.

It is only natural that the reform approaches will need to be updated to support these changes.

This report offers a fresh frame of reference shaped by context, driven by outcomes, and informed by alternatives. It has three clear messages for policy makers and industry practitioners. First, reform approaches must be shaped by the political and economic contexts of individual countries. Second, reform approaches should be tailored to achieve desired policy outcomes. Finally, multiple institutional pathways to achieve the desired outcomes must be possible. There is no one-size-fits-all framework, and the particular needs and challenges of low-income and fragile environments deserve special consideration.

Our hope is that this report can refresh the thinking on power sector reform in the developing world; help deliver electricity access to those who need it most; and ultimately result in a clean, green, and financially sustainable power sector.

Riccardo Puliti
Global Director, Energy and
Extractive Industries and
Regional Director, Infrastructure (Africa)
The World Bank

Acknowledgments

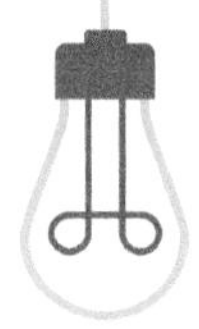

This report would not have been possible without generous funding provided by the Energy Sector Management Assistance Program (ESMAP) and the Public-Private Infrastructure Advisory Facility (PPIAF).

Valuable guidance was received from the report's peer reviewers: Pierre Audinet, Clive Harris, Dejan Ostojic, Sheoli Pargal, Mike Toman, and Maria Vagliasindi (World Bank) and Tooraj Jamasb (Durham University). The report also benefited from the comments and suggestions of Sudeshna Ghosh Banerjee (World Bank) throughout the process.

The wide scope of the report meant that a very large team of sector experts was involved in preparing each chapter, including many World Bank specialists complemented by external consultants.[1] These were based on the main themes of the project, namely, political economy of power sector reform, economic regulation of the power sector, wholesale power markets, utility reform, cost recovery, and disruptive technology.

- For the political economy theme, original research was done by a team led by Ashish Khanna and comprising Anton Eberhard, Catrina Godinho, Alan David Lee, Brian Levy, Zainab Usman, and Jonathan Walters.
- For the regulation theme, Katharina Gassner and Joseph Kapika led the team, which comprised Martin Rodriguez Pardina and Julieta Schiro of Macroconsulting Argentina, as well as Kagaba Paul Mukibi.
- For the power market theme, original research was conducted by a team led by Debabrata Chattopadhyay and comprising Hugh Rudnick and Constantin Velasquez from the University of Chile, as well as Tatyana Kramskaya and Martin Schroder.
- For the utility restructuring and governance theme, the team was led by Vivien Foster and comprised Joeri de Wit, Victor Loksha, and Anshul Rana, with advisory support from Pedro Antmann, Elvira Morella, Mariano Salto, and Pedro Sanchez.
- For the cost recovery theme, original research was conducted by a team led by Ani Balabanyan and comprising Joern Huenteler, Arthur Kochnakyan, Tu Chi Nguyen, Arun Singh, and Denzel Hankinson and Nicole Rosenthal of DH Infrastructure. The analysis is based on a methodologically consistent set of financial models prepared for 25 utilities across 14 countries and 3 Indian states. The financial analysis was led by Arthur Kochnakyan,

[1] All names listed are World Bank staff unless otherwise stated.

supported by a team of independent consultants, including Emiliano Lafalla, Adrian Ratner, Vazgen Sargsyan, and Martin Tarzyan. The chapter was drafted by Joern Huenteler and Tu Chi Nguyen.

- For the disruptive technologies theme, original research was done by Kelli Joseph and Jonathan Walters, both independent consultants, under the guidance of Gabriela Elizondo Azuela. The team also benefited from feedback provided by Pierre Audinet.

Econometric modeling for the report was done by Nisan Gorgulu of The George Washington University under the guidance of Jevgenjis Steinbucks.

Although the report's scope is global, it benefited immensely from detailed analysis undertaken for the 15 countries in the Rethinking Power Sector Reform Observatory. The unprecedented data collection and analysis would not have been possible without a combination of independent local consultants and World Bank colleagues from the respective country teams covering the energy sector.

- For Colombia, data collection and stakeholder interviews were done by Edison Giraldo, Jorge Giraldo, and Isaac Dyner Rezonzew, all of whom worked as independent consultants. Gabriela Elizondo Azuela, Janina Franco, Elvira Morella, and David Reinstein provided valuable support and guidance.
- For the Dominican Republic, George Reinoso was responsible for collecting data and conducting stakeholder interviews as an independent consultant. Valuable support and feedback were provided by Pedro Antmann and Elvira Morella.
- For the Arab Republic of Egypt, data collection and stakeholder interviews were conducted by Hafez el Salmawy as an independent consultant. Marwa Mostafa Khalil and Ashish Khanna provided valuable guidance and feedback.
- For India, Deloitte India was responsible for the data collection and conducting stakeholder interviews in the country. The team benefited from the expert guidance and support of Mani Khurana.
- For Kenya, data collection and stakeholder interviews were conducted by David Mwangi, an independent consultant, under the guidance of Anton Eberhard and Catrina Godinho and the University of Cape Town. Laurencia Karimi Njagi and Zubair K. M. Sadeque provided valuable support and feedback.
- For Morocco, Tayeb Amegroud, an independent consultant, was responsible for collecting data and conducting stakeholder interviews. The project benefited greatly from the guidance provided by Moez Cherif and Manaf Touati.
- For Pakistan, data collection and stakeholder interviews were carried out by Fariel Salahuddin, an independent consultant. The team benefited from the guidance of Anjum Ahmad, Defne Gencer, Rikard Liden, Anh Nguyen Pham, Mohammad Saqib, and Richard Spencer.
- For Peru, Eduardo Zolezzi, an independent consultant, was responsible for collecting data and conducting stakeholder interviews. Valuable guidance and support were provided by Janina Franco.
- For the Philippines, the data collection and stakeholder interviews were undertaken by Rauf Tan, an independent consultant. Yuriy Myroshnychenko provided valuable support and feedback, while Wali del Mundo acted as a reviewer.
- For Senegal, data collection and stakeholder interviews were conducted by Assane Diouf, an independent consultant, under the guidance of Anton Eberhard, Catrina Godinho, and Celine Paton of the University of Cape Town. Valuable support and guidance were provided by Manuel Berlengiero, Alioune Fall, Manuel Luengo, and Chris Trimble.

- For Tajikistan, data collection and stakeholder interviews were conducted by Jamshed Vazirov, an independent consultant. The team benefited from the guidance and support of Arthur Kochnakyan and Takhmina Mukhamedova.
- For Tanzania, Anastas Mbawala, an independent consultant, was responsible for collecting data and conducting stakeholder interviews under the guidance of Anton Eberhard and Catrina Godinho of the University of Cape Town. Joern Huenteler and Vadislav Vucetic provided valuable guidance.
- For Uganda, data collection and stakeholder interviews were conducted by Geofrey Bakkabulindi, independent consultant, under the guidance of Anton Eberhard and Catrina Godinho of the University of Cape Town. Raihan Elahi and Mbuso Gwafila provided valuable support and feedback.
- For Ukraine, Svetlana Golikova, independent consultant, was responsible for collecting data and conducting stakeholder interviews. The team benefited from the guidance and support of Fabrice Karl Bertholet and Dmytro Glazkov.
- For Vietnam, Nguyen Trinh Hoang Anh, Adam Fforde, Eric Groom, and Tran Dinh Long, all independent consultants, were responsible for data collection and conducting stakeholder interviews in the country. Franz Gerner and Tran Hong Ky provided valuable guidance, while Sebastian Eckardt and Madhu Raghunath acted as reviewers.

The report also benefited immensely from global databases including the World Bank Private Participation in Infrastructure Database (PPI 2018), the S&P Global Platts World Electric Power Plants Database, *Regulatory Indicators for Sustainable Energy (RISE)*, *Tracking SDG7: The Energy Progress Report*, and other databases managed by the International Energy Agency (IEA) and the U.S. Energy Information Administration (EIA).

The report was edited by Steven Kennedy. Final design was provided by Critical Stages, and typesetting was undertaken by Datapage.

About the Authors

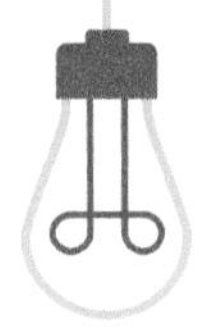

Vivien Foster is the Chief Economist for the Infrastructure Vice Presidency of the World Bank. Throughout her 20 years at the World Bank, she has played a variety of leadership roles and contributed to client dialogue, as well as advisory and lending engagements, in more than 30 countries across Africa, Asia, Europe, Latin America, and the Middle East. She has spearheaded several major policy research initiatives, including: *Water, Electricity, and the Poor* (2005), examining the distributional impact of utility subsidies; *Africa's Infrastructure* (2009), analyzing the continent's network infrastructure challenges; *Building Bridges* (2009), detailing China's growing role as infrastructure financier for Africa; *Tracking SDG7: The Energy Progress Report* (2013–18), a global dashboard for tracking progress toward the achievement of SDG7 goals for energy; and *Regulatory Indicators for Sustainable Energy (RISE)* (2016, 2018), monitoring worldwide adoption of good-practice policies to support sustainable energy. She is a graduate of Oxford University; she holds a master's degree from Stanford University and a PhD from University College London, both in economics.

Anshul Rana is a consultant in the Office of the Chief Economist for the Infrastructure Vice Presidency at the World Bank. He specializes in institutional reform in the power sector. He has also worked on the global dashboard to track progress toward SDG7 goals and on RISE, an Energy Sector Management Assistance Program (ESMAP) product used to monitor policy frameworks to support sustainable energy globally. Prior to joining the World Bank, he taught at the School of Advanced International Studies at Johns Hopkins University, focusing on the political economy of infrastructure development and energy policy in the developing world. He has also worked as a reporter for major newspapers and television networks in India and the United States. He holds a master's degree in international economics from the School of Advanced International Studies at Johns Hopkins University.

Background Papers

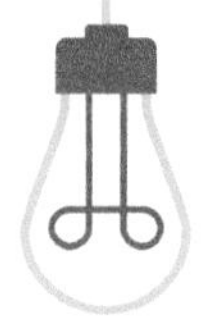

The report draws from 27 background papers that are being published on a rolling basis in the World Bank's Policy Research Working Paper series and can be accessed on the project website at http://www.esmap.org/rethinking_power_sector_reform.

The background papers include some that aim to paint a broad-brush picture of power sector reform across the developing world:

- Foster, V., S. Witte, S. Ghosh Banerjee, and A. Moreno. 2017. "Charting the Diffusion of Power Sector Reforms across the Developing World." Policy Research Working Paper No. 8235, World Bank, Washington, DC.
- Foster, V., and S. Witte. Forthcoming. "Evaluating Electricity Tariff Structure Design in the Developing World." Policy Research Working Paper, World Bank, Washington, DC.

Considerable emphasis was placed on conducting thorough reviews of existing literature covering the main themes of this report, namely, utility reform, political economy of power sector reform, economic regulation of the power sector, wholesale power markets, and cost recovery. These reviews are published in the following set of background papers.

- Bacon, R. W. 2018. "Taking Stock of the Impact of Power Utility Reform in Developing Countries: A Literature Review." Policy Research Working Paper No. 8460, World Bank, Washington, DC.
- Huenteler, J., I. Dobozi, A. Balabanyan, and S. Ghosh Banerjee. 2017. "Cost Recovery and Financial Viability of the Power Sector in Developing Countries: A Literature Review." Policy Research Working Paper No. 8287, World Bank, Washington, DC.
- Lee, A. D., and Z. Usman. 2018. "Taking Stock of the Political Economy of Power Sector Reform in Developing Countries: A Literature Review." Policy Research Working Paper No. 8518, World Bank, Washington, DC.
- Pardina, M. R., and J. Schiro. 2018. "Taking Stock of Economic Regulation of Power Utilities in the Developing World: A Literature Review." Policy Research Working Paper No. 8461, World Bank, Washington, DC.
- Rudnick, H., and C. Velasquez. 2018. "Taking Stock of Wholesale Power Markets in Developing Countries: A Literature Review." Policy Research Working Paper No. 8519, World Bank, Washington, DC.

Detailed power market case studies were prepared for four countries from the Rethinking Power Sector Reform Observatory that were most advanced in the creation of wholesale power markets, namely, Colombia, India, Peru, and the Philippines.

- Rudnick, H., and C. Velasquez. 2019. "Learning from Developing Country Power Market Experiences: The Case of the Philippines." Policy Research Working Paper No. 8721, World Bank, Washington, DC.
- Rudnick, H., and C. Velasquez. 2019. "Learning from Developing Country Power Market Experiences: The Case of Colombia." Policy Research Working Paper No. 8771, World Bank, Washington, DC.
- Rudnick, H., and C. Velasquez. 2019. "Learning from Developing Country Power Market Experiences: The Case of Peru." Policy Research Working Paper No. 8772, World Bank, Washington, DC.
- Rudnick, H., and C. Velasquez. Forthcoming. "Learning from Developing Country Power Market Experiences: The Case of India." Policy Research Working Paper, World Bank, Washington, DC.

The background papers also include detailed country case studies that provide a narrative of the reform dynamics for each country in the Rethinking Observatory and evaluate the impact of reforms on key dimensions of sector performance.

- Bacon, R. W. 2019. "Learning from Power Sector Reform: The Case of Pakistan." Policy Research Working Paper No. 8842, World Bank, Washington, DC.
- Bacon, R. W. 2019. "Learning from Power Sector Reform: The Case of the Philippines." Policy Research Working Paper No. 8853, World Bank, Washington, DC.
- Bacon, R. W. Forthcoming. "Learning from Power Sector Reform: The Case of Ukraine." Policy Research Working Paper, World Bank, Washington, DC.
- Godinho, C., and A. Eberhard. 2019. "Learning from Power Sector Reform: The Case of Kenya." Policy Research Working Paper No. 8819, World Bank, Washington, DC.
- Godinho, C., and A. Eberhard. 2019. "Learning from Power Sector Reform: The Case of Uganda." Policy Research Working Paper No. 8820, World Bank, Washington, DC.
- Godinho, C., and A. Eberhard. Forthcoming. "Learning from Power Sector Reform: The Case of Tanzania." Policy Research Working Paper, World Bank, Washington, DC.
- Khurana, M. Forthcoming. "Learning from Power Sector Reform: The Case of Andhra Pradesh." Policy Research Working Paper, World Bank, Washington, DC.
- Khurana, M. Forthcoming. "Learning from Power Sector Reform: The Case of Rajasthan." Policy Research Working Paper, World Bank, Washington, DC.
- Khurana, M. Forthcoming. "Learning from Power Sector Reform: The Case of Odisha." Policy Research Working Paper, World Bank, Washington, DC.
- Lee, A. D., and F. Gerner. 2019. "Learning from Power Sector Reform: The Case of Vietnam." Policy Research Working Paper, World Bank, Washington, DC.
- Paton, C., C. Godinho, and A. Eberhard. Forthcoming. "Learning from Power Sector Reform: The Case of Senegal." Policy Research Working Paper, World Bank, Washington, DC.
- Rana, A. 2019. "Learning from Power Sector Reform: The Case of the Dominican Republic." Policy Research Working Paper, World Bank, Washington, DC.
- Rana, A., and A. Khanna. 2019. "Learning from Power Sector Reform: The Case of the Arab Republic of Egypt." Policy Research Working Paper, World Bank, Washington, DC.

- Rudnick, H., and C. Velasquez. Forthcoming. "Learning from Power Sector Reform: The Case of Peru." Policy Research Working Paper, World Bank, Washington, DC.
- Rudnick, H., and C. Velasquez. Forthcoming. "Learning from Power Sector Reform: The Case of Colombia." Policy Research Working Paper, World Bank, Washington, DC.
- Usman, Z., and Amegroud, T. 2019. "Lessons from Power Sector Reforms: The Case of Morocco." Policy Research Working Paper No. 8969, World Bank, Washington, DC.

Abbreviations

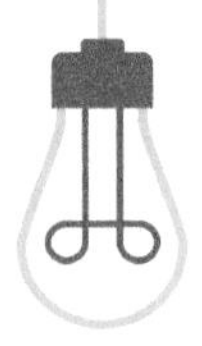

BOO	build, own, operate
BOOT	build, own, operate, transfer
BOT	build, operate, transfer
DER	distributed energy resources
EE	energy efficiency
EIA	Energy Information Administration
ESMAP	Energy Sector Management Assistance Program
EVN	(national utility of Vietnam)
FY	fiscal year
GNI	gross national income
GW	gigawatt
HHI	Herfindahl-Hirschman Index
IEA	International Energy Agency
IPP	independent power producers
ISO	independent system operator
IT	information technology
km	kilometer
KPI	key performance indicator
kWh	kilowatt-hour
LACE	levelized avoided cost of energy
LCOE	levelized cost of energy
MW	megawatt
MWh	megawatt-hour
OECD	Organisation for Economic Co-operation and Development
PJM	Pennsylvania–New Jersey–Maryland

PPA	power purchase agreement
PPI	Private Participation in Infrastructure database
PPIAF	Public-Private Infrastructure Advisory Facility
PSP	private sector participation
PSRI	Power Sector Reform Index
QFD	quasi-fiscal deficit
RE	renewable energy
REV	Reforming the Energy Vision (New York)
RIIO	revenue using incentives to deliver innovation and outputs (United Kingdom)
RISE	Regulatory Indicators for Sustainable Energy
SAIDI	System Average Interruption Duration Index
SAIFI	System Average Interruption Frequency Index
SDG7	Sustainable Development Goal 7
SO	system operator
SOE	state-owned enterprise
T&D	transmission and distribution
TSO	transmission system operator

Key Messages

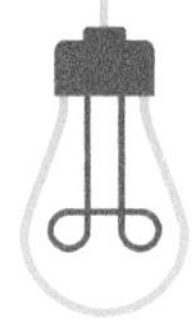

During the 1990s, a new paradigm for power sector reform was put forward that emphasized the restructuring of utilities, the creation of regulators, the participation of the private sector, and the establishment of competitive power markets. Twenty-five years later, only a handful of developing countries have fully implemented these Washington Consensus policies. Across the developing world, reforms were adopted rather selectively, resulting in a hybrid model in which elements of market orientation coexist with continued state dominance of the sector.

This book aims to revisit and refresh the thinking on power sector reform approaches for developing countries. The approach relies heavily on evidence from the past, drawing both on broad global trends and deep case material from 15 developing countries. It is also forward looking, considering the implications of new social and environmental policy goals, as well as emerging technological disruptions.

A nuanced picture emerges. Regulation has been widely adopted, but practice often falls well short of theory, and cost recovery remains an elusive goal. The private sector has financed a substantial expansion of generation capacity. Yet, its contribution to power distribution has been much more limited, and its performance on efficiency can sometimes be matched by well-governed public utilities. Restructuring and liberalization have been beneficial in a handful of larger middle-income nations but have proved too complex for most countries to implement.

Based on these findings, the report points to three major policy implications.

- ***Context dependence.*** First, reform efforts need to be shaped by both the political and economic context of the host country. The 1990s reform model was most successful in countries that had reached certain minimum conditions of power sector development and offered a supportive political environment. When these same reforms were adopted in more challenging environments, the risk of policy reversal was high, while successful outcomes were by no means guaranteed. The 1990s approach to power sector reform is more compatible with political systems that are based on a market-oriented ideology and contestable power structures. Economic preconditions include a relatively large power system at a high level of electrification with good operational and financial data and a well-functioning framework of tariff regulation.

  The report proposes a two-track approach, with countries in more challenging environments focusing on governance reforms and and the achievement of financial viability, while more ambitious structural reforms are deferred until systems are at a more mature stage of development.

- ***Outcome orientation***. Second, reform efforts should be driven and tailored to desired policy outcomes and less preoccupied with following a predetermined process. Since the 1990s, countries' policy objectives for the power sector have widened beyond security of supply and fiscal sustainability to encompass important social and environmental goals, such as universal access and power sector decarbonization. The evidence indicates that Washington Consensus reforms alone will not deliver on twenty-first-century policy objectives; they need to be complemented by more targeted policy measures.
- ***Pluralist approaches***. Third, countries found alternative institutional pathways to achieving good power sector outcomes. Among the best-performing power sectors in the developing world are some that decisively implemented the 1990s reform model and others that retained a dominant and competent state-owned utility, guided by strong policy objectives, and combined this with a more gradualist and targeted role for the private sector. This evidence makes a case for greater pluralism of approaches going forward.

Overview: Key Findings and Policy Implications

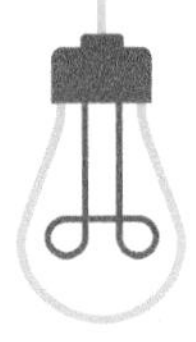

INTRODUCTION

During the 1990s, a new paradigm for power sector organization grew out of the wider "Washington Consensus" on development and was spearheaded by multilateral institutions. The new paradigm came on the heels of growing dissatisfaction with state-owned utilities (Bacon and Besant-Jones 2001). These vertically integrated monopolies had successfully supported the rollout of national infrastructure networks in many countries during the 1960s–80s but had begun to show limitations in the form of inefficient operations, burgeoning subsidies, and financing constraints. The 1990s power sector reform model comprised a package of four structural reforms:

- Regulation (through the creation of an autonomous regulatory entity)
- Restructuring (entailing corporatization and full vertical and horizontal unbundling of the utility)
- Private sector participation (particularly in generation and distribution)
- Competition (ultimately in the form of a wholesale power market).

The 1990s reform model was based on the idea that reforms would lead to beneficial behavior change among the key sector actors, resulting in improved sector performance. Behavior changes when private management is introduced. Private management reorients enterprises from bureaucratic and political incentives to profit-seeking, cost-control, and customer orientation. Market pressure or regulatory incentives would discipline any potential for private management abuses. The private sector and the regulator would prevent day-to-day political interference. The combination of stronger commercial incentives, competitive pressures, and regulatory oversight was expected to improve the efficiency and cost recovery of power utilities. The resulting decline in state subsidies and increase in financial viability would make possible the major investment programs needed to achieve security of supply in fast-growing power systems (World Bank 1993). This thought process is presented as a theory of change underpinning the 1990s reform model in figure O.1. The theory of change is used as a conceptual framework for evaluating the model's efficacy in this study. By 2015, the adoption of Sustainable Development Goal 7 (SDG7) and the Paris Climate Accord had broadened the policy objectives for the power sector, bringing a new focus on electrification and decarbonization, goals that had not been envisaged in the 1990s.

The aim of this study is to revisit, refresh, and update thinking on power sector reform in developing countries in

FIGURE O.1 **The 1990s model was based on an underlying theory of change**

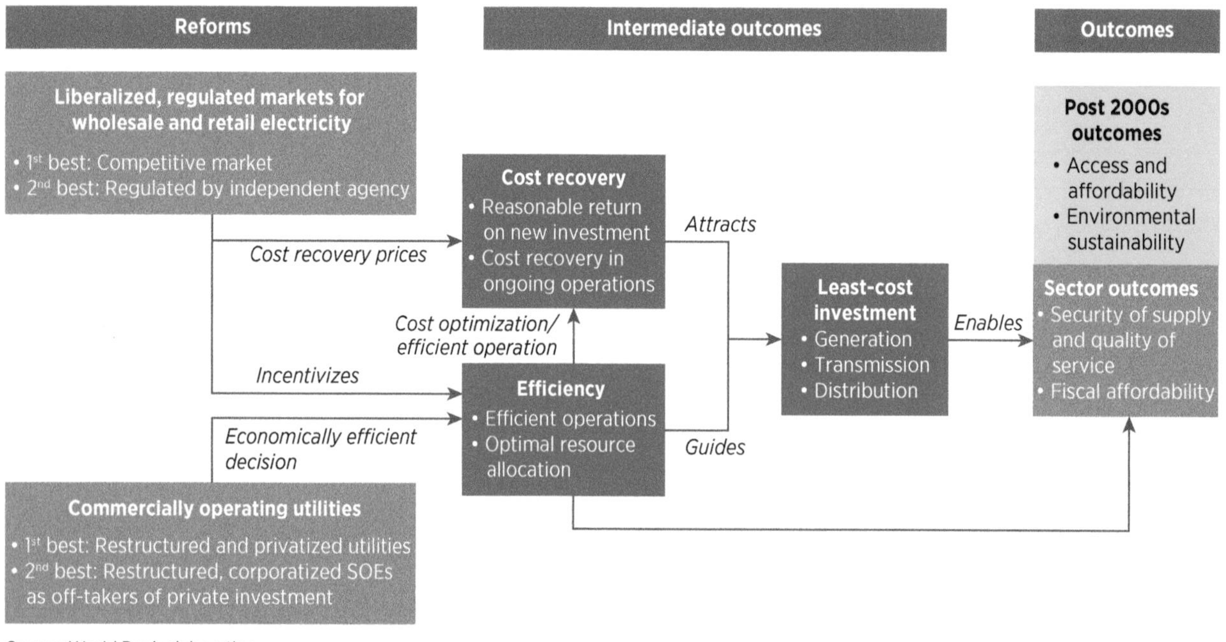

Source: World Bank elaboration.
Note: SOEs = state-owned enterprises.

the light of historical evidence and future trends. The prescriptions of the 1990s reform model were primarily derived from economic theory and principles. By the early 2000s, it had become clear that the model was not universally applicable in practice (Besant-Jones 2006). We now have a quarter-century of empirical evidence against which to evaluate the approach. The case for such an evaluation hinges both on the practical difficulties encountered with the application of the model in the developing world, as well as on the significant changes in policy objectives. At the same time, the emergence of disruptive technologies raises questions about how the recommendations of the 1990s model may need to be adapted going forward.

Relying on a rich new evidence base, the study looks back over 25 years of experience with power sector reform across the developing world. The approach is heavily evidence-based, drawing on reform efforts and performance in 88 developing countries, complemented by a Power Sector Reform Observatory that provides deep-dive studies of 15 countries.[1] Countries are not judged for the reforms they have undertaken but rather for the results they have delivered. Sector outcomes are evaluated along multiple dimensions, including traditional objectives such as security of supply, as well as the new policy agenda focusing on electrification and decarbonization.

At the same time, the study looks ahead to the technological disruptions sweeping the power sector, developments that are challenging conventional wisdom about sector organization and structure. Traditionally, power systems have developed around centralized infrastructure designed to reap economies of scale and achieve simultaneous balancing of supply and demand through the one-way flow of power to passive consumers. However, the current wave of innovations—including decentralized renewable energy, battery storage, and digitalization—empowers consumers and other decentralized actors to participate in the production of electricity and in so-called demand-response services,[2] generating reverse flows along power networks and introducing the possibility of trade at the retail level. Moreover, as large-scale battery storage becomes increasingly flexible and cost-effective, the need for power systems to simultaneously match supply and demand will recede.

The purpose of this overview is to summarize the lessons from the study and reflect upon their implications for future practice. Ten key findings are followed by the policy implications of those findings. The comprehensive analysis contained in the main report begins with a survey of the uptake of the 1990s power sector model by developing countries, considering both the economic and political drivers of reform. Attention then turns to the implementation of each of the fundamental building blocks of the reform model: sector restructuring and governance; private sector participation; regulation; and market liberalization. Thereafter, reform measures are evaluated in terms of their impacts both on intermediate sector outcomes (such as efficiency and cost recovery) and on final sector outcomes (such as security of supply, access and affordability, environmental sustainability).

The study suggests that future reforms should be shaped by context, driven by outcomes, and informed by alternatives. The 1990s reform model is sometimes misconstrued as a universally applicable policy prescription. However, the findings reported here suggest instead that the 1990s model contains valuable insights that can support improvements in efficiency, cost recovery, and security of supply when deployed in the right circumstances and for the right reasons. However, economic and political preconditions are found to be important determinants of the success of reforms; these deserve closer consideration when determining the appropriate reform path for each country. Reform choices also need to be guided by desired sector outcomes, notably, with respect to decarbonization and universal access objectives. Fortunately, good sector outcomes can be achieved in a variety of institutional settings, as the experience of developing countries around the world has shown. Those settings will be tested, as new business models emerge in response to the technological disruptions that are reshaping the economic logic of the sector.

KEY FINDINGS

This section summarizes the most relevant and interesting results of the study in the form of 10 key findings.

- *Finding #1.* Uptake of power sector reform in the developing world did not evolve according to the textbook model.
- *Finding #2.* Power sector reforms were more likely to gain traction if they were consistent with the country's political system and ideology and led by champions enjoying broad stakeholder support.
- *Finding #3.* The private sector made an important contribution to expanding power generation capacity in the developing world, albeit with significant challenges along the way.
- *Finding #4.* Wholesale power markets helped improve efficiency in the minority of countries that was ready for them; many others found themselves stuck in transition.
- *Finding #5.* Good corporate practices, particularly with respect to human resources and financial discipline, were associated with better utility performance; these were more prevalent among privatized utilities.
- *Finding #6.* Private sector participation in power transmission and distribution delivered good outcomes in favorable settings; elsewhere, it was susceptible to reversal.
- *Finding #7.* Regulatory frameworks have been widely adopted, but implementation

has often fallen far short of design, particularly when utilities remained under state ownership.

- *Finding #8.* Cost recovery has proved remarkably difficult to achieve and sustain; the limited progress made owes more to efficiency improvements than to tariff hikes.
- *Finding #9.* The outcomes of power sector reform were heavily influenced by the starting conditions in each country.
- *Finding #10.* Good sector outcomes were achieved by countries adopting a variety of different institutional patterns of organization for the sector.

Finding #1: Uptake of power sector reform in the developing world did not evolve according to the textbook model

The diffusion of power sector reform in the developing world was strongly affected by contextual factors. The 1990s power sector reform model spread rapidly across both the developed and developing worlds. A quarter-century later, however, the patterns of adoption are quite different. Organisation for Economic Co-operation and Development (OECD) countries have adopted (on average) close to 80 percent of the 1990s policy prescriptions, although with some notable exceptions. The degree of adoption in the developing world is much lower at under 40 percent. The level of uptake differs systematically according to the geographical, economic, and technical characteristics of countries (map O.1).[3] Specifically, reform adoption is twice as high in Latin America relative to the Middle East, in middle-income relative to low-income groups, and in countries with larger power systems relative to smaller ones. Moreover, the momentum of reform slowed markedly over time, with uptake more limited during the decade from 2005–15 than from 1995–2005.

MAP O.1 Power sector reform spread unevenly across the developing world

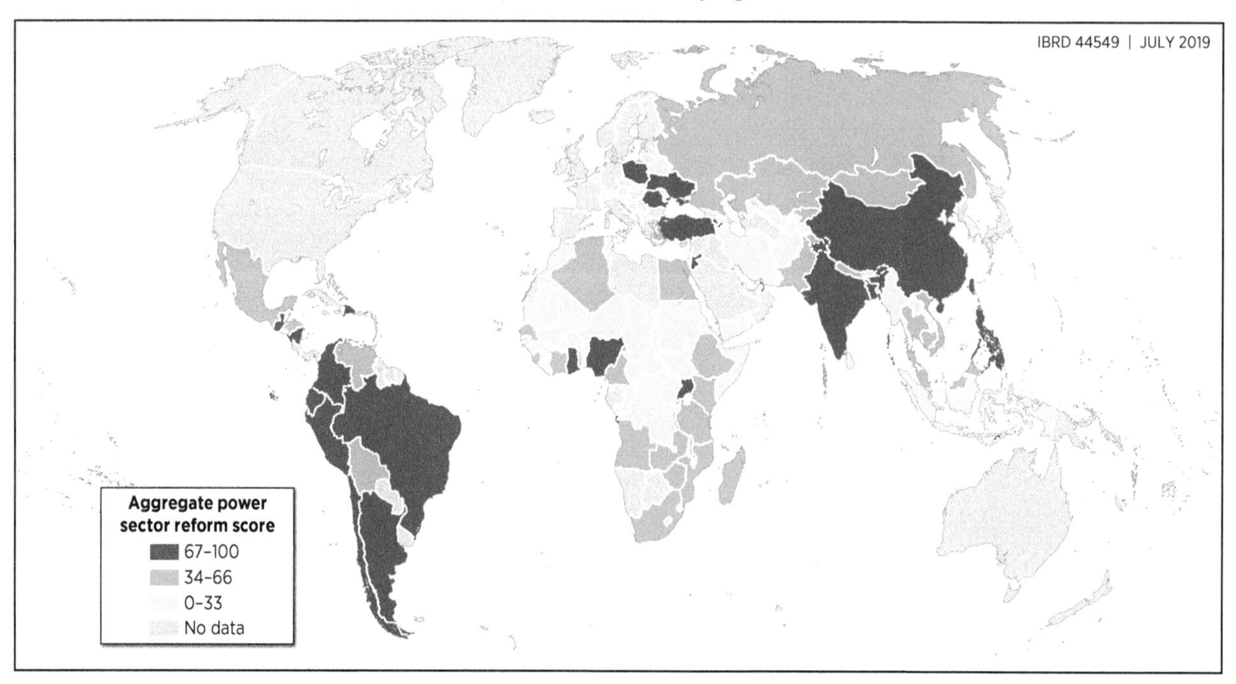

Sources: World Bank elaboration based on Rethinking Power Sector Reform utility database 2015; Regulatory Indicators for Sustainable Energy 2016.
Note: PSRI score based on existing legislation (as of 2015), which may be different from practice.

As a result, reform implementation diverged from the theoretical paradigm. Overall, barely a dozen developing countries were able to implement the 1990s model in its entirety. Instead, most are stuck at an intermediate stage of implementation, sometimes referred to as the "hybrid model" (Eberhard and Gratwick 2008). A further quarter of developing countries—including many small, low-income, and fragile states—have barely begun to reform their power systems. Underlying this partial implementation has been a tendency to cherry-pick components of the 1990s model that were easier to implement, while leaving others aside. Creation of a regulatory entity and private sector participation in generation through independent power projects (IPPs) were, by far, the two most popular reforms, adopted by more than 70 percent of developing countries; the uptake of other reforms was much lower. This à la carte approach to reform does not sit well with the original conception of the 1990s model as a coherent package of mutually supportive reform measures. It meant that countries ended up with contradictory reform combinations, such as private sector participation in distribution without a regulator—or, more frequently, the other way around.

Finding #2: Power sector reforms were more likely to gain traction if they were consistent with the country's political system and ideology and led by champions enjoying broad stakeholder support

The 1990s reform model drew heavily on economic first principles, with no explicit attention to the political dynamics of the reform process. Yet, the reality is that the power sector is highly politicized across much of the developing world. Power utilities—with their significant employment rolls and contracting volumes, as well as their ability to direct valued electricity services to different communities—are a natural focus for patronage politics. Moreover, the cost and quality of electricity supply has the potential to become an electoral issue that can mobilize public unrest and topple governments.

Power sector reforms almost always take place in the context of a crisis and often as part of a wider national transformation process. There are few examples of countries that reformed in the absence of a crisis or of countries that failed to reform when beset by crisis. The triggering events sometimes originated within the power sector, such as a drought or oil price shock or a situation of unsustainable utility debt. However, in many cases, the power sector was implicated in a wider national crisis, either linked to fiscal stabilization (such as tariff reforms in the Arab Republic of Egypt) or to socioeconomic transition (such as privatization in Ukraine). This finding underscores the fact that power sector reform does not take place in a vacuum; it needs to be understood in terms of the wider political and economic context.

The trajectory of reform varies substantially across countries, with reform announcements providing no guarantee of sustained implementation. The reform process typically begins with the public announcement of a reform program. Some countries then move rapidly toward implementing the full suite of reforms announced, as in the case of Peru (figure O.2a). In other cases, reforms rapidly lose momentum, with delivery falling well short of original aspirations and even being susceptible to reversal over time, as in the case of Senegal (figure O.2b). Overall, the gap between reform announcements and implementation can be quite considerable (figure O.3).

Reform trajectories reflect the political dynamics around the power sector in each country, as well as the strategy adopted for reform implementation. Although reforms are announced by countries

FIGURE O.2 **The trajectory of power sector reform followed different paths across countries**

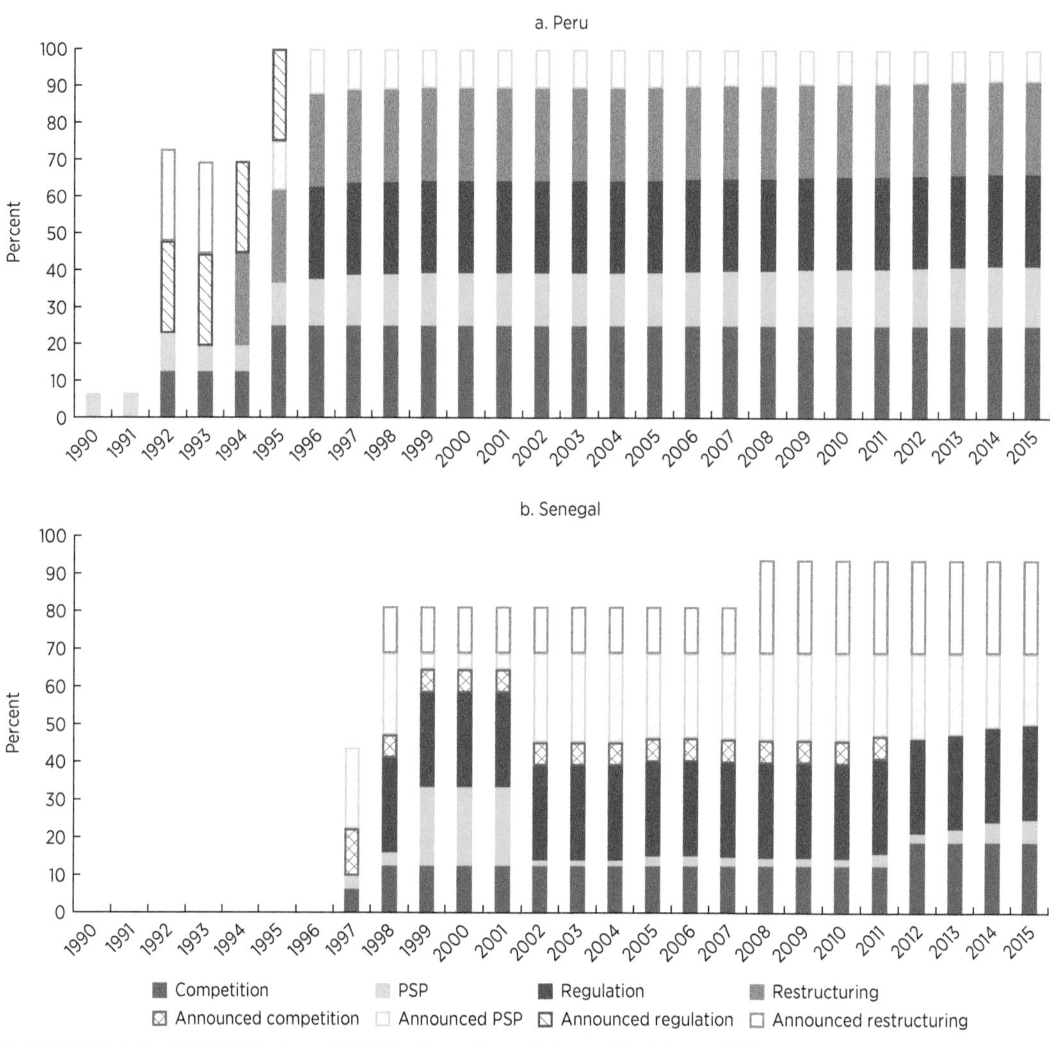

Source: World Bank elaboration based on Rethinking Power Sector Reform utility database 2015.
Note: PSP = private sector participation.

across the ideological spectrum, evidence suggests that those with a stronger market orientation are more likely to make meaningful progress with implementation. Similarly, reforms tend to proceed further in countries that have contestable or multipolar political systems, as opposed to those where power is more centralized. This is consistent with the observation that the reform process typically involves the delegation and decentralization of power by breaking up national monopolies, delegating responsibility to regulators and private operators, and allowing new entry to competitive markets. The strategy for reform implementation at the sector level is also important. Countries that can mobilize a strong reform champion, ideally supported by a stable and competent bureaucracy, generally go further with sector reform. However, unless wider stakeholder alignment is achieved through outreach efforts and ultimately legislative support, reforms may prove

FIGURE O.3 **For some countries, the gap between reform announcement and implementation has been considerable**

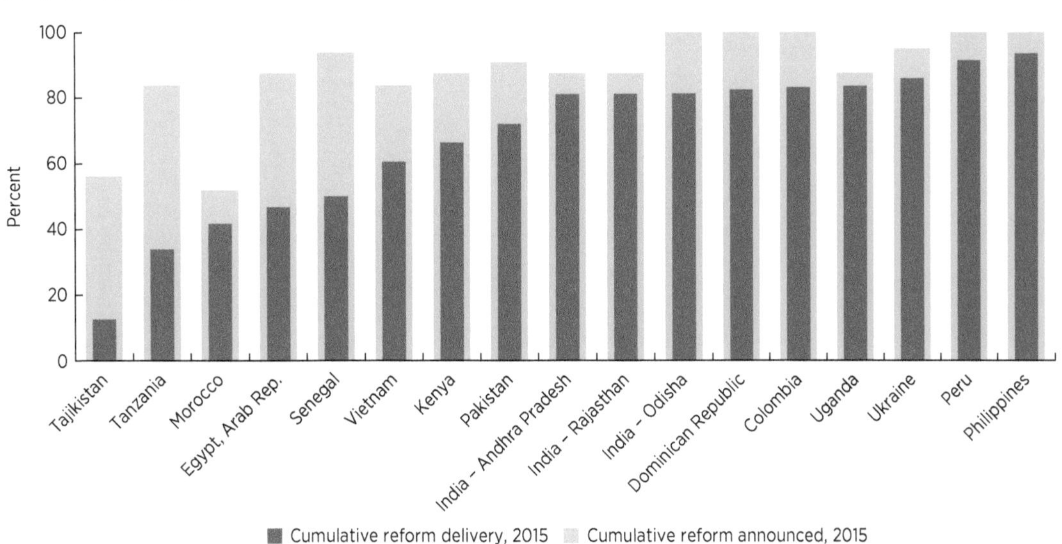

Source: World Bank elaboration based on Rethinking Power Sector Reform utility database 2015.

difficult to sustain and vulnerable to reversals of various kinds. Finally, while donors play an important role in introducing reform ideas and supporting their implementation, they do not seem to have much influence on a country's overall reform trajectory, which is rather shaped by local political factors.

Finding #3: The private sector made an important contribution to expanding power generation capacity in the developing world, albeit with significant challenges along the way

The private sector has contributed just over 40 percent of new generation capacity in the developing world since 1990, a share that has been remarkably consistent across country income groups. The absolute amount of private investment in Africa has been relatively low, but it still represents about 40 percent of total investment, similar to other regions. Across income groups, the share of private sector investment in capacity additions hovers around 40 percent (figure O.4a). For modern renewable energy technologies—now in the ascendancy—the share was almost twice as high, at around 70–80 percent (figure O.4b). Nevertheless, only a handful of countries was able to rely exclusively on the private sector for almost all new generation capacity. Foreign sponsors have been a major source of private investment in power generation, particularly in the Middle East and Sub-Saharan Africa (figure O.5). South Asia stands out as the only region where the majority of private investment has been domestically sourced.

Nevertheless, private investments in generation have not always been guided by principles of least-cost planning. During the 1990s, little attention was paid to power system planning, leaving many developing countries without strong technical capacity in this critical area. This was unfortunate at a time when demand for electricity was growing so quickly across the developing world that the scale of the system had to double every decade in many countries. Even when plans were made, they were seldom enforced. Only one in five countries makes

FIGURE O.4 The private sector's contribution to new generation capacity was steady across income groups but heavily skewed by generation technology

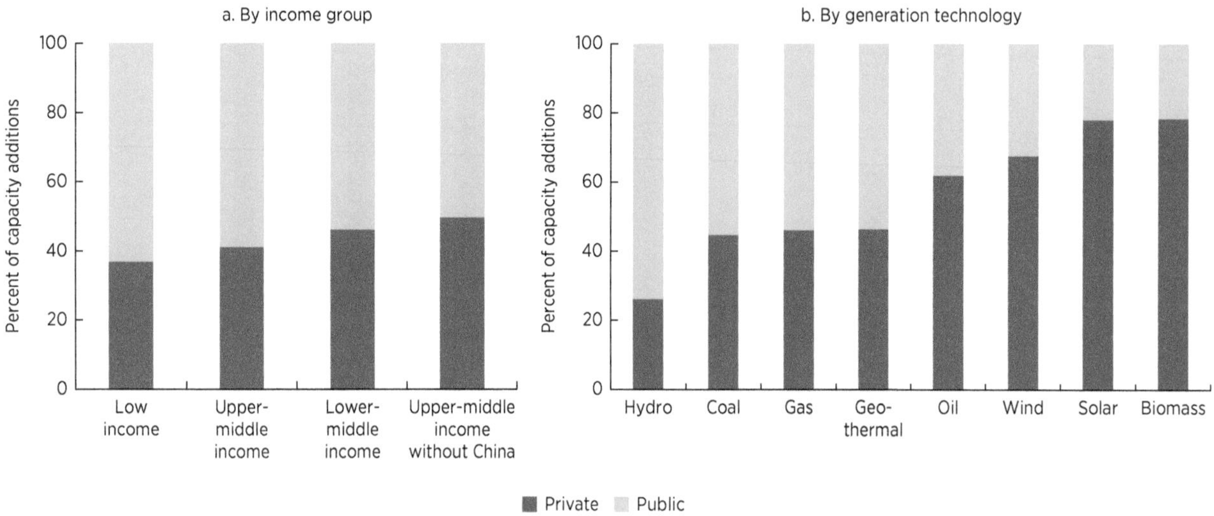

Source: World Bank elaboration based on World Bank–PPIAF Private Participation in Infrastructure Database 2018; UDI World Electric Power Plants database 2017.

FIGURE O.5 The bulk of private investment in generation came from foreign sponsors

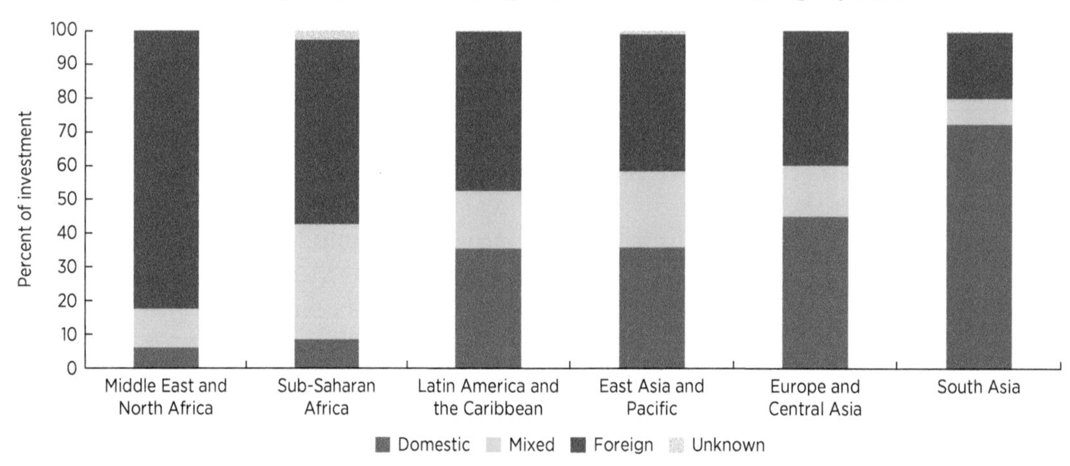

Source: World Bank elaboration based on World Bank–PPIAF Private Participation in Infrastructure Database 2018; UDI World Electric Power Plants database 2017.

power system plans mandatory, which often leaves important decisions about plant capacity vulnerable to the vagaries of political interference or unsolicited bids. In contrast to Latin America and the Middle East, where competitive tendering is more established (although the number of deals in the latter region is limited), direct negotiation of deals for IPPs remains widespread across Sub-Saharan Africa and South Asia (figure O.6). Such nontransparent procurement processes jeopardize value for money in generation and invite allegations of corruption, which has bedeviled IPP programs in some countries—Tanzania being a prominent example. Countries with strong planning and

FIGURE O.6 **Direct negotiation of IPPs remains widespread in South Asia and Sub-Saharan Africa**

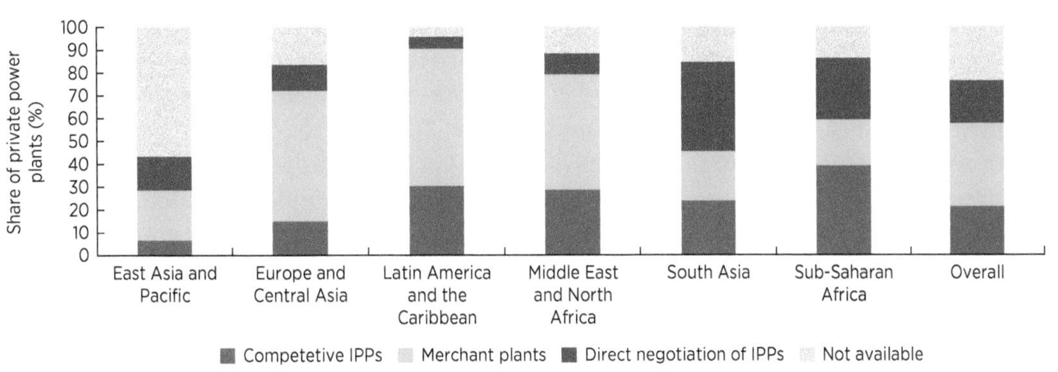

Source: World Bank elaboration based on World Bank–PPIAF Private Participation in Infrastructure Database 2018.
Note: IPPs = independent power producers.

procurement frameworks were more likely to be able to expand generation capacity to keep pace with growth in peak demand. The available evidence suggests that the features of the planning and procurement framework most closely associated with good outcomes for security of supply are the existence of institutional capacity for planning, the use of a transparent and participatory process for developing plans, and the adoption of competitive bidding for new generation.

Striking the right balance of risk between the public and private sector in power generation has proved challenging. IPPs face a plethora of risks, including demand risk, fuel price risk, exchange rate risk, and termination risk. All can weaken investor interest, particularly in untested markets, until a reliable track record has been established. In response, governments provide contractual protections of various kinds. Oil price and currency fluctuations are often passed through directly to the tariff specified in the power purchase agreement (PPA). "Take-or-pay" clauses prevalent in many African IPP contracts guarantee purchase of power even in the absence of demand; elsewhere, capacity charges at least ensure that fixed capital costs can be covered. Sovereign guarantees often need to be provided to compensate investors in case of premature termination. At one end of the spectrum, IPP programs have sometimes stalled when private sector demands for risk mitigation were not matched by the willingness of governments to provide them. Examples include Egypt's first IPP program in the early 2000s and Vietnam's program in the 2010s. At the other end of the spectrum, when governments have assumed excessive risk, IPP programs have sometimes triggered financial crises. Large-scale IPP programs have left governments exposed to currency or oil price risks, as happened during the Asian financial crisis of the late 1990s in Pakistan and the Philippines, where the power sector became a major contributor to public debt.

Finding #4: Wholesale power markets helped improve efficiency in the minority of countries that were ready for them; many others found themselves stuck in transition

Only one in five developing countries has introduced a wholesale power market, reflecting the formidable list of preconditions that must be met before such markets become possible or meaningful. Power markets are for the most part found in middle-income countries whose power systems are relatively large, financially viable, and unbundled (both vertically and horizontally)—and where regulatory governance is sound. However, regional power markets at varying stages of development are also allowing smaller countries in Africa, Central America, and South Asia to capture some of the benefits of power trade.

Close to half of the developing countries have adopted the single buyer model as a (sometimes indefinite) step toward wholesale competition. After some vertical and horizontal unbundling of the sector, IPPs compete alongside incumbent generators to supply power to the publicly owned single buyer, which is typically the transmission (and sometimes also distribution) utility. Although often conceived as a transitional model toward a competitive market, in practice most countries have remained stuck at this stage. A key concern is that the long-term take-or-pay arrangements that are often required to induce IPP investments in emerging markets can introduce distortions into power dispatch and build contractual rigidity into the power system—both of which significantly limit the scope for competition when a wholesale market is eventually introduced.

FIGURE O.7 **Electricity spot prices have shown wide variation across developing country markets**

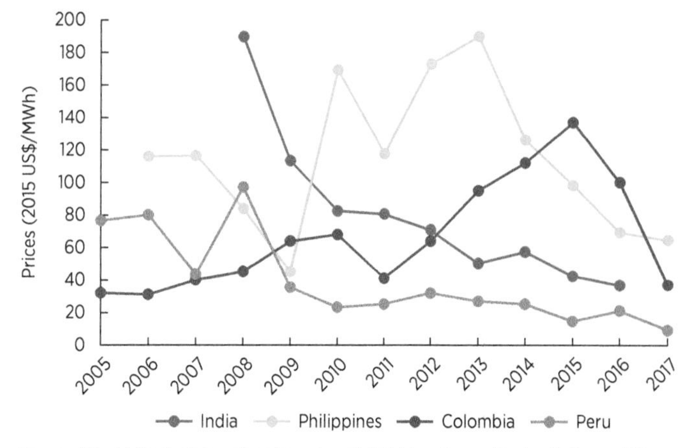

Source: World Bank elaboration based on Rethinking Power Sector Reform utility database 2015.
Note: MWh = megawatt-hours.

Effective functioning of wholesale markets requires a high-resolution, short-term pricing mechanism, as well as a sound and adaptive governance structure. The main function of wholesale power markets is to provide efficient short-term price signals to guide dispatch and inform investment. Prices across developing country spot markets have varied widely, ranging between US$20 to US$200 per megawatt-hour, with price trends conveying the evolution of local market conditions, such as expanding investment in India or drought conditions in Colombia (figure O.7). High spatial resolution of prices—such as the nodal prices used in Peru—is important to signal transmission constraints. Close monitoring of market prices and performance by an independent watchdog, such as the system operator or regulator, has proved important to detect abuses of market power often attributable to inadequate restructuring of generation assets prior to the launch of the market (Jamasb, Newberry, and Pollitt 2005; Jamasb, Nepal, and Timilsina 2015; Nepal and Jamasb 2012). This has been particularly challenging in the Philippines, but it has improved over time owing to new entries and the interconnection of segmented markets, reflected in tumbling wholesale market prices (figure O.7). Good governance of the system operator is critical for the impartial and effective dispatch practices that underpin price formation. Some countries have chosen to combine this function with that of transmission system operator, which is a viable option as long as conflicts of interest can be avoided. The functions of system and market operator have also proved possible to combine.

Despite expectations, spot market prices have not provided adequate incentives for investment in new generation capacity across the developing world. There has been relatively little entry by merchant plants[4] and limited willingness of regulators to allow spot market prices to spike during scarcity periods to the levels needed to incentivize new investment. Accordingly, several countries have adopted regulated capacity payments, which, although effective in incentivizing new investment, have led to concerns about excess capacity—for example, in Chile. Capacity markets have also been tried, though without success, in Colombia. Increasingly, supply auctions are proving to be an effective model for ensuring security of supply across several Latin American countries. In supply auctions, potential generators compete for the right to supply power to distribution companies on a long-term basis, but they do so without take-or-pay provisions.

More recently, decarbonization of the generation mix has emerged as a new policy objective to be pursued, creating further challenges for wholesale power markets. With few exceptions, decarbonization was not a major policy objective pursued through least-cost generation plans during the period under study. Generation investments were largely driven by concerns over security of supply, which coincidentally pushed hydro-dominated countries toward greater carbon intensity and oil-dominated countries toward lower carbon intensity. Nevertheless, these experiences illustrate that such policy-directed investment decisions can materially move the dial on carbon intensity once that becomes the objective. More recently, some Latin American countries, as well as India, have adapted their supply auctions to explicitly support the transition to renewable energy by targeting certain generation technologies. The growing share of variable renewable energy has created even further challenges for capital cost recovery in the generation segment, since the presence of resources such as wind and solar—which are characterized by zero marginal cost—can lead to periods of zero and even negative spot prices in some markets. Also, the variability of wind and solar resources increases the need for fast-ramping flexible resources to balance the system as needed, yet many markets lack mechanisms for appropriately incentivizing such ancillary services.

Finding #5: Good corporate practices, particularly with respect to human resources and financial discipline, were associated with better utility performance; these were more prevalent among privatized utilities

Corporatization of public utilities was conceived as a way to put the power sector on a more commercial footing. Prior to 1990, many public power utilities operated as administrative departments of their respective line ministries without any separate corporate existence. Doing so left them subject to the vicissitudes of public administration and unable to adopt a commercial orientation. For this reason, the first step to power sector reform in many countries was to separate out the operational functions associated with service provision into a distinct state-owned corporation, typically operating under company law. In doing so, many important decisions were made regarding the governance of the company and the establishment of management processes.

There is a significant governance gap between corporatized public utilities and privatized ones. A well-established literature on corporate governance of state-owned enterprises provides a clear frame of reference for good practice in this domain. For those jurisdictions where power utilities are entirely state-owned, corporate governance tends to reflect about 55 percent of good-practice measures, suggesting considerable room for improvement.[5] Governance scores tend to be systematically higher for private utilities, falling in the 60–90 percent range, a level only occasionally matched by public utilities. Boards of private utilities enjoy almost complete decision-making autonomy, whereas those of public utilities have limited freedom on critical matters of finance and human resources—particularly with respect to raising capital and appointing the chief executive officer. Public utilities also suffer considerable interference in the appointment and removal of board members. Overall, public utilities tend to be less rigorous in staff hiring, with more limited use of standard good practices, such as advertising, shortlisting, interviewing, and checking of references. Owing to public sector employment restrictions, they also have less ability to reward employees through performance bonuses or to fire those who perform poorly. Public utilities also tend to fall particularly short with respect to basic accounting practices that are universal in the private sector. When it comes to adoption of information technology, by contrast, there seems to be relatively little difference between public and private utilities.

Good practice on corporate governance is strongly correlated with good utility performance in terms of cost recovery and distribution efficiency—irrespective of public or private management. Surprisingly little has been documented to date regarding the extent to which corporatized power utilities pursued good governance practices and the resulting performance impact. New evidence presented in this study suggests that the quality of managerial practices related to human resources and financial discipline are strongly associated with better performance on distribution efficiency and operating cost recovery (figure O.8). The correlation holds irrespective of whether utilities are publicly or privately managed, since the best-performing public utilities exhibit somewhat better management practices than their peers. Board autonomy and accountability, however, are not so clearly linked to performance. Some of the dimensions of corporate governance that are most strongly associated with efficient utility performance are the publication of accounts consistent with international financial reporting standards, the explicit definition of public service obligations, the ability to fire employees for nonperformance, the use of transparent hiring processes for selecting employees, the adoption of modern information technologies, and the board's freedom to appoint and remove the chief executive officer.

Finding #6: Private sector participation in power transmission and distribution delivered good outcomes in favorable settings; elsewhere, it was susceptible to reversal

Private sector participation in transmission has not been widespread, but some successful examples exist in Latin America and Asia. The reform model of the 1990s was primarily concerned with establishing private sector participation in generation and distribution. The transmission segment was regarded as a natural monopoly, exercising system-coordination functions best handled under public ownership. Nevertheless, the experience of some countries in Latin America has illustrated that new transmission lines can readily be bid out under build-operate-transfer structures where the investment climate is adequate. These contracts are similar to those used for IPPs, but more straightforward, to the extent that there are no fuel costs or dispatch issues to consider, and remuneration is reduced to a simple annuity payment covering capital and operating costs over the life cycle. Cases of system-wide transmission concessions or even divestiture are much rarer.

Some of the early-reforming countries introduced widespread private sector participation in their distribution sectors. The financial health and operational strength of distribution utilities is a key driver of overall power sector performance. A financially precarious distribution utility can undermine the entire payment chain, while operational weaknesses in the local grid can prevent power from reaching customers even when it is available. For precisely these reasons, the 1990s model

FIGURE O.8 Certain aspects of corporate governance are strongly associated with improved efficiency performance for distribution utilities

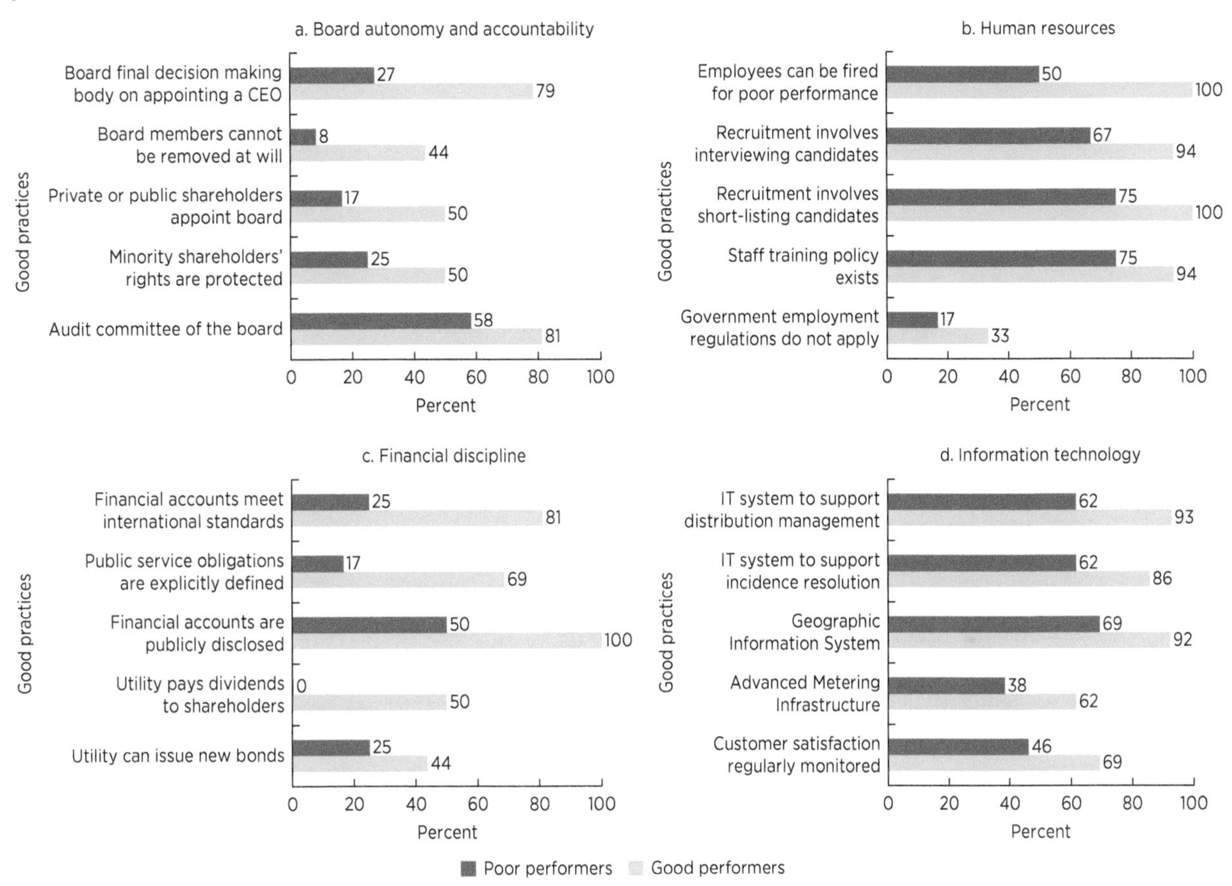

Source: World Bank elaboration based on Rethinking Power Sector Reform utility database 2015.
Note: IT = information technology.

prescribed private sector participation in the distribution tier as one of the first measures to be taken to turn around an ailing power sector. This is reflected in the surge of private sector participation in distribution that took place during the 1990s (figure O.9). Divestiture of distribution utilities was prevalent among early-reforming countries in Latin America, Central Asia, and Eastern and Central Europe, although it was comparatively rare in Africa and in East and South Asia. Nevertheless, even among countries undertaking privatization of power distribution utilities, relatively few privatized the entire distribution sector. More typically, public and private distribution utilities have coexisted within the same country, with private operators often serving capital cities or larger commercial centers. The decision to privatize only some distribution utilities may reflect differences in the commercial viability of the service areas, or variations in the local political environment, particularly in countries where electricity distribution remains a subnational responsibility.

Private sector participation in distribution has proved susceptible to reversals, and appetite for the reform subsided in the 2000s. Overall, 32 distribution transactions in 15 developing countries have been reversed (in the case of divestitures) or

FIGURE O.9 **Private sector participation in distribution peaked in the late 1990s before declining**

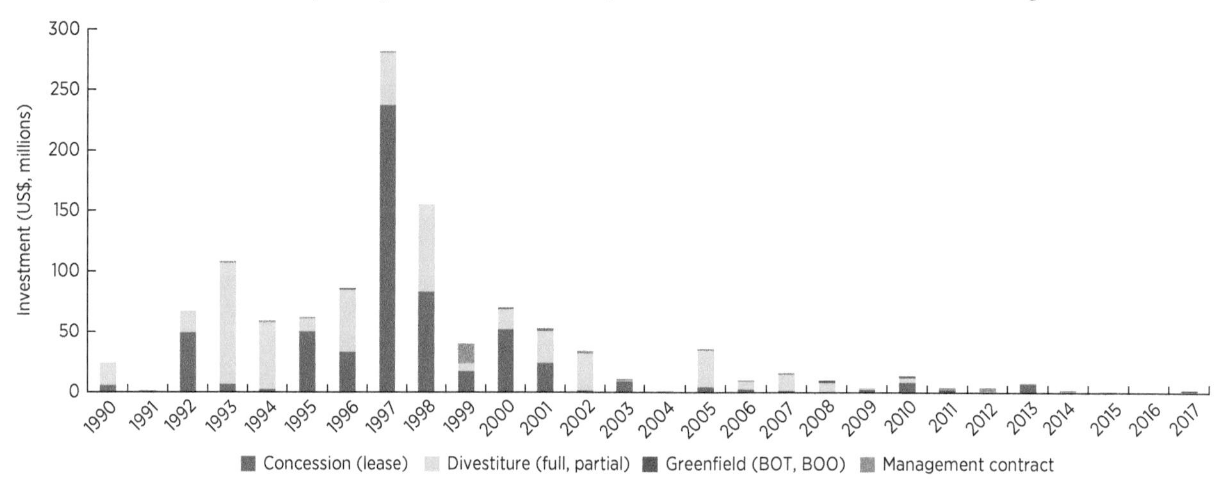

Source: World Bank elaboration based on World Bank–PPIAF Private Participation in Infrastructure Database 2018.
Note: BOO = build, own, operate; BOT = build, operate, transfer.

prematurely terminated (in the case of concessions and other contractual instruments), particularly during the first decade of reform. The probability of reversal was particularly high in Sub-Saharan Africa, affecting more than 20 percent of transactions. Sub-Saharan Africa's experiments with utility management contracts, in particular, have been checkered, encountering difficulties in recruiting and retaining qualified managers and suffering from tense labor relations and inadequate transfer of skills to local staff. Privatization reversals were most often associated with defective operational data (for example, serious underestimation of system losses) that led to unsustainable bids (for example, in the Indian state of Odisha), or with the government's unwillingness to apply tariff regulation as laid down in the legal framework (as in the Dominican Republic). Stakeholder opposition has also been a serious issue in some cases (as in Senegal, where the labor unions vehemently opposed utility privatization). Customers, in particular, often bear the brunt of tariff hikes associated with privatization, without always seeing an immediate impact on the quality of service, and this can sometimes lead to public disaffection (as in the Pakistani city of Karachi or Uganda). Such concerns led to a dramatic tail-off in private sector participation in electricity distribution after the early 2000s (figure O.9).

Private sector participation in distribution is strongly associated with full cost recovery. Private sector participation is the only reform that is associated with higher levels of full capital cost recovery, as opposed to recovery of operating costs alone. Among the countries reviewed that have undertaken significant and sustainable privatization of the distribution segment, it is exceedingly rare for tariffs to fall below full cost recovery levels. This partly reflects the fact that countries achieving higher levels of cost recovery are more likely to attract private sector participation; it also indicates that the presence of the private sector obliges the government to follow through on tariff regulations that call for cost recovery pricing.

With respect to efficiency, the performance of privatized distribution utilities is on par with the top half of performers among public utilities. Many of the privatized utilities studied perform to a high degree of operational efficiency (figure O.10). However, a group of publicly owned utilities (in the Indian

FIGURE O.10 Private sector participation is associated with much higher levels of cost recovery, while performance on efficiency is within the range observed for public utilities

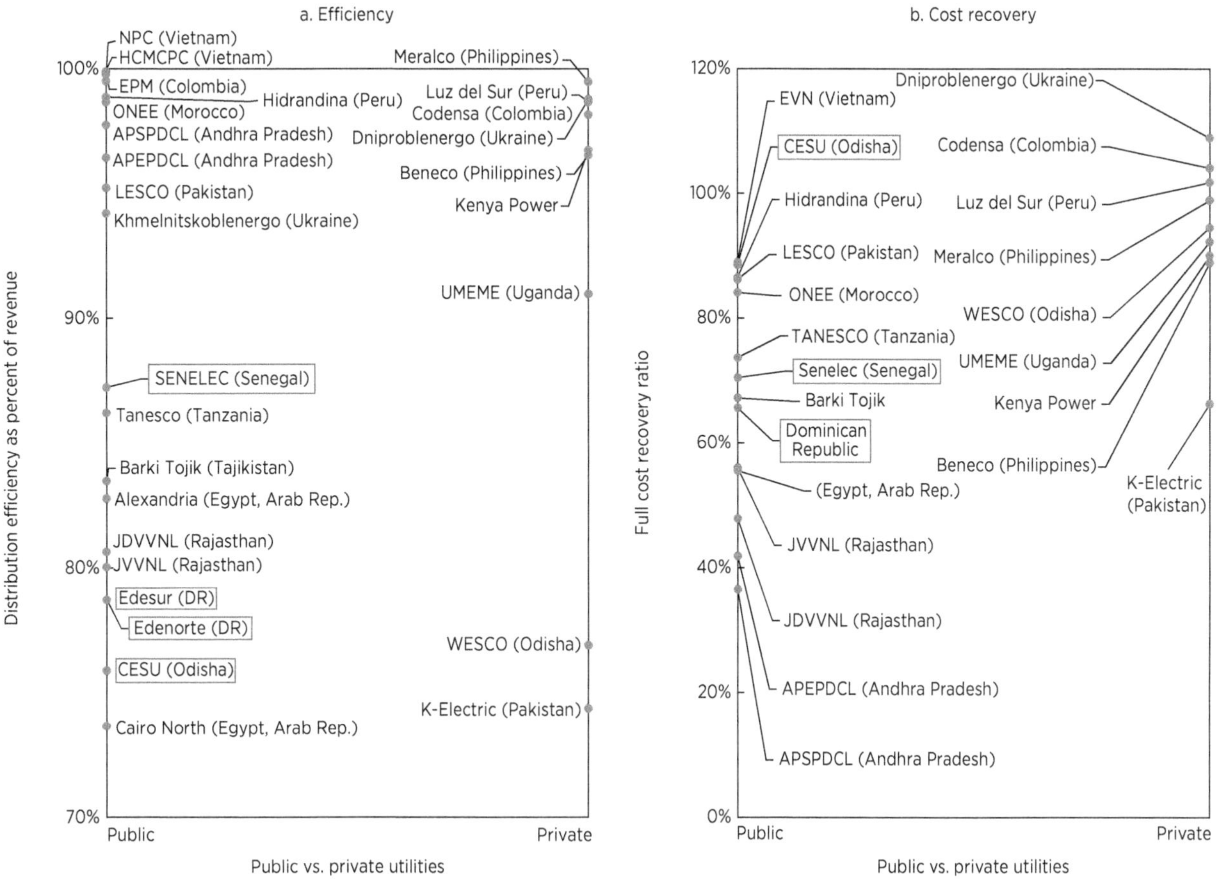

a. Efficiency

NPC (Vietnam)
HCMCPC (Vietnam)
Meralco (Philippines)
EPM (Colombia)
Hidrandina (Peru)
Luz del Sur (Peru)
ONEE (Morocco)
Codensa (Colombia)
APSPDCL (Andhra Pradesh)
Dniproblenergo (Ukraine)
APEPDCL (Andhra Pradesh)
Beneco (Philippines)
LESCO (Pakistan)
Kenya Power
Khmelnitskoblenergo (Ukraine)
UMEME (Uganda)
SENELEC (Senegal)
Tanesco (Tanzania)
Barki Tojik (Tajikistan)
Alexandria (Egypt, Arab Rep.)
JDVVNL (Rajasthan)
JVVNL (Rajasthan)
Edesur (DR)
Edenorte (DR)
WESCO (Odisha)
CESU (Odisha)
K-Electric (Pakistan)
Cairo North (Egypt, Arab Rep.)
100% 90% 80% 70%
Distribution efficiency as percent of revenue
Public Private
Public vs. private utilities

b. Cost recovery

Dniproblenergo (Ukraine)
EVN (Vietnam)
CESU (Odisha)
Codensa (Colombia)
Hidrandina (Peru)
Luz del Sur (Peru)
LESCO (Pakistan)
Meralco (Philippines)
ONEE (Morocco)
WESCO (Odisha)
TANESCO (Tanzania)
Senelec (Senegal)
UMEME (Uganda)
Barki Tojik
Kenya Power
Dominican Republic
Beneco (Philippines)
(Egypt, Arab Rep.)
K-Electric (Pakistan)
JVVNL (Rajasthan)
JDVVNL (Rajasthan)
APEPDCL (Andhra Pradesh)
APSPDCL (Andhra Pradesh)
120% 100% 80% 60% 40% 20% 0%
Full cost recovery ratio
Public Private
Public vs. private utilities

Source: World Bank elaboration based on Rethinking Power Sector Reform utility database 2015.
Note: Red boxes indicate utilities that have seen privatization rollback.

state of Andhra Pradesh, Morocco, and Vietnam) performs as efficiently as the privatized utilities. There are also some privatized utilities facing difficult operating environments (such as in the Pakistani city of Karachi or the Indian state of Odisha) that perform no better than some of the worst public utilities. At the same time, some of the worst-performing public utilities are cases of failed privatization (as in the Dominican Republic and Senegal).

There is also evidence that private sector participation is associated with good sector outcomes. Ultimately, the impact of reform is best evaluated in terms of results. Analysis undertaken for this study suggests that private sector participation has a significant positive impact on generation capacity and electricity access in low-income countries and that it supports the expansion of renewable energy in middle-income countries.

However, by far the strongest driver of electrification is income per capita, rather than any structural reform. The substantial progress on electrification made in many countries approaching middle-income status from 1990–2015 primarily took the form of utility-driven, grid-based electrification programs backed by clear political targets and public investment. In some cases (such as India, Morocco, and Vietnam), these efforts predated the sector reform process in the country. In other cases (such as Kenya, Tanzania, and Uganda), they were adopted long after the power sector reform, usually in response to the limited dynamism of electrification in the post-reform period. Grid electrification may be loss-making for the utility at the margin, meaning that it cannot be left to commercial incentives alone. With the advent of solar technology, decentralized private sector actors are playing an increasingly important role in the electrification process, although the jury is still out on the question of whether the most remote populations can be served on a purely commercial basis.

Finding #7: Regulatory frameworks have been widely adopted, but implementation has often fallen far short of design, particularly when utilities remained under state ownership

The creation of regulatory agencies was widely embraced and supported by sound regulatory frameworks in many countries. As of 2015, over 70 percent of developing countries had created a power sector regulator. On paper, the associated regulatory frameworks were relatively well-designed, incorporating provisions to balance the autonomy and accountability of the regulatory framework. In addition to the central functions of all such entities—regulation of tariffs and service quality (based on detailed methodologies laid down in the regulatory framework), regulators are widely responsible for licensing market entry, including negotiation of the terms of PPAs (85 percent) and competitive procurement (60 percent). They may also play a role in other important policy areas, such as clean energy (80 percent), power market design (65 percent), and electrification (55 percent).

In practice, however, it has proved very difficult to apply regulatory frameworks as written, and this has adversely affected the efficacy of regulation. Regulatory frameworks are to varying degrees overlooked or contradicted in practice (Andres, Guasch, and Diop 2007; Gilardi and Maggetti 2011). Whereas, on average, countries meet about 47 percent of good practice regulatory standards on paper, this score drops to 30 percent in practice.[6] The gap between regulation on paper and regulation in practice can be relatively narrow (as in Peru and Uganda, where the gap is less than 10 percentage points) or extremely wide (as in the Dominican Republic and several Indian states, where the gap can be 30–50 percentage points) (figure O.11a). One critical area is the authority of regulators to determine electricity tariffs, which is legally granted in 94 percent of countries but actually honored in only 65 percent—with a lot of caveats. Not surprisingly, the achievement of operating cost recovery is significantly related to the quality of regulation as practiced rather than as written.

Although originally conceived as an enabler of privatization and competition, regulation was often introduced into sectors still dominated by monopolistic state-owned actors. Many countries that fit this description adopted legal frameworks based on the principles of incentive regulation, according to which the regulator harnesses the utility's profit motive to incentivize efficient delivery of high-quality services. Such incentives are not typically effective unless regulated utilities operate according to strong commercial principles, making them responsive to incentives. Regulation does seem to have worked quite well, however, in countries with largely privatized distribution sectors. Moreover, evidence indicates that the presence of private actors in the sector is associated with much closer adherence by regulators to the established legal framework. The reason may simply be that it is more difficult for the government to deviate from enacted regulations when third-party private actors are involved.

Where utilities remain in public hands, the Ministry of Finance can become an

FIGURE O.11 **Significant divergence exists between regulation on paper and regulation in practice**

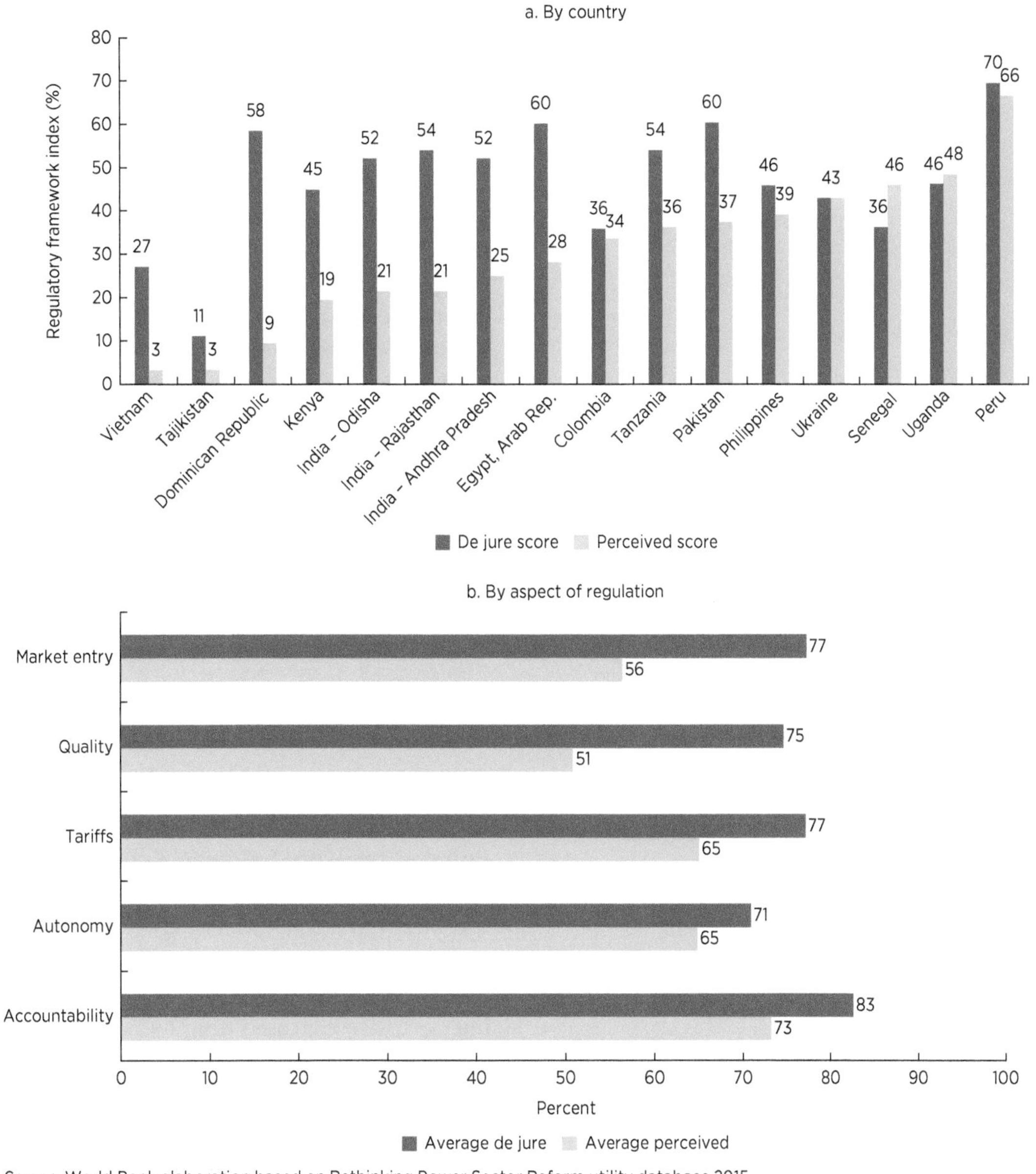

Source: World Bank elaboration based on Rethinking Power Sector Reform utility database 2015.

important player in the tariff-setting process. Countries where utilities remain publicly owned are often characterized by weak regulatory authority over tariff-setting and a soft budget constraint overall. When tariffs are not allowed to keep pace with costs (figure O.12), a degree of fiscal liability is created bringing the Ministry of Finance into the frame. Several countries, such as Egypt and Senegal, have explicitly recognized this in their tariff-setting frameworks, committing to fiscal transfers that exactly compensate for any shortfall in cost recovery from tariffs. This approach acknowledges that sector costs must ultimately be covered by a combination of taxes and user charges and provides a coherent framework for making such trade-offs. Nevertheless, the Senegalese experience illustrates the challenges of meeting such commitments during periods of fiscal stress.

FIGURE O.12 **Regulatory tariff recommendations are not always respected in practice**

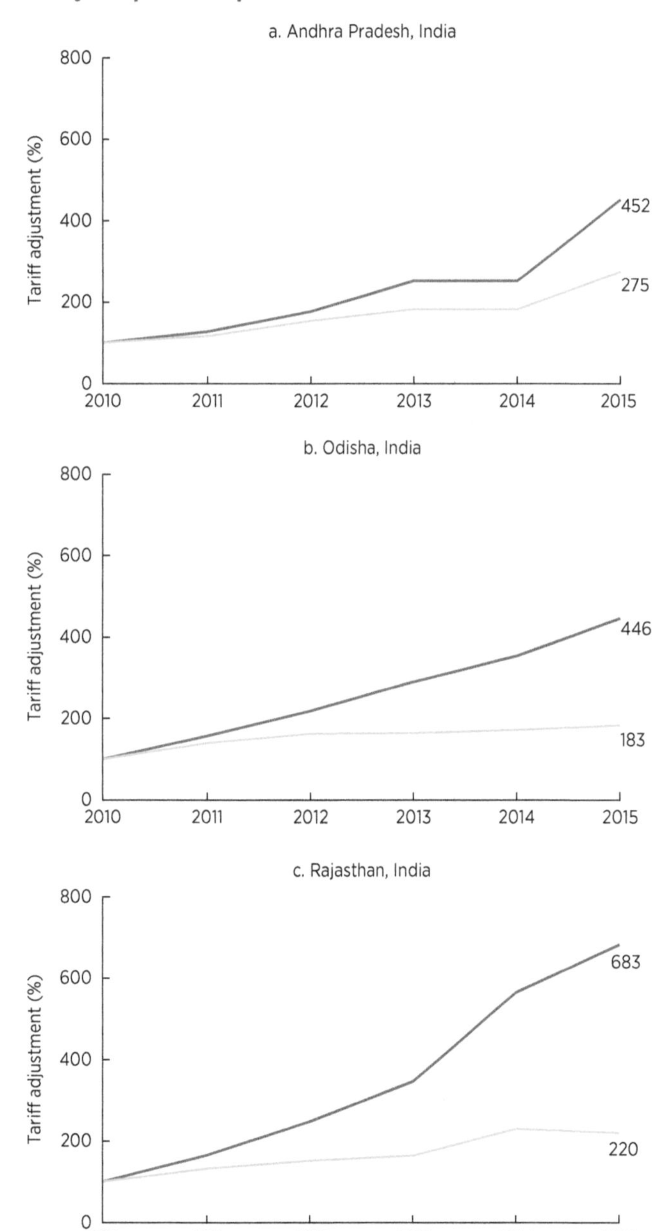

Source: World Bank elaboration based on Rethinking Power Sector Reform utility database 2015.

While regulators have struggled with tariff-setting challenges, quality-of-service regulation has not received the attention that it deserves and is too often observed in the breach. The shortfall in practice is particularly large for regulations pertaining to quality of service and market entry (figure O.11b). Indeed, few countries were found to have a meaningful system in place for regulating quality of service. (Colombia and Peru are among the few that do.) On the one hand, legal requirements to develop and monitor quality-of-service standards and penalize noncompliance are not always observed by regulators. On the other hand, utilities may lack the information systems to fully comply with such a framework and to manage reliability issues adequately. This is a serious deficiency, given the importance of service reliability for customers.

Finding #8: Cost recovery has proved remarkably difficult to achieve and sustain; the limited progress made owes more to efficiency improvements than to tariff hikes

Full cost recovery has been a challenge for power utilities. Only about half of them can be considered financially viable. Over the 25-year period under review, the extent to which end-user tariffs covered the full capital cost of supplying electricity increased from 69 percent to 79 percent, and about as many countries saw their performance on cost recovery deteriorate as improve (figure O.13a). Strikingly, even countries with relatively low cost of service sometimes struggle to achieve full capital cost recovery. In fact, full capital cost recovery is almost exclusively confined to utilities that have been privatized. Experience shows that progress toward cost recovery is subject to sudden erosion by exogenous factors, such as droughts, devaluations, and oil price shocks. Although full capital cost recovery has proved difficult to attain, almost all of the utilities have achieved operating cost recovery. Moreover, about half of the

FIGURE O.13 **More countries made progress on efficiency than on cost recovery, 1990–2015**

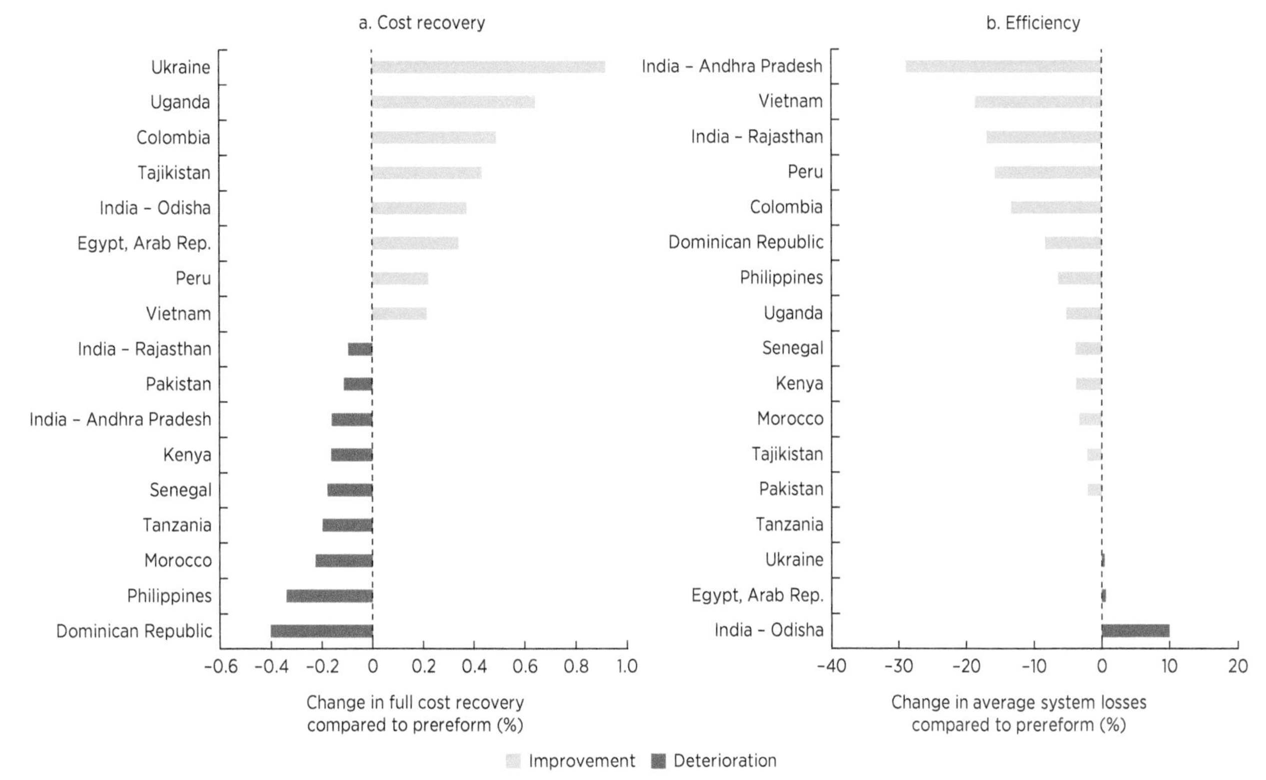

Source: World Bank elaboration based on Rethinking Power Sector Reform utility database 2015.

utilities can be considered financially viable in the sense of covering both operating costs and historic debt service and repayment obligations, albeit without providing a full rate of return on their asset base.

Where progress on full cost recovery was made over time, cost reductions played a greater role than tariff adjustments in bringing utilities closer to this goal. Specifically, average system losses across the study sample of countries fell from 24 to 17 percent between 1990 and 2015, and improvements were observed in more than 80 percent of jurisdictions (figure O.13b). Indeed, some countries would already be able to fully recover costs based on current tariffs if they could raise their commercial and operational efficiency to industry benchmarks. Tariff adjustments, however, have proved hard to apply as some regulators have seen their recommended adjustments aggressively scaled back or even completely overturned by the political authorities.

Utilities with revenue shortfalls are seldom fiscally compensated. The quasifiscal deficit across the study sample remains high, averaging close to one percentage point of gross domestic product, with underpricing being the major contributor in most cases. Financial analysis of the utilities showed that such shortfalls are not typically compensated by fiscal transfers from the state. Instead, utilities are forced to adopt a range of suboptimal coping strategies that often include taking on high-cost short-term commercial debt to cover cashflow shortfalls or simply falling into arrears with upstream suppliers of bulk fuel or electricity.

Cross-subsidies among customer groups and across consumption levels have long been the norm for electricity

tariff structures and may further undermine cost recovery. About three-quarters of developing countries practice cross-subsidies between commercial and residential customers, with the former paying on average more than twice as much as the latter for each unit of electricity purchased. A similar share of countries makes use of increasing block tariffs for residential customers, which typically provide sizable discounts at low or even average consumption levels and then step up tariffs for higher levels of consumption without ever reaching full cost recovery even in the highest consumption brackets. Deeper analysis shows that while modest amounts of cross-subsidy have been accommodated historically without seriously prejudicing the achievement of cost recovery, cross-subsidization can seriously undermine the financial equation of the utility if even the highest-paying customers are not paying at the cost recovery level.

Finding #9: The outcomes of power sector reform were heavily influenced by the starting conditions in each country

The 1990s power sector reform model was largely derived from principles believed to apply universally, independent of context. In practice, numerous preconditions—both economic and political—have emerged as important in shaping the applicability of the approach. Across the developing world, systematic differences can be observed in the uptake of the 1990s reform model across countries, based on factors such as income group, system size, and political system. Contextual factors also seem to have played a role in shaping the outcome of reforms (table O.1). The analysis distinguishes among "comprehensive reformers," which applied at least 70 percent of the prescriptions of the 1990s model; "limited reformers"; "stronger performers," which scored above average on outcome variables capturing progress on security of supply, electrification, and decarbonization; and the remaining "weaker performers."

A first group of countries largely applied the full policy prescriptions of the 1990s reform model and went on to see a range of positive outcomes as a result, experiencing improved operational efficiency and cost recovery, as well as enhanced security of supply. Foremost among these were Colombia, Peru, and the Philippines. In all these cases, the reform package was adopted comprehensively and relatively rapidly during the 1990s without major implementation setbacks. A continuous

TABLE O.1 Overview of preconditions among groups of countries at the time of reform

	Sector preconditions						Country preconditions	
	Cost of electricity ($/kWh)	Full cost recovery (%)	System losses (%)	Access to electricity (%)	Electricity consumption (kWh pc pa)	System size (GW)	Income level (GDP pc)	Quality of governance (index)
Comprehensive reformers								
Stronger performers	0.15	69	19	82	1,413	20	1,405	-0.43
Weaker performers	0.17	70	30	53	315	15	756	-0.49
Limited reformers								
Stronger performers	0.13	55	21	77	804	22	737	-0.55
Weaker performers	0.23	84	27	27	172	2	428	-0.40

Source: World Bank data.
Note: The Rethinking Power Sector Reform observatory countries are assigned their categories, specified in the table. All figures relate to the pivotal reform year for each country during the decade of the 1990s or the nearest data point available in some cases. India and Ukraine are excluded from system size calculations. GDP = gross domestic product; GW = gigawatt; kWh = kilowatt-hour; pa = per annum; pc = per capita.

process of second-generation reforms fine-tuned the operation of the model. Each of these countries faced its own challenges, but these could be accommodated, by and large, within the parameters of the new institutional framework.

A second group of countries also adopted comprehensive reforms but did not experience the same comparatively smooth implementation and positive outcomes. In Pakistan, for example, the unbundled power sector has been plagued by a chronic circular-debt crisis that undermines the payment chain; the only privatization in the distribution sector continues to be disputed in the courts after more than a decade of litigation. In the meantime, the country struggles to achieve security of supply and universal access to electricity. Other illustrative cases include the Dominican Republic and the Indian state of Odisha, where an extensive power reform was undertaken, including privatization of distribution utilities. However, in both cases, it proved difficult in practice to apply the prescribed framework of tariff regulation, leading to a subsequent renationalization and persisting concerns about security of supply, as well as weak performance on both intermediate and final outcomes.

Some insight into these disparate experiences can be gained by comparing the preconditions that existed in these two groups of comprehensive reformers at the time of the reform in the 1990s (compare the first two rows of table O.1). In particular, those countries where reforms proved to be successful started out from a much more advantageous national and sectoral position than the others. In terms of country context, the strong performers had already achieved an income level (around US$1,400 per capita) that was approximately double that of the weak performers, and they also enjoyed a better institutional environment, as captured by the World Bank Governance Index. In terms of sector context, the strong performers started out with much better operational performance in terms of system losses (19 percent versus 30 percent), much higher levels of electrification (82 percent versus 53 percent), and a much more developed energy system with significantly higher installed capacity (20 gigawatts versus 15 gigawatts). Their per capita electricity consumption was about four times as high. Even among the group of countries that made only limited reforms, the stronger performers enjoyed significantly better preconditions than those with weaker performance.

Finding #10: Good sector outcomes were achieved by countries adopting a variety of different institutional patterns of organization for the sector

Although the 1990s reform model started out with a unified reform blueprint, that blueprint was adapted to widely varying degrees. A significant minority of countries remains with a traditional vertically integrated national utility model, while the majority finds itself under an assortment of hybrid models.

Countries where adoption of reforms has been slower or more limited have, in some cases, performed as well, in terms of sector outcomes, as those that went further with the reform agenda. Comparing across a wide range of postreform outcomes covering security of supply, social inclusion, and environmental sustainability shows that the stronger performers divide into two equal groups comprising both comprehensive and limited reformers (table O.2). The performance differences are remarkably small between these two groups of countries; the limited reformers do slightly better on reliability, access, and affordability, and slightly worse on overall adequacy of capacity and carbon intensity. In a similar fashion, the weaker performers are also evenly split between countries that took a more comprehensive or limited approach to reform.

TABLE O.2 Comparison of country performance according to reform strategy

	Security of supply				Social inclusion		
	SAIFI	Normalized capacity (in MWs per million population)	Capacity diversification (HHI)	Meeting demand (ratio)	Access to electricity (%)	Affordability (% of GNI)	Carbon intensity (gCO_2/kWh)
Stronger performers							
Comprehensive reformers	8.8	551	0.4	1.1	94	4.6	357
Limited reformers	4.3	429	0.6	1.0	99	1.4	387
Weaker performers							
Comprehensive reformers	61.3	166	0.2	0.9	62	4.7	330
Limited reformers	30.3	45	0.5	1.0	35	13.0	419

Sources: IEA data; World Bank data.
Note: All figures relate to postreform performance as measured in 2015. The outcomes are judged based on a detailed framework provided in table 9.2 in chapter 9. Green signifies good outcome, yellow signifies moderate outcome, and red signifies poor outcome. GNI = gross national income; HHI = Herfindahl-Hirschman Index; MWs = megawatts; SAIFI = System Average Interruption Frequency Index.

Of particular interest, then, are the institutional paths taken by limited reformers that achieved stronger performance outcomes. Salient in this group are countries like Morocco and Vietnam, as well as the Indian state of Andhra Pradesh. What these cases appear to have in common is a continued role for a competent state-owned utility, with a more targeted role for the private sector.

Morocco kept a vertically integrated, publicly owned monopoly at the core of the sector, while opening to the private sector for certain generation plants and city-level distribution concessions. Rather than focusing on structural reform and the creation of regulatory capacity, Morocco's energy policy was characterized by the articulation of clear and ambitious social and environmental objectives at the highest political level. Those objectives were accompanied by clear institutional responsibility and accountability for delivery and supported by adequate investment finance, capturing both public and private sources as appropriate.

In Vietnam, the sector continues to be dominated by the incumbent utility operating as an unbundled public sector holding company with weak regulatory oversight. Vietnam's power sector journey prioritized the achievement of universal access through a sustained and well-financed program spearheaded by the national utility of Vietnam (EVN). The country is moving toward the staged implementation of a wholesale power market, in which a minority of privately owned generators competes alongside publicly owned subsidiaries of EVN.

In the Indian state of Andhra Pradesh, the state government completed unbundling and regulatory reforms but stopped short of privatizing the distribution segment. Instead, considerable efforts were made to sharpen incentives for managerial performance through the establishment of clear performance indicators relating to revenue collection, combined with frequent monitoring by senior management and financial reward for good outcomes. This approach was combined with legal reforms to make power theft a prosecutable criminal offense.

Finally, although Kenya does not feature among the stronger performers globally, it does present the best overall range of sector outcomes among the Sub-Saharan African case studies considered. Kenya's approach to reform was also incremental and distinctive. In particular, majority public ownership was retained in the distribution sector, but an almost equal share of equity floated on the Nairobi Stock Exchange provided an additional discipline on corporate governance of Kenya Power.

POLICY IMPLICATIONS

The 10 policy implications that follow draw on the review of historical evidence provided and on a forward look at disruptive technology trends in the power sector. The momentous technological changes underway—notably, increasingly cost-effective decentralized technologies—are posing fundamental questions about the viability of the traditional centralized utility and promising to change the structure of the power sector. In some frontier markets, the wave of change takes the form of distribution utilities splitting into a wires business and a distribution system operator, whose primary role is to provide a platform that consumers and businesses can use to trade energy both within the distribution segment and into the wholesale power market. In other cases, the new technologies are seen primarily as opportunities to improve the efficiency and effectiveness of the traditional utility.

As these debates play out into an uncertain future, at least two things seem clear.

First, power consumers will no longer be captive to underperforming utilities. The technological disruption in OECD member countries is taking place against a backdrop of universal access to a relatively high-quality and reasonably priced grid service. In contrast, across the developing world, many utility customers are faced with a costly and unreliable supply. Historically, the only alternative for unsatisfied customers was to supply their own electricity using expensive diesel generators. As rooftop solar power becomes cheaper and approaches grid parity, self-generation will become increasingly attractive where utility service is deficient, particularly once battery storage becomes more cost-effective. This development will start to contest the monopoly power of the incumbent utility, potentially providing incentives for improved performance. At the same time, there is the risk that already precarious utilities may be exposed to further financial distress resulting from grid defection.

Second, the speed and coherence of the technological transition will depend critically on the design of the regulatory framework, which shapes the incentives for innovation. Incentives for utilities to innovate depend on the regulatory regime under which they operate, since it is this that determines whether and how investments and operational savings can be turned into profits. Incentives for customers to innovate will depend on how much freedom they are given by the regulatory framework to engage in decentralized energy production and storage activities, as well as the associated impact on tariffs. Incentives for new players to enter the market and innovate will similarly depend on the flexibility of the regulatory licensing regime. In view of this, it is clear that the design of the regulatory framework will give countries a certain amount of discretion to accelerate or impede the uptake of disruptive technologies.

The following policy implications draw on lessons learned from past experience and also identify how disruptive technologies are likely to affect aspects of the power sector reform agenda.

- *Policy implication #1*. The design of power sector reforms should be informed by the enabling conditions of each country and oriented primarily toward achieving better sector outcomes.
- *Policy implication #2*. The design of power sector reform needs to be thoroughly grounded in the political realities of each country.
- *Policy implication #3*. Greater emphasis should be placed on building institutional capacity for power sector planning and associated implementation.
- *Policy implication #4*. Generation plants should be procured through a transparent and competitive process, with as much contractual flexibility as the context allows.
- *Policy implication #5*. Unbundling should not be the highest priority where more

fundamental financial and governance challenges persist; it should be undertaken primarily to facilitate deeper reforms.

- *Policy implication #6.* Wholesale power markets remain a viable option for countries that have put in place all the foundational measures; others may derive greater benefit from regional trade.
- *Policy implication #7.* Greater efforts should be made to strengthen the corporate governance and managerial practices of state-owned utilities.
- *Policy implication #8.* The regulatory framework needs to be adapted to reflect the institutional context and to accommodate emerging technological trends.
- *Policy implication #9.* Private sector participation in distribution should be considered only when enabling conditions are met.
- *Policy implication #10.* Delivering on the twenty-first century agenda of universal access and decarbonization calls for additional reform measures targeted explicitly at these objectives.

Policy implication #1: The design of power sector reforms should be informed by the enabling conditions of each country and oriented primarily toward achieving better sector outcomes

The 1990s power sector reform model was derived from economic first principles believed to apply universally, independent of context. As a result, it lacks a framework for customizing reform to the country context. In practice, numerous enabling conditions—both economic and political—have emerged as important in shaping its applicability. Across the developing world, systematic differences can be observed in the uptake of the model across countries, depending on their income group, system size, political system, and other factors. Drawing on the case studies that have informed this study, contextual factors also seem to have played a role in shaping the outcome of reforms.

Experience suggests that it may be helpful to think about power sector reform engagements in two phases, depending on the nature of the country environment. This overall framework is depicted in table O.3, which presents the reform measures likely to be applicable in more challenging versus more mature environments, as well as the enabling conditions that signal a country's readiness for various aspects of the reform package.

In more challenging environments, a basic set of preliminary reform measures is proposed. This applies to countries that may be challenged by low incomes, fragile settings, small scale, or other limiting factors. The priority in these environments should be to work toward a foundation of good sector governance and basic financial viability, without embarking on overly complex structural reforms.

The policy implications are as follows:

- *Regulation.* Critical at this juncture is to adopt a transparent and well-founded tariff-setting methodology and to apply it each year. This could be done by a regulatory agency or, at this stage, by a competent unit within the Ministry of Energy or the Ministry of Finance. An adequate initial aspiration for tariff-setting would be to ensure financial viability through recovery of enough capital costs to service and repay existing debt. Equally important would be for the Ministry of Energy to lay the foundations for monitoring the quality of service. The process of tariff and quality regulation should be integrated with other processes for overseeing state-owned enterprises (relating, for example, to performance contracts or fiscal transfers).
- *Restructuring.* This is unlikely to be a high priority at this stage. A vertically integrated power system may be easiest to manage while putting in place strong foundations for the sector. However, the entry of the private sector into generation—through

TABLE O.3 Customizing power sector reforms to country context

	More challenging environments	Enabling conditions	More mature environments
Regulation	Establish clear tariff-setting methodology with oversight from ministry of energy or finance. Aim for achievement of limited capital cost recovery (that is, financial viability). Establish clear quality-of-service framework with oversight from Ministry of Energy.	• *Cost recovery ratio exceeds 70 percent.* • *Revenue collection ratio exceeds 90 percent and is enforced by disconnection.* • *System losses are below 15 percent.* • *Electrification rate exceeds 80 percent.* • *Regular audited financial accounts are compliant with international financial reporting standards.* • *Modern IT systems are in place and deliver good operational data.* • *Regular tariff adjustments are in line with regulatory methodology.* • *The political context is supportive, in terms of ideology, leadership, and stakeholders.* • *Generation capacity reaches 1–3 GW.* • *No major bottlenecks exist on the transmission grid or in fuel supply.*	Create separate regulatory entity. Aim for full capital cost recovery. Ensure enforcement of quality-of-service regulation.
Restructuring	Retain vertically integrated utility, and selectively introduce private investment for new plants.		Restructure the power sector to separate out the transmission system operator and ensure adequate degree of competition in generation.
Privatization	Focus on establishing sound corporate governance arrangements and good managerial practices for power distribution, with special focus on human resource management and measures to promote financial discipline. Prioritize electrification through carefully planned parallel efforts with reach of the grid and off-grid, backed up by strong political commitment and adequate public funding.		Strengthen commercial incentives in distribution through measures such as: credit-rating and bond issues; stock market listing; and/or private sector participation.
Competition	Ensure adequate technical capacity for power system planning directly linked to competitive procurement of generation. Introduce economic dispatch of generation plants administered by utility.		Open the grid to third-party access and allow bilateral contracting between generators and large customers. Create wholesale power market. Conduct supply auctions for investment in new plant.

Source: World Bank.
Note: GW = gigawatts; IT = information technology.

supply contracts with the utility—can play a valuable role in expanding capacity.

- *Private sector participation.* It may be best at this stage to limit private involvement to generation. For the distribution segment, the emphasis should be on building good governance and managerial practices, particularly with respect to financial discipline and human resource management.
- *Competition.* The only relevant form of competition at this stage is likely to be competition for the right to build new generation plants. Particularly critical is the development of the technical capacity required to conduct least-cost planning to determine what plants to build, with mandatory links to a competitive procurement process. Furthermore, some of the benefits of a competitive market can be mimicked through the administrative practice of economic dispatch.

In more mature environments, it becomes feasible to contemplate a more sophisticated package of reforms, as long as these improve sector outcomes. This applies particularly to middle-income countries with stable political environments and large power systems, where progress has been made toward good governance and financial viability for the sector. Given that reform is a means to an end, the priority in these environments should be to identify where power sector performance continues to fall short of expectations and to pursue more advanced reform measures geared to delivering results in these specific areas.

The policy implications are as follows:

- *Regulation.* Thought should be given to establishing a separate regulatory entity if one does not already exist. It now becomes more important to set tariffs to achieve full capital cost recovery, as well as to tighten enforcement of quality-of-service regulation. Strengthening the regulatory framework is particularly critical if the policy objective is for the sector to repay investment finance at market rates.
- *Restructuring.* This is the right juncture at which to consider vertical unbundling to create a separate transmission system operator that will support impartial third-party access to the grid. At the same time, it becomes important to break up generation assets to provide for sufficient competitive pressure among market players.
- *Private sector participation.* Countries moving toward a wholesale power market should ideally divest at least part of their generation assets to the private sector to ensure some diversity of ownership among competing companies. In the distribution tier, countries experiencing operational inefficiencies may wish to consider private sector participation. Where public utilities are performing efficiently, the case for private sector participation is weaker; the need to raise additional capital, however, may make it necessary for the utility to obtain a credit rating to support access to bond finance, or a minority stock exchange listing, both of which will also have the desirable effect of tightening the utility's financial discipline.
- *Competition.* Countries at this stage are ready to consider the transition to a wholesale power market. This should be accompanied by parallel supply auctions or an equivalent measure to ensure timely development of adequate new generation capacity.

The transition from challenging to mature environments can be gauged in terms of certain key enabling conditions. In practice, it may not be necessary or feasible for countries to meet every one of these enabling conditions; however, the more conditions that are met, the better are the prospects for implementation of the more sophisticated reforms. Most of these enabling conditions are related to readiness for the introduction of private participation in distribution. This is more likely to succeed when certain minimum thresholds of financial viability and commercial efficiency have been passed, and when the challenge of electrification is at a reasonably advanced stage. Good financial and operational data systems will also help to reduce information asymmetries and increase confidence among private participants, as will a good track record of regulatory tariff-setting and a conducive political environment. Other enabling conditions are more directly related to the establishment of wholesale power markets. In particular, the power system should be large enough to support at least five competing generation firms (at least 3 gigawatts) and to generate enough turnover to justify the fixed costs of establishing market platforms (at least US$1 billion in annual revenues).

Policy implication #2: The design of power sector reform needs to be thoroughly grounded in the political realities of each country

Commitments to power sector reform should reflect a sober assessment of the country's political economy. The 1990s reform model drew heavily on economic first principles, with no explicit attention to the political dynamics of the reform process. Yet, the reality is that the power sector is highly politicized across much of the developing world. Understanding a country's political dynamics and how they impinge on stakeholder interactions in the power sector should be the starting point for any power sector reform. Rather than overlooking the political dimension, a smart reform process should be adapted to fit the political context, harnessing potential reform champions and explicitly engaging in consensus-building with contrarian groups.

The policy implications are as follows:

- *Undertake a political economy analysis before engaging in reform.* The analysis should aim at discovering how the power sector touches upon the country's vested interests and political groupings to identify potential winners and losers from reform. It should also consider whether the proposed direction of reform is compatible with the country's ideological orientation and broader political system. The findings of the political economy analysis should explicitly guide the design of the reform program to be adopted.
- *Integrate outreach and communication efforts to engage all relevant stakeholders.* The communications campaign should be based on messages that can be used by the reform champions to articulate the value proposition associated with the reform. Those messages can be disseminated through a variety of channels. Communications should be complemented by outreach that directly engages with all stakeholders, particularly those most threatened by the reform process. In addition to an intensive effort at the outset of a reform process, there is a need to monitor the state of public opinion throughout implementation, as sudden changes in the political environment can easily lead to reform reversals.

Policy implication #3: Greater emphasis should be placed on building institutional capacity for power sector planning and associated implementation

The 1990s model had little to say on the issue of planning. The implicit assumption was that the advent of a wholesale power market would somehow circumvent the need for planning. The ultimate goal of the 1990s model was to create a competitive market. At the time, it was assumed that private investments in power generation would be adequately guided by price signals. The role of the state was seen primarily as the regulator of a privately owned and operated competitive sector, and great emphasis was placed on the creation of a capable regulatory institution and associated legal framework. Central planning functions were overlooked or downplayed. Indeed, in some countries, the planning function traditionally housed in national power utilities or line ministries fell through the cracks as power sector reform processes worked to unbundle the incumbent utilities and to build technical capacity in regulatory agencies operating outside of line ministries. In practice, power markets proved difficult to establish in all but a handful of developing countries; even there, price signals have not provided an adequate basis for investment decisions.

The policy implications are as follows:

- *Create strong technical capacity for planning and empower the planning function.* The development of a strong planning capacity for the development of new generation and transmission infrastructure should be prioritized as a critical component of power sector reform. Various alternative institutional models have been successfully used around the world to locate the planning function, including the line ministry, the transmission utility, the system operator, or a dedicated technical agency. Regulators can play a valuable role in the technical review of investment plans as part of the process of setting revenue requirements for capital expenditure.
- *Make sure the power system plan is actually implemented.* As important as the planning process itself is a strong link between the

power system plan and the procurement of new generation and transmission plant, so that procurement is aligned with the plan and contracted in a timely and cost-effective manner that keeps pace with demand. Without such a clear linkage, governments are vulnerable to unsolicited proposals that may not represent the most cost-effective option for the power system.

- *Incorporate new technologies in power system planning.* Technologies such as distributed energy resources, together with storage and demand response, have the potential to reduce the costs of reaching supply-demand balance. However, the incorporation of such resources is not considered in traditional power system planning, in part because they introduce significant complexity into standard planning methodologies, but also because they would not necessarily be undertaken by the incumbent utility. Storage—in particular—can play multiple roles in the power system, potentially substituting for conventional investments in generation, transmission, and distribution assets. There is a need to modernize planning tools and techniques to integrate such considerations.

Policy implication #4: Generation plants should be procured through a transparent and competitive process, with as much contractual flexibility as the context allows

Although IPPs have proved a popular and effective means of bringing private capital into power generation, much room for improvement remains in the way such projects are implemented. Direct negotiation of projects, often in response to unsolicited proposals, remains widespread across Africa and Asia, raising concerns about value for money and the potential for corruption. At the same time, the need to mitigate risk to reassure investors entering uncharted waters has left many countries with rigid take-or-pay contracts and extensive guarantee clauses that both constrain the efficiency of dispatch and saddle the utility and the government with onerous liabilities.

The policy implications are as follows:

- *Mandate the use of competitive procurement for generation projects.* Competitive bidding of new generation plants should be the default modality for procurement. If unsolicited proposals are considered—only in clearly defined and exceptional cases and when their prefeasibility and compatibility with existing investment plans can be established—they should also be subjected to a competitive process.
- *Maximize the flexibility of contractual provisions.* Risk-mitigation mechanisms will inevitably be needed in unproven environments, but these should be carefully scrutinized and limited to the minimum required to meet investors' legitimate expectations of return. Doing this could mean, for instance, scaling back the volume or duration of take-or-pay clauses or making use of two-part pricing mechanisms that separate capacity and energy charges.
- *Consider the adoption of supply auctions wherever possible.* The foregoing challenges have been successfully addressed by countries that have moved toward the adoption of supply auctions, ensuring a pipeline of regular, well-structured offerings of batches of new generation plant. These are linked to long-term contracts with distribution utilities that give generators first right of supply without committing to take-or-pay arrangements. A growing number of countries are adopting such mechanisms to procure variable renewable energy, and these could readily be extended to cover other technologies.

Policy implication #5: Unbundling should not be the highest priority where more fundamental financial and governance challenges persist; it should be undertaken primarily to facilitate deeper reforms

In the past, power sector restructuring has, at times, been treated as a panacea for reform and prioritized as an early reform measure. However, in and of itself, power sector restructuring does little to tackle the fundamental issues of weak governance and financial fragility that plague the power sector in many developing countries. Moreover, restructuring a sector that suffers from weak governance and financial fragility may only exacerbate the challenges of technical coordination and financial payment along the supply chain.

In reality, unbundling was never intended as an isolated reform measure but rather as a necessary precursor for a competitive market. Unless the latter is a realistic possibility in the medium term, restructuring the sector may not be a pressing matter. Unbundling entails significant transaction costs, as well as the potential loss of economies of scale and scope, which should not be underestimated (Pollitt 2008; Vagliasindi 2012). For these reasons, the relevance of unbundling to smaller power systems is particularly questionable. There is a well-established minimum size threshold of 1 gigawatt before countries should even consider embarking on sector restructuring, and a further threshold of 3 gigawatts before they definitely need to unbundle should they be preparing for the establishment of a wholesale power market.

The policy implications are as follows:

- *Consider unbundling when there is a clear purpose for doing so and where enabling conditions are in place.* The purpose behind unbundling might be to establish a wholesale power market in the not-too-distant future or to introduce private sector participation in a specific segment of the industry but not elsewhere. The enabling conditions would include (1) a minimum system size of at least 1 gigawatt to avoid the loss of economies of scale and (2) adequate institutional governance, including strong payment discipline and technical coordination along the supply chain.

Policy implication #6: Wholesale power markets remain a viable option for countries that have put in place all the foundational measures; others may derive greater benefit from regional trade

The 1990s power sector reform model held up a competitive power market as the endpoint of reform. The aspiration remains legitimate, but it has proved to be farther out than originally envisaged. The difficulty of fulfilling the many enabling conditions that a wholesale power market requires has deferred indefinitely the introduction of such markets across much of the developing world. Nevertheless, their attainment remains a valuable and legitimate aspiration, provided that the enabling conditions can be met. Indeed, the present wave of technological disruption only increases the value of wholesale power markets, which, when properly designed, can support the discovery of rapidly evolving costs and foster the integration into the power system of variable renewables, ancillary services, battery storage, and demand response.

The policy implications are as follows:

- *Ensure that the enabling conditions for a wholesale power market are in place.* Countries should not consider developing such a market until a wide range of preconditions have been met. These include the following: (1) a fully restructured power sector that has created at least five competing generators with diversified ownership, (2) an absence of significant constraints in transmission or fuel availability, (3) a financially viable sector with a solid payment chain, (4) solid regulatory practices, and (5) sufficient system size. A wholesale power market entails certain fixed costs that are unlikely to be justified by the potential efficiency gains until the market is large enough. As a rule of thumb, power markets are not likely to become very interesting until a country reaches a national market turnover of around US$1 billion, which is equivalent to a power system size of some 3 gigawatts.

- *Avoid getting locked into transitional arrangements.* Countries that are ready to move to a competitive market should consider carefully whether transition mechanisms are really needed, since experience suggests there is a relatively high risk of getting stuck in intermediate stages, in particular, the single-buyer model.
- *Establish a strong transmission system operator.* The transmission utility plays a critical role in a competitive power market, ensuring equitable access of third parties to the grid infrastructure, and potentially also playing a leading role in power sector planning, system planning, and sometimes market operation.
- *Monitor and adapt the design of the wholesale power market based on implementation experience.* Wholesale power markets may not always function according to design. Proactive monitoring for potential abuses of market power is very important, particularly in the early stages, as is the flexibility to learn from this experience and adapt market design accordingly.
- *Provide a parallel mechanism for incentivizing investment in generation.* Short-term market price signals alone are not always adequate to provide incentives for investment in new capacity. Parallel capacity mechanisms are needed, with supply auctions proving to be particularly efficient and effective. Such auctions can be adapted to target low-carbon forms of energy (with associated storage) and can increasingly be used to contract for adequate ancillary services to balance variable renewable energy.
- *Modernize wholesale power markets to accommodate new resources.* Conventional power market designs are not adapted for the presence of variable renewable energy resources, battery storage, or increasingly sophisticated demand response. Integrating them calls for the development of new pricing mechanisms that are able to remunerate the ancillary services required for the successful integration of variable renewable energy, provide suitable price signals to incentivize efficient investment in utility-scale battery storage, and allow demand-response aggregators to participate in the process of dispatch.
- *Participate in regional and cross-border trading arrangements wherever possible.* Regional power markets also offer significant benefits for arbitrage based on differential generation costs and load profiles among neighboring countries. Other benefits include shared reserve margins and greater flexibility to accommodate variable renewable energy. For countries not yet ready to develop wholesale power markets domestically, regional markets can provide an important first step. Nevertheless, even regional markets entail certain basic minimum enabling conditions that cannot always be taken for granted—in particular, creditworthiness on the part of power importers and security of supply on the part of power exporters.
- *Move toward economic dispatch of power plants.* Deviations from principles of economic dispatch are widespread in the developing world, leading to major generation inefficiencies. Countries not yet ready to develop wholesale power markets should consider having their system operator move toward the practice of economic dispatch based on the marginal costs of operating different plants.

Policy implication #7: Greater efforts should be made to strengthen the corporate governance and managerial practices of state-owned utilities

The 1990s reform model focused on privatization of distribution utilities, but the reality is that most remain publicly owned. The creation of corporatized public utilities out of traditional ministerial departments was viewed as a short transitional measure toward eventual privatization, which would lead to a full overhaul of managerial practices. However, given the relatively limited uptake of privatization in the distribution segment, it has become very important to address enduring weaknesses in the corporate governance of public utilities. The evidence shows that there is wide variation in the performance of public utilities; a substantial minority reaches efficiency levels comparable to private utilities, while the majority continues to flag. Better-performing public utilities share many aspects of good corporate governance with each other and with private utilities.

The policy implications are as follows:

- *Improve human resource management of public utilities.* Public utilities should take care to apply aspects of human resource management that are strongly associated with improved performance. These relate primarily to the quality of the selection process for hiring employees—in particular, the application of standard good practices, such as advertising vacancies, shortlisting and interviewing candidates, and conducting reference checks. The liberty to fire employees for underperformance is also found to be important, although this is often difficult to enforce in public sector environments.
- *Strengthen financial discipline of public utilities.* Similarly, public utilities should adopt certain aspects of financial discipline that are strongly associated with improved utility performance. Again, these comprise standard measures, such as the publication of externally audited financial accounts that are prepared in conformity with international financial reporting standards. Another good practice is the explicit identification and costing of public service obligations that cannot be justified on commercial grounds.

Policy implication #8: The regulatory framework needs to be adapted to reflect the institutional context and to accommodate emerging technological trends

The creation of sector regulators has been a popular reform, but many of these entities find themselves regulating public rather than private utilities. The power sector reform model of the 1990s envisaged the creation of a regulatory entity as a prerequisite for introducing private sector participation, particularly in power distribution. The regulator was supposed to play the dual role of protecting private investors from opportunistic government meddling, while also protecting consumers from abuses of privately held monopoly power. The evidence suggests that regulation has functioned much more effectively where the private sector entered power distribution than where utilities remained state-owned.

Moreover, the regulatory regimes of the 1990s did not anticipate the current wave of technological disruption in the power sector. The power sector has seen momentous technological change since the development of the 1990s power sector reform model. The changes are challenging the traditional approach to tariff regulation, which is based on ensuring that the utility collects enough revenue to enable it to roll out new infrastructure. It also raises questions about the traditional design of tariff structures that were often motivated by social policy concerns in a context where consumers were largely captive.

The policy implications are as follows:

- *Ensure that the instruments of price regulation are consistent with the governance of the utility.* There is little value in applying the instruments of incentive regulation—designed to harness the profit motive of private utilities—to state-owned utilities that are not driven by profit maximization and may not even operate under hard budget constraints. In these cases, it makes more sense to use traditional cost-of-service regulation and focus on creating supportive managerial performance incentives. Even the creation of a separate regulatory entity may be less of a priority when the sector remains state-owned, because, in practice, both the utility and the regulator are likely to be closely overseen by the line ministry, making regulatory independence somewhat illusory. Nevertheless, irrespective of which institution is responsible for regulation, a clear, well-grounded methodology for tariff-setting, applied on an annual basis, is of tantamount importance.
- *Aim for limited capital cost recovery initially.* Most regulatory tariff methodologies are based on principles of full capital cost recovery, including remuneration of the full asset base at the market cost of capital. Where utilities have been privatized, this principle is critical for financial sustainability. However, in the case of state-owned utilities, which often benefit from significant capital grants, it is not essential to remunerate the full asset base at the market cost of capital. Rather, the concern should be to ensure that the utility is able to cover the costs associated with the loans that are carried on its books. This limited capital cost recovery, which ensures the financial viability of the enterprise, is a reasonable interim tariff-setting objective.
- *Integrate regulation with other key public sector processes for state-owned utilities.* In some countries, regulatory frameworks coexist with other forms of state oversight. Utilities may be held accountable through performance contracts with the Ministry of Energy, for example, while tariff-setting is inextricably linked with financial oversight and subsidy decisions that lie in the hands of the Ministry of Finance. Rather than creating parallel tracks, regulation should build upon and integrate these complementary processes. Quality-of-service regulation should be reflected in the key performance indicators determined under performance contracts. Tariff and subsidy decisions should be taken simultaneously in a coordinated manner, ensuring that the overall revenue requirements of public utilities are met through a combination of both sources.
- *Give greater attention to creating a credible regulatory framework for quality of service.* With regulatory attention focused primarily on tariff-setting, efforts to provide a credible framework for monitoring quality of service and enforcing the achievement of the prescribed standards have been inadequate. Such a framework is of critical importance to ensure that regulatory reforms yield tangible benefits for electricity consumers.
- *Test the "future-readiness" of the regulatory framework.* The regulatory pricing regime for power utilities can affect the incentives for adoption of new technologies. For instance, traditional cost-of-service regulation will not encourage a utility to adopt technologies that may reduce demand for energy or meet demand at a lower investment cost. The regulatory licensing regime may also create barriers to the entry of new actors, such as providers of distributed energy resources or demand aggregators. There is therefore a need to review existing regulatory frameworks to evaluate whether they offer adequate incentives for innovation.
- *Ensure that the economics of decentralized electricity supply are reflected in tariff structures.* Electricity tariff structures have traditionally been designed under the premise that consumers have limited alternatives to grid

electricity, so pricing can be guided primarily by considerations of fairness and equity rather than economic efficiency. This practice has led to tariff structures under which costs are recovered primarily through volumetric charges, with extensive embedded cross-subsidies across consumption bands and consumer groups. Because such tariff structures fail to recognize the fixed-cost nature of the power grid, they overreward customers choosing to self-supply and fail to convey time-of-use price signals that would incentivize customers to participate more actively in demand response. Future tariff structures will have to give greater weight to fixed charges that take into account customer load. Volumetric charges will have to reflect time of use and be designed in combination with structures to remunerate prosumers injecting power into the grid.[7]

Policy implication #9: Private sector participation in distribution should be considered only when enabling conditions are met

Privatization of distribution utilities has delivered good outcomes in suitable environments, but it has proved risky where conditions were not right. Private sector participation in power distribution was widely adopted in Latin America and parts of Europe and Central Asia, with outcomes that were quite encouraging. Nevertheless, it has also been associated with disappointing performance and dramatic reversals in cases where the utility was not yet functioning at a basic level or the authorizing environment was weak. Some countries that eschewed utility privatization found other ways to incorporate the benefit of private sector discipline through financial market channels.

The policy implications are as follows:

- *Determine whether the economic preconditions for distribution privatization are in place.* Private sector participation is more likely to be successful in circumstances where (1) there is reasonably accurate information about the operating performance of the utility and the condition of its assets; (2) retail tariffs are relatively close to full (capital) cost recovery (at least 70 percent); (3) it is accepted that customers can be disconnected for nonpayment of bills; and (4) a competent regulator possesses the power to adjust tariffs as needed and the technical competence to monitor quality of service.
- *Evaluate whether the political preconditions for privatization of distribution are in place.* Even when the economic preconditions for private sector participation are in place, political impediments may remain. Private sector participation is more likely to be politically feasible in circumstances where (1) there is a broad, established tradition of private sector–led economic activity; (2) domestic actors can be involved in the privatization; (3) the value of private sector participation is clear; and (4) positive outcomes can be arranged for key stakeholder groups.
- *Explore alternative modalities for engaging the private sector.* The 1990s model considered private sector participation primarily in terms of private ownership, or at least management, of the utility. However, financial markets can provide another channel through which private sector discipline can be introduced into power distribution. This can be done through mechanisms such as listing minority shares of a state-owned utility on a local stock exchange or having the utility secure a credit rating and raise its own bond finance.
- *Maintain a proper focus on energy access.* Strengthening the utility's commercial orientation should sharpen its incentive to expand its market through electrification. However, in many developing countries, unserved customers are unprofitable owing to high incremental costs and relatively low consumption. This underscores the need

to complement distribution reforms with a sound electrification planning process comprising clear targets, an associated public funding program, and a suitable monitoring framework. At the same time, off-grid rural electrification can be advanced by creating a suitable enabling environment for private provision of off-grid solar power.

Policy implication #10: Delivering on the twenty-first century agenda of universal access and decarbonization calls for additional reform measures targeted explicitly at these objectives

Universal electrification eventually comes into conflict with a utility's commercial incentives and requires parallel policy and financial supports. Strengthening utilities' commercial orientation through private sector participation or other means can drive a rapid expansion of connections in urban areas. However, extending access to electricity to the periurban and rural periphery often leads a utility into diminishing and even negative marginal returns on investment, particularly if the power consumption of poor households remains very low. Thus, universal electrification cannot be achieved purely by allowing a utility to pursue commercial incentives. It requires complementary policy action to set access targets, provide sustained public subsidies to offset the associated financial losses, and exploit the opportunities offered by solar technology for off-grid electrification. Looking back over the past 25 years, progress on electrification was not typically synchronized with power sector reform (figure O.14a); rather, it reflected policy commitments that became increasingly likely as a country's per capita income grew. In some countries, the big push on electrification preceded sector reform; in others, it came more as an afterthought.

Power sector reform provides certain enabling conditions for decarbonization, but additional policy and planning measures must be taken to direct investors toward cleaner energy options. Private sector investment in generation can make a significant contribution to expanding renewable energy capacity. In addition, a wholesale power market, particularly when complemented by supply auctions, can provide a useful mechanism for price discovery related to new technologies, as well as a solid economic framework for pricing services ancillary to variable renewable energy and for remunerating demand response. Nevertheless, the evidence suggests that significant progress toward decarbonization over the past 25 years has been primarily driven by policy targets rather than by institutional reforms per se (figure O.14b). For most countries over this period, the overriding policy goal for generation was security of supply rather than decarbonization, leading oil-dependent countries to become less carbon-intensive as they diversified into gas, and hydro-dependent countries to become more carbon-intensive as they diversified into fossil fuels.

The policy implications are as follows:

- *Advance electrification on multiple fronts.* Countries making the most rapid progress toward electrification have done so by making simultaneous progress on- and off-grid, based on an integrated spatial master plan. They typically make long-term commitments to ambitious electrification targets, supporting them with public and donor finance and providing a suitable enabling environment. A critical issue is to ensure that both the upfront and ongoing costs of electricity are affordable for the target populations.
- *Determine explicit policy targets for decarbonization.* Achieving decarbonization goals requires explicit government direction of investment decisions in power generation, as well as incentives for the adoption of low-carbon technologies and more efficient consumption of energy.

FIGURE O.14 Progress on twenty-first-century policy objectives for electrification and decarbonization, 1990–2015, countries ranked in descending order of reform effort

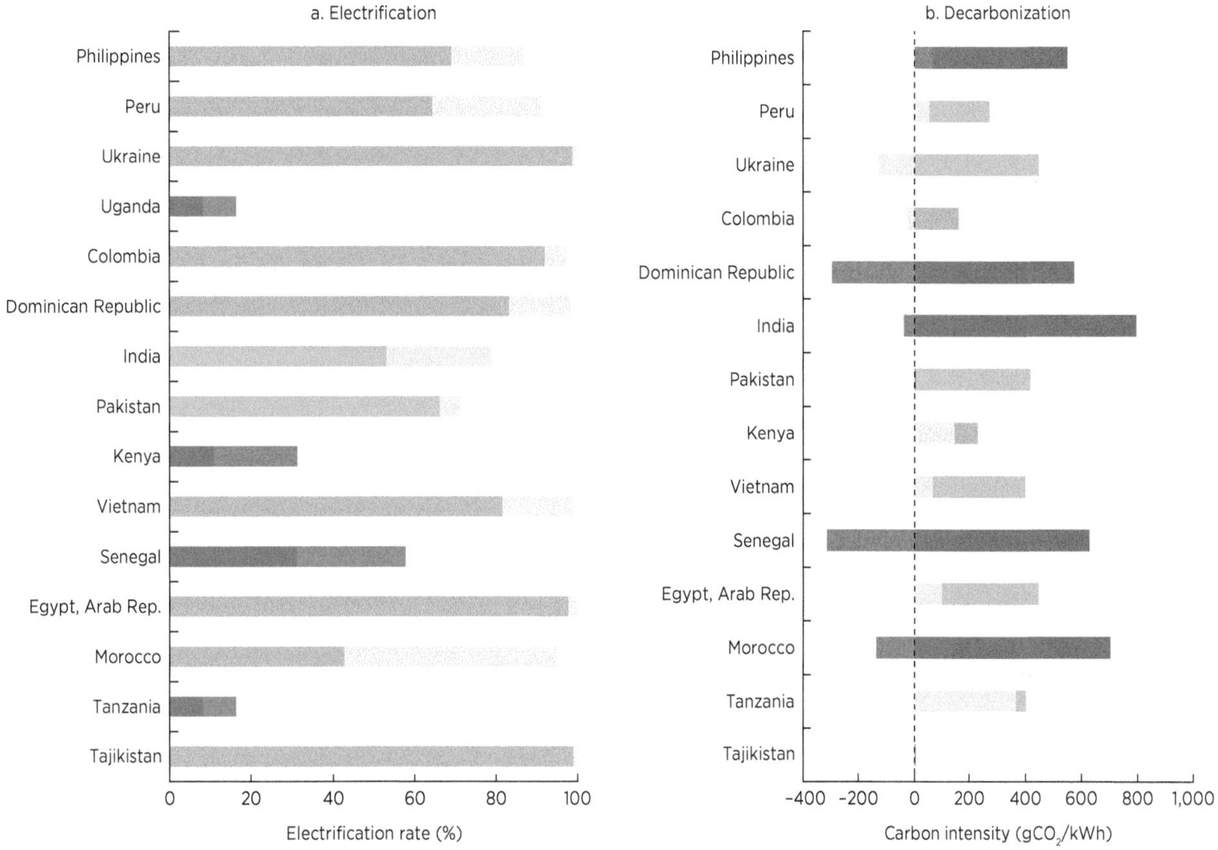

a. Electrification

Source: Based on data from Tracking SDG7 report and IEA.
Note: Dark shaded bars represent prereform electrification; the light shaded bars represent the change since then. For all panels the color of the bar follows the evaluation framework set up in table 9.2. IEA = International Energy Agency; SDG7 = Sustainable Development Goal 7.

b. Decarbonization

Source: Based on data from Tracking SDG7 report and IEA.
Note: Dark shaded bars represent average value in 2010–15; light shaded bars represent the change in values from prereform era. For all panels the color of the bar follows the evaluation framework set up in table 9.2. gCO_2/kWh = grams of carbon dioxide produced per kWh; IEA = International Energy Agency; SDG7 = Sustainable Development Goal 7.

CONCLUSIONS

Overall, it is recommended that future reforms be increasingly shaped by context, driven by outcomes, and informed by alternatives.

First, there is a need to shift from a context-neutral approach to reform to one that is shaped by context. An overarching message is that the design of reforms should be sensitive to country conditions. The 1990s power sector reform model was largely derived from economic first principles and first tested in relatively sophisticated environments. As a result, it lacks a framework for adapting reform to the country context. In practice, numerous preconditions—both economic and political—have emerged as important in shaping its applicability. A more structured approach to mapping out such prerequisites should figure prominently in future efforts along the lines offered in this report.

Second, there is a need to shift from process-oriented reform to outcome-oriented reform. The 1990s model focused primarily on a particular package of institutional reforms, which, it was argued, would lead in time to better overall sector outcomes. Rather, it is important to design a reform process by identifying the most critical outcomes and working backward from there to identify the measures most likely to remove key bottlenecks and roadblocks preventing achievement of the desired outcomes.

Third, there is a need to shift to a more pluralistic range of institutional models. Although the 1990s power sector reform blueprint has demonstrated its ability to deliver in certain country contexts, the results have been quite disappointing in other settings. Moreover, some countries that adopted only limited reforms have achieved outcomes at least as good as those achieved by countries that went further with the reform agenda. These findings make the case for a more pluralistic approach to power sector reform going forward, recognizing that there is more than one route to success.

NOTES

1. The Rethinking Power Sector Reform Observatory includes Colombia, Dominican Republic, the Arab Republic of Egypt, India (states of Andhra Pradesh, Odisha, and Rajasthan), Kenya, Morocco, Pakistan, Peru, the Philippines, Senegal, Tajikistan, Tanzania, Uganda, Ukraine, and Vietnam.
2. Demand response is defined as when the end user changes their electricity usage in response to price signals or incentives payments.
3. A simple Power Sector Reform Index was constructed to aggregate data across the four dimensions of power sector reform considered in this study. The index gives each country a score in the range 0–100 on each dimension of reform. The scores give equal weight to each step of each dimension on the reform continuum. The simple average of the four 0–100 scores is used to summarize the extent of reform. The index is purely descriptive and has no normative value. This index is described in greater detail in Chapter 2, and full technical definitions are provided in the annex of the chapter.
4. Merchant plants are typically nonutility power generation plants that compete to sell power. They usually do not have long term power purchase agreements and are mostly found in competitive wholesale power market places.
5. A Utility Governance Index measures the extent to which specific utilities conform to good practices. It is difficult to say exactly when and how good governance and management practices have been adopted over time, because such measures are usually implemented within institutions and do not necessarily involve major legal or structural changes that can readily be tracked in the public record. Nevertheless, it is possible to measure the current rate of adoption of such practices. Based on a sample of 19 state-owned and 9 privatized utilities from the 15 observatory countries, the Utility Governance Index measures the existence of best practices in utility rules and regulations. For example, a utility may, on paper, allow managers to hire and fire employees based on performance—and the index captures this—however it is unable to tell whether the manager actually does so. This index is described in greater detail in chapter 4 and the full technical definitions are provided in the annex of the chapter.
6. The survey conducted in each of the 15 Observatory countries included 355 categorical and quantitative questions on the regulatory system. The questions were both descriptive and normative. Normative questions aimed to capture regulatory best practices based on the literature. To synthesize the normative data in a convenient and intelligible format, a Regulatory Performance Index was created. Two versions of the same index were calculated for each country. First, a de jure index derives from the country's regulatory framework as captured on paper in laws, regulations, and administrative procedures. Second, a perception index determines whether the paper provisions are applied in practice. The local consultant in each country provided the perception index; his or her professional opinion was informed by some 20 interviews with key stakeholders in the reform process. The perception index was also reviewed by the World Bank country energy team knowledgeable about local context. Despite best

efforts, this second index is more subjective than the first. This index is described in further detail in chapter 6 with technical definitions in the annex of the chapter.

7. Prosumers are entities that consume as well as produce electricity.

REFERENCES

Andres, L., J. Guasch, and M. Diop. 2007. *Assessing the Governance of Electricity Regulatory Agencies in the Latin American and Caribbean Region: A Benchmarking Analysis.* Washington, DC: World Bank.

Bacon, R. W., and J. Besant-Jones. 2001. "Global Electric Power Reform, Privatization and Liberalisation of the Electric Power Industry in Developing Countries." *Annual Review of Energy and the Environment* 26 (November): 331–59.

Banerjee, S. G., A. Moreno, J. Sinton, T. Primiani, and J. Seong. 2017. *Regulatory Indicators for Sustainable Energy: A Global Scorecard for Policy Makers.* Washington, DC: World Bank.

Besant-Jones, J. 2006. "Reforming Power Markets in Developing Countries: What Have We Learned?" Energy and Mining Sector Board Discussion Paper No. 19, World Bank, Washington, DC.

Eberhard, A., and K. Gratwick. 2008. "Demise of the Standard Model of Power Sector Reform and the Emergence of Hybrid Power Markets." *Energy Policy* 36 (10): 3948–60.

Foster, V., S. Witte, S. G. Banerjee, and A. Moreno. 2017. "Charting the Diffusion of Power Sector Reforms across the Developing World." Policy Research Working Paper 8235, World Bank, Washington, DC.

Gavin, M., and D. Rodrik. 1995. "The World Bank in Historical Perspective." *American Economic Review* 85 (2): 329–34.

Gilardi, F., and M. Maggetti. 2011. "The Independence on Regulatory Authorities." In *Handbook on the Politics of Regulation,* edited by D. Levi-Faur, 201–14. Edward Elgar Publishing.

IEA (International Energy Agency), IRENA (International Renewable Energy Agency), UN (United Nations), WBG (World Bank Group), and WHO (World Health Organization). 2018. *Tracking SDG7: The Energy Progress Report.* Washington, DC: World Bank Group.

Jamasb, T., R. Nepal, and G. R. Timilsina. 2015. "A Quarter Century Effort Yet To Come of Age: A Survey of Power Sector Reforms in Developing Countries." Policy Research Working Paper 7330, World Bank, Washington, DC.

Jamasb, T., D. Newberry, and M. Pollitt. 2005. "Core Indicators for Determinants and Performance of the Electricity Sector in Developing Countries." Policy Research Working Paper No 3599, World Bank, Washington, DC.

Jayarajah, C., and W. Branson. 1995. *Structural and Sectoral Adjustment: World Bank Experience, 1980–92.* Washington, DC: World Bank.

Nepal, R., and T. Jamasb. 2012. "Reforming the Power Sector in Transition: Do Institutions Matter?" *Energy Economics* 34 (5): 1675–82.

Pollitt, M. 2008. "The Arguments for and against Ownership Unbundling of Energy Transmission." *Energy Policy 36 (2)*: 704–71.

PPI Database. 2018. Private Particicipation in Infrastructure Database. https://ppi.worldbank.org.

Vagliasindi, M. 2012. "Power Market Structure and Performance." Policy Research Working Paper 6123, World Bank, Washington, DC.

Williams, J., and R. Ghanadan. 2006. "Electricity Reform in Developing and Transition Countries: A Reappraisal." *Energy 31*: 815–44.

World Bank. 1993. "The World Bank's Role in the Electric Power Sector: Policies for Effective Institutional, Regulatory, and Financial Reform." Policy Paper, World Bank, Washington, DC.

Setting the Stage

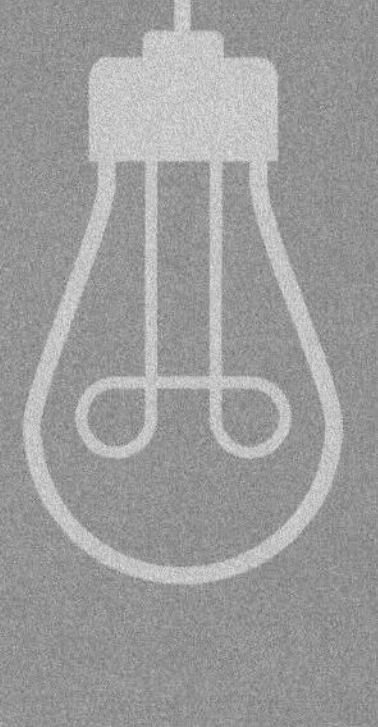

PART

I

What Do We Mean by Power Sector Reform?

1

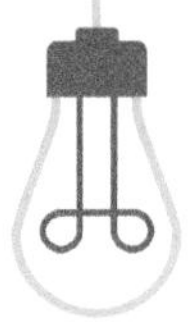

MOTIVATION

During the 1990s, a new paradigm for power sector organization emerged from the wider "Washington Consensus," a term coined in 1989. Multilateral institutions spearheaded the new paradigm across the world, and it rapidly took hold just as the Soviet Union was unraveling. Marked by 10 neoliberal policy recommendations, the era featured two policies that were particularly relevant to the power sector—namely, the privatization of state-owned enterprises (SOEs) and the abolition of regulations restricting competition. International financial institutions, notably the World Bank (box 1.1) and the International Monetary Fund, played prominent roles in diffusing market ideas throughout the developing world. They encouraged structural and sectoral adjustment programs based on market-oriented macroeconomic and fiscal policies (Jayarajah and Branson 1995). The World Bank explicitly recommended enforcing conditions to make governments commit to market reforms in the power sector (World Bank 1993). Other multilateral development banks and bilateral donors adopted similar approaches (Williams and Ghanadan 2006). Financial support was accompanied by extensive research, advocacy, technical assistance, and capacity building, working off the landmark World Development Report of 1994 (Gavin and Rodrik 1995; World Bank 1994).

By the early 2000s it had become clear, however, that power sector reform was not universally applicable in the developing world. Despite widespread uptake in Latin America and Eastern Europe, implementation proved more complicated than anticipated. The reforms required many distinct implementation phases that often called for second-generation reforms (as in Brazil and Turkey). In addition, the associated political and social challenges could be significant. Sometimes, the announced reforms could not be implemented, as, for example, in Lebanon or Zambia. At other times, reforms were implemented only to be ultimately reversed, as with the renationalization of privatized utilities in Bolivia and Kazakhstan. Many developing countries were selective about the 1990s policy recommendations, an approach that produced various hybrid models and outcomes (Gratwick and Eberhard 2008). It became evident that the 1990s reform model had been developed largely in the context of member countries of the Organisation for Economic Co-operation and Development (OECD), where, with universal access achieved and demand stagnant, boosting efficiency was the primary concern.

BOX 1.1 The World Bank and three decades of power sector reforms

Because the World Bank is often associated with the 1990s reform model, it is important to understand how the institution's own understanding and implementation of power sector reform have evolved, through the lens of official guidance documents published since the 1990s.

1993 policy paper

A World Bank policy paper issued in 1993 stands out as an early articulation of the 1990s reform model. The policy paper clearly articulates a shift in World Bank policy, which had once "largely supported the state-owned monopoly power utilities," (World Bank 1993, 11) toward a new lending focus that would "aggressively pursue the commercialization and corporatization of, and private sector participation in, developing-country power sectors" (World Bank 1993, 16). The paper further clarified that "a requirement for all power lending will be an explicit country movement toward the establishment of a legal framework and regulatory process satisfactory to the Bank" (World Bank 1993, 59). Thus, the paper clearly advocates the establishment of an independent regulator and a movement toward more commercial and corporate principles governing the sector, with emphasis on increased private sector participation. Unbundling or restructuring was suggested as a means to increase efficiency and reduce costs. Although this was the main thrust of the paper, there was also some passing recognition that "there can be no one standard approach ... for all countries" (World Bank 1993, 56). Rather, "individual countries should be encouraged to review and select the options, mechanisms, and pace of reform most appropriate to their needs and circumstances" (World Bank 1993, 22).

1994 World Development Report

The *World Development Report 1994: Infrastructure for Development* further underlines the shift in policy toward the 1990s reform model across all infrastructure sectors—including power—entailing greater competition, stronger commercialization, and a more market-oriented approach. The report stresses that infrastructure should be managed "like a business not a bureaucracy" and with more private sector involvement in management, financing, or ownership to ensure a commercial orientation (World Bank 1994, 2). The report calls for introducing competition directly (open entry) or indirectly (through competitive bidding) to increase efficiency and accountability. It views unbundling as promoting "new entry and competition in segments that are potentially competitive" but cautions against higher transaction costs (World Bank 1994, 53).

2003 World Bank Group review

A 2003 review of the World Bank Group's experience in promoting private sector participation in the power sector states that the Bank "underestimated the complexity and time required" for reforms to mature and achieve lasting and equitable sector outcomes (World Bank 2003, 26). It goes on to underline that "no single blueprint" is suitable for all sector reforms and private sector participation and that "an evolving menu of options for the combinations and sequences of reforms" had developed over time (World Bank 2003). The report mentions that the 1993 policy's emphasis on "learning by doing" had led to inadequate weight being placed on the political economy of reform (World Bank 2003, vi). The report mentions that reforms to date were "complex, time consuming, resource intensive and require[d] sequencing," although it provides few details on their timing and sequencing (World Bank 2003, ix).

2004 guidance note

Within a decade of the 1993 policy paper, the World Bank's operations were moving toward a more nuanced application of the 1990s reform model, captured in the World Bank's 2004 guidance note for "Public and Private Sector Roles in the Supply of Electricity Services." This note clearly states that staff should consider the full range of reform options, from pure public interventions to public-private arrangements and pure private arrangements, and "assess the credibility and realism of proposed Government strategies while developing interventions" (World Bank 2004, 4). It also "cautions against prescriptive, one-size-fits-all recommendations" (World Bank 2004, 1).

2006 board discussion paper

More than ten years after the publication of the 1993 policy paper, one of its authors, John Besant-Jones (2006), compiled "lessons learned" from reforms' implementation in developing countries. His paper served as a follow-up to the 2004 guidance note. Entitled "Reforming Power Markets in Developing Countries: What Have We Learned?" the paper emphasizes that a one-size-fits-all approach was "ruled out by the extensive range of economic and institutional endowments of these countries" (Besant-Jones 2006, 1). It highlights the importance of starting conditions and country context (political, social, and so forth). The paper recognizes that "the economic case for breaking up

(Box continued next page)

BOX 1.1 The World Bank and three decades of power sector reforms *(Continued)*

a vertically integrated utility rests on various factors" and may be worthwhile when the benefits exceed the costs (Besant-Jones 2006, 3). It advocates intermediate reform options (such as power-trade areas or the single-buyer model) for countries too small for a competitive power market. The paper argues for a strategy for sequencing reforms as being less risky and more sustainable, though it recognizes other approaches. It reasons that most developing countries whose power sector reforms had made substantial progress followed a logical sequence and "passed primary legislation for power market reform, established sector regulation, transacted with [independent power producers], and privatized some of the power supply industry" (Besant-Jones 2006, 111).

2016 Independent Evaluation Group evaluation

More than two decades after the landmark 1993 policy, the emphasis came to rest explicitly on the context and timing of reforms, particularly the political economy. The Independent Evaluation Group's 2016 evaluation of World Bank interventions in the power sector emphasizes the need to address the political economy of reforms by "aligning program timelines with government reform programs" (World Bank 2016, 23) and matching the "scale of the [World Bank Group] support to the scope of reforms and political risk" (World Bank 2016, 24). It cautions against "overambitious agendas and excessive conditionalities" (World Bank 2016, 25).

As a result, the model did not pay heed to the policy objectives most critical to developing countries—namely, meeting growing demand for electricity and completing the electrification process.

The policy objectives of the power sector have since evolved and expanded to encompass environmental and social concerns. Climate change emerged as a major global concern in the 2000s, with the concomitant need to decarbonize the energy sector. This need led to extensive public policy interventions promoting the scale-up of new renewable energy technologies, while also providing stronger incentives for energy efficiency. The climate change debate also surfaced the need to balance environmental action with social objectives. This imperative was enshrined in 2015 as Sustainable Development Goal (SDG) 7, committing international financial institutions to universal access to energy that is modern, affordable, reliable, and sustainable. Yet some 1 billion of the world's people continue to live without access to electricity (IEA and others 2018), and many more make do with inadequate and unreliable supply. These key environmental and social objectives did not figure into the 1990s paradigm of power sector reform.

More recently, the power sector has faced profound technological changes. The plummeting cost of solar photovoltaic power and new developments in battery storage, combined with digitized power grids, are creating possibilities for the decentralization of energy services, with power provided by a wider variety of actors. These technological trends are disrupting frontier markets where some are even calling into question the need for a traditional, centralized utility. Again, these technologies were absent from the 1990s reform landscape.

In light of the historical evidence and likely future trends, this study aims to revisit, refresh, and update the thinking on power sector reform in developing countries. The prescriptions of the 1990s reform model derived primarily from economic theory and principles. A quarter century of experience with applications of the model allows us to reevaluate this approach. The case for reevaluation hinges both on the practical difficulties encountered in the developing world and on the significant changes in policy objectives and technological opportunities that have unfolded during the intervening period. At the same time, the emergence of disruptive technologies raises questions as to how the recommendations of the 1990s model might need to be adapted going forward.

This study presents the findings of a research exercise that documents and evaluates power sector reforms across the developing world. Over a three-year period, the Rethinking Power Sector Reform initiative took stock of accumulated knowledge, compiling the latest research evidence and data on global power sector reform trends.[1] In particular, deep-dive case studies across 15 developing countries examine both the political dynamics and technical evidence regarding the process of power sector reform and its implications for sector performance. This introductory chapter provides a brief history of power sector reform, outlines the 1990s model, and sketches out the theory of change used to analyze and evaluate the impacts of reform. Finally, it articulates the nature of technological disruption in the sector and its potential relevance for future reforms.

A BRIEF HISTORY OF POWER SECTOR REFORM

From the 1870s to 1920s, electric power services were largely unregulated, elite, and private.[2] This period is regarded as the first of four institutionally normative phases in the power sector (Bhattacharyya 2011). In most countries, supply began as a fragmented market of local power providers owned by decentralized private companies or municipal governments (Besant-Jones 2006). Beyond those benefiting from public street lighting, users of electricity were mostly private—firms and privileged households.

From the 1920s onward, as electricity became a mass public good, governments took increasing control over the sector, and utilities grew from oligopolies to monopolies. The electricity sector started out in the form of private companies serving local systems. As demand for the new service increased, the need arose for larger, integrated supply systems to capture economies of scale and scope; costs and prices declined. Many governments came to consider the entire sector as a natural monopoly, whereby integration would minimize the costs of coordination between supply chain functions and finance (IEA 1999). States could also capture economies of scale by funding large projects whose high capital costs were less easily financed by private investors. State control in the sector was thus justified on grounds of economic efficiency, in addition to the public policy objectives of consumer welfare, national security, and industrial growth (Besant-Jones 2006; Brown and Mobarak 2009). To avoid a monopoly's negative outcomes, such as excessive profits, solutions included public ownership and regulation. This second phase continued through the 20th century.

From the 1940s to the 1960s, developing countries established state-owned monopoly utilities in a wave of consolidation and nationalization. These efforts received external support, including from the World Bank.[3] Public monopolies in the power sector were considered "generally satisfactory in most developing countries, in an environment of low inflation and low debt levels, and with governments allowing utilities a significant degree of managerial autonomy" (World Bank 1993, 34). More broadly, public investment in infrastructure and management of markets for economic stability were consistent with the Keynesian economics that dominated many developed countries from the 1940s to 1970s (Jahan, Mahmud, and Papageorgiou 2014). In parallel, public monopolies aligned well with the socialist and nationalist ideologies that prevailed in many newly independent developing countries. The United States' own model of regulated, private-investor-owned monopoly utilities was "widely admired and exported abroad in postwar years, though few developing countries had the capacity to duplicate the public-private checks and balances inherent in the American system" (Williams and Dubash 2004).

By the 1970s and 1980s, however, various economic and political factors came together to trigger a shift away from the paradigm of state control. Countries exhausted economies of scale and scope in the power sector, depending on the fuel and technology used and the legacy of prior policies (Victor and Heller 2007). The 1970s oil crises made countries aware of their vulnerability to fuel imports. This awareness contributed to growing consciousness of the benefits of energy conservation, especially in the United States, where experience with nuclear power eroded trust in utilities. The oil crises contributed to global economic recession; Latin America's debt crisis in the 1980s also played a role. Indebted countries turned to the International Monetary Fund and other foreign sources for finance. At the same time, technological innovation opened up new possibilities. Newly developed combined cycle gas turbine (CCGT) power plants were more efficient, at a smaller scale, than fossil fuel power plants, dramatically reducing the capital requirements as well as the marginal cost of power generated from new plants. Advances in information and communication technology made it easier to coordinate grid operation and integrate independent plants. These gains, however, were offset by several challenges. Average costs of power generation reflected sunk capital investment in aging assets that became uneconomic before the end of their expected life. As high-income countries' demand growth slowed, their power companies expanded business to developing country markets. Subsequent poorly managed international capital flows contributed to the 1997 Asian financial crisis, prompting further support from the International Monetary Fund and the World Bank.

Market-oriented power sector reforms began as experiments based on economic theory, taken up by political leaders. Beginning in 1978, Chile was the first to pursue comprehensive market reforms in its power sector (Bacon 1995), fusing elements from existing arrangements in Belgium, France, and the United Kingdom (Pollitt 2004). The ideological foundations, however, can be traced to the United States, in particular, to Milton Friedman and Friedrich Hayek at the University of Chicago. A generation of Chileans known as the "Chicago Boys" studied economics there in the 1950s. When Augusto Pinochet came to power in Chile in 1975, he empowered newly appointed officials from this group to pursue a "revolutionary market society," including in the power sector (Clarke 2017). By the late 1960s, Friedman had asserted the success of free market ideas as a counterrevolution to Keynesian economics. The influence of these ideas reached new heights when Prime Minister Margaret Thatcher and President Ronald Reagan took office in 1979 and 1981, respectively. Thatcher's pursuit of economy-wide market reforms included enacting full reform of the U.K. power sector by 1989, with a wholesale market of unprecedented complexity (Erdogdu 2014). In the United States, the 1978 Public Utilities Regulatory Policy Act allowed relatively efficient independent power producers to serve the grid and so conserve energy in response to the oil crises. By facilitating competition with incumbent monopolies, however, the act paved the way for subsequent broader restructuring in line with the liberal economic agenda of President Reagan and his successor, President George H. W. Bush.

By the 1990s, market-oriented reforms in the power sector had crystalized into a global norm, albeit with alternative formulations. The market-oriented experiences of early reformers involved a loose set of ideas. The reform literature assembled them into what was subsequently called a "blueprint for action" (Bacon 1995, 124), a "standard model" (Littlechild 2001, 1), a "standard prescription" (Hunt 2002, 8), or "textbook architecture" (Joskow 2008, 11) of sector norms. Other characterizations in the literature are normative, including those in Conway and Nicoletti (2006), EBRD (2010), and ESMAP (1999). The core elements of the 1990s reform model (described in further detail in the next section) entailed restructuring the incumbent utility; creating an independent regulatory entity; introducing private sector ownership (or at least commercial orientation); and opening to competition where relevant. This full suite of structural reforms was further underpinned by pricing reforms aimed at achieving cost-reflective tariffs, to be guaranteed by the regulatory entity, and upon which private sector investment and market competition were essentially premised.

Nevertheless, by the 2000s, government intervention in the power sector reemerged amid evolving policy concerns and observations about the perceived limitations of market-oriented reforms. This renewed intervention may be considered the fourth phase of power sector reform, with a "new debate ... about the need for intervention in the market" (Bhattacharyya 2011, 720). One concern involved the security of supply, including in high-income countries such as the United Kingdom, as old plants reached the end of their life. California's 2000–01 electricity crisis revealed striking failures of both market and sector regulation (Hunt 2002). Although these failures are not intrinsic to model reforms, the crisis nevertheless contributed to broad concern, first, over the influence of private actors in the sector and, second, about the global slowdown in the pace of subsequent market reforms (Bhattacharyya 2011). More recently, international interest in universal access and clean energy has reached new heights, as reflected in the SDGs and the Paris Agreement. These accords set a global policy agenda for the energy sector likely to be sustained for decades. Concurrently, technology is rapidly advancing in solar photovoltaics, wind turbines, energy storage, microgrids, distributed resources, electric vehicles, and related information and communication technologies.

THE 1990s POWER SECTOR REFORM MODEL

Power sector reforms in the 1990s were driven by a range of internal and external factors that varied from country to country (Bacon 1995; Bacon and Besant-Jones 2001; ESMAP 1999; Jamasb, Nepal, and Timilsina 2015; Joskow 2008; Wamukonya 2003).

In developing countries, the principal driver of reform was the poor performance of SOEs and the associated fiscal consequences (Bacon 2018). Developing countries faced surging demand for electricity, as well as low levels of electrification. Marked by high costs and low operational efficiency, SOEs proved unequal to demand, leading to a yawning supply–demand deficit, static electrification, and widespread blackouts and brownouts. Subsidies to the sector mushroomed as governments held tariffs below cost-recovery levels to avoid politically unpopular increases. The mounting fiscal burden starved the sector of the capital needed to invest in generation capacity and grid extension. To put this situation in perspective, the quasi-fiscal deficit of the power sector as a result of inefficient operations and underpricing of electricity became macroeconomically significant; it has been estimated to have a median value of about 1 percent of gross domestic product across a range of developing countries in Africa and the Middle East (Alleyne 2013; Briceño-Garmendia, Smits, and Foster 2009; Camos and others 2017; Trimble and others 2016). Governments' willingness to permit the worsening of financial performance provided no incentive for utilities to cut costs and improve efficiency. The further worsening of the fiscal situation through an external cost shock, such as an oil price hike or currency devaluation, sometimes provided the final impetus for reform. In other cases, the immediate catalysts were broader debt crises and ensuing power sector reforms as a precondition for access to multilateral finance.

In transition economies, power sector reform was part of a broader effort to create a market economy (Bacon 2018). The sector faced different challenges in these countries, which tended to have excess generation capacity and high rates of access. Nevertheless, they still charged prices far below replacement costs and struggled to maintain their aging infrastructure. Over and above these sector issues was the general reform strategy of creating a market economy based on private ownership and competition. A further incentive was provided by the desire of certain countries to join the European Union, membership in which would require them to bring the power sector in line with European Union directives.

In developed countries, power sector reform was seen as a means to reduce prices for consumers while raising proceeds for the national treasury (Bacon 2018). Developed countries did not typically suffer from many of the severe operational and financial challenges facing the power sector elsewhere. Accordingly, the motivation for reform was different. The advent of new technologies, particularly CCGT, that could lower generation costs and did not require very large units in order to reap the benefits of economies of scale, reduced the strength of the case for allowing a vertically integrated monopoly. Without economies of scale, plants and firms could be smaller, opening the field for multiple players and competition, once private capital could enter the industry. The hope was that competition would deliver lower prices, better quality, and greater choice for consumers, while the associated privatization of state-owned assets would strengthen the financial position of the state. An unintended consequence of energy deregulation in member countries of the OECD was the creation of a new breed of energy multinational. This new kind of multinational would eventually go looking for investment opportunities abroad as power sector reform took root in the emerging economies.

The 1990s power sector reform model comprises a package of four structural reforms (Foster and others 2017):

- The first of the structural actions is *regulation*, which requires an autonomous regulatory entity able both to provide a degree of political independence and to hold utilities accountable for their operational and financial performance.
- The second is *restructuring*, which involves steps toward the eventual full vertical and horizontal unbundling of the incumbent state-owned monopoly.
- The third is *private sector participation*, which brings private management and capital into the sector to boost operational efficiency and investment (which often involved a preceding step of corporatizing SOEs).
- The fourth element is *competition*, which initially allows generators to compete to supply a monopoly utility and eventually allows customers to negotiate their supply contracts directly with power producers and traders supported by a power exchange; in some countries, retail competition for small customers was facilitated through suppliers of alternative energy.

Regulatory reform is defined as the establishment of an autonomous entity with responsibility for regulatory oversight and with some role in decision making (Foster and others 2017). The power sector provides policy-making, regulatory, and service provision functions. Policy making charts the sector's strategic direction. Regulation oversees the sector to ensure that it follows and enforces the strategic direction, while service provision is the implementation of that strategic direction. Traditionally, all three functions have been combined within the line ministry; however, with widespread corporatization of the service provider to form an enterprise distinct from the line ministry, as well as the growing delegation of these activities to the private sector, the need for a clearer regulatory function has been felt. A widely adopted model, drawing heavily on experience in the United Kingdom and the United States, has been to create an entity dedicated to regulation. Independence is considered important. It isolates the regulator (and, ultimately, the service provider) from short-term, opportunistic political interference, and provides balance between the investor's right to a fair return on capital and the consumer's right to value. Independence, although always relative, has been defined in terms of an institutional existence, governance structure, and budget line that are separate from the line ministry itself. In practice, however, genuine independence has proved difficult in many political systems and cultures. Governments are generally reluctant to relinquish their discretionary powers over political patronage such as electricity tariffs, power sector investment plans, and utility employment. The roles of the regulator include setting tariffs to recover efficient costs, monitoring and enforcing service standards, and overseeing market entry.

Restructuring reform is defined as movement along a spectrum toward full vertical and horizontal unbundling of the sector (Foster and others 2017). The starting point for restructuring reform is typically a vertically integrated national monopoly utility, and its theoretical endpoint is a fully restructured sector, with restructuring entailing both the vertical unbundling of generation, transmission, and distribution, and the horizontal unbundling of the generation and distribution tiers to create multiple companies operating in parallel. Such a process paves the way for competition, according to the theoretical rationale. First, vertical unbundling aims at removing any conflict of interest that may arise when a single utility has more than one function along the electricity supply chain. For example, a transmission company that also engages in generation may have the incentive to prioritize grid access for its own generation capacity as opposed to that of competitors. It is relevant to distinguish partial vertical unbundling—in which, for example, generation is broken out but transmission and distribution remain joined—from full vertical unbundling, whereby separate entities undertake all three segments of the electricity supply chain. Second, horizontal unbundling aims at diluting market power, which is particularly relevant for generation. For example, a country with five or six similar generation companies will likely experience stronger competitive pressure than a country with only two companies—a large one and a small one. Although full unbundling entails separate ownership of the different entities created in the process, in practice unbundling may proceed gradually—beginning with defining the distinct management units and then separating accounts and constituting separate entities under distinct ownership.

Competition is defined as the coexistence of multiple service providers in the same market (Foster and others 2017). Competition among service providers promotes efficiency and innovation. When multiple companies compete head to head for consumers, a market discipline emerges, along with pressure to keep costs down to efficient levels and to improve service quality. The large economies of scale in the power sector mean that key activities (for example, transmission) are traditionally considered natural monopolies, making it inefficient to have more than one supplier. Even under a natural monopoly, however, it is still possible to have different companies compete for the right to supply the market on a monopoly basis for a certain period of time. The liberalization of the power sector therefore often proceeds in incremental stages, beginning with the opening up of generation to independent power producers that compete for the market. Eventually, it may transition to a full single-buyer model where generation is fully divested from the incumbent utility, with the latter acting as the single buyer of generation on behalf of end consumers. The next stage—once the transmission segment has been fully unbundled—is to allow third-party access to the power grid so large customers can purchase power directly from generators on a bilateral negotiated basis. In due course, it may evolve into a wholesale power market, with a centralized price-setting mechanism and a variety of contracts and products being exchanged. In some instances, a final step would unbundle the distribution and retail functions of the utility, allowing the latter to be open to competition for energy supply.

Private sector participation is defined as the introduction of private sector management and investment, whether through temporary contractual arrangements or permanent asset sales (Foster and others 2017). The 1990s power sector reform model presupposed that the power utility had already been corporatized—that is, separated from a ministerial department and constituted as a free-standing SOE operating with a commercial orientation and governed by company law. Thereafter, private sector participation can be implemented to varying degrees along a number of dimensions. First, the *scope* of participation varies according to the extent of the electricity supply chain affected. Private sector participation may initially be undertaken in one segment of the supply chain, often generation, but not necessarily in another, such as distribution. Second, it may affect some, but not all, companies in a particular supply segment. For example, a country may privatize some of its generation plants, but leave the others under public ownership. The *coverage* of private sector participation may be gauged for generation according to the percentage of capacity under private control, and for distribution by the percentage of distribution companies under private control. Third, the *depth* of private sector participation may range from, for example, a management contract, where the private sector has neither responsibility for investment nor any exposure to commercial risk; to a concession, where the private sector has time-bound responsibility for investment and exposure to commercial risk; to a divestiture, where the private sector permanently takes over all responsibilities and risks associated with electricity supply. Although the 1990s reform model encouraged private sector participation in generation and distribution, no concomitant recommendation existed for the transmission segment, whose strong natural monopoly characteristics and strategic character seemed to justify continued public ownership.

It is important to note that the various measures constituting the 1990s power sector reform model were seen as mutually supportive and intended for implementation as a package (Bacon 2018). Some policies have a direct effect, whereas others act as facilitators, without which the direct policies cannot be effective. For example, privatizing the SOE without prior restructuring would simply create a private monopoly in lieu of a public one. In the absence of regulation and competition, private monopolies tend to use their monopoly power to maximize profits by restricting output via excessive prices. Hence, they absorb much of the gain from efficiency improvement and produce little benefit for consumers even though production costs have been cut. Competition serves to discipline abusive behavior and is made possible by a prior unbundling of the sector; for natural monopolies, regulation can redistribute the gains between consumers and the producer. Sector unbundling without privatization simply multiplies the number of SOEs in the sector without fundamentally affecting their performance. Or again, regulation without privatization essentially duplicates the control of whichever government department was responsible for the actions of the SOE. In short, all the elements of the reform package are needed for the full performance impacts to materialize.

A THEORY OF CHANGE

This study uses the theory of change underpinning the 1990s reform model as a conceptual framework for evaluating the model's efficacy. This theory of change proposes that reforms would lead to beneficial behavior change among the key sector actors, resulting in improved sector performance. Behavior changes when private management is introduced. Private management reorients enterprises from bureaucratic and political incentives toward profit seeking, cost control, and customer orientation. Potential for private management abuses would be disciplined either by market pressures in competitive segments of the supply chain or by regulatory incentives in natural monopoly segments of the supply chain. The presence of the regulator, as well as the engagement of the private sector, would meanwhile prevent opportunistic interference in the day-to-day operation of the sector for political ends. Thus, overall, the adoption of the 1990s reform package was expected to improve cost recovery and operational efficiency. Over time, the utility becomes more financially viable and permits greater investment. These intermediate outcomes would then feed through into greater security of supply. Security of supply was the main outcome envisaged in the 1990s paradigm (figure 1.1).

More recently, however, the range of outcomes sought for the sector has been expanded to include social and environmental benefits. These outcomes are enshrined in SDG7 and comprise universal access to electricity as well as renewable energy and energy efficiency, both important contributors to a clean and low-carbon energy future, helping to reduce global warming and improve local air quality. These objectives were never explicitly addressed by the 1990s model; however, to the extent that they rely on increased investment for their

FIGURE 1.1 **1990s model sector reforms: Inferred simple theory of change**

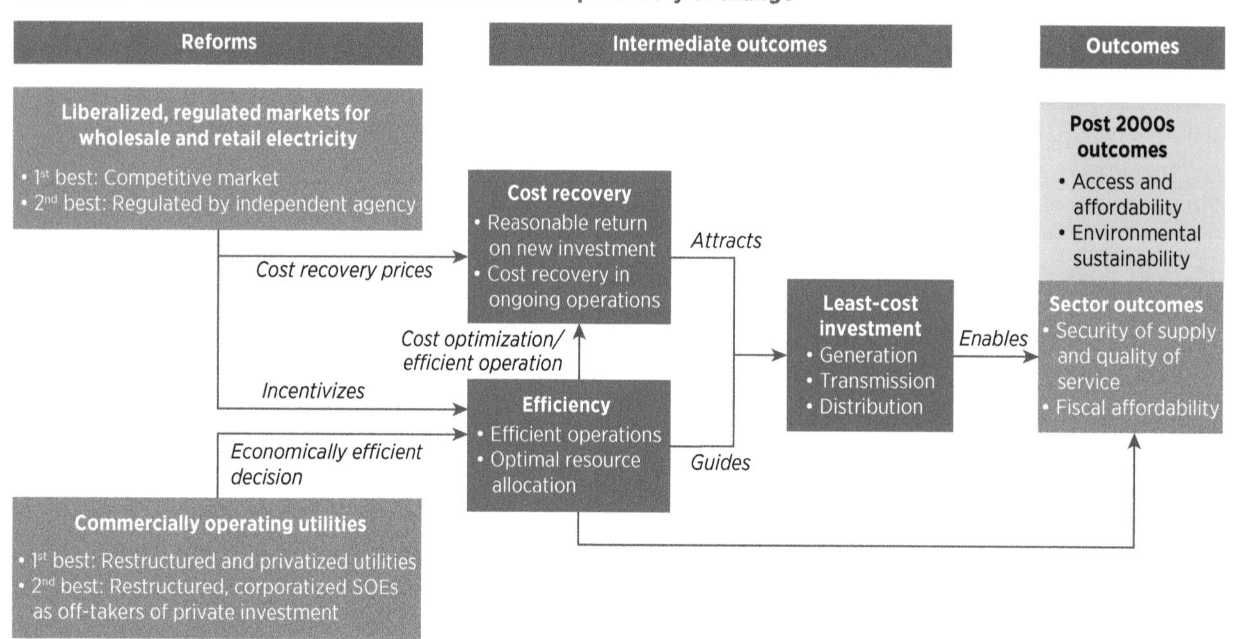

Source: World Bank elaboration.
Note: SOE = state-owned enterprise.

achievement, they could (at least in principle) benefit from the kinds of reforms supported by the 1990s model. On the one hand, it is unfair to judge the performance of the 1990s model against environmental and social policy objectives it was never explicitly designed to achieve. On the other hand, given the importance of these policy objectives, it is relevant to ask if the 1990s model was incidentally helpful in advancing these newer agendas.

This study uses a hybrid methodology—quantitative cross-country analysis combined with in-depth qualitative country case studies. Much of the academic literature on power sector reform focuses on econometric analysis of large sample cross-country panel data (Erdogdu 2011; Nepal and Jamasb 2012; Sen, Nepal, and Jamasb 2016; Zhang, Parker, and Kirkpatrick 2005), with a view to estimating whether various reforms affect sector and economic performance outcomes (Bacon 2018). Although wide-ranging and precise, these studies are constrained by the availability of panel data for suitable indicators and cannot provide insights on the process and dynamics of reforms. Much of the policy literature on power sector reform focuses instead on in-depth case studies of country-level reforms (Newberry and Pollitt 1997; Pineau 2009). Although those studies are rich in content, their narrow scope does not allow for wider generalizations. Following the approach of Vagliasindi and Besant-Jones (2013), this study adopts a hybrid approach—power sector reforms in 15 countries are studied in depth (with India providing three state-level case studies). These case studies can then be both tapped for qualitative insights and pooled to provide cross-country data patterns. It is true, however, that a sample of 15 countries remains relatively small, and cross-country analysis cannot be considered representative in any statistical sense. Nevertheless, the sample contains a range of geographies, income groups, political systems, and power sector conditions, allowing cross-country comparisons to be illustrative of wider relationships between power sector measures and outcomes (box 1.2).

Because power sector reform is but a means to an end, the important thing is whether countries are able to achieve good performance outcomes. As noted, the 1990s power sector reform model is motivated by the desire to obtain good sector performance outcomes; it offers one particular theory of change regarding how these outcomes can be reached. This report is based on the premise that the implementation of the 1990s sector reform model—or, for that matter, of any other institutional model for the power sector—is valuable insofar as it improves performance outcomes. This report will focus on evaluating the degree to which countries improved their performance outcomes in their power sectors and understanding the extent to which the reform measures may have contributed.

Power sector performance can be evaluated along three key dimensions: security of supply, social inclusion, and environmental sustainability. For the power sector to fulfill its role as an enabler of economic activity, it must expand generation capacity to keep pace with peak demand so consumers have a reliable supply of electricity. For the power sector to fulfill its role as a platform for social inclusion, a country must attain universal electrification and provide basic service that is affordable across all population tiers. For the power sector to contribute to environmental sustainability, it must reduce the carbon intensity of electricity generation and diminish the impact on local air quality.

Whereas the 1990s reform model focused primarily on the economic dimension of the power sector, since 2010 the social and environmental dimensions of the power sector have grown in significance. The three dimensions of power sector reform described represent a 21st-century understanding of the sector.

BOX 1.2 The Rethinking Power Sector Reform Observatory: An introduction

At the heart of this study is a Power Sector Reform Observatory, comprising data from 15 countries. Although not representative in the statistical sense, this sample illustrates a range of country and power sector settings. As illustrated in table B1.2.1, the observatory countries are spread evenly across the six developing regions of the world and include a range of income levels, political systems, system scales, and degrees of power system development.

Furthermore, the sample was designed to include countries at an advanced reform stage (for example, Peru and the Philippines) in addition to countries that have made slower progress (for example, Tajikistan and Tanzania). It also aimed to include countries with strong and weak power sector performances. Of particular interest were countries at an advanced stage of reform yet with relatively weak performance (for example, the Dominican Republic) and countries with more modest reforms yet relatively strong performance (for example, Vietnam).

In each of the 15 observatory countries, a local consultant was hired to undertake an ambitious data collection exercise. The exercise detailed 90 quantitative performance indicators for the power system overall and for the one or two major distribution utilities in that jurisdiction; a questionnaire seeking qualitative data across 460 categorical variables produced an exhaustive description of power sector reform measures affecting policy, planning, regulation, and utility governance. A series of in-depth open-ended interviews with some 20 key stakeholders documented their experiences and perspectives on the country's power sector reform journey.

Following the collection phase, data were validated by the project team with input from the World Bank energy team for each country. The information collected across all 15 countries was consolidated into a single cross-country power sector database. Quantitative data were cleaned and analyzed according to a consistent rubric. Qualitative data were classified between descriptive variables and those that captured normative features of the institutional framework. Researchers developed a series of normative indexes to aggregate this wealth of information into a more tractable form; these indexes cover good practices in power system planning, power sector regulation, and utility governance. The indexes will be introduced later, in their corresponding chapters. Interview transcripts were also compared across countries and summarized according to a qualitative framework seeking to identify any differences and commonalities.

TABLE B1.2.1 Power Sector Reform Observatory countries and selected characteristics

Country	Region	Income group	Political system[a]	Installed capacity (GW)[b]	Electrification rate (%)[c]
Colombia	Latin America and Caribbean	Upper middle	Multipolar	17.3	99
Dominican Republic	Latin America and Caribbean	Upper middle	Multipolar	3.8	100
Egypt, Arab Rep.	Middle East and North Africa	Lower middle	Unipolar	42.1	100
India	South Asia	Lower middle	Multipolar	291.3	85
Kenya	Sub-Saharan Africa	Lower middle	Transition	2.4	75[d]
Morocco	Middle East and North Africa	Lower middle	Unipolar	7.8	100
Pakistan	South Asia	Lower middle	Transition	29.3	72[d]
Peru	Latin America and Caribbean	Upper middle	Unipolar	12.2	95
Philippines	East Asia and Pacific	Lower middle	Transition	21.7	91
Senegal	Sub-Saharan Africa	Lower middle	Transition	0.9	65
Tajikistan	Europe and Central Asia	Lower middle	Unipolar	5.4	100
Tanzania	Sub-Saharan Africa	Low	Transition	1.5	33
Uganda	Sub-Saharan Africa	Low	Unipolar	0.9	27
Ukraine	Europe and Central Asia	Lower middle	Multipolar	49.9	100
Vietnam	East Asia and Pacific	Lower middle	Unipolar	37.9	100

Source: World Bank elaboration based on data collected for Rethinking Power Sector Reform Project.
Note: GW = gigawatt.
a. At time of reform implementation.
b. Platts database up to 2017 for operational installed capacity.
c. Access data from IEA, IRENA, UN Statistics Division, WBG, and WHO 2018.
d. Based on individual country statistics.

They are particularly informed by recent political agreements such as SDG7, which articulates a target for universal access to affordable electricity by 2030 alongside substantial progress toward a cleaner power sector. The Paris Climate Accord, too, calls for energy sector measures that cap global warming at 1.5 degrees Celsius. In the 1990s, the key concern was to achieve conditions necessary to support investment in new infrastructure aimed at achieving security of supply. Because the urgency of environmental and social concerns was not as evident in the 1990s when the power sector reform model was developed, it would be anachronistic to evaluate the outcomes of power sector reforms against these more recently articulated goals. Nevertheless, given their prominence, it is relevant to ask whether the 1990s power sector reform model might also play a role in advancing these agendas.

AN UNCERTAIN FUTURE

The traditional power sector reform agenda is already being affected by technological disruptions sweeping through the sector. Technological change—particularly when it is rapid—fundamentally affects the cost structure of industries and may reshape a sector's institutional organization. The electricity sector is undergoing precisely such changes, which are increasingly calling into question many of the traditional paradigms and reshaping both the generation and distribution tiers.

The power sector reform paradigm of the 1990s emerged during a period of relative technological stability. The technical principles governing the electricity industry did not change fundamentally during its first century of existence. For many years, technology change in the power sector has been largely incremental. Open cycle gas turbine plants gradually gave way to CCGT plants, meters got progressively smarter, and renewable energy costs gradually fell. A few technologies may have slightly affected institutions—for example, modular CCGTs made the unbundling and privatization of generation a bit easier, and prepayment meters helped alleviate financial distress along the electricity supply chain—but not much could really be described as truly disruptive of institutions. It was against this relatively stable technological backdrop that the 1990s power sector reform model emerged, developing under the implicit assumption that technology was largely a given.

In contrast, the electricity sector in the 21st century has found itself at the center of momentous technological change. Like cars displacing horses, or cell phones displacing landlines, or photography going digital, a new tipping point—that is, the point at which a new technology becomes so inexpensive it is nearly ubiquitous—may be here. In the power sector, this means that almost all new investments in generation will soon be renewable, with implications for grid operations, system planning, and market design. It means that battery storage will become increasingly economic and prevalent, breaking the longstanding constraint that supply and demand for electricity must match precisely and instantaneously at every moment. It means that power grids are becoming increasingly digitized and intelligent, enabling more sophisticated coordination of supply, demand, and storage activities across a growing number of actors. It means electricity consumers in many countries are playing roles that are much different from those they played in the past. Exceeding the old passive consumption roles, they can engage in their own decentralized generation, sell surplus energy or storage capacity back into the grid, participate in remunerated demand-reduction measures, and even trade electricity on the distribution grid.

These new technologies also challenge existing models of planning, regulation, and institutional structure. The integration of renewables, storage, and consumer-sited distributed energy resources has profound implications for patterns of electricity consumption, approaches to power sector reform, and pathways toward climate goals. When power no longer flows one way, but instead needs to be coordinated across multiple consumer-sited locations on the distribution system, then regulatory models designed for one-way power flow need to be reconsidered, sector planning and grid operations become much more complicated, and traditional business models are challenged. As consumers have the option to produce some of their own energy or reduce some of their own consumption, they alter the total system peak demand, and may even alter how the grid system is (or should be) planned.

Politicians and technocrats, policy makers and system planners, investors and civil society, and financiers and regulators are all attempting to understand the implications of that change. They are trying to figure out what will happen to investment decision making and to sector institutions in the disruptive technology maelstrom, and that outcome may be unclear for some time. As these new disruptive technologies are adopted, there will be big winners and big losers, and the inevitable tensions between them will affect the likely pace at which new technologies are adopted. Which energy future countries adopt has a lot to do with the choices they have already made to enable electricity production. Thus, policy makers and practitioners will have to balance their act delicately to adapt to these coming changes.

Finally, these technological disruptions will clearly have implications for the future application of the 1990s reform model. The technological disruption is still at an early stage, and it would be foolhardy to attempt to predict where it will ultimately lead. Moreover, because of large differences in starting conditions, policy objectives, and institutional capacities, the direction taken by technological disruption could be expected to differ substantially across developed and developing countries. What is clear is that the technological disruption could have implications for the prescriptions of the 1990s reform model, affecting their degree of relevance and ease of implementation. Each key chapter of this report ends with some preliminary reflections on how technological disruption will affect the various dimensions of reform covered by the study.

This report is organized as follows. Part I (chapters 1–3) sets the stage for power sector reform. Chapter 2 outlines the uptake, diffusion, packaging, and sequencing of power sector reforms around the world. These reforms' underlying political economy dynamics are examined in chapter 3, which also considers variations in implementation.

Part II (chapters 4–7) examines the four building blocks of power sector reform. Chapter 4 looks at utilities and the extent to which unbundling and governance reforms were able to turn around operational performance and what strategies were adopted to address financial shortfalls. Chapter 5 investigates the role of the private sector in determining the performance of the power sector. Chapter 6 considers the practice of regulation, evaluating the quality of regulatory governance and practice and the extent to which they have been successful in insulating the sector from political interference and helping utilities achieve cost recovery. Chapter 7 considers developing country experience with the creation of wholesale power markets, clarifying the conditions in which they may be helpful and elucidating a range of market design and transition issues.

Part III (chapters 8–9) looks at the impact of power sector reforms both on improving utility performance and on enhancing outcomes for consumers. Chapter 8 examines the extent of utilities' progress toward cost recovery and operational efficiency. Intermediate outcomes are seen to improve utilities' financial viability and permit greater investment. Finally, chapter 9 evaluates the extent to which power sector reforms have been able to deliver on the performance outcomes they were intended to achieve.

NOTES

1. The 20 background papers for this project are being gradually released and posted on the Energy Sector Management Assistance Program website, https://www.esmap.org/rethinking_power_sector_reform.
2. This section is based on Lee and Usman (2018).
3. See Kapur, Lewis, and Webb (1997) for a fascinating account of the World Bank's history, and Collier (1984) for its early, prominent support of the power sector of developing countries.

REFERENCES

Alleyne, Trevor. 2013. *Energy Subsidy Reform in Sub-Saharan Africa: Experiences and Lessons.* Washington, DC: International Monetary Fund.

Bacon, Robert W. 1995. "Privatization and Reform in the Global Electricity Supply Industry." *Annual Review of Energy and the Environment* 20 (1): 119–43.

———. 2018. "Taking Stock of the Impact of Power Utility Reform in Developing Countries: A Literature Review." Policy Research Working Paper 8460, World Bank, Washington, DC.

Bacon, Robert W., and John Besant-Jones. 2001. "Global Electric Power Reform, Privatization, and Liberalization of the Electric Power Industry in Developing Countries." *Annual Review of Energy and the Environment* 26 (1): 331–59.

Besant-Jones, John E. 2006. "Reforming Power Markets in Developing Countries: What Have We Learned?" Energy and Mining Sector Board Discussion Paper No. 19, World Bank, Washington, DC.

Bhattacharyya, Subhes C. 2011. *Energy Economics: Concepts, Issues, Markets, and Governance.* London: Springer.

Briceño-Garmendia, C., K. Smits, and V. Foster. 2009. *Financing Public Infrastructure in Sub-Saharan Africa: Patterns and Emerging Issues.* Washington DC: World Bank Group.

Brown, David S., and Ahmed Mobarak. 2009. "The Transforming Power of Democracy: Regime Type and the Distribution of Electricity." *American Political Science Review* 103 (2): 193–213.

Camos, D., R. Bacon, A. Estache, and M. M. Hamid. 2017. *Shedding Light on Electricity Utilities in the Middle East and North Africa: Insights from a Performance Diagnostic.* Washington, DC: World Bank.

Clarke, Timothy David. 2017. "Rethinking Chile's Chicago Boys: Neoliberal Technocrats or Revolutionary Vanguard?" *Third World Quarterly* 38 (6): 1350–65.

Collier, Hugh. 1984. *Developing Electric Power: Thirty Years of World Bank Experience.* Washington, DC: World Bank.

Conway, Paul, and Giuseppe Nicoletti. 2006. "Product Market Regulation in the Non-Manufacturing Sectors of OECD Countries: Measurement and Highlights." Economics Department Working Paper No. 530, Organisation for Economic Co-operation and Development, Paris.

EBRD (European Bank for Reconstruction and Development). 2010. *Transition Report 2010: Recovery and Reform.* London: EBRD.

Erdogdu, Erkan. 2011. "The Impact of Power Market Reforms on Electricity Price-Cost Margins and Cross-Subsidy Levels: A Cross-Country Panel Data Analysis." *Energy Policy* 39 (3): 1080–92.

———. 2014. "The Political Economy of Electricity Market Liberalisation: A Cross-Country Approach." *Energy Journal* 35 (3): 91–128.

ESMAP (Energy Sector Management Assistance Program). 1999. *Global Energy Sector Reform In Developing Countries: A Scorecard.* Washington, DC: World Bank.

Foster, Vivien, Samantha Witte, Sudeshna Ghosh Banerjee, and Alejandro Moreno. 2017. "Charting the Diffusion of Power Sector Reforms Across the Developing World." Policy Research Working Paper 8235, Rethinking Power Sector Reform, World Bank, Washington, DC.

Gavin, Michael, and Dani Rodrik. 1995. "The World Bank in Historical Perspective." *American Economic Review* 85 (2): 329–34.

Gratwick, Katharine N., and Anton Eberhard. 2008. "Demise of the Standard Model of Power Sector Reform and the Emergence of Hybrid Power Markets." *Energy Policy* 36 (10): 3948–60.

Hunt, Sally. 2002. *Making Competition Work in Electricity.* New York: Wiley.

IEA (International Energy Agency). 1999. *Electricity Market Reform: An IEA Handbook.* Paris: Organisation for Economic Co-operation and Development.

IEA (International Energy Agency), IRENA (International Renewable Energy Agency),

UN (United Nations) Statistics Division, WBG (World Bank Group), and WHO (World Health Organization). 2018. *Tracking SDG7: The Energy Progress Report.* Washington, DC: World Bank Group.

Jahan, Sarwat, Ahmed Saber Mahmud, and Chris Papageorgiou. 2014. "What Is Keynesian Economics?" *Finance and Development* 51 (3): 53–54.

Jamasb, Tooraj, Rabindra Nepal, and Govinda R. Timilsina. 2015. "A Quarter-Century Effort Yet to Come of Age: A Survey of Power Sector Reforms in Developing Countries." Policy Research Working Paper 7330, World Bank, Washington, DC.

Jayarajah, Carl, and William Branson. 1995. *Structural and Sectoral Adjustment: World Bank Experience, 1980–92.* A World Bank operations evaluation study. Washington, DC: World Bank Group.

Joskow, P. L. 2008. "Lessons Learned from the Electricity Market Liberalization." *Energy Journal* 29 (special issue): 9–42.

Kapur, Devesh, John P. Lewis, and Richard C. Webb. 1997. *The World Bank: Its First Half Century.* Volumes 1 and 2. Washington, DC: World Bank.

Lee, Alan David, and Zainab Usman. 2018. "Taking Stock of the Political Economy of Power Sector Reforms in Developing Countries: A Literature Review." Policy Research Working Paper 8518, Rethinking Power Sector Reform, World Bank, Washington, DC.

Littlechild, S. C. 2001. "Electricity: Regulatory Developments from Around the World." Beesley Lectures on Regulation series XI, Institute of Economic Affairs/London Business School, London, October 9.

Nepal, Rabindra, and Tooraj Jamasb. 2012. "Reforming the Power Sector in Transition: Do Institutions Matter?" *Energy Economics* 34 (5): 1675–82.

Newberry, David M., and Michael G. Pollitt. 1997. "The Restructuring and Privatisation of Britain's CEGB—Was It Worth It?" *Journal of Industrial Economics* 45 (3): 269–303.

Pineau, Pierre-Olivier. 2009. "Electricity Sector Reform in Cameroon: Is Privatization the Solution?" *Energy Policy* 30 (11–12): 2249–61.

Pollitt, Michael G. 2004. "Electricity Reform in Chile: Lessons for Developing Countries." *Competition and Regulation in Network Industries* 5 (3–4): 221–62.

Sen, Anupama, Rabindra Nepal, and Tooraj Jamasb. 2016. "Reforming Electricity Reforms? Empirical Evidence from Asian Economies." OIES Paper: EL 18, Oxford Institute for Energy Studies, Oxford, UK.

Trimble, Christopher Philip, Masami Kojima, Ines Perez Arroyo, and Farah Mohammadzadeh. 2016. "Financial Viability of Electricity Sectors in Sub-Saharan Africa: Quasi-Fiscal Deficits and Hidden Costs." Policy Research Working Paper 7788, World Bank, Washington, DC.

Vagliasindi, Maria, and John Besant-Jones. 2013. *Power Market Structure: Revisiting Policy Options.* Directions in Development Series. Washington, DC: World Bank.

Victor, David G., and Thomas C. Heller. 2007. *The Political Economy of Power Sector Reform: The Experiences of Five Major Developing Countries.* New York: Cambridge University Press.

Wamukonya, Njeri. 2003. "Power Sector Reform in Developing Countries: Mismatched Agendas." *Energy Policy* 31 (12): 1273–89.

Williams, James, and Navroz K. Dubash. 2004. "Asian Electricity Reform in Historical Perspective." *Pacific Affairs* 77 (3): 411–36.

Williams, James, and Rebecca Ghanadan. 2006. "Electricity Reform in Developing and Transition Countries: A Reappraisal." *Energy* 31 (6–7): 815–44.

World Bank. 1993. *The World Bank's Role in the Electric Power Sector: Policies for Effective Institutional, Regulatory, and Financial Reform.* Washington, DC: World Bank.

———. 1994. *World Development Report 1994: Infrastructure and Development.* New York: Oxford University Press.

———. 2003. "Private Sector Development in the Electric Power Sector: A Joint OED/OEG/OEU Review of the World Bank Group's Assistance in the 1990s." World Bank, Washington, DC.

———. 2004. "Public and Private Sector Roles in the Supply of Electricity Services." Operational Guidance for World Bank Group Staff, Energy and Water Department, World Bank, Washington, DC.

———. 2016. "Financial Viability of the Electricity Sector in Developing Countries: Recent Trends and Effectiveness of World Bank Intervention." World Bank, Washington, DC.

Zhang, Yinfang, David Parker, and Colin Kirkpatrick. 2005. "Competition, Regulation, and Privatization of Electricity Generation in Developing Countries: Does the Sequencing of the Reforms Matter?" *Quarterly Review of Economics and Finance* 45 (2–3): 358–79.

How Far Did Power Sector Reform Spread in the Developing World?

2

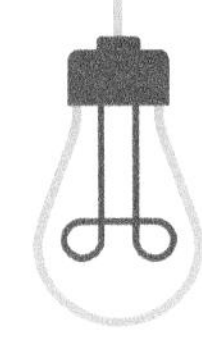

Guiding Questions

- *What factors have affected the uptake of power sector reform over time?*
- *Was power sector reform implemented as originally envisaged?*

Summary

- *The uptake of power sector reform in the developing world lags well behind that in the member countries of the Organisation for Economic Co-operation and Development.*
- *Following the rapid diffusion of power sector reform from 1995 to 2005, uptake notably slowed in the subsequent decade, 2005–15.*
- *A country's geographical, economic, and power system characteristics strongly affect the uptake of reform.*
- *Power sector reform packages show clear signs of cherry-picking of recommended measures: easier ones were implemented more frequently than more difficult ones.*
- *Many countries are settled at an intermediate stage of power sector reform.*
- *Power sector reforms were often packaged and sequenced in ways unrelated to the original reform logic.*

This chapter examines how extensively developing countries adopted the policy prescriptions of the 1990s power sector reform model.[1] Building on a global survey of reform measures and their diffusion and uptake, the chapter tracks the model's four components: regulation, restructuring, private sector participation (PSP), and competition. Considering each of these measures separately allows researchers to characterize how the reform measures were sequenced and packaged. The guiding questions for this chapter are the following: How rapidly did the 1990s model of power sector reform spread globally, and has the initial momentum been sustained? Did the model adopted in developing countries go as far it did in developed countries? Were reforms sequenced and packaged in line with the original policy recommendations? Finally, how was the adoption of power sector reforms influenced by the particular characteristics of countries' economies and power sectors?

A Power Sector Reform Index (PSRI) was developed as part of this study to measure the extent to which countries adopted the four structural measures of the 1990s model (box 2.1). The index aims only to describe a country's reform efforts; it makes no judgment on whether more reform is better or worse than less. By quantifying the extent of implementation in every country and every year in the sample, however, the index is useful for characterizing and analyzing patterns. The main findings are presented as a series of observations in the next section of this chapter.

KEY FINDINGS

The analysis set out in this chapter is based on a survey of 22 developed and 88 developing countries and their implementation of power sector reform measures. Undertaken for the Regulatory Indicators for Sustainable Energy (RISE) project for the period from 1990 to 2015, this survey provides a rich source of evidence regarding global patterns of power sector reform. Six main findings are described below.

Finding #1: The uptake of power sector reform in the developing world lags substantially behind that in the member countries of the Organisation for Economic Co-operation and Development

The 1990s power sector reform model has been more widely adopted among Organisation for Economic Co-operation and Development (OECD) countries than in the developing world. The median PSRI score is 78 for OECD countries; it is 37 for developing countries. This disparity is evident in comparisons of the PSRI frequency distribution for both groups of countries (figure 2.1). For the OECD country group, the values are strongly skewed toward scores in the 80–90 range; and only about 1 in 10 of these countries scores below 50. For the developing country group, the distribution is much flatter and concentrated in the lower range of the PSRI, with two-thirds of countries

BOX 2.1 Defining the Power Sector Reform Index

A simple Power Sector Reform Index was constructed to aggregate data across the four dimensions of power sector reform considered in this study. The index gives each country a score in the range 0–100 on each dimension of reform. The scores give equal weight to each step of each dimension on the reform continuum (see table B2.1.1). The simple average of the four 0–100 scores is used to summarize the extent of reform, although one could also weight the four according to their relative difficulty and importance. The index is purely descriptive and has no normative value.

TABLE B2.1.1 Power Sector Reform Index

<table>
<tr><td>Regulation</td><td colspan="3">No regulator = 0</td><td colspan="3">Regulator = 100</td></tr>
<tr><td>Restructuring</td><td colspan="2">Vertically integrated = 0</td><td colspan="2">Partial vertical unbundling = 33</td><td>Full vertical unbundling = 67</td><td>Vertical and horizontal unbundling = 100</td></tr>
<tr><td>Competition/ liberalization</td><td>Monopoly = 0</td><td colspan="2">Independent power producers = 25</td><td>Single buyer model = 50</td><td>Bilateral contracts = 75</td><td>Competitive market = 100</td></tr>
<tr><td>Private sector participation</td><td colspan="6">0.5 × (percentage of generation capacity with private sector participation)
+ 0.5 × (percentage of distribution utilities with private sector participation)</td></tr>
</table>

FIGURE 2.1 **OECD countries score systematically higher on the Power Sector Reform Index**

PSRI Scores: OECD vs. developing countries

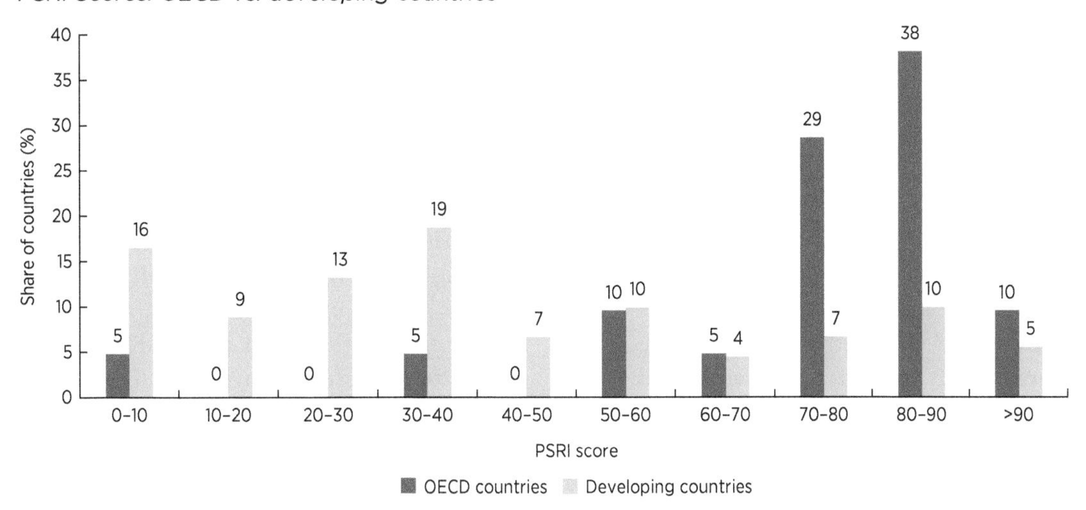

Sources: Based on Regulatory Indicators for Sustainable Energy 2017; World Bank Private Participation in Infrastructure Database 2016 (https://ppi.worldbank.org/); and industry data.
Note: OECD = Organisation for Economic Co-operation and Development; PSRI = Power Sector Reform Index.

scoring below 50. Yet adoptions of the full set of measures envisaged by the 1990s model have been far from universal, even in the OECD country group. For example, Japan and the Republic of Korea both retain vertically integrated utilities, and France's electricity supply chain remains almost entirely state owned. Overall, about 85 percent of OECD countries has fully unbundled power sectors, vertically and horizontally; however, only about half of them (40 percent) has introduced PSP in generation and distribution.

The gap in reform uptake between OECD and developing countries is particularly pronounced for some elements of reform. Although the aggregate reform scores differ significantly, this disparity conceals important variations in the elements of reform that are implemented (figure 2.2). For instance, the prevalence of PSP is relatively similar between OECD and the developing world (about 85 to 95 percent). Both groups have introduced at least a degree of PSP in their power sectors. The depth of privatization is greater, meanwhile, in the OECD countries, affecting on average 50 percent of the sector; the developing world has managed to privatize only 33 percent of its power sector. A large

FIGURE 2.2 **OECD countries are more likely to have adopted restructuring and liberalization reforms**

Advancement on four reform elements: OECD vs. developing countries

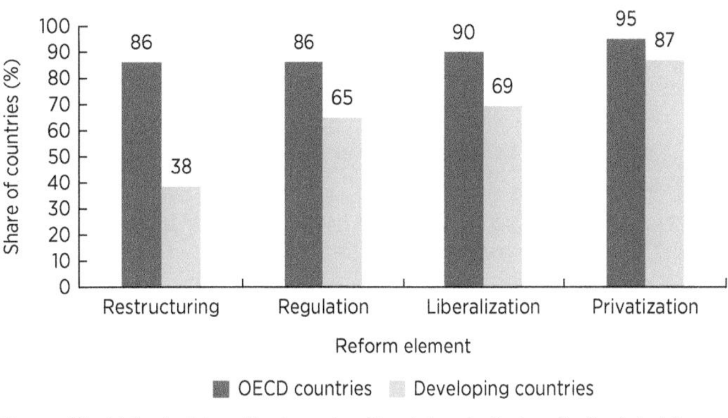

Source: World Bank elaboration based on Regulatory Indicators for Sustainable Energy 2017; World Bank Private Participation in Infrastructure Database 2016 (https://ppi.worldbank.org/); and industry data.
Note: OECD = Organisation for Economic Co-operation and Development.

gap exists, however, between the two groups on the prevalence of regulatory agencies (65 percent in developing versus 86 percent in OECD countries), and liberalization (69 percent in developing and 90 percent in OECD countries). The most striking disparity in reform uptake is seen in restructuring. The share of OECD countries embarking on these reforms is about 50 percentage points higher than it is for developing countries. The discrepancy is even greater in the depth of reform. Compared with developing countries, OECD countries are twice as likely to have fully unbundled their power sectors (vertically and horizontally); they are 10 times more likely to have introduced a wholesale power market.

Finding #2: The rapid diffusion of power sector reform from 1995 to 2005 was followed by a notable slowdown from 2005 to 2015

The diffusion of power sector reforms across the developing world has slowed since 2005. The year 1990 marks the baseline—when early reforms in Chile and the United Kingdom began spreading to the developing world. From 1990 to 1995, reformers experimented primarily with contract-based forms of PSP in generation and, to a lesser extent, distribution. Divestitures were relatively rare, and less than 10 percent of countries had established a regulatory entity. By 1995, the global average PSRI score for developing countries stood at just 12 (figure 2.3). The subsequent decade saw rapid diffusion of power sector reforms, boosting the global index almost threefold to a PSRI score of 35 in 2005. These measures established regulatory entities and introduced PSP or the divestiture of generation and distribution. Between 2005 and 2015, the uptake of these measures slowed considerably, with the global score climbing only 8 points (from 35 to 43) in 10 years.

Furthermore, power sector reform since 2005 has been limited to certain relatively straightforward measures. Reform uptake is slowing down across a range of key measures, with adoption rates varying by a factor of two or three across the two decades (figure 2.4). Most of the reform from 2005 to 2015 centered on the creation of regulatory entities and partial vertical unbundling—seen in about 15–20 percent of surveyed countries. Major reform measures—for example, full vertical unbundling, privatization of generation or distribution, and creation of the single-buyer model—were adopted by only about 5 percent of the surveyed countries.

The 1990s power sector reform model has been promoted tirelessly for 25 years. Still, the developing world is less than halfway to full adoption. The slowdown since 2005 does not, however, reflect saturation. We know that about 70 percent of the developing countries analyzed has cherry-picked the most prevalent reform measures; meanwhile, other countries have yet to even embark on power sector reforms. The rate and extent of reform suggest that several factors may be inhibiting uptake.

FIGURE 2.3 The uptake of power sector reform measures has been slowing since 2005

Uptake of power sector reform measures in the developing world, by type of measure, 1995–2015

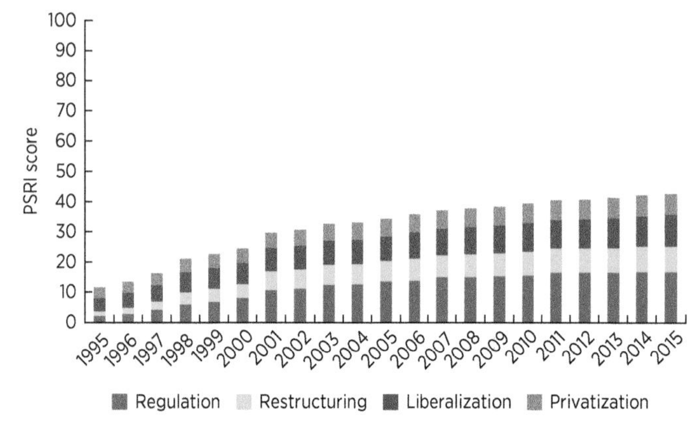

Source: Foster and others 2017.
Note: PSRI = Power Sector Reform Index.

FIGURE 2.4 A slowdown is evident across a wide range of different reform measures

Reform measures undertaken, 1995–2005 vs. 2005–15

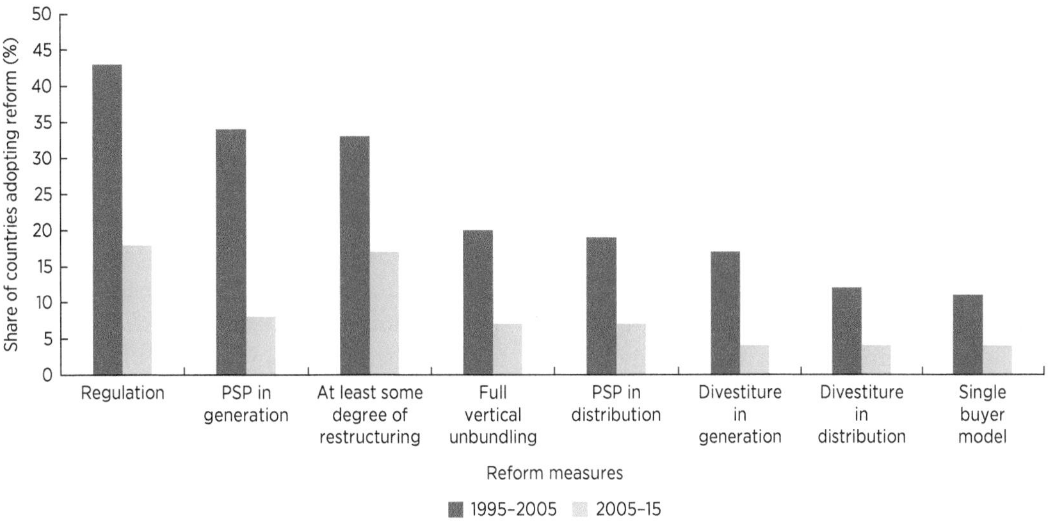

Source: Foster and others 2017.
Note: PSP = private sector participation.

Finding #3: Geographical, economic, and power system characteristics all strongly affect the uptake of power sector reform

The uptake of various reform measures is closely related to a country's geography, income level, power system size, and political economy. As a general rule, reform uptake is much stronger in Latin America and the Caribbean and in Europe and Central Asia compared to other regions. A country's income group typically has a strong positive effect on the prevalence of reform; the physical size of the power system has an even stronger positive effect. There is weaker evidence that oil importers go further with their reforms than do oil exporters. Uptake is more prevalent in multipolar (with competing power centers across jurisdictions) political systems relative to unipolar (where political power is concentrated) ones. The magnitude and statistical significance of these effects vary according to the dimension of reform.

With regulatory reform measures, the country-specific effects are negligible, perhaps because these measures are relatively easy to adopt. The strongest and only statistically significant differences observed in the uptake of regulation are seen across income groups. Specifically, the prevalence of regulatory agencies is highest for the upper-middle-income group. Furthermore, the leading geographic group, Latin America and the Caribbean, was 48 percent more likely to have introduced regulatory reforms than the lagging group, the Middle East and North Africa; overall results by geographic region were not, however, statistically significant.

Large differences were seen in the adoption of restructuring measures, particularly across geographic regions. For unbundling, the gap between the leading region, Europe and Central Asia, and the lagging region, Sub-Saharan Africa, is as much as 60 percentage points and statistically significant. Moreover, a high-income country, an oil-importing state, or a country with a power system producing more than 10 gigawatts (GW) can each add between 14 and 46 percentage points to the probability of unbundling its power sector.

With respect to competition, the scale of a country's power system proves critical for the adoption of wholesale power markets. Looking at liberalization overall, we see the most statistically significant differential across regions. The leading region, Latin America and the Caribbean, and the lagging region, Middle East and North Africa, are separated by a gap of up to 74 percentage points. Also striking is that countries with multipolar political systems are not necessarily more inclined to adopt competitive power markets. Countries with a large power system are about 50 percent more likely to introduce competition into their power systems. The effect of size is particularly stark for wholesale power markets. No country with a national power system of less than 1 GW of capacity has adopted a wholesale power market. Their systems are likely too small to support enough generators for a competitive market to be meaningful. Among countries with systems in excess of 10 GW, about 50 percent has a wholesale power market—more than twice the ratio for countries with smaller systems in the 1–5 GW range.

FIGURE 2.5 Latin America and the Caribbean led the way on power sector reform with many other regions lagging behind

Reform progress, by region, 1995–2015

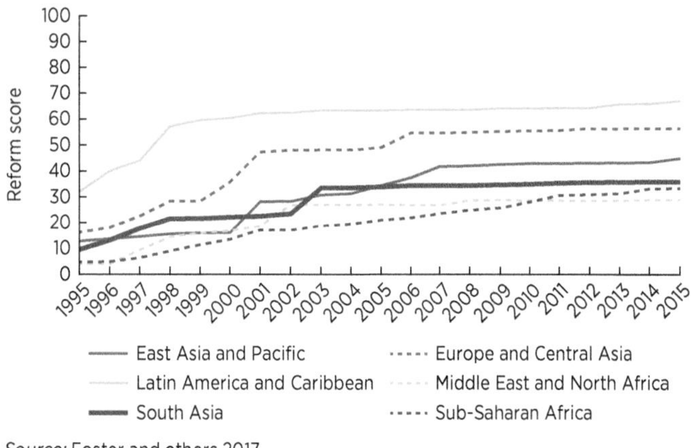

Source: Foster and others 2017.

Regarding PSP, differences persist across country groupings, with little prospect of convergence. The gap between the uptake of PSP in the leading region, Latin America and the Caribbean, and the lagging region, Sub-Saharan Africa, is 43 percentage points. Countries in higher-income groupings are more likely to have PSP by a differential of 23 percentage points. Again, a country with a large power system is about 30 percent more likely to undertake PSP in its system. By and large, countries in the different size groupings show no signs of converging. Differences in uptake are therefore becoming more pronounced over time. Countries producing more than 1 GW are twice as likely to have PSP in generation as those with smaller systems. In the case of distribution, the threshold is higher, and only in systems with more than 5 GW of capacity does the probability of PSP double relative to countries with smaller systems.

The persistence of such pronounced differences in reform uptake over two decades and across country groupings suggests that country characteristics constrain reform. The 1990s reform model would appear more readily applicable in certain environments. The fact that regional variations are the largest of any observed suggests a couple of things. First, some of the other effects are aggregated and accentuated by geography—for example, by the concentration of small, low-income, fragile countries in Africa. Second, region-specific drivers might be at work, including perhaps an intraregional bandwagon or domino effect.

On virtually every type of reform measure, Latin America and the Caribbean has been a pioneer, whereas the Middle East and North Africa region has lagged. Chile's pioneering reforms of the 1980s spread rapidly across Latin America and the Caribbean, doubling the average reform score from about 30 to 60 in the five years from 1995 to 2000 (figure 2.5). Europe and Central Asia also reformed at a brisk pace, reaching a score of about 60 in 2005, taking an additional five years to get there. In other developing regions, reforms have been adopted gradually, reaching an average score of just 30 to 40 by 2015. Noteworthy, nonetheless, is Sub-Saharan Africa's reform journey, which showed the largest increase in the PSRI from 1995–2005, eventually outstripping the Middle East and North Africa and registering scores similar to those observed in South Asia.

Overall reforms have had relatively little traction in lower-income countries and those with smaller power systems. Country characteristics such as geography, income group, power system size, and political system seem to have had a statistically significant influence on uptake (figure 2.6). One of the greatest influences has been system size: countries with installed capacity above 10 GW scored more than twice as high on the PSRI as those with systems below 1 GW. Similarly, countries in the middle-income bracket and those with relatively competitive political dynamics scored much higher than others. Countries with unipolar political systems also score about 15 percentage points behind those with multipolar systems. Equally striking is the fact that reform uptake is about the same for fragile states and nonfragile states (although countries that have not implemented any reform are almost all fragile), and between countries with stronger and weaker rules of law.

Finding #4: Countries cherry-pick their power sector reform measures

The 1990s power sector reform model was originally conceived of as a package of measures. In practice, developing countries have been selective in what measures they adopt. Some elements of the package have proved much more popular than others. The different reform measures can be divided into three broad groups according to their popularity (figure 2.7).

The most popular reform, undertaken by two-thirds of developing countries, is the introduction of regulatory agencies and independent power producers (IPPs). Creating a regulatory entity is politically straightforward—the reform simply adds to the sector rather than changes it. A regulatory entity may, in principle, represent a challenge to the sector's political hegemony. In practice, however, regulators don't always possess the power to provide such challenges. PSP in generation is the second-most-widespread reform measure, particularly through build, operate, transfer arrangements (called BOTs) or IPPs, although divestiture is also common. PSP in generation has proved less challenging to implement than other measures in part because it can often be applied to new plants. This application avoids the need to alter existing institutions, involves few employees, is relatively straightforward to manage, and is remote from the customer interface. Furthermore, it lends itself to a simple ring-fenced power purchase arrangement, either with a creditworthy utility or one whose creditworthiness can be enhanced with a state guarantee. Meanwhile, about one-third of developing countries was unwilling or unable to undertake even these relatively tractable measures over 20 years of tireless promotion in policy circles.

About half of developing countries made progress in the vertical unbundling of their utilities and in introducing PSP in distribution. Both measures disrupt the status quo. Unbundling means that the incumbent utility has to be restructured into multiple entities, creating governance and human resource challenges. When the private sector enters into distribution, it changes the point of interface with the public—a politically sensitive change that likely imposes tougher payment discipline, higher tariffs, and more stringent oversight of labor practices in a larger workforce.

FIGURE 2.6 **Reform uptake has been strongly associated with income group and scale of system**

Reform uptake by various country characteristics, 1995–2015

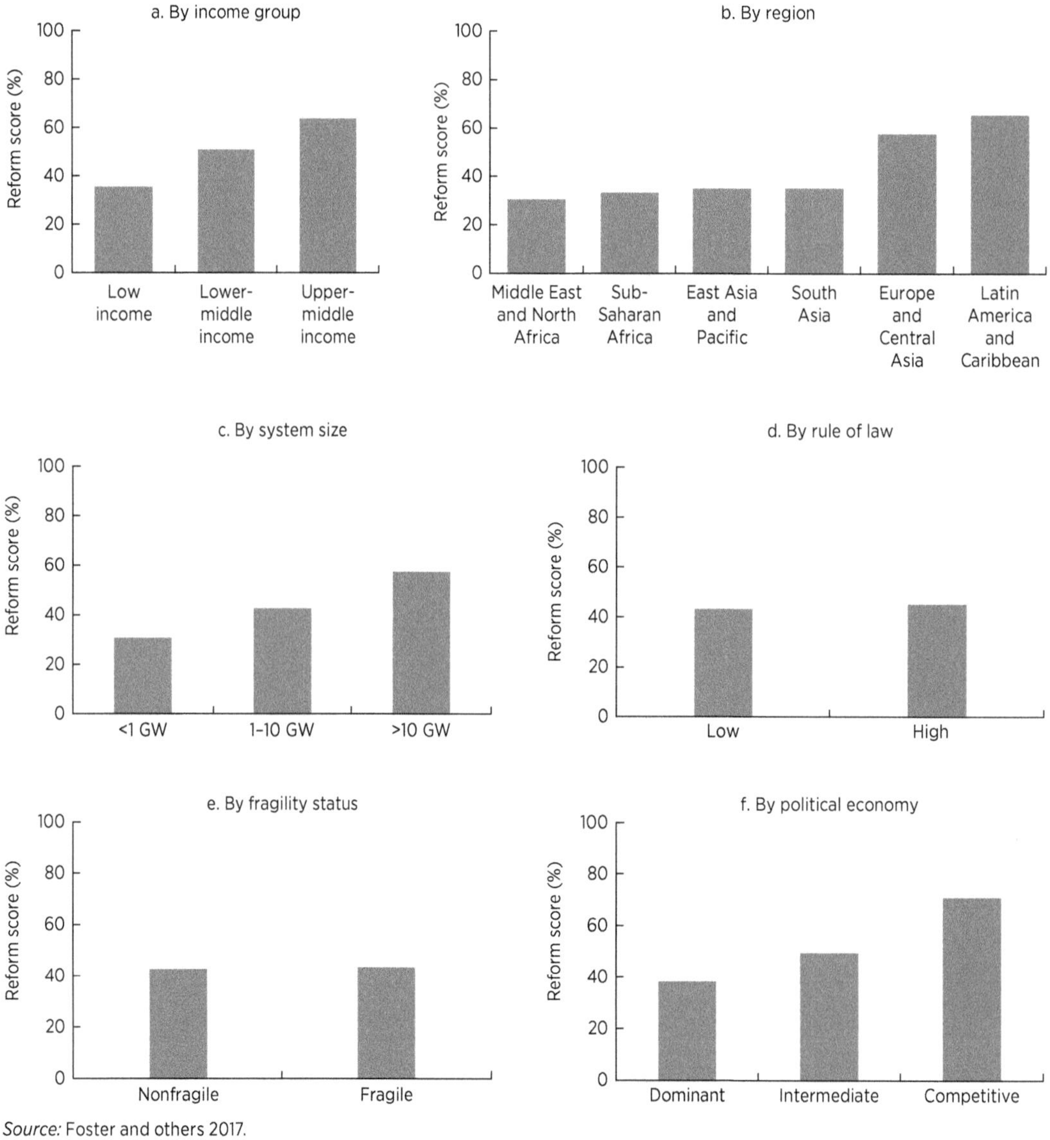

a. By income group

Low income | Lower-middle income | Upper-middle income

b. By region

Middle East and North Africa | Sub-Saharan Africa | East Asia and Pacific | South Asia | Europe and Central Asia | Latin America and Caribbean

c. By system size

<1 GW | 1–10 GW | >10 GW

d. By rule of law

Low | High

e. By fragility status

Nonfragile | Fragile

f. By political economy

Dominant | Intermediate | Competitive

Reform score (%)

Source: Foster and others 2017.

Note: All panels are based on data from developing countries and do not include any high income countries. GW = gigawatt. For panels a, b, and c, differences are statistically significant at the 5 percent level. For panel f, differences are statistically significant at the 10 percent level.

Only one in five developing countries achieved full vertical and horizontal unbundling of the sector and created a wholesale power market. It makes sense that the creation of power markets would be limited, because their establishment is relevant only in larger power systems, remains contingent on the prior adoption of many other reform measures, and asks states to relinquish much of their influence over the sector.

In addition, the comparative rarity of PSP in transmission partly reflects that this element was not a strong policy prescription of the reform model, because of the many public good functions provided by the power grid. Finally, management contracts are not widely adopted in the power sector, except in Sub-Saharan Africa, where experience has been mixed. Although these contracts are relatively easy to introduce from a political and technical standpoint, the impacts have been limited.

Whereas momentum is slowing for reforms aimed at restructuring and competition, it remains robust for regulation and PSP (figure 2.8). The adoption of utility unbundling and creation of wholesale power markets has been almost flat since the late 1990s. By contrast, the creation of regulatory agencies and adoption of PSP have continued to enjoy some uptake throughout the period.

FIGURE 2.7 Some reform measures proved a lot more popular than others

Relative uptake of various reform measures, by type, 1995–2015

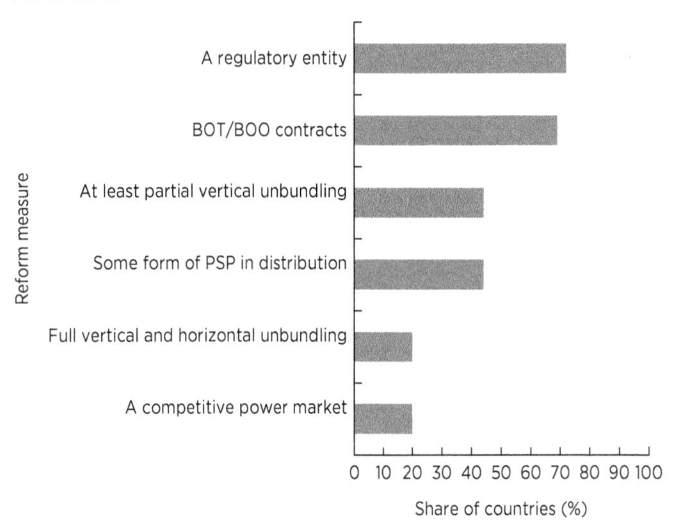

Source: Foster and others 2017.
Note: BOO = build, own, operate; BOT = build, operate, transfer; PSP = private sector participation.

FIGURE 2.8 Some types of reforms diffused more rapidly than others

Reform year, by share of countries, 1995–2015

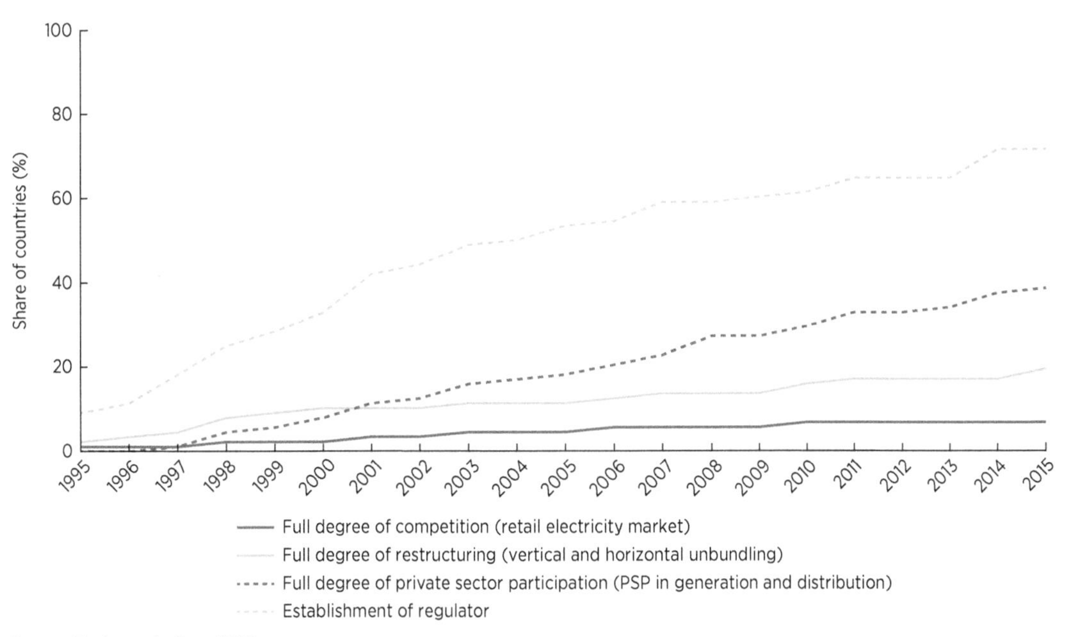

Source: Foster and others 2017.
Note: PSP = private sector participation.

FIGURE 2.9 Barely a dozen developing countries managed to implement the full 1990s reform package

Reform package adoption, by country and reform element, 1995–2015

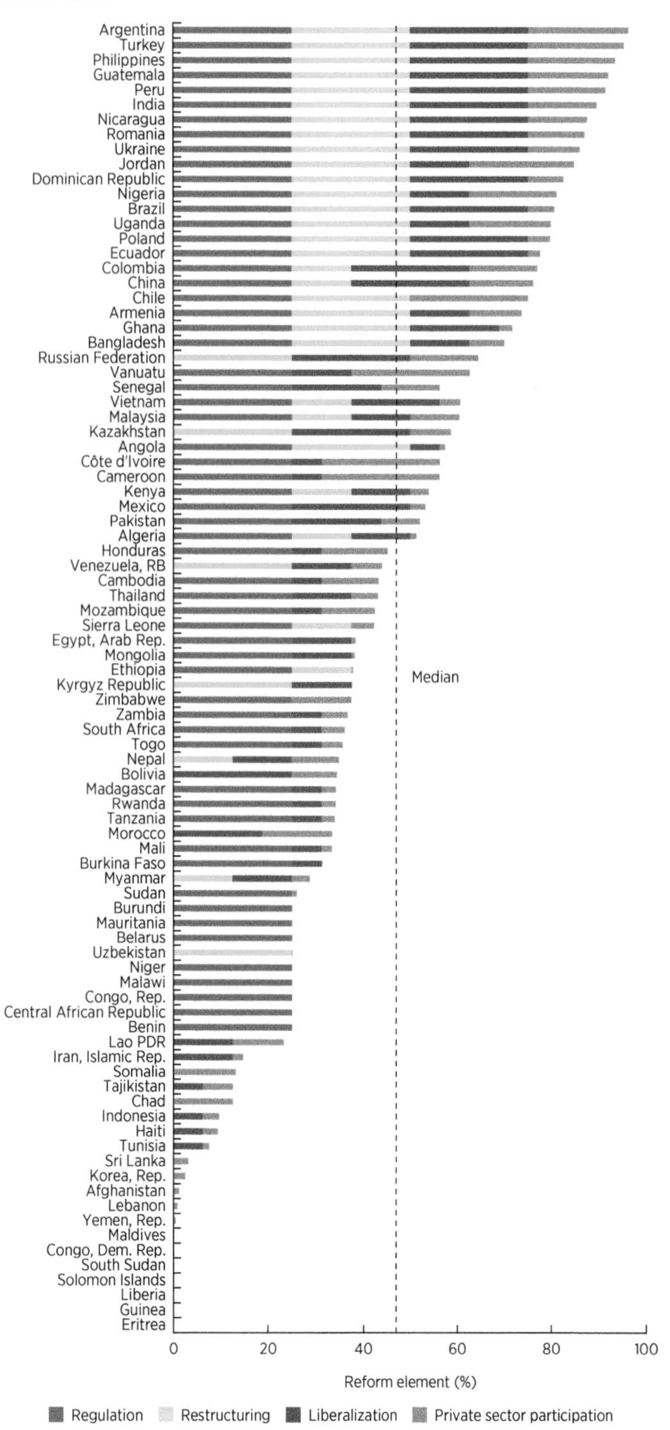

Source: Foster and others 2017.

Finding #5: Many countries find themselves stuck at an intermediate stage of power sector reform

Barely a dozen developing countries have managed to implement the full 1990s reform package, with most countries stuck at an intermediate stage of reform. Overall, the median score on the PSRI is 43, and three-quarters of developing countries score below 72 on the index (figure 2.9). This score illustrates clearly how most developing countries find themselves at the early or intermediate stage of the power sector reform agenda. Not a single developing country has a PSRI score of 100, and only about a dozen have scores above 80. The most aggressive reformers are found in Latin America and the Caribbean (Argentina, Brazil, Guatemala, Nicaragua, and Peru) as well as Europe and Central Asia (Romania and Turkey). Also, outliers are found in every region (map 2.1). Jordan, Nigeria, and the Philippines stand out as aggressive reformers in regions where bold reform has not been the norm; India also figures in this group. At the other extreme, only about a dozen countries have taken no steps toward implementation of the 1990s reform model, scoring close to zero on the index. They are primarily fragile (Afghanistan, Democratic Republic of Congo, Eritrea, Lebanon, Liberia, South Sudan, and Republic of Yemen) or small island states (Maldives and Solomon Islands). One striking feature is that the clear majority of developing countries has taken partial power sector reform measures but remains stuck at an intermediate stage: about half of countries score between 20 and 60 on the index.

When countries have adopted holistic power sector reforms, the time it took for implementation ranged from 3 to 18 years, with a median value of 12 years. Countries with high PSRI scores undertook reforms at differing speeds (figure 2.10). To understand

MAP 2.1 Outlier countries on power sector reform exist in every region

Power sector reforms, by country, 1995–2015

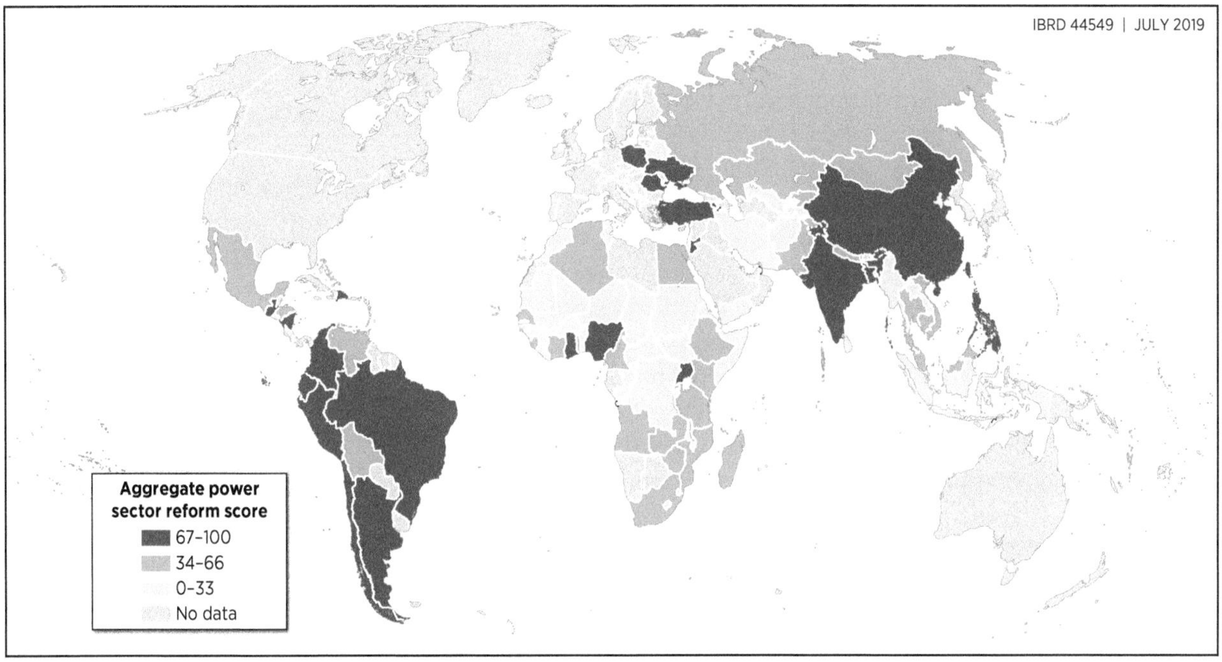

IBRD 44549 | JULY 2019

Aggregate power sector reform score

67–100

34–66

0–33

No data

Sources: World Bank elaboration based on Rethinking Power Sector Reform utility database 2015; Regulatory Indicators for Sustainable Energy 2016.
Note: PSRI score based on existing legislation (as of 2015), which may be different from practice.

the time periods needed to accomplish major reforms, we confined our attention to those countries that completed the bulk of the reform agenda by introducing a regulatory entity, unbundling vertically and horizontally, introducing a wholesale power market, and privatizing at least 50 percent of their generation and distribution sectors. Eight countries fall in this category. In Latin America countries enacted reforms relatively rapidly in under five years. Peru took three years, Argentina four, and Guatemala five. Reforms in Europe and Asia took substantially longer, in the 10–20 year range: Turkey 12 years, India 13 years, the Philippines 15 years, and Romania 16 years. Ukraine took 18 years. These long implementation time frames indicate that considerable time is needed to adopt a complex reform package.

FIGURE 2.10 Countries with high PSRI scores undertook reforms at differing speeds

Time elapsed between first and most recent reform step (in years)

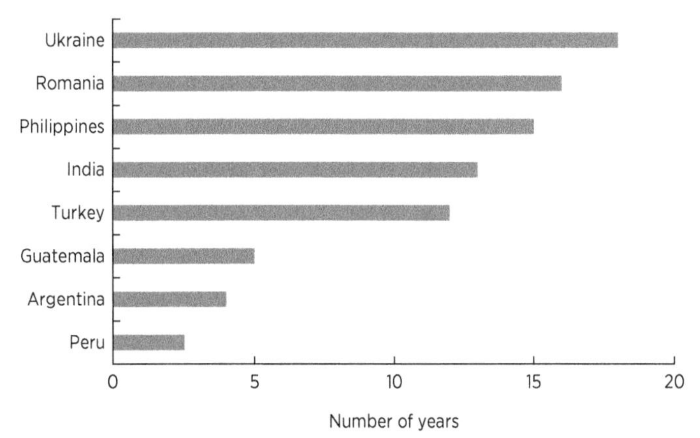

Source: Based on Regulatory Indicators for Sustainable Energy 2017; World Bank Private Participation in Infrastructure Database 2016 (https://ppi.worldbank.org/); and industry data.
Note: PSRI = Power Sector Reform Index.

Finding #6: Countries often packaged and sequenced power sector reforms in ways unrelated to original reform logic

Power sector reform measures were originally conceived as a coherent package, and it was envisaged that they would be implemented according to a certain logical sequence (figure 2.11). Regulation would typically be the starting point, so that the rules and incentives of the sector would be specified before any changes were made to the institutional actors. This step would be followed by unbundling to create the desired sector architecture, which, once completed, would pave the way for potential changes in ownership. Given that the distribution segment represents the beginning of the payment chain for electricity, it was generally thought that it made sense to begin PSP with distribution to improve the sector's operational performance and commercial viability. Improvements here would then generate a solid revenue stream able to provide the basis for remunerating PSP in generation. Only after all these elements were in place would it make sense to move toward a wholesale power market, which relies on multiple actors and a strong set of commercial incentives—both of which would need to be achieved through prior unbundling and PSP measures. Some competition could be introduced at earlier stages, however. IPPs could precede unbundling, for example, whereas single-buyer or third-party access could follow unbundling.

Most countries reversed the order of reform implementation, contravening the sequence originally envisaged. There is no common sequence for the adoption of reform measures across countries; rather, all possible permutations can be found (figure 2.12). Overall, 34 percent of countries introduced regulation before undertaking any PSP in generation or distribution. This pattern is overshadowed by the 52 percent of countries that did things the other way around. Similarly, some 20 percent of countries introduced PSP into distribution ahead of generation; they were greatly outnumbered by the 40 percent of countries that began with generation instead. This practice may reflect the fact that, although it is more logical, from an economic perspective, to start PSP in distribution, it is politically and practically more feasible to introduce PSP to the generation segment, where employment, affordability, and customer interface are lesser issues.

About one-third of developing countries ended up with partial and incoherent combinations of the reform package. We have seen how countries have been selective in their implementation of reforms, and their cherry-picking was not always informed by the synergies produced among the different reform combinations. By examining the reform combinations that countries ended up with in 2015, we were able to identify imbalances; for example, one dimension of reform proceeded further than was warranted by achievements along a related dimension.

FIGURE 2.11 **Power sector reform was conceived as a coherent package of measures implemented according to a logical sequence**

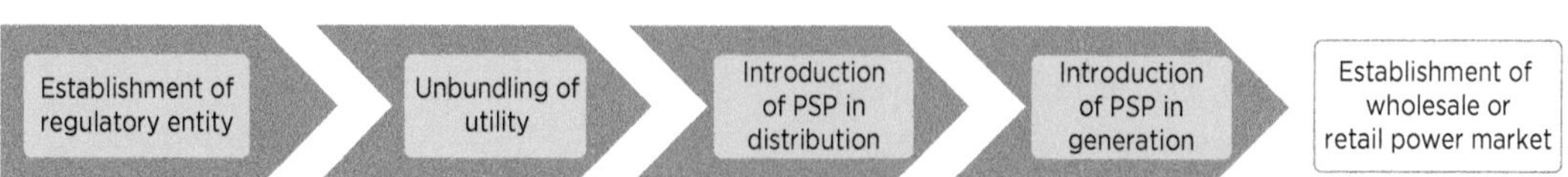

Source: World Bank elaboration.
Note: PSP = Private sector participation.

FIGURE 2.12 A significant percentage of countries adopted reforms in unorthodox ways

Variations on the initially recommended reform sequence, by share of countries

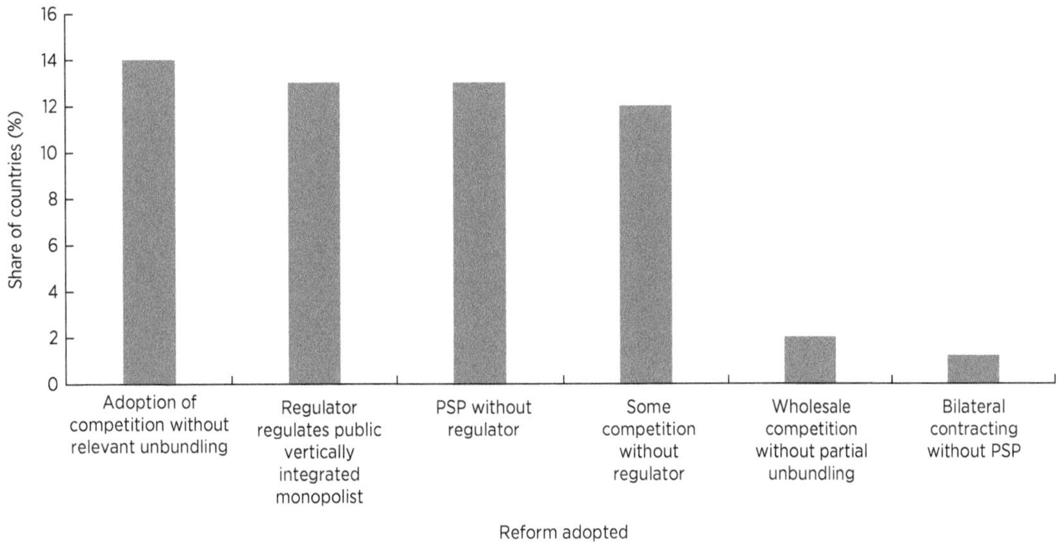

Source: Foster and others 2017.
Note: PSP = private sector participation.

The measures to restructure and privatize utilities are strongly linked with those that introduce competition, and these measures have proved challenging to progress that is balanced. Vertical unbundling distinguishes natural monopolies from activities that are potentially competitive; PSP then places potentially competitive segments on a commercial footing, and these segments are allowed to interact through measures that introduce competition. Over 30 percent of countries demonstrates imbalances among these three elements of structural reform. For example, almost 10 percent of countries introduced power markets before fully unbundling their power sectors. Conversely, just over 10 percent of countries completed full vertical and horizontal unbundling of the sector, but they still practice a single-buyer model when they look structurally prepared for a power market. Furthermore, almost 17 percent of countries has been able to introduce PSP in both generation and distribution yet has not introduced any competition beyond allowing IPPs. Another anomalous situation is found in countries (11 percent) that have introduced PSP in a utility that remains a vertically integrated national monopoly.

Another area of incoherence is seen in the creation of regulation entities. On the one hand, strictly speaking, a regulatory entity is not essential when the sector is organized as a publicly owned vertically integrated utility, because the whole rationale for state ownership is to act in the public interest, and no wider market relations require regulation. Moreover, when the regulator and the public utility report directly to the line ministry, the regulator has difficulty exercising authority. Nonetheless, 13 percent of developing countries created regulators when their utility remained an integrated state-owned monopoly. On the other hand, a regulator is critical once private ownership or operation has been introduced to any part of the sector, and as the reforms of competition continue. Still, 13 percent of developing countries lacks any regulatory entities despite having introduced PSP while 12 percent of developing countries lacks regulatory entities despite having commenced the process of increasing competition.

CONCLUSION

In summary, these patterns in the uptake of power sector reform illustrate that, although the Washington Consensus has been commonplace across the OECD, implementing recommended reforms in their entirety remains challenging to developing countries. Policy prescriptions for power sector reform diffused rapidly around the world during the decade 1995–2005; uptake in stable, middle-income countries with large power systems was significant during this period. Even among this group, however, only a handful of countries was able to fully implement the reform model as originally envisaged and typically managed to do so over a long period. A much larger group of countries implemented reforms selectively (typically emphasizing regulation and IPPs) and often ended up at an intermediate stage of reform without much impetus to introduce further reform measures. In addition, a significant minority of countries has barely started on reform and does not appear to present suitable conditions for doing so.

NOTE

1. This chapter is based on a background paper by Foster and others (2017).

REFERENCES

Banerjee, Sudeshna Ghosh, Francisco Alejandro Moreno, Jonathan Edwards Sinton, Tanya Primiani, and Joonkyung Seong. 2017. "Regulatory Indicators for Sustainable Energy: A Global Scorecard for Policy Makers." World Bank, Washington, DC.

Foster, Vivien, Samantha Helen Witte, Sudeshna Ghosh Banerjee, and Alejandro Vega Moreno. 2017. "Charting the Diffusion of Power Sector Reforms across the Developing World." Policy Research Working Paper 8235, Rethinking Power Sector Reform, World Bank, Washington, DC.

How Did Political Economy Affect the Uptake of Power Sector Reform?

3

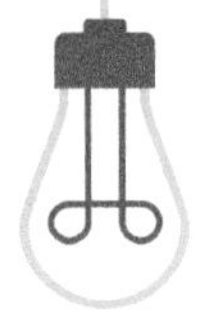

Guiding questions

- *How rapidly did countries implement reforms, and how long did they manage to sustain them?*
- *How did the nature of a country's political system affect its experience of reform?*
- *How important were reform champions and stakeholder alignment in delivering and sustaining reform?*
- *How will technological disruption affect the political economy of power sector reform?*

Summary

- *Crises typically provide political windows of opportunity for reform, yet the ambition and implementation of the reform have little to do with the depth of the crisis.*
- *Donor influence can be a catalyst and provide support for reform, but local ownership is what determines the boldness of reform and its ultimate sustainability.*
- *Countries espousing market-oriented ideologies are more likely to implement and sustain bold reforms.*
- *Reforms can be introduced across a range of political systems but are more likely to progress in countries where power is decentralized and contestable.*
- *Reform champions are often pivotal figures who help to ensure reform momentum, particularly when they are supported by stable bureaucracies.*
- *Unless champions orchestrate durable stakeholder alignments, reforms can become unsustainable.*
- *Legislation enshrines a level of political commitment to reform that supports longer-term sustainability.*
- *Looking ahead, the pace of technological disruption will reflect the interplay between existing vested interests in the power sector and new decentralized actors entering the sector.*

INTRODUCTION

This chapter examines the extent to which power sector reform in the developing world has been shaped by the political dynamics of each country.[1] It first lays out the implementation stages for power sector reform. It then charts the progress made by different countries through these stages and explores the influence of diverse factors on the evolution of reforms. The guiding questions for this chapter are the following: To what extent did countries implement and then sustain reform? What factors affected the speed of implementation? How was the reform process affected by the ideology and structure of a country's political system? Finally, how important were reform champions and stakeholder alignment in delivering and sustaining a reform agenda?

The 1990s model of power sector reform was born in an age of relative political innocence. The Cold War was ending, political left and right seemed to be converging, political and economic liberalism felt itself triumphant, and some even declared the "end of history" was nigh. Former socialist countries were joining the "transition to market economies," and developing countries were downplaying the economic role of the state as they entered periods of "structural adjustment." Many countries were moving toward greater political pluralism. From this historical context emerged the 1990s model of power sector reform, which emphasized accountability through the separation of powers and the transparent interplay of political forces. A technocratic and avowedly apolitical vision of the future power sector emerged.

In theory, the 1990s model aimed to improve sector efficiency and attract investment, in part by depoliticizing key decisions. It envisaged a power sector that would pursue public purpose through the motivator of private profit, with competition or antimonopoly regulation providing checks and balances. A key objective of reform measures was to depoliticize sector decision making (Lee and Usman 2018). This objective was to be achieved by initially corporatizing and eventually privatizing service providers, taking business decisions out of the hands of bureaucrats. At the same time, the creation of an autonomous regulatory entity—operating at arm's length from the executive branch of government with its own source of financing and independent leadership—was intended to insulate tariff and licensing decisions from direct political interference. The global diffusion of reforms from the 1990s to 2002 was for the most part unprecedented, and so could not be accompanied by evidence-based forecasts of the possible impacts on efficiency or the political feasibility of implementation. Geopolitical factors made policy makers willing to pursue reforms despite risks and costs, especially in response to crisis. For reform advocates of the period, the intrinsically political objectives of equitable access to modern and clean energy were not generally high on the agenda, but least-cost "efficient" solutions to power supply were most certainly paramount.

In practice, it proved impossible to contain political influence over the sector. Various commentators have underscored the power sector's inherently political nature. As noted by Kofi Annan (APP 2015, 16), "Governments often view [electricity] utilities primarily as sites of political patronage and vehicles for corruption." Similarly, "Power market reform is an inherently political process ... often an arena of conflict between competing interests that are of fundamental importance to society" (Besant-Jones 2006, 14). The experience of power sector reform has reflected this reality, with political factors determining the extent to which reform is feasible—beyond the original technocratic vision. This experience has led to the rise of explicitly political interpretations of power sector reform dynamics, and eventually to approaches that explicitly take the political economy into account. Though still emerging, such approaches seem to offer a sturdy framework for understanding reform as it takes shape on the ground (Lee and Usman 2018).

This chapter examines power sector reform through a political economy lens; it aims to understand how political forces shaped the reforms. Drawing on extensive stakeholder interviews (see box 3.1), the discussion examines how people interact politically, in pursuit of specific interests, within the context of their ideas, history, and perceptions. The narrative considers how the means of influence matter to the formulation and

BOX 3.1 Methodology for political economy stakeholder analysis

Using original, in-depth qualitative analysis of 15 countries constituting the Power Sector Reform Observatory, this study examines the role of the political economy in power sector reform (discussed in chapter 1). The countries were drawn from across the ideological spectrum—from state- to market-oriented political economies and hybrids of both. They are unipolar (where political power is concentrated) to multipolar (with competing power centers across jurisdictions). During the period under consideration, some of the countries evolved from unipolar to multipolar (for example, Peru) or moved among different political systems (for example, Pakistan). The countries perform differently on the relevant World Governance Indicators (for example, on rule of law, corruption, regulatory quality, and government effectiveness). The Latin American countries, whose annual income per capita is above US$5,000, perform relatively well on governance. But governance scores vary wildly among countries with an annual income per capita under US$3,000 (see figure B3.1.1).

In each of the 15 observatory countries, local experts conducted interviews with some 20 or more stakeholders involved in the reform process, often over long periods or during key episodes.[a] Some of the stakeholders had occupied positions of authority, whereas others were close observers. Interviews were structured partly around specific questions about reform episodes and processes in that country, but they were also allowed to flow in an open-ended way in order to be more informative and frank. Interviews were written up individually, and also summarized into a narrative of events supported by a timeline of reforms and a stakeholder mapping exercise. The interview-driven methodology can support deeper analysis of political economy processes in particular countries at particular times. Interviews make it possible, first, to get behind the scenes to see what happened and why, and, second, to obtain a range of perspectives and interpretations. It is essentially like triangulating history, providing considerable country-specific detail to illustrate and support broad global generalizations.

Apparent trends across the case studies were systematically verified through tabulation. This in turn allowed for qualitative comparisons of political factors and their impact on reform outcomes.

FIGURE B3.1.1 No strong relationship between income group and quality of governance

World Governance Indicator, by income group and GDP per capita, 2010–15

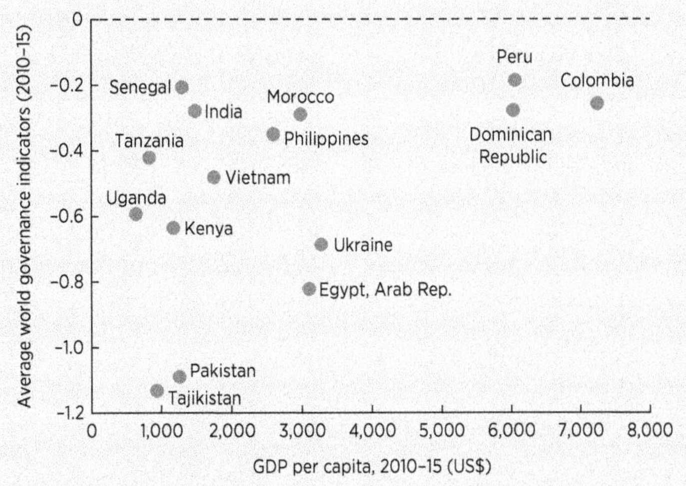

Source: Based on Rethinking Power Sector Reform utility database 2015 and World Bank World Development Indicators database 2018.
Note: GDP = gross domestic product.

a. All interviewees gave generously of their time and experience; most were candid and insightful. Interviews were confidential, so speakers are identified only generically (for example, as a senior power sector official or manager of a private power company).

implementation of policy, and how real-life institutions shape events and outcomes. Power sector reform is a complex political process that may take many years to implement. It is dependent on unfolding circumstances and often nonlinear in nature. Political history matters as much as economic concepts, but history in pursuit of economic objectives can, to a greater or lesser extent, be managed politically.

For the purpose of this analysis, the stages of the power sector reform process involve steps forward and potentially backward. The stages of the reform process can be characterized as, first, the announcement of the reform; second, the process of implementation; third, the delivery endpoint; and, finally, longer-term sustainability (figure 3.1). The Power Sector Reform Index (PSRI) developed for the study[2] (see annex 3A) can be used to track the extent of reform that each country announced and implemented year on year from 1990 to 2015.

The first stage of reform is usually a public announcement of the intention to reform. Announcements generally feature a sector strategy, road map, or policy paper. In some countries, such announcements (made over a consecutive three-year period) were ambitious, signaling an intention to undertake more than 60 percent of the measures entailed by the 1990s reform model. Four countries—Colombia, Pakistan, Senegal, and Uganda—pledged to undertake more than 80 percent of the 1990s reforms. A minority of countries—Morocco, Tajikistan, and Tanzania—made modest announcements committing to implement less than 50 percent of the 1990s reform agenda. On average, across the 15 observatory countries, reform announcements made over a three-year period mentioned 68 percent of the measures set out in the 1990s reform agenda.

The second stage of reform is implementation. Countries implement reform measures with varying speeds. Some countries not only announced ambitious reforms but also went ahead to implement them boldly, completing at least 60 percent of the 1990s reform agenda within three years. The three boldest reformers—Colombia, Peru, and Ukraine—each implemented as much as 70 percent of the 1990s reform package within three years. Other countries, even those that had expressed similar ambition, were more incremental in their reform process. Figure 3.2 illustrates the reform trajectories of a bold reformer, like Ukraine, versus an incremental reformer, like Vietnam. A third and smaller group of countries—including Morocco and Tajikistan—took only limited action toward the implementation of reform. On average, across the 15 observatory countries, countries were able to implement 50 percent of the 1990s reform agenda in the space of three consecutive years.

The third stage of reform is delivery, which denotes the maximum extent of reform that the country was able to implement. Irrespective of the speed at which they reform, countries differ in the extent to which they were finally able to follow through on the implementation of their overall reform announcement.[3] Although none of the countries succeeded in fully implementing the reforms they announced over 1990–2015, the degree of

FIGURE 3.1 Power sector reform process: Announcement, implementation, delivery, and sustainability

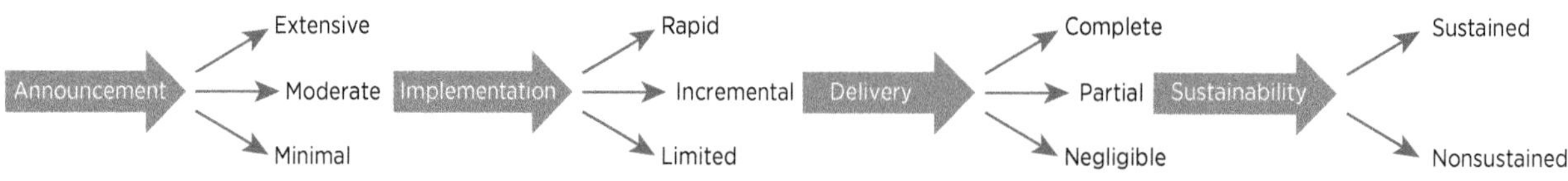

Source: World Bank 2019.

FIGURE 3.2 The contrasting reform trajectories of a bold reformer, Ukraine, and an incremental one, Vietnam

PSRI, by reform measures, 1990–2015

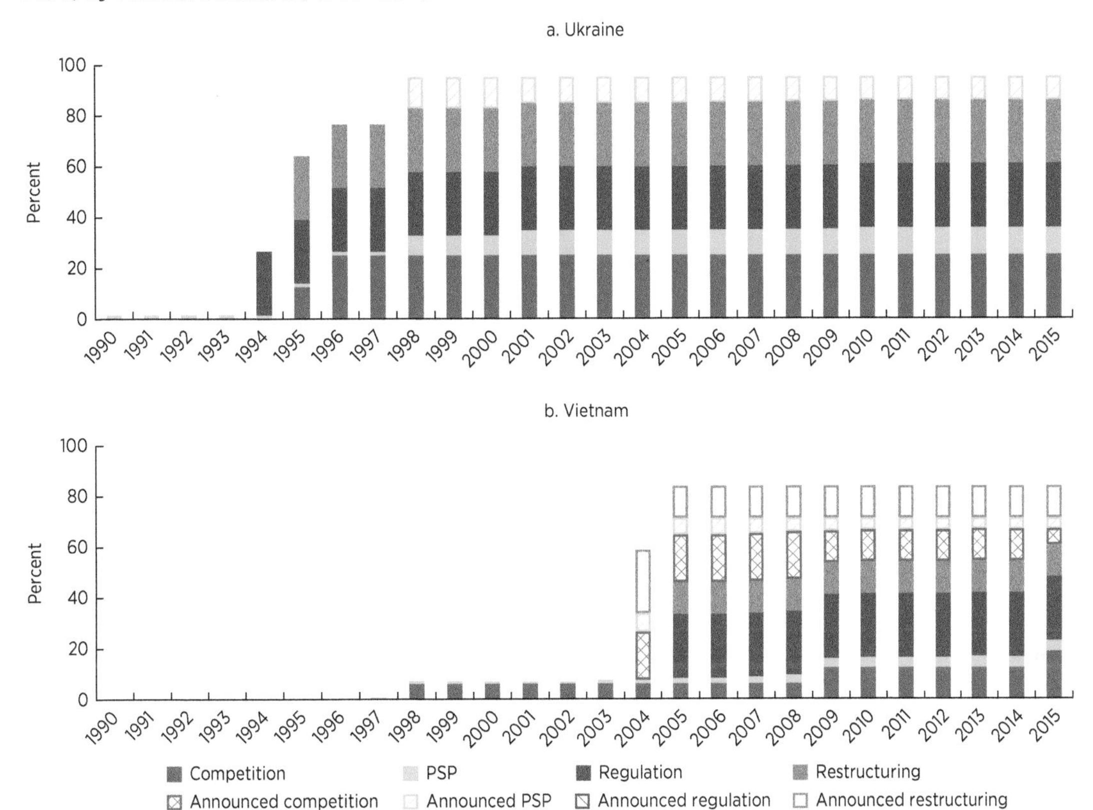

Source: Based on Rethinking Power Sector Reform utility database 2015.
Note: PSP = private sector participation; PSRI = Power Sector Reform Index.

implementation varied from those that *substantially* achieved their announced reform goals, to those that did so *partially*, or those whose progress was only *negligible*. Two groups of countries stand out. The first group includes those whose reform delivery closely aligned with their announcements: Peru, the Philippines, Uganda, and Ukraine. The second group implemented reforms that fell short of their reform announcement—short by as much as 40 percentage points. This group included the Arab Republic of Egypt, Senegal, Tajikistan, and Tanzania (figure 3.3). On average, across the 15 observatory countries, reform delivery as of 2015 amounted to 67 percent of the 1990s reform agenda, or about 20 percentage points short of what was announced cumulatively from 1990 to 2015 (figure 3.3).

The final stage of reform is sustaining the measures implemented. Reforms can always be undone by contrary political forces. Although most countries were able to sustain the structural reforms they implemented during the study period, a significant minority reversed enacted privatization reforms, as in the Dominican Republic, the Indian state of Odisha, and Senegal (figure 3.4). These reversals reduced their reform index scores by 11 percentage points on average.

FIGURE 3.3 Contrast between reforms announced and reforms delivered in observatory countries and states

PSRI, announced vs. delivered, 2015

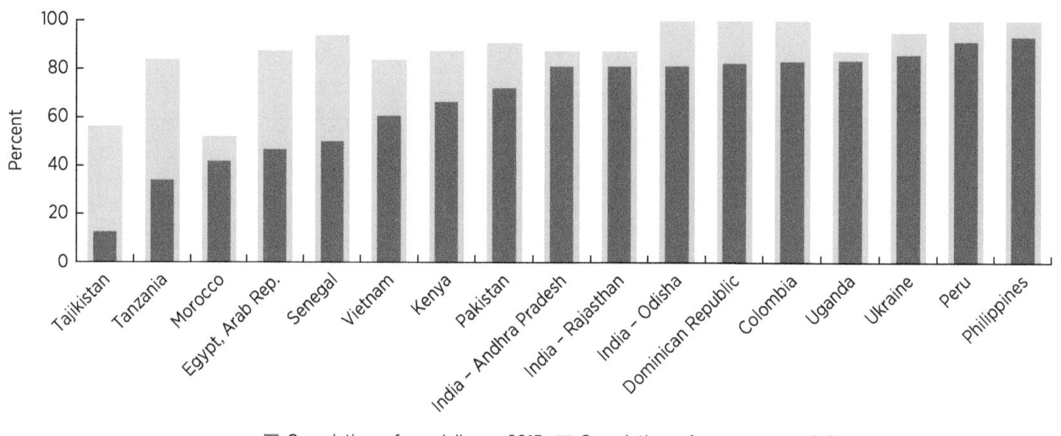

Source: Based on Rethinking Power Sector Reform utility database 2015.
Note: PSRI = Power Sector Reform Index.

FIGURE 3.4 Some countries experienced reversals of reform once implemented

Reform reversals: Dominican Republic and Senegal

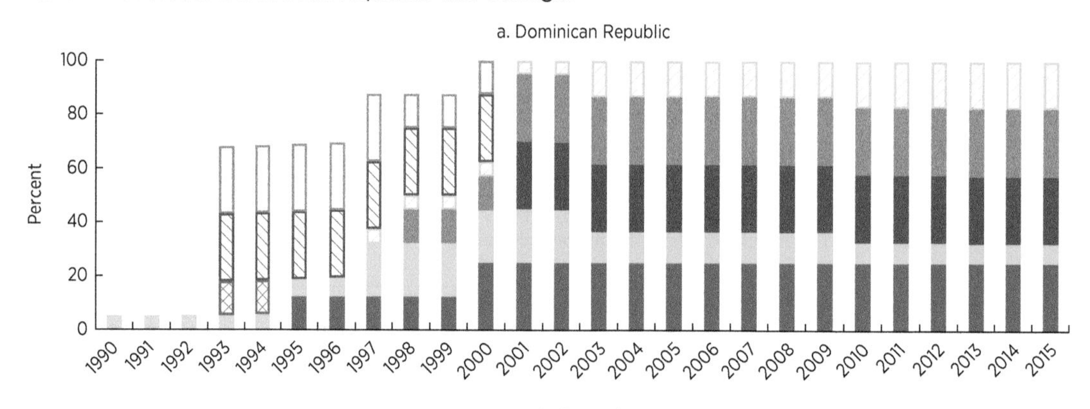

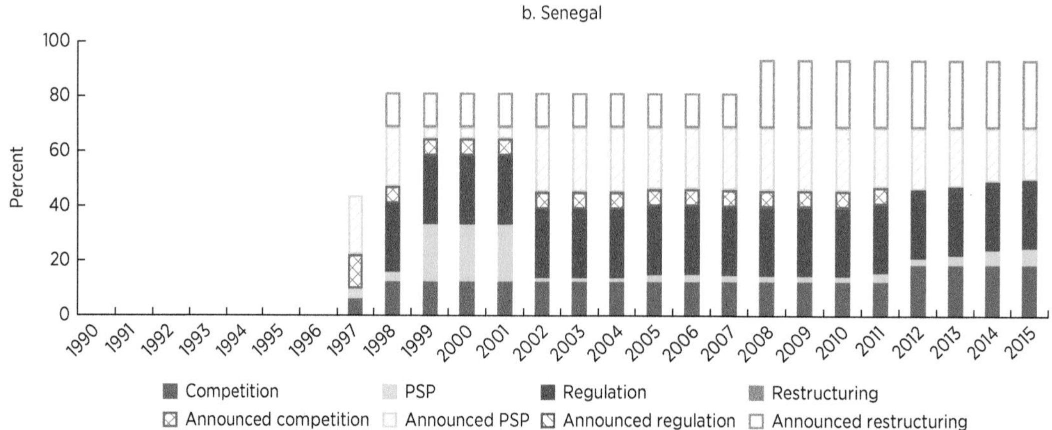

Source: Based on Rethinking Power Sector Reform utility database 2015.
Note: PSP = private sector participation.

Countries display very different patterns in their reform processes (table 3.1). Only two countries (Peru and Ukraine) announced ambitious reforms, implemented them boldly, substantially delivered on them, and sustained them over time. Four entities (the Indian states of Andhra Pradesh and Rajasthan, as well as the Philippines and Uganda) announced ambitious reforms, implemented them incrementally, substantially delivered them, and subsequently sustained them. At the other end of the spectrum were countries that both promised and delivered little. The most extreme case is Tajikistan, which delivered only 13 percent of the 1990s reform agenda. Overall, bold reform implementation correlates with the extent of reform ultimately delivered (with a correlation coefficient of 0.87). Put differently, countries approaching reform boldly tended to have less of a shortfall between announcement and implementation (with a correlation coefficient of minus 0.52). Moreover, the correlation between the bold reform and likelihood of reversal is low (with a correlation coefficient of 0.18). In fact, the probability of reform reversal at 29 percent was slightly higher among incremental reformers than at 25 percent for bold reformers.

Characterizations of a country's reform process are purely descriptive and not in the least normative. As described in chapter 2, the reform measure index developed for the study does not intend to pass judgment on whether more reform is better or worse. It does, however, prove helpful in measuring how much reform took place in a standardized way; it allows meaningful comparisons of the extent of reform across countries and over time.

Many factors shape the dynamics of a country's power sector reform. To understand them, we needed a qualitative comparison of the political economy narratives across the

TABLE 3.1 The pace of power sector reform in the observatory countries and states

	Announcement	Implementation	Delivery	Sustainability
Peru	Extensive	Rapid	Complete	Sustained
Ukraine	Extensive	Rapid	Complete	Sustained
Colombia	Extensive	Rapid	Partial	Sustained
India - Odisha	Extensive	Rapid	Partial	Nonsustained
India - Andhra Pradesh	Extensive	Incremental	Complete	Sustained
India - Rajasthan	Extensive	Incremental	Complete	Sustained
Philippines	Extensive	Incremental	Complete	Sustained
Uganda	Extensive	Incremental	Complete	Sustained
Pakistan	Extensive	Incremental	Partial	Sustained
Dominican Republic	Extensive	Incremental	Partial	Nonsustained
Senegal	Extensive	Incremental	Negligible	Nonsustained
Vietnam	Extensive	Limited	Negligible	Sustained
Kenya	Cautious	Incremental	Partial	Sustained
Egypt, Arab Rep.	Cautious	Limited	Negligible	Sustained
Morocco	Minimal	Limited	Partial	Sustained
Tajikistan	Minimal	Limited	Negligible	Sustained

Source: Based on Rethinking Power Sector Reform utility database 2015.
Note: All announcements made over a period of three years with an intention to implement more than 60 percent of the 1990s reform model are cumulatively classified as "ambitious." Those between 50 and 60 percent are "cautious," and those below 50 percent are classified as "minimal." Economies managing to implement more than 60 percent of the 1990s model reforms over a period of three years are classified as "rapid," those implementing between 60 and 50 percent are "incremental," and those below 50 percent are called "limited." In economies where the difference between total announced and total implemented reforms is less than 15 percent, delivery is classified as "complete." Where the difference is between 16 and 25 percent, it is classified as "partial." Where the difference is greater than 25, the delivery of reform is classified as "negligible."

FIGURE 3.5 Nested model of political influences in power sector reform

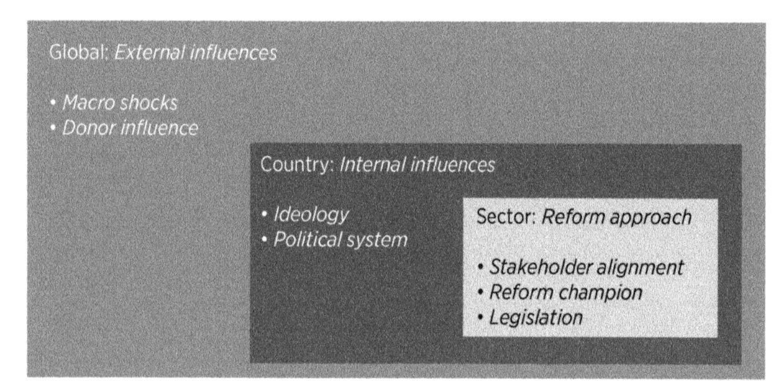

Source: Based on Rethinking Power Sector Reform utility database 2015.

15 observatory countries. This exercise surfaced several factors that recurred with regularity across all cases. Conceptually, they belong in three nested categories (figure 3.5), the broadest being exogenous factors, such as the eruption of a crisis or influence exerted by foreign donors. Factors pertaining to the national political sphere fall into the middle nest and include a country's political system and ideology. Finally, factors affecting reform at the level of the power sector include reform champions, stakeholder coalitions, and legislation.

KEY FINDINGS

Analysis in this chapter draws on the power sector reform process characterized in table 3.1 and on the nested model of political influences on power sector reform in figure 3.5. Drawing on the extensive qualitative material culled from stakeholder interviews, we examined the relevance of each political influence at each stage of reform. Seven key findings are set out below.

Finding #1: Crises typically provide political windows of opportunity for reform, yet the ambition and implementation of the reform have little to do with the depth of the crisis

What we must understand is that the story started in the 1990s, when we were living in a desperate situation with no investments. The sector had been managed very poorly, politically speaking, with illogical prices, and so forth. During the 1990s we had no choice—changes had to be made. These changes were not necessarily well planned, but we had to act in the urgency of the moment to cover the deficit. The first reaction was to bring back private investments: it was logical and correct. Now, how it was done and why it was done the way it was, was simply due to desperation. We looked at what had been done in other countries, did a quick balance, and got some ideas.

—*Energy sector official*

What were the goals of the reform then? It cannot address efficiency. It cannot address additional capacity. It cannot address higher cost of electricity. It cannot. Only [thing] to address was: stop the bleeding at that time. So it was purely reactional—and then you put all that into the law: long-term objectives. Oh my gosh.

—*Energy sector official*

The initial impetus for power sector reform among the 15 country case studies almost always sprang out of economic crisis. Crisis certainly presents opportunities for action. It reduces the strength of opposition to reform and highlights the benefits of reform. These opportunities create critical junctures in the history of a country so that things that looked politically unlikely suddenly become eminently possible. Some sort of crisis is, in fact, almost a prerequisite for reform. A glance at the observatory country cases is informative. It is unusual for a country to pursue power sector reform outside of an economic crisis. Vietnam is the only possible exception, although it too struggled to meet the rapid growth in demand for electricity. At the other extreme, Tajikistan, alone among the countries, introduced no significant reform measures despite finding itself in a deep sector crisis.

The triggering crisis can be macroeconomic or specific to the power sector (table 3.2). Macroeconomic crises take a variety of forms. Government budget pressures on the power sector can create such a crisis. Another kind of crisis might take the form of accumulated non-performing loans made to the power sector by the financial sector. Crises could emerge from the external balance of payments, as external power sector debt accumulates. A depreciating currency increases costs such that shortages of foreign exchange make fuel and spare parts unaffordable. Macroeconomic triggers like these were seen in power sector reform episodes in Egypt, Kenya, Peru, and Tanzania. In the case of Ukraine, although power sector reform was triggered by macroeconomic crisis, it was also part of a wider structural shift toward a market economy. Alternatively, the crisis can manifest itself more directly in the power sector through blackouts and power rationing, as seen in Colombia, Pakistan, or the Philippines. Or the impending threat of electricity shortages might stem from the financial fragility of the sector, as happened in the Dominican Republic and India. Electricity shortages may themselves signal an underlying financial crisis in the sector or, in hydro-dominated systems, occur because of drought. In other cases, crises are brought on after countries emerge from periods of armed conflict or political turmoil, as with Uganda.

The case of Uganda illustrates how a wider economic and political crisis wore down opposition to power sector reform. When Yoweri Museveni took power in Uganda in 1986, the economy was barely functioning after years of economic mismanagement. There was political oppression, untrammeled corruption, and war. Inflation was rampant, production was minimal, the exchange rate was under extreme pressure, and the government budget was in disarray. The power sector was likewise shattered—losses were running to at least 40 percent, generation was down to 60 megawatts with poor system reliability. The Uganda

TABLE 3.2 Crisis at the time of power sector reform

Economy	Macroeconomic crisis	Utility financial crisis	Power supply crisis
Colombia	S	S	S
Dominican Republic	S	S	S
Egypt, Arab Rep.	M	M	—
India - Andhra Pradesh	—	M	—
India - Odisha	M	M	—
India - Rajasthan	M	S	—
Kenya	M	—	—
Morocco	—	M	M
Pakistan	—	S	S
Peru	S	S	S
Philippines	—	M	M
Senegal	S	S	S
Tajikistan	—	S	S
Tanzania	S	S	S
Uganda	S	S	S
Ukraine	S	—	—
Vietnam		No Crisis	

Source: Based on Rethinking Power Sector Reform utility database 2015.
Note: Vietnam had no sectoral crisis at reform time. M = moderate crisis; S = severe crisis; — = not available.

Electricity Board was in chronic operational and financial crisis. By this stage, the public was exhausted by the failures of the previous regimes and open to almost any option—including neoliberal ones—to fix both the economy and the power sector. This pragmatism made Uganda's thoroughgoing reforms possible—more so than in most other African countries at that time. By the early 1990s, the government was implementing bold reforms of its power sector: a fully unbundled utility, private sector participation in generation and distribution, a competent and relatively independent regulator, and tariffs adjusted to reflect costs. The implementation of so many structural changes speaks to the power of crisis as a political catalyst for reform.

Colombia's power sector contributed greatly to the country's macroeconomic crisis, and it benefited from the wider political and institutional reforms introduced in response. By 1990, Colombia was in a deep macroeconomic crisis. The power sector was responsible for 30 percent of total external debt and 33 percent of the public sector deficit—an unsustainable burden for the state. Support from financing agencies was drying up. The situation was exacerbated by drought in 1992. A year of blackouts and power rationing revealed the sector's deep structural failings and threatened Colombia's economic recovery. The government declared a social and economic emergency. The new 1991 constitution incorporated institutional and governance reforms emphasizing decentralization and transparency. Building on this legal foundation, the government began to restructure the energy sector and its regulatory institutions. By adopting new public service and electricity laws, the government was able to reform tariff regulation, open the sector to private investment, and initiate a competitive wholesale electricity market.

Nevertheless, although countries mostly responded to crisis with ambitious reform announcements, subsequent implementation and delivery bear little relationship to the severity of the crisis. Crisis is the standard backdrop for power sector reform, but there is little evidence on the relationship between crisis and ultimate power sector reform outcomes in the wider literature; the limited evidence is equivocal about the reform value of a crisis (Lee and Usman 2018). These reservations are borne out by data on the 15 observatory countries, where we found no statistically significant relationship between the severity of the crisis, the ambition of the reform, and the trajectory of its implementation, including the boldness of implementation and how substantially reforms were delivered and ultimately sustained (table 3.3). Indeed, in some cases, reversals of model reforms have occurred during crisis conditions, such as when utilities were renationalized in the Dominican Republic, the Indian state of Odisha, and Senegal. Crisis can therefore be understood as creating a space for response; the choice of response and its ultimate sustainability depend on other factors.

In the Indian state of Rajasthan, a severe crisis prompted the announcement of a deep reform of the power sector, but political resistance prevented full adoption and sowed the seeds for further crisis. In 1998, the year before reforms began, system losses were at 42 percent and collection rates had fallen to 76 percent.

TABLE 3.3 Crisis severity and patterns of power sector reform

		Announcement			Implementation			Delivery			Sustainability	
		Extensive	Cautious	Minimal	Rapid	Incremental	Limited	Complete	Partial	Negligible	Sustained	Nonsustained
Crisis	Severe											
	Moderate											
	None											
Statistical significance		—			—			—			—	

Source: Based on Rethinking Power Sector Reform utility database 2015.
Note: This table shows the extent to which the severity of the crisis affects the subsequent process of reform across the sample 15 country cases. The darker the shading of the box, the more countries from the sample fall into each combination of circumstances. The Chi-squared test was performed to reveal any statistically significant relationship between the severity of the crisis and the process of reform, with the results reported as follows: **** = 95–99 percent significance (significant); *** = 90–95 percent significance (less significant); ** = 85–90 percent significance (somewhat significant); * = 75–85 percent significance (low significance); — = not significant. Full statistical results are reported in table 3B.1 in annex 3B.

The state electricity board was in critical condition, so the state enacted the Rajasthan Power Sector Act in 1999 to restructure the sector and improve operational and financial performance. The goal was to carry out tariff reform, introduce private sector participation in distribution, and make the sector more efficient. Although the unbundling succeeded, opposition to privatization emerged early in the reform implementation, halting any attempts to move ahead with the measure. Left with the option of raising tariffs or reducing system losses, the government focused on the latter and managed to keep losses in check for several years. Success was short-lived, however, and losses soon mounted. Combined with a tariff freeze between 2004 and 2011, these losses left the sector in a deep financial crisis with cumulative losses of over US$3 billion by 2008–09.

Finding #2: Donor influence is often a catalyst for reform and provides ongoing support, yet local ownership shapes the boldness of the reform and ultimately determines its sustainability

As an economy, we were the strongest in the region. I think because of that the donors … pushed us a lot harder than they pushed other countries. Sometimes our political leaders … could not understand why the country was being pushed so hard.… The push from outside, over a long period of time, made the leaders realize it was the only way to go and decided to see if it would work. Over time, we found that it did work for us, and we have embraced it since.

—Utility manager

The current Electricity Sector Reform Strategy is a result of the poor performance of the power sector and perhaps an indication that government has given up on supporting an inefficient utility. Government thought they had done so much to help the utility, but it is still not working. These reforms are not due to external pressure but come from within government.

—Academic

For policy reform to succeed, governments need to exhibit leadership and ownership of the process. Local ownership is sometimes questioned when external development agencies exert influence, or where there are other external influences. Donors have sometimes applied policy-based loan conditions or provided technical assistance. Donor-supported knowledge exchange and reform advocacy can also be influential, particularly if there is a dominant global reform blueprint. Development partners aim to provide advice that is technocratic and nonpolitical, but their counterparts in government do not necessarily perceive their purported neutrality. Moreover, political opinion in some countries is shaped by adverse historical experience with foreign investment or debt dependence.

During the 1990s, many countries were heavily dependent on donors for financial assistance and subject to donor influence on power sector reform. Almost all the 15 observatory countries were dependent on foreign assistance at the time of reform, with aid accounting for about 5 percent of gross domestic product (GDP) on average (figure 3.6). Even for countries with historically low reliance on official development assistance, donor influence could be high when macroeconomic crisis occurred, as in Colombia, the Dominican Republic, and Peru. International financial institutions were instrumental in diffusing market ideas during the 1980s and 1990s, reflecting the resurgence of neoclassical economic thinking. A notable feature of World Bank and International Monetary Fund lending in the 1980s was structural and sectoral adjustment programs based on market-oriented macroeconomic and fiscal policies (Jayarajah and Branson 1995). The term

FIGURE 3.6 Dependency on foreign aid was substantial at the time of power sector reform

Official development assistance at the time of power sector reform

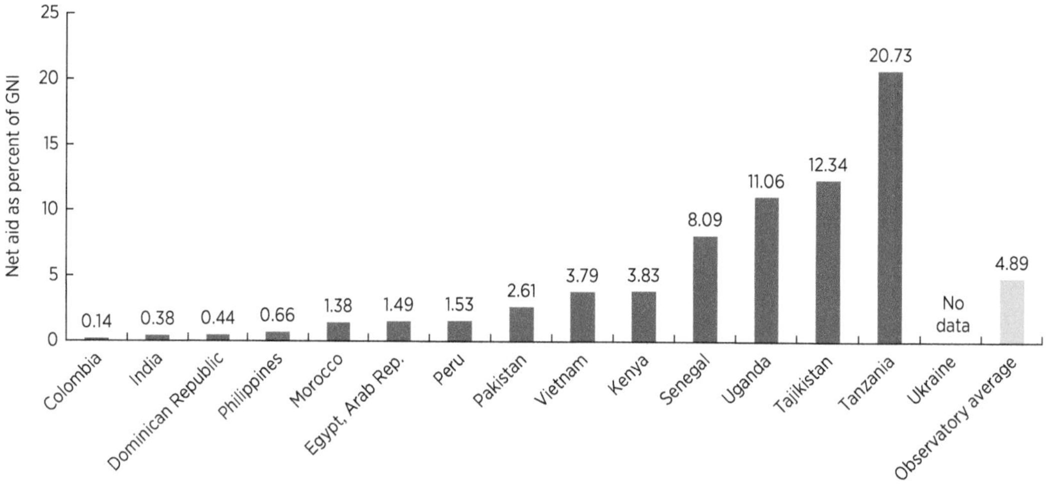

Source: Based on World Development Indicators database 2018.
Note: The net assistance as a percentage of gross national income (GNI) is calculated as a three-year moving average for the year in which substantial reform was first announced.

"Washington Consensus" was coined in 1989, as noted at the outset of chapter 1, to describe 10 such policies supported by the United States, International Monetary Fund, and World Bank (Williamson 2005). Among these policies, two in particular are relevant to the power sector: the privatization of state-owned enterprises and the abolition of regulations restricting competition. The World Bank (1993) explicitly recommended enforcing conditions to make governments commit to power sector market reforms. Other multilateral development banks and bilateral donors adopted similar approaches (Williams and Ghanadan 2006). In addition, reforming countries continued to rely on international financial institutions for significant investment projects. World Bank support for the 15 observatory countries throughout the reform process is shown in annex 3C.

Governments motivated by the need for finance are more likely to follow the norms of their financiers. One study (Henisz, Zelner, and Guillén 2005) suggests that privatization and the establishment of regulatory agencies are relatively easy to observe and enforce compared with other reforms such as liberalization and independent regulatory practice, and thus are more tractable as targets of foreign aid. Although donor influence over the adoption of reform agendas is evident in many countries, it is also clear that countries that do not feel ownership for such reforms can be adept at paper reforms (through a policy paper or even legislation) that are never enacted, or take symbolic measures (such as the creation of a regulator) (Pritchett, Woolcock, and Andrews 2010) that may have a lower political cost but do not address the sector's underlying problems (Levy 2014).

These kinds of cosmetic reform measures have been termed "isomorphic mimicry" whereby countries mimic reform activity without implementing any measures (Pritchett, Woolcock, and Andrews 2010).[4] In Tajikistan, for example, the country's vertically integrated utility, Barki Tojik, has a history of poor operational and financial performance. Finally, the government's need to lessen its financial burden together with donor pressure led in

2004 to the formation of a state commission. The commission recommended a cautious restructuring, unbundling the utility over a long time period. In 2008 the country finally adopted a resolution to restructure, and several subsidiaries and companies, all controlled by Barki Tojik, were formed. No meaningful unbundling occurred, however, and Barki Tojik remains a fully state-controlled, vertically integrated utility. Prior to the 2008 resolution, Tajikistan passed several laws and resolutions referring to privatizing state-owned enterprises, including Barki Tojik, but no steps were taken in that direction either.

Kenya's power sector had two successive waves of sector reform. What began in 1996 as a perfunctory, donor-driven reform process gave way, in 2006, to a second wave of deeper, home-grown reform. An aid embargo was introduced in 1991 in response to Kenya's weak macroeconomic performance and governance failures. Donor organizations insisted on sector reform. Reluctantly, Kenya complied. The measures taken at that time nevertheless laid important foundations—policy and regulatory functions were separated from commercial activities, and generation was unbundled from transmission and distribution. The government introduced cost-reflective tariffs, liberalizing generation. The sector reform process also benefited from significant technical assistance and capacity building. Key sector officials were sent abroad on study tours to observe progress in other countries. In time, Kenya's political and technocratic leadership was persuaded of the relevance of power sector reform. A second wave of reform, initiated in 2006, was led entirely by these converts to domestic reform, champions who were able to build on foundations laid down in the 1990s. They strengthened independent regulation, partially privatized the national generation company, and created agencies to address renewable energy and energy access goals. Donor conditionality may have helped trigger power sector reform in Kenya, but local leadership gave reform its deep roots and momentum.

In Ukraine, repeated reform efforts mirror the influences of competing outside influences—principally the Russian Federation and the European Union. Since Ukraine's independence in 1991, following the Soviet Union's breakup, the country's geographic location has meant significant and countervailing political dynamics. Ukraine tried early on, in the 1990s, to establish a wholesale competitive spot market in electricity. Although broadly reflective of Ukraine's movement toward a market economy, this liberalization of the power sector remained deeply at odds with the inherently oligarchic economy lacking in any kind of competition. The reform failed. Twenty years later, the same tensions are in place. Nevertheless, the plan to introduce a wholesale electricity market resurfaced. This time the agreements are tied to integration with the European Union, which in turn provides momentum and incentive. Ukraine's electricity market and regulatory reform can converge with the European Union's *acquis communautaire*.[5] The reform seems to be moving ahead.

Although donors have been important catalysts of reform and providers of ongoing support, the dynamics of the reform process are largely driven by domestic political factors (table 3.4). A glance at the experience of the 15 observatory countries shows that, although donor influence has been ubiquitous, it has not definitively shaped a country's pace of reform or subsequent delivery. No statistically significant relationship can be found between aid dependency and the depth of reforms undertaken. This suggests that reform implementation may be driven primarily by factors internal to the country, something we consider in the following sections. Still, once a country begins moving ahead with reform, donor support remains critical to implementation. For example, the World Bank has remained involved in nearly all of the reforms undertaken in the observatory countries (see annex 3C).

TABLE 3.4 Donor influence on patterns of power sector reform

		Announcement			Implementation			Delivery			Sustainability	
		Extensive	Cautious	Minimal	Rapid	Incremental	Limited	Complete	Partial	Negligible	Sustained	Nonsustained
Donor influence	High											
	Moderate											
Statistical significance		—			—			—			—	

Source: Based on Rethinking Power Sector Reform utility database 2015.
Note: This table shows the extent of donor influence on the process of reform across the sample of 15 country cases. The darker the shading, the higher the number of countries from the sample that fall into each combination of circumstances. The Chi-squared test was applied to uncover statistically significant relationships between ideology and reform, with the results reported as follows: **** = 95–99 percent significance (significant); *** = 90–95 percent significance (less significant); ** = 85–90 percent significance (somewhat significant); * = 75–85 percent significance (low significance); — = not significant. Full statistical results are reported in table 3B.2 in annex 3B.

Finding #3: Countries espousing market-oriented ideologies are more likely to implement and sustain bold reforms

> *The donors recommended that we undertake reforms.... There was this idea that government should not be involved in the business.... A neighboring country came to us and said that we should not accept the donor conditions ... the most that should be done was unbundling. We listened, but we kept quiet. Our neighbor had a different background to us. They had an interest in government providing services such as electricity because of their socialist history.*
>
> —*Utility official*

Ideology plays an important role in a political economy. Some countries had a longstanding orientation toward either state dominance (Tajikistan, Tanzania, and Vietnam) or a market economy (the Philippines) (table 3.5). Others might lean more heavily in one direction or another over time and are best characterized as hybrids (India and Senegal). Despite the importance of ideology, political parties across the developing world do not necessarily divide along ideological lines.

The relationship between power sector reform and political ideology is not easy to define. The 1990s power sector reform model, grounded as it is in neoliberal thinking, appears to sit more easily within a market-oriented ideology. This tendency, however, is not clear cut. Some left-wing or even communist governments espouse the role of the market, and the private sector, in power sector reform (Senegal and Vietnam in the late 1990s), whereas some right-wing governments rely on the state (Rajasthan in India, and Senegal, in the 2000s). The academic literature presents mixed evidence on ideology as a driver in power sector reform (Erdogdu 2013, 2014). The same mixed results can be found in the case study literature, which finds that ideology has shaped power sector reform in Africa (Gore and others 2018), but not among Indian states (Cheng and others 2016). The reform narratives of the 15 observatory countries likewise show that a subtle blend of ideology, historical legacy, and political perceptions explains power sector reform, albeit not in a deterministic way or in ways that fit into a simple left–right spectrum.

Tanzania's strong brand of African socialism has asserted itself by limiting power sector reforms, notwithstanding its

TABLE 3.5 Observatory countries and states and their ideological orientation

Ideological leaning	Country/State
Market-oriented	Colombia, Peru, Philippines, Uganda
Hybrid	Dominican Republic, India (Andhra Pradesh, Odisha, and Rajasthan), Kenya, Morocco, Pakistan, Senegal, Ukraine
State-oriented	Arab Republic of Egypt, Tajikistan, Tanzania, Vietnam

Source: World Bank 2019.

official policy pronouncements. The spirit of President Julius Nyerere's *Ujamaa*—Tanzanian African socialism—and the Arusha Declaration of 1967 remain powerful influences. The 1967 declaration outlined a vision of economic justice, a socialist command economy, nationalization, self-reliance, and independence from foreign private investment or foreign aid. Those principles set the foundation for a degree of Tanzania's postindependence national unity, ethnic cohesion, and political stability, not universal at the time in Sub-Saharan Africa. The country's lackluster economic performance, however, led to dwindling donor support in the 1980s, prompting loan conditionality around the liberalization of key economic sectors that, by the 1990s, extended to the power sector. Public commitments were made in 1997 to privatize the national utility, Tanzania Electric Supply Company Limited (TANESCO), and in 1999 to fully unbundle the power sector. Nevertheless, 20 years later, the only major reforms enacted in Tanzania are opening the market to independent power producers (IPPs) and creating an independent regulator. The power utility, TANESCO, remains a vertically integrated state-owned monopoly, characterized by weak financial and technical performance and tariffs that are not cost-reflective. The successive governments have manifested varying approaches to the liberalization and restructuring of the power sector. Without momentum or policy continuity, public support for power sector reform appears neither deep nor wide.

In Vietnam, the country's longstanding state-orientation has shaped a more cautious and incremental approach to power sector reform. The state still plays a leading role in the economy in general and the power sector in particular, under communist party guidance. The introduction of the role of markets has been deliberately steady, with the 2004 Electricity Law, the basis for sector reform, taking more than eight years to draft. The 2006 road map set out a careful rollout of power sector reform—designing each stage of the process as a pilot to test, improve, and learn from, followed by full implementation. The approach has been gradual and consensus-driven (within the framework of the political and sector leadership), as well as sensitive to social harmony. Nevertheless, this approach has meant that reform in Vietnam has come slowly. In essence, markets are a technocratic device rather than part of any ideological or political transformation. Implementation has been slow and much delayed.

In short, ideology affects both the pace at which reform is implemented and the extent to which announcements are followed through to full implementation. This observation is somewhat at variance with the academic literature cited earlier in the chapter. A systematic look at the 15 observatory countries suggests, however, that ideology is only weakly significant (at 20 percent)[6] in explaining reform announcements but becomes more important as the reform process unfolds (table 3.6). Market-oriented ideology emerges as a strong driver for bold reform implementation (significant at the 1 percent level): half of the market-oriented countries implemented bold reforms compared with none of the state-oriented countries. Ideology also appears to be strongly related (again at the 1 percent significance level)

TABLE 3.6 Ideology and patterns of power sector reform

		Announcement			Implementation			Delivery			Sustainability	
		Extensive	Cautious	Minimal	Rapid	Incremental	Limited	Complete	Partial	Negligible	Sustained	Nonsustained
Ideology	Market											
	Hybrid											
	State											
Statistical significance		*			****			****			*	

Source: Based on Rethinking Power Sector Reform utility database 2015.
Note: This table shows the extent to which a country's prevailing ideology influences the subsequent process of reform across the sample of 15 country cases. The darker the shading of the box, the more countries from the sample that fall into each combination of circumstances. The Chi-squared test was applied to uncover statistically significant relationships between ideology and reform, with the results reported as follows: **** = 95–99 percent significance (significant); *** = 90–95 percent significance (less significant); ** = 85–90 percent significance (somewhat significant); * = 75–85 percent significance (low significance); — = not significant. Full results are reported in table 3B.3 in annex 3B.

to whether a country fully delivers on its original reform announcements: two-thirds of market-oriented countries substantially followed through on promised measures compared with none in the case of state-oriented countries. Ideology does not seem to be closely related, however, to the sustainability of reforms.

Finding #4: Reforms can be introduced across a range of political systems but are more likely to progress in countries where power is decentralized and contestable

> *There were huge fights, but they managed to make way for reforms because of the president's mode of work. He would engage people for long hours, talk to the opposition to explain why liberalization and privatization made sense—the meetings could run for a whole day and a whole night. In this way, he wore the opposition down to accepting the approach. The president would go through rounds and rounds of discussions—with industry, members of parliament, and others. The success of the reform was probably because the president took this role upon himself.*
>
> —*Former senior official*

> *We haven't really done any power sector reforms at all even since 1998. All that happened in 1998 was that they unbundled the national utility. So they unbundled it and made several generation and distribution companies, but they were all still populated by people from the national utility and still owned by the government. They were treated as one organization.*
>
> —*Government official*

The nature of a country's political system is relevant to power sector reform. Countries vary according to the extent to which political power is sustained by a centralized decision maker (a unipolar system) or oscillates among a more dispersed set of actors (multipolar) (table 3.7). The 15 observatory countries present a range of political systems. Some have had sustained unipolar leadership of varying kinds—for example, the king in Morocco, President Museveni in Uganda, and the communist party in Vietnam. Others have been consistently multipolar, such as the democracies of Colombia or India. Still others went through major transitions in their political systems during the reform process. Peru transitioned from a unipolar system under President Alberto Fujimori to a multiparty democracy. The Philippines followed a similar

TABLE 3.7 Political systems of observatory countries

Country	Concentration of political power: First significant reform	Concentration of political power: 2015
Arab Republic of Egypt, Morocco, Tajikistan, Uganda, Vietnam	Unipolar	Unipolar
Kenya, Pakistan,[a] Senegal, Tanzania	Transition	Multipolar
Colombia, Dominican Republic, India, Philippines, Ukraine	Multipolar	Multipolar
Peru	Unipolar	Multipolar

Source: Based on Rethinking Power Sector Reform utility database 2015.
a. Pakistan transitioned through military and democratic governments during this period.

path after Ferdinand Marcos was ushered from power. Pakistan, in contrast, has periods of democratic government punctuated with episodes of military rule. The reform stories of all of these countries illustrate how shifts in political systems affect implementation of power sector reform.

In Pakistan, distribution was privatized during a period of martial law, but the legitimacy of this action remains in question. Pakistan's reform journey began in 1992 with the publication of a Power Sector Strategic Plan that envisaged opening generation, unbundling the national utility, privatization, and regulation. Almost all these measures were implemented between 1994 and 1997 by a democratically elected government; privatization proved a sticking point. After the introduction of military rule in 1999, a decision was taken in 2005 to privatize Karachi Electric. The emergency conditions in force at the time put all privatization rules and processes into abeyance, and the process therefore happened without prior deliberation and approval by the Council of Common Interest. The unions, opposition political parties, and consumer representatives have contested the privatization of Karachi Electric ever since. As a result, the political and investment environment has not been conducive to further privatization of electricity distribution; subsequent governments have not been able to make progress on this issue.

Following major structural reforms under centralized rule in Peru, the subsequent democratic government focused efforts on second-generation reforms. In the late 1980s, Peru faced a severe macroeconomic crisis that was mirrored by a power sector crisis. The Fujimori government that came to power in 1990 initiated a drastic program of stabilization and structural reform; it encompassed a thoroughgoing reform of the power sector, including profound unbundling, privatization, regulation, and liberalization measures. Following promulgation of a sector law in 1991, 100 percent of transmission, 70 percent of generation, and 45 percent of distribution were transferred to private ownership, management, and operations in the space of a few years. When democratic government returned in 2002, attempts to privatize the remaining distribution companies in the provinces met with civil unrest and were eventually dropped. Attention turned to several second-generation reforms designed to revive flagging investment levels in the sector. A new sector law in 2006 reorganized the system operator, reformed planning and regulation functions, improved administration of the electricity market, and introduced regular supply auctions. This proactive management of incremental reforms has proved successful in sustaining sector performance.

In Senegal, a change in the political regime led to a reversal of the utility privatization process. It was the socialist regime (in power for 40 years, from independence to 2000) that first opened the power sector to private investment and decided in 1997 to begin privatizing the vertically integrated utility, Senelec. The Senelec privatization eventually resulted in the award of a 25-year concession to a consortium of foreign private companies, but this decision was followed by widespread labor unrest, pervasive blackouts, and the arrests of some labor leaders. Elections held in the backdrop of privatization and the consequent unrest led to the first-ever defeat of the Socialist party and brought the liberal Senegalese Democratic party to power. One of the first actions of this new dispensation was to cancel Senelec's concession contract. A second attempt at privatization subsequently failed, in part because of weak private sector interest but also because political will was lacking. The rights of labor and public perceptions about the role of the foreign private sector in providing a public service played important roles, producing intense ideological cleavages that themselves determined Senegal's reform path.

The manner in which a country's political system might shape its power sector reform process is conceptually ambiguous. On the one hand, unipolar systems may find it more straightforward to implement power sector reforms should the central decision maker be convinced of the merits of this course of action. On the other hand, given that the nature of power market reform is to depoliticize the sector by ceding control to autonomous bodies (the regulator and nongovernment actors such as the private sector), this course of action might be inherently less attractive to a unipolar leader.

In practice, countries with unipolar systems seem more likely to drag their feet on implementing reform. The evidence from the 15 observatory countries suggests that foot-dragging is dominant among unipolar systems (table 3.8). The nature of the political system appears to have no material impact on a country's announced reform ambitions. There is, however, a strong relationship between multipolar political systems and bold reform implementation (statistically significant at 6 percent), with multipolar countries almost twice as likely to implement bold reforms compared with unipolar countries. Furthermore, countries with multipolar systems are almost twice as likely to fully deliver on their original reform announcements relative to unipolar countries (statistically significant at the 15 percent level). These findings highlight nuances

TABLE 3.8 Type of political system and patterns of power sector reform

		Announcement			Implementation			Delivery			Sustainability	
		Extensive	Cautious	Minimal	Rapid	Incremental	Limited	Complete	Partial	Negligible	Sustained	Nonsustained
Political system	Unipolar											
	Transition											
	Multipolar											
Statistical significance		—			***			**			—	

Source: Based on Rethinking Power Sector Reform utility database 2015.
Note: This table shows to what extent a country's political system influences the process of reform across the sample of 15 country cases. The darker the shading of the box, the higher the number of countries from the sample that fall into each combination of circumstances. The Chi-squared test was applied to uncover statistically significant relationships between ideology and reform, with the results reported as follows: **** = 95–99 percent significance (significant); *** = 90–95 percent significance (less significant); ** = 85–90 percent significance (somewhat significant); * = 75–85 percent significance (low significance); — = not significant. Full statistical results are reported in table 3B.4 in annex 3B.

beyond the results of studies involving larger sample sizes, which find no correlation between competitive political systems and power sector market reform elements in developing countries (Erdogdu 2014; Foster and others 2017).

Finding #5: Reform champions often emerge as pivotal figures who help to ensure momentum toward implementation, particularly when supported by stable bureaucracies

The president has an interest in the power sector since he came in a wake of a shortage in supply, which turned into a political problem to the former president. He does care about tariff reform, as a part of the subsidy reform policy. However, [on] the other reform dimensions and details the president has less role. He has less clear impact on these aspects as he may look to them as details which may not attract his attention. Regarding the cabinet … they deal with the program case-by-case and not in a comprehensive plan with an integrated vision.

—Government official

Political class was not much involved; reform was essentially a one-man decision. It did not trickle down to political class and civil servants. Once the political decision was taken at a high level, civil servants formed groups to take care of reforms. Two years after the reform agenda was adopted, the leader lost elections. Even then, top political class was committed but at bureaucracy level no personal commitment was there owing to changes made there (by the new government) like energy secretary, etc.

—Former regulator

When a director general [(DG) of the utility] is there for 18 months, or a maximum of two years, he can't really do anything. The political authorities are very susceptible [to public opinion]. So when the street complains, the minister or DG gets sacrificed. Another one comes without that translating into any significant improvements. In reality, it's the political authorities at the highest level, who should have refrained from interfering in the management of the company, by putting in place the people that were really needed.

—Senior government official

In 2004, the relationship between the position of PS [permanent secretary] and president changed. Previously they had been appointed by the minister but, from 2004, the president took on the responsibility of these appointments. In the past, the minister could have worked with the minister of public service to "fire" the PS. This created a stronger and protected relationship between the technical experts in the ministries and the president, which the president used to get the technical advice he needed.

—Electricity company official

Reform champions sometimes emerge and play a pivotal role in directing the reform process (see table 3.9). They are able to articulate a clear vision regarding the endpoint of the reform, motivate action by critical players, and use their political capital to address various roadblocks that may emerge along the way. Few studies are available on the role of individual leaders in power sector reforms across a large number of countries.[7] About half of the 15 countries in the observatory had a clearly identifiable reform champion. The nature of these champions varies from the highest authority—the king in Morocco, president in Uganda, or chief minister in some of the Indian states to technocrats in the line ministries responsible for the sector. The reform narratives illustrate that champions are most effective when they can rely on strong and stable leadership at senior

TABLE 3.9 Overview of reform champions

Country/State	Champion(s)	Role
Dominican Republic	President Fernandez	With reform legislation stuck in congress, carried out almost the entire reform process through decrees and ministerial orders.
India – Andhra Pradesh	Chief Minister Naidu	Despite strong political opposition, pushed through with reforms, driving adoption and implementation.
India – Odisha	Chief Minister Patnaik	First political leader to buy in to 1990s model reforms for power sector in India. Implemented full spectrum of the reform including private sector participation in distribution, which several later reformers did not.
Kenya	Senior ministry and utility officials	Convinced other actors and orchestrated their actions on basis of their conviction, technical credibility, and closeness to the president.
Morocco	Monarch	Articulated a vision of reform and directed efforts of key institutions.
Pakistan	PML(N) and bureaucrats	Party enacted comprehensive reform law in 1992 but lost power following year. On coming back in 1997, moved fast to implement various aspects of the law—regulatory, restructuring WAPDA.
Peru	President Fujimori	Bought into the 1990s model reforms and implemented all aspects of restructuring, competition, and private sector participation in a span of a few years.
Philippines	President Arroyo	Put her weight behind the EPIRA legislation to ensure it would get the votes necessary to be adopted as law.
Uganda	President Museveni	Bought into reform, drove implementation, and persuaded skeptical stakeholders to collaborate.

Source: Based on Rethinking Power Sector Reform utility database 2015.
Note: EPIRA = Electric Power Industry Reform Act; PML(N) = Pakistan Muslim League (Nawaz); WAPDA = Water and Power Development Authority.

technical levels—permanent secretaries, directors general, chief executive officers, regulators, and sometimes ministers. In some instances, such as Kenya, the technocrats themselves have championed reform. The power sector reform is, in other words, highly technical and highly political at the same time.

In Uganda, reform momentum was maintained by strong technocrats in key ministry positions enjoying strong support from political leadership. Circa 1990, the Ministry of Finance, Planning, and Economic Development laid the legal and policy groundwork for public enterprise reform and divestiture (which formed the basis for later power sector reform), because that was deemed to be critical to macroeconomic stabilization and recovery. Top bureaucrats in the ministry, who remained in office for more than two decades (1992–2013), were critical to the overall reform process in the country. As institutional leadership of power sector reform passed more to the Ministry of Energy and Mineral Development, the role of the top bureaucrat there (who had been in office for 20 years) was pivotal. The long-term close relationships between such technocrats and the president clearly determined both the delivery of Uganda's power sector reform and the way in which it progressively deepened over the years and adjusted to evolving circumstances.

In the Philippines, President Gloria Arroyo used her scant political capital to ensure that the power sector reform legislation would pass the congress. When Arroyo took over in January 2001, the country was still reeling under the effects of the Asian financial crisis of 1997–98, with a depreciating currency and exodus of capital. On the political side, her party leader and former president Joseph Estrada had been impeached, leaving her with limited space to maneuver. The Estrada government had presented a comprehensive power sector reform bill in the congress, but the changing political environment meant it was going nowhere. Arroyo prioritized the bill when she entered office, using all her political capital to get it passed in the congress despite several attempts to destabilize her government and impeach her. As a result, the Electric Power Industry Reform Act (EPIRA) was signed into law and became effective in July 2001.

In Morocco, the king provides the political vision for the sector as well as ensuring the stability of the technical cadre. Morocco is an active albeit constitutional monarchy that provides a degree of autonomy for the prime minister, cabinet of ministers, and parliament. The monarchy typically exercises strategic leadership and has played a central role in articulating a long-term vision for the power sector. In general, the most senior leaders of the power sector are appointed by the king and require his full support to remain in office. As such, the average tendency is for key technical leaders and staff to have long tenures (even if some individuals can be rapidly removed if judged ineffective), and to enjoy high-level political support. Key initiatives in the power sector—such as its initial opening to the private sector, a drive to achieve universal electricity access, and a strategic shift to renewables—are initiated at the highest level. Follow-up implementation is the responsibility of the technical leaders, but royal oversight helps remove roadblocks and keeps the overall process on schedule. The implementation of Morocco's renewable energy strategy by the Moroccan Agency for Sustainable Energy (MASEN) is a classic example of that process. The president of MASEN has a strong professional background in financial sector leadership; he was appointed by the king to get financing for challenging projects, has held office for almost 10 years, and has enjoyed strong political support in removing obstacles to implementation of the strategy.

The Indian civil service, in contrast, sees much more rapid turnover of critical positions, making it more challenging to follow through on reform implementation at the state level. Senior bureaucrats at the state level are part of a civil service; members are recruited nationally and can be seen as a permanent executive branch of the government. These officers are generalists and move from one ministry to another, separate from the political system, but all bureaucratic appointments, from the power secretary and the chief executive officer of the state-owned utility to the top police officers in every city, are made by the state government. Consequently, officers are usually moved from one ministry to the other after a period of two to three years. Each new political dispensation shuffles the top bureaucrats at the ministries. This discontinuity limits the benefits of long-term technical staff who can buy into and push reforms.

The impact of reform champions is evident in the pace of reform implementation and helps to ensure that reform announcements are fully delivered (table 3.10). The presence of a champion does not affect a country's publicly avowed level of ambition on reform. It matters significantly, however, for the success of implementation. Countries lacking a reform champion were four times as likely to stall on the implementation of reforms (significant at the 15 percent level). The impact is even starker when it comes to delivery. More than half the countries lacking a reform champion largely failed to deliver on any of the reform announcements, whereas all those with reform champions made considerable progress (significant at the 2 percent level). Champions do not seem to be so helpful, however, when it comes to the longer-term sustainability; perhaps because they are not around.

TABLE 3.10 Reform champions and patterns of power sector reform

		Announcement			Implementation			Delivery			Sustainability	
		Extensive	Cautious	Minimal	Rapid	Incremental	Limited	Complete	Partial	Negligible	Sustained	Nonsustained
Reform champion	Strong											
	Weak											
Statistical significance		—			**			****			—	

Source: Based on Rethinking Power Sector Reform utility database 2015.
Note: This table shows to what extent the strength of the reform champion influences the subsequent process of reform across the sample of 15 country cases. The darker the shading of the box, the higher the number of countries from the sample that fall into each combination of circumstances. The Chi-squared test was applied to uncover statistically significant relationships between ideology and reform, with the results reported as follows: **** = 95–99 percent significance (significant); *** = 90–95 percent significance (less significant); ** = 85–90 percent significance (somewhat significant); * = 75–85 percent significance (low significance); — = not significant. Full results are reported in table 3B.5 in annex 3B.

Finding #6: Unless champions can orchestrate wider stakeholder alignment, the sustainability of reforms may be called into question

> *There are vested interests that take advantage of the situation in the power industry. When you talk of reforms, you mean change. However, there are people who benefit from the status quo, especially the utility's situation (such as politicians, suppliers—especially fuel suppliers). These vested interests do not support reform. Unfortunately, they are also placed in positions of influence.*
>
> *—Former utility official*

> *The entity responsible to do the reform hesitated due to the fear of public/employees' reaction. Furthermore, they are not convinced about the timeframe for the reform. Therefore, they are normally on the side of postponing the reform. There is reluctance on the leadership level from information disclosure.*
>
> *—Government advisor*

> *Regarding the media, on one side it is loyal to the government since a substantial part is owned by the State and more than that, it [is] being subsidized by the State. On the other hand, it does not have sufficient information to address the reform deeply. In general, and due to its inability to convey the message of reform they are more tending to status quo, which is more understood to them. The reform is not a priority to the media. Also, it shows the lack of capacity of the state to communicate to the public regarding the reforms.*
>
> *—Government advisor*

Reform champions alone cannot ensure the success of a reform process. Power sector reforms are often initiated by a political leader or technical-level reform champion, or by one particular institution; and they can achieve quite a lot and impart momentum to the reform program. To deepen and widen the reforms, and to sustain the momentum, a wider range of stakeholders and interests needs to be engaged; compromises reached between contesting views, a coalition formed, and levels of communication heightened to directly involve stakeholders and often also the public. The degree of interaction between people and institutions needs to intensify to keep the reforms moving (Lee and Usman 2018). The public may come to oppose electricity privatization or foreign private investment because of various negative associations (Sen, Nepal, and Jamasb 2016; Zelner, Henisz, and Holburn 2009;). It is also evident, however, that public participation in power sector policy processes has been limited in many recent developing country cases, even where civil society is active, such as in India (Dubash 2002; Nakhooda, Dixit, and Dubash 2007).

Power sector reform touches on many competing interests and inevitably creates losers as well as winners. Regulatory tariff adjustments designed to achieve cost recovery will make electricity less affordable for key constituencies. Gains in the commercial and operational efficiency of utilities improve their financial performance but may also threaten jobs or prevent pilferers from availing themselves of "free" electricity through network theft. Reform of power markets may spell the loss of business for those supplying fuel to private generators. Privatization of power utilities may create new commercial opportunities, but those new opportunities may also spell the loss of political patronage and bureaucratic influence and ultimately benefit foreign commercial interests. It follows that friction can and does emerge among different stakeholder groups, which can sometimes threaten to derail reform attempts. These problems may be at least partially mitigated by aligning stakeholder interests through consensus-building dialogue and potentially compensatory measures, such as labor protection or generous redundancy packages, or safety net subsidies to protect the poorest from tariff increases, or by adjusting the design of privatization programs to create greater opportunities for the domestic private sector.

As a result, an integral part of power sector reform is managing friction among stakeholder groups. Stakeholder groups can have different impacts on the reform process depending on their power and influence. Labor unions fear loss of jobs, competing ministries fight for influence, consumer groups may oppose any tariff hikes, and regional utilities or subnational leaders want to avoid losing control (table 3.11). How these frictions are resolved may determine the direction of the reform. In some cases, such as Colombia and Peru, local governments wielded enough power to shape the national reform process by forcing the central government to compromise on some measures. In other cases, such as the Indian states of Andhra Pradesh and Rajasthan, consumers were powerful enough to stop tariff reform undermining utility financials.

In Tajikistan, vested interests among key stakeholders have consistently blocked the path of reform. One key political economy issue has involved the national utility, Barki Tojik, and its inability to recover costs from two critical and energy-intensive state-owned entities, the aluminum company, TALCO, and the Agency for Land Reclamation and Irrigation, which provides irrigation services to farmers. Neither entity is charged a cost-recovery tariff for electricity, and their substantial accumulated debts are repeatedly forgiven. There are at least two political economy explanations for these large implicit subsidies. One is that the government gains legitimacy through supporting employment in these entities, because any downsizing might expose them to market conditions. The other explanation is that those entities, although nominally state-owned, exhibit beneficial ownership by the ruling elite and their political and business allies. As such, key political interests would be threatened by any deep structural reform in the power sector that would seek financial discipline, cost recovery, operational autonomy for unbundled sector entities, or private participation. If deep power sector reform involved the state giving up control over the levers that provide the large subsidies, the beneficial owners would lose the influence that ensures their very benefits. This dynamic in Tajikistan's political

TABLE 3.11 Stakeholder friction in power sector reforms: An overview

Country/State	Stakeholders	Nature of conflict
Colombia	Municipal utilities	Successfully opposed restructuring and struck a compromise to perform under new guidelines of market power. Result is a mixed ownership sector.
Dominican Republic	Political actors	Reluctance in implementing reforms due to loss of patronage opportunities led to prolonged delays in implementing reform legislation.
Egypt, Arab Rep.	Political actors	Reluctance to implement tariff reforms associated with power sector reform due to fear of civil unrest.
India - Andhra Pradesh	Farmers	Strong political force, successfully lobbied for free electricity undermining sector financials.
India - Odisha	Utility employees	State electricity board employees opposed privatization but were won over with employment guarantees.
India - Rajasthan	Consumers	Successfully lobbied for a tariff freeze undermining utility financials.
Kenya	Political actors	Exerted influence over regulator to control tariffs during politically sensitive times such as elections.
Morocco	Municipal utilities	Successfully opposed reform legislation and a full market model because they use revenue from electricity businesses to balance losses from sales of water.
Pakistan	PEPCO	The body established to implement restructuring and privatization in sector in 1996 still exists and wants to maintain its role, thereby slowing the reform process.
Peru	Provincial governments	Regional utilities and politicians were opposed to privatization and successfully blocked attempts to privatize provincial power utilities.
Philippines	Oligarchs	Oligarch industry players with congressional affiliations exercised preference for hybrid state/market approach with restricted competition.
Senegal	Labor unions	Actively opposed privatization of the power utility.
Tajikistan	Economic elite	Oppose any reform linked to tariff subsidy because it hits their economic interests in industry and agriculture.
Tanzania	Political actors	Large number opposed to private sector involvement in power sector.
Uganda	Incumbent utility	Incumbent staff and senior executives were opposed to restructuring and privatization but failed to stop process.
Ukraine	Economic elite	Economic and political elite acquired controlling stakes of utilities through the privatization process.
Vietnam	Party members	Serious disagreement between factions in party on reform path led to prolonged debate and delays in implementation.

Source: Based on Rethinking Power Sector Reform utility database 2015.
Note: PEPCO = Pakistan Electric Power Company.

economy has so far been sufficient motivation to block deep power sector reform in the country.

Across a number of Indian states, reforms promoted by influential chief ministers often flounder at later stages under their successors, because of a lack of wider support. This pattern repeats itself in the states of Andhra Pradesh and Rajasthan.

In Andhra Pradesh, the then–chief minister eschewed privatization but pursued tariff reforms and information technology investments aimed at enhancing the efficiency and financial viability of the utilities. Extensive stakeholder engagement was undertaken with unions and employees but did not extend to electricity consumers. Farmers, in particular, retain a deep sense of entitlement to free power. The issue of electricity subsidies for agriculture became politically charged, overriding the reform process and leading to the government's electoral defeat in 2004–05. The policy of free power for farmers persists to this day and has debilitated the financial standing of power utilities in the state.

In Rajasthan, reforms of the power sector were initiated at the turn of the century. The objective was the financial turnaround of the sector using tariff reforms and privatization of the distribution business. The reforms were driven by a chief minister motivated by fiscal concerns, who nevertheless did not lay the necessary groundwork of stakeholder engagement. After a new government came to power, the reforms stalled in 2004–05. Tariffs were frozen for the next eight years, with an accumulated debt of up to US$14 billion by 2014.

Colombia's experience shows how reforms can be adjusted to accommodate the competing interests of national and municipal governments, thereby achieving the necessary consensus to progress. Colombia's power sector has long combined national-level power utilities, controlled by central government along with powerful municipal utilities—notably, Empresa de Energia Electrica de Bogota (EEEB) in Bogotá and Empresas Públicas de Medellín (EPM) in Medellín—that together accounted for about 40 percent of electricity distribution. There had long been tension over the allocation of roles between central government and municipal actors in the sector, and these came to a head when the sector reform laws of 1994 called for unbundling and privatizing the utilities. The municipal governments objected. A compromise allowed EPM and EEEB to remain vertically integrated public utilities as long as they separated the accounting of their generation and distribution activities and abided by new restrictions on market shares in these activities. EPM flourished under the new regulatory framework and remains a vertically integrated publicly owned utility and one of the main actors in the power sector. EEEB, in contrast, was unable to turn its performance around and eventually underwent vertical separation and privatization.

Egypt employed an effective public communications campaign to explain the need for electricity tariff reforms. It preempted consumer opposition. In 2014 Egypt experienced major electricity supply outages that led to civic unrest. The government moved swiftly to address concerns about security of supply by contracting for short-term rental power, importing liquefied natural gas, and building new gas-fired plants. Once these fundamentals were in place, the government restored the financial equilibrium of the power sector through a five-year rising tariff trajectory. The initial price increases of electricity and major categories of fuels (40 to 78 percent) were intended to reduce its subsidy burden. The magnitude of the reform and consumer sensitivity to energy price hikes meant the government had to develop an effective communication strategy. The campaign highlighted the inherently inequitable nature of the subsidies, explaining how they benefited the rich disproportionately (Moerenhout 2018). At the same time, the government communicated that fiscal savings from energy subsidy reform would be at least partially redirected to social programs, which up until that time received fewer resources than energy subsidies. The emphasis on social spending was in line with the country's new constitution. This strategy helped to forge a social consensus around tariff reform and staved off unrest. Other countries' experience with communication campaigns is summarized in box 3.2.

Finally, in some instances, stakeholders can lend support to reforms that provide opportunities to advance their own interests, as happened in Ukraine. The period of 2000–02 saw several dubious privatization efforts in Ukraine stall the entire reform process. Large business interests used their clientelist relationship with senior government officials to secure ownership of generating assets and distribution companies. In a sector plagued by nonpayment by distributors in the wholesale electricity market, the issue of dues was huge at every level. The wholesale electricity market owed money to generators, which owed money to fuel suppliers. Meanwhile, the government's own fuel-supply companies took the state-owned generation companies to court, and the court promptly forced the sale of company assets to settle the debt. These assets were sold in a

BOX 3.2 Importance of public communication strategy to support reforms

International experience has shown that communication is critical to the success of major economic reforms. If an effective communication program is not implemented before, during, and after reform measures go into effect, it is difficult to earn the public's trust and foster understanding of the political decisions that underpin the reform. A well-researched communication program with informational, attitudinal, and behavioral objectives can enhance the effectiveness of reform.

An effective communication campaign involves mapping key stakeholders, using outreach and two-way dialogue with citizens, conducting opinion research, consulting with stakeholders, creating and testing compelling messages that build awareness of reform benefits, assigning credible messengers, identifying good channels of communication, coordinating within government, setting strategic goals, and communicating consistently with evidence-based messages.

Public reactions to reform programs are highly contextual and dynamic. A well-informed public understands the rationale for reform and greatly improves the likelihood of success. Some successful examples are listed below.

In **Vietnam**, public acceptance of an electricity tariff reform was facilitated by a communication strategy that focused on raising awareness about the rationale for tariff hikes. A capacity gap analysis was conducted to identify ways to improve Vietnam Electricity's (EVN) communication efforts, and opinion research was conducted to understand public sentiments.

Iraq undertook a qualitative assessment of key consumer perceptions and awareness levels of tariffs and subsidies. The assessment mapped stakeholders and analyzed the audience; surveys and focus groups gathered information on citizens' views. The government strengthened its internal coordination to deliver consistent and convincing messages.

Ukraine rolled out a communication strategy in support of stiff tariff hikes. The rollout included a 30-second public service announcement that aired on 19 TV stations across Ukraine and appeared on 15 government websites. The announcement was rooted in a detailed understanding of public perceptions; the key messages reflected the public's concern. The effort included involving local media in major cities and making them aware of the energy sector status.

In **Belarus,** focus group discussions and a stakeholder mapping found that opposition to tariff reforms stemmed from the lack of knowledge consumers had about tariff-setting policy and reform processes. A communication strategy was designed to support efforts to better engage consumers in the governance of district heating providers. Workshops were held with local service providers to build capacity for improved public communication with consumers; well-designed graphics helped to explain reforms and the benefits of energy efficiency. These efforts—and clearer and more transparent heating bills—helped mitigate resistance to higher tariffs.

Source: Worley, Pasquier, and Canpolat 2018.

nontransparent manner, leading to allegations of corruption in a process termed "asset stripping." Auctions were announced in random fashion. In one case in 2001, about 4,000 megawatts of thermal units were sold for as little as US$38 million. The same interests that had acquired the generation assets went on to take controlling stakes when the distribution companies were privatized—the result of weak corporate laws that accorded private sector control of the companies with shareholding as low as 26 percent. This corrupt process became known as shadow privatization.

Achieving stakeholder alignment helps safeguard the sustainability of reform measures (table 3.12). Stakeholder alignment does not appear to be critical at the early stages of reform. It has no material impact on the initial level of reform ambition, or on the pace of reform implementation, or on the extent of delivery of reform measures. Where it appears to make a difference is in the sustainability of reform. Among countries without stakeholder alignment, about one-third undergo privatization reversals, compared with none among countries that achieved stronger alignment (significant at the 11 percent level).

Finding #7: Legislation is usually a necessary statement of political commitment to reform that helps to support longer-term sustainability

There was the Sector Policy Paper closely followed by the Energy Act, with clear guidelines

TABLE 3.12 Stakeholder alignment and patterns of power sector reform

		Announcement			Implementation			Delivery			Sustainability	
		Extensive	Cautious	Minimal	Rapid	Incremental	Limited	Complete	Partial	Negligible	Sustained	Nonsustained
Stakeholder alignment	Strong											
	Weak											
Statistical significance		—			—			—			**	

Source: Based on Rethinking Power Sector Reform utility database 2015.
Note: This table shows to what extent the degree of stakeholder alignment influences the subsequent process of reform across the sample of 15 country cases. The darker the shading of the box, the higher the number of countries from the sample that fall into each combination of circumstances. The Chi-squared test was applied to uncover statistically significant relationships between ideology and reform, with the results reported as follows: **** = 95–99 percent significance (significant); *** = 90–95 percent significance (less significant); ** = 85–90 percent significance (somewhat significant); * = 75–85 percent significance (low significance); — = not significant. Full results appear in table 3B.6 in annex 3B.

on what needed to be implemented—that is why things went the way they did.

—*Former senior government official*

The reform has had achievements that can show that the … general law of electricity was a major accomplishment, it is a good law, although there are some things that need to be modified; but it is a good law. The country received good investment through the reform process, but a new wave of reform has to be made; but above all, we must change the climate of legal security, law enforcement, and improve the institutional framework for new investments coming, and let them flow. And the state should stop interfering in everything and stop politicizing everything. If you are investing in new companies, or if you are staying with your companies; it has to let institutions work, you cannot be a judge and part of your institutions; do not obstruct investment, because the state may take up to a year to grant a concession to a plant, or two years. That's crazy.

—*Sector official*

Legislation is often a critical step in crystalizing political commitment to reform and paving the way for implementation. Drafting a legal framework is important for working out the implementation details of a reform process, while the process of approving legislation necessarily engages a range of political actors and becomes one important vehicle for consensus building. Global-level analysis suggests that legislation enabling the introduction of IPPs, for example, helps to attract private investment in the power sector from domestic and foreign sources (Urpelainen and Yang 2017). Among the observatory countries, most of them introduced legislation as a foundation of reform. The process was not always straightforward, however. In some countries, the sector law was not enacted until many years after the policy commitment, as in the cases of the Dominican Republic and Egypt (see table 3.13).

The benefits of legislation are felt primarily in safeguarding the longer-term sustainability of reforms (table 3.14). The presence of strong sector legislation does not seem to affect the ambition of reform announcements, or the speed or efficacy of implementation

TABLE 3.13 Legislation and power sector reform: An overview

Country	Foundational legislation	Years elapsed[a]	Comments
Colombia	Laws 142 and 143	0	Severity of macroeconomic crisis and power shortages in 1990s prompted immediate legislative action.
Dominican Republic	Electricity Law 2001	8	Law introduced in congress in 1993 but faced opposition on privatization and languished in the legislature for eight years.
Egypt, Arab Rep.	Electricity Law 2015	15	Committed to reform in 2000, but political momentum subsided, and full legislation was deferred until 2015.
India - Andhra Pradesh	Andhra Pradesh Electricity Reforms Act 1999	4	State government instituted a committee to recommend a reform path in 1995 feeding into eventual legislation.
India - Odisha	Orissa Electricity Reform Act 1995	2	Government began reform process in 1993 with assistance from the World Bank and U.K. Department for International Development.
India - Rajasthan	Rajasthan Power Sector Reform Act 1999	1	Because other states had drafted similar legislation for reforms, Rajasthan took limited time to do the same.
Kenya	Electricity Power Act 1997, Energy Act 2006	5, 14	Relatively weak legislation was passed five years after reform commitment; the legislative foundation took nine years to materialize.
Morocco	Law No 57-2009, Law No-37-2016	n.a.	No conventional sector reform laws; focus on role of MASEN in promoting renewable energy.
Peru	Law for Power Concessions 1992	1	The country committed to overhauling its economic policies to come out of the 1991 macro crisis.
Philippines	Electric Power Industry Reform Act 2001	5	Original legislation tabled in 1996 did not pass but was adopted after refiling in 1998.
Uganda	Electricity Act 1999	1	The sector strategy was created in 1998 and in 1999; the government followed up with legislation.
Vietnam	Electricity Law 2004	8	The original draft law was created in 1996 and went through 25 versions before being enacted as law in 2004.

Source: Based on Rethinking Power Sector Reform utility database 2015.
Note: MASEN = Moroccan Agency for Sustainable Energy; n.a. = not applicable.
a. Elapsed time between announcement of power sector reform and enactment of the legislation on which the reform is based.

TABLE 3.14 Legislation and patterns of power sector reform

		Announcement			Implementation			Delivery			Sustainability	
		Extensive	Cautious	Minimal	Rapid	Incremental	Limited	Complete	Partial	Negligible	Sustained	Nonsustained
Legislation	Strong											
	Weak											
Statistical significance		—			—			—			*	

Source: Based on Rethinking Power Sector Reform utility database 2015.
Note: This table shows to what extent the strength of the legislative framework influences the subsequent process of reform across the sample of 15 country cases. The darker the shading of the box, the higher the number of countries from the sample that fall into each combination of circumstances. The Chi-squared test was applied to uncover statistically significant relationships between ideology and reform, with the results reported as follows: **** = 95–99 percent significance (significant); *** = 90–95 percent significance (less significant); ** = 85–90 percent significance (somewhat significant); * = 75–85 percent significance (low significance); — = not significant. Full statistical results are reported in table 3B.7 in annex 3B.

and delivery. Countries with strong legislation, however, have a much higher chance of sustaining those reforms (significant at 21 percent).

In the Dominican Republic, difficulties in passing new sector legislation meant that major reforms were undertaken on a fragile legal basis that was vulnerable to reversal. Following the decision to reform the power sector, a new reform bill was prepared in 1992 incorporating the standard reform prescriptions of unbundling, privatization, and liberalization. The bill was tabled in the legislature in 1993 but was opposed by several members who wanted to maintain partial state ownership of sector assets, and it did not pass the body till 2001. In the absence of comprehensive legislation, the government went ahead with reforms through decrees and ministerial resolutions. In 1998, decrees were used to unbundle the national utility into seven companies and create a sector regulator with an associated regulatory framework. In 1999, three distribution companies and two generation companies were privatized with 50 percent stakes, with power purchase agreements extending to 2003; by 2001, the wholesale electricity market had begun operation. Owing to weaknesses in the legal framework, political interference remained strong and the government refused to allow tariff adjustments under the new regulatory system. All this led to reversing the privatization of two of the distribution utilities by 2003 and the third one in 2009.

In Vietnam, the final version of the Electricity Law took years to materialize, but the long gestation ensured that various stakeholders were on board in the end. Work on drafting a new comprehensive electricity law began in 1996; over the next eight years, 25 different versions would be prepared. The Ministry of Industry led the drafting of the law and the team included representatives of Vietnam Electricity (EVN) and various other government departments and ministries. The lengthy discussion over the bill reflected deep disagreements within Vietnam's leadership on the respective roles of state-owned enterprise and the party in the 1990s and early 2000s. Although this debate was not fully settled by 2004, the balance of opinion among stakeholders had shifted toward more competition in the sector, and the 2004 Electricity Act was passed. This act resulted in the restructuring of EVN, increased private sector participation in power generation, and the initiation of a path toward a wholesale market.

LOOKING AHEAD

Adoption of disruptive technologies will depend on the political interplay of winners and losers. As countries prepare themselves for an uncertain future in the face of rapid technological change in the power sector, there are questions about so-called disruptive technologies and their impact on the sector's institutions and structures. As with historic experience of reform, the speed and scope of innovation in any given country's power sector will likely be shaped by political dynamics and lobbying by potential winners and potential losers.

New technology will disrupt not only industry cost structures, business models, and regulatory instruments, but also the political economy of the sector. New winners and new losers will emerge from the change process, developments that go to the heart of political economy dynamics. A lot will depend on how each player sees its interests affected by disruption, and whether each makes common cause (forms coalitions) with those it perceives as fellow stakeholders (table 3.15). Some players will see the future as theirs. They will support the energy transition. These players could include storage providers, prosumers, renewable energy IPPs, mini/microgrid operators, and electric vehicle owners/providers. In contrast, conventional IPPs and fuel suppliers could oppose the transition (particularly the IPPs) out of fear, anticipating that their assets could become stranded and perhaps inadequately compensated.

TABLE 3.15 Political economy analysis: How key stakeholders respond to technological disruption

	Change in interests	Change in influence/authority	Reasons to resist change	Reasons to support change	Possible coalition partners
Regulator	More defensive	Reduced role as competition increases (seen as out of date)	Erodes power	Carve out new role	RE IPPs, utilities, prosumers, fractalists, storage providers
Incumbent utility	More defensive	Diminish somewhat	Stranded assets, threat to revenue	If opportunity to control profit from disruptive techs	Conventional IPPs, fuel suppliers, regulator if threat dominates
Conventional IPPs	Become less conventional	Depends on strategy, pro- or antichange	Loss of business opportunities, loss of revenue from dispatch	Switch to RE, storage, fractalism	Depends on strategy
Renewable IPPs	Not so interested in pursuing subsidies	Increases as they become more valuable/needed	Unused to market risk	Bigger opportunities	Incumbent utility, regulator
Renewable minigrids	Much bigger market opportunity	Changes with market growth	None	Bigger opportunities	Fractalists, telecom, mobile money, desperate fuel suppliers
Consumers/prosumers	Sell electricity, consume for transport, higher stake in reliable electricity	Gain market clout, more motivated as citizens/voters	Some might lose subsidies or other privileges	Opportunity for better/cheaper electricity, maybe to sell also	Minigrids, fractalists, regulator

Source: World Bank 2019.
Note: IPP = independent power producer; RE = renewable energy.

Particularly intriguing will be the stance of incumbent utilities; some might see the energy transition as an insuperable threat to be resisted, whereas others may see it as a huge opportunity to be pursued. Utilities that are able to remain relevant need not necessarily be the losers and may even transition out of their legacy roles into entirely new identities. Much of this will depend on existing electricity production models and the role utilities may or may not play in owning generation resources. Utilities may also resist the adoption of disruptive technologies, seeing them as a threat to their viability. Unless regulators can adapt rate designs to support new forms of cost recovery, utilities are unlikely to innovate on their own. Depending on the pace of adoption of distributed energy resource solutions, utilities may only be able to resist for so long.

Countries are likely to follow different energy transition paths, because their institutional receptivity to disruption will vary considerably. Much of this will be shaped by the existing market and regulatory structures set up to enable the existing electricity production model. Declining technology costs could create a viable alternative to the utility structure, and the ability to integrate renewable technologies, while maintaining a reliable electricity system, will be an ongoing challenge. How countries respond to these challenges will determine how competitive they will be economically, their attractiveness as an investment destination, and their credibility as government-managers of technology change and innovation. One thing seems clear: how a country handles its energy transition, or fails to do so, will be a really big deal.

CONCLUSION

By examining relationships among the power sector actors, one can detect the range of political factors driving power sector reform (table 3.16). Power sector reform takes a

TABLE 3.16 Political drivers of each stage of the reform process: An overview

Political drivers	Stage of reform			
	Announcement	Implementation	Delivery	Sustainability
Crisis	—	—	—	—
Donors	—	—	—	—
Ideology	*	****	****	*
Political system	—	***	**	—
Reform champion	—	**	****	—
Stakeholder alignment	—	—	—	**
Legislation	—	—	—	*

Source: Based on Rethinking Power Sector Reform utility database 2015.
Note: The Chi-squared test examines whether it is possible to reject the null hypothesis of no statistically significant relationship between the existence of each of the following factors: (1) the extent of the reform announcement; (2) the boldness of the reform implementation; (3) the completeness of the reform delivery; and (4) the longer-term sustainability of reform. A value below X percent denotes statistical significance at the X percent level. **** = 95–99 percent significance (significant); *** = 90–95 percent significance (less significant); ** = 85–90 percent significance (somewhat significant); * = 75–85 percent significance (low significance); — = not significant.

long time. It typically begins with ambitious announcements of reform intentions. Implementation progresses at varying paces that may or may not fully deliver on original promises. Delivery is sometimes subject to reform reversals later on. Most countries embarked on power sector reform in response to a crisis and typically in response to donor pressure. Although these external factors undoubtedly played a catalytic role in the decision to reform, the subsequent trajectory of reform primarily reflects internal factors. Of these factors, ideology appears to have by far the strongest effect, particularly on the pace of implementation and the extent of ultimate delivery. The nature of the political system also becomes relevant at the implementation stage, in that countries with centralized power structures tend to make less progress with reform. Finally, the approach to reform within the power sector itself also turns out to be important. Reform champions make a significant difference in following through on implementation. Longer-term sustainability appears to rest, however, on stakeholder alignment and actual legislation.

Looking ahead, technological disruptions will be reshaping the political economy dynamics of the power sector. The recent wave of innovation in the power sector is introducing a new set of actors and eroding the position of institutions that have been the traditional protagonists of the sector. Although the net effect of these changes remains uncertain, it is clear that these new influences will alter the political landscape of reform.

ANNEX 3A. GLOBAL POWER SECTOR REFORM INDEX

The standard package of reforms prescribed by international donors in the 1990s included four principal components: restructuring (vertical and horizontal unbundling of power utilities), private sector participation, creation of an independent regulator, and competition in power generation.

In order to aggregate across the four dimensions of power sector reform considered in this study, a simple Power Sector Reform Index is constructed. The index gives each country a score on an interval of 0 to 100 on each dimension of power sector reform. The scores are based on giving equal weight to each step on each dimension of the reform continuum (see table 3A.1). The average of the four 0–100 scores is used to provide an overall summary of the extent of reform.

TABLE 3A.1 Power Sector Reform Index

Regulation	No regulator = 0			Regulator = 100	
Restructuring	Vertically integrated = 0	Partial vertical unbundling = 33		Full vertical unbundling = 67	Vertical and horizontal unbundling = 100
Competition	Monopoly = 0	Indepent power producers = 25	Single-buyer model = 50	Bilateral contracts = 75	Competitive market = 100
Private sector participation	0.5 × (percentage of generation capacity with private sector participation) + 0.5 × (percentage of distribution utilities with private sector participation)				

ANNEX 3B. CHI-SQUARED CONTINGENCY TABLES

TABLE 3B.1 Crisis and patterns of power sector reform

		Announcement			Implementation			Delivery			Sustainability	
		Extensive	Cautious	Minimal	Rapid	Incremental	Limited	Complete	Partial	Negligible	Sustained	Nonsustained
Crisis	Severe	0.47	0	0.12	0.18	0.29	0.12	0.24	0.18	0.18	0.47	0.12
	Moderate	0.18	0.12	0.06	0.06	0.18	0.12	0.12	0.18	0.06	0.29	0.06
	None	0.06	0	0	0	0	0.06	0	0	0.06	0.06	0
Chi-squared test P-value		34%			55%			52%			88%	

Source: Based on Rethinking Power Sector Reform utility database 2015.
Note: Fractions denote the share of the country sample in each category and sum to 1.00 within each of the boxes. The Chi-squared test examines whether it is possible to reject the null hypothesis of a statistically significant relationship between the severity of the crisis and (1) the extent of the reform announcement; (2) the boldness of the reform implementation; (3) the completeness of the reform delivery; and (4) the long-term sustainability of reform. A value below *X* percent denotes statistical significance at the *X* percent level.

TABLE 3B.2 Donor impact on patterns of power sector reform

		Announcement			Implementation			Delivery			Sustainability	
		Extensive	Cautious	Minimal	Rapid	Incremental	Limited	Complete	Partial	Negligible	Sustained	Nonsustained
Donor influence	High	0.59	0.12	0.18	0.24	0.35	0.29	0.29	0.29	0.29	0.71	0.18
	Moderate	0.12	0	0	0	0.12	0	0.06	0.06	0	0.12	0
Chi-squared test P-value		62%			28%			62%			49%	

Source: Based on Rethinking Power Sector Reform utility database 2015.
Note: Fractions denote the share of the country sample in each category and sum to 1.00 within each of the boxes. The Chi-squared test examines whether it is possible to reject the null hypothesis of a statistically significant relationship between the severity of the crisis and (1) the extent of the reform announcement; (2) the boldness of the reform implementation; (3) the completeness of the reform delivery; and (4) the long-term sustainability of reform. A value below *X* percent denotes statistical significance at the *X* percent level.

TABLE 3B.3 Ideology and patterns of power sector reform

		Announcement			Implementation			Delivery			Sustainability	
		Extensive	Cautious	Minimal	Rapid	Incremental	Limited	Complete	Partial	Negligible	Sustained	Nonsustained
Ideology	Market	0.24	0	0	0.12	0.12	0.00	0.18	0.06	0	0.24	0
	Hybrid	0.41	0.06	0.06	0.12	0.35	0.06	0.18	0.29	0.06	0.35	0.18
	State	0.06	0.06	0.12	0	0	0.24	0	0	0.24	0.24	0
Chi-squared test P-value		19%			1%			1%			20%	

Source: Based on Rethinking Power Sector Reform utility database 2015.
Note: Fractions denote the share of the country sample in each category and sum to 1.00 within each of the boxes. The Chi-squared test examines whether it is possible to reject the null hypothesis of a statistically significant relationship between the severity of the crisis and (1) the extent of the reform announcement; (2) the boldness of the reform implementation; (3) the completeness of the reform delivery; and (4) the long-term sustainability of reform. A value below *X* percent denotes statistical significance at the *X* percent level.

TABLE 3B.4 Political system and patterns of power sector reform

		Announcement			Implementation			Delivery			Sustainability	
		Extensive	Cautious	Minimal	Rapid	Incremental	Limited	Complete	Partial	Negligible	Sustained	Nonsustained
Political system	Unipolar	0.18	0.06	0.12	0.06	0.06	0.24	0.12	0.06	0.18	0.35	0
	Transition	0.12	0.06	0.06	0	0.18	0.06	0	0.12	0.12	0.18	0.06
	Multipolar	0.41	0	0	0.18	0.24	0	0.24	0.18	0	0.29	0.12
Chi-squared test P-value		27%			6%			15%			37%	

Source: Based on Rethinking Power Sector Reform utility database 2015.
Note: Fractions denote the share of the country sample in each category and sum to 1.00 within each of the boxes. The Chi-squared test examines whether it is possible to reject the null hypothesis of a statistically significant relationship between the severity of the crisis and (1) the extent of the reform announcement; (2) the boldness of the reform implementation; (3) the completeness of the reform delivery; and (4) the long-term sustainability of reform. A value below *X* percent denotes statistical significance at the *X* percent level.

TABLE 3B.5 Reform champions and patterns of power sector reform

		Announcement			Implementation			Delivery			Sustainability	
		Extensive	Cautious	Minimal	Rapid	Incremental	Limited	Complete	Partial	Negligible	Sustained	Nonsustained
Reform champion	Strong	0.41	0.06	0.06	0.12	0.35	0.06	0.24	0.29	0	0.41	0.12
	Weak	0.29	0.06	0.12	0.12	0.12	0.24	0.12	0.06	0.29	0.41	0.06
Chi-squared test P-value		74%			15%			2%			60%	

Source: Based on Rethinking Power Sector Reform utility database 2015.
Note: Fractions denote the share of the country sample in each category and sum to 1.00 within each of the boxes. The Chi-squared test examines whether it is possible to reject the null hypothesis of a statistically significant relationship between the severity of the crisis and (1) the extent of the reform announcement; (2) the boldness of the reform implementation; (3) the completeness of the reform delivery; and (4) the long-term sustainability of reform. A value below *X* percent denotes statistical significance at the *X* percent level.

TABLE 3B.6 Stakeholder alignment and patterns of power sector reform

		Announcement			Implementation			Delivery			Sustainability	
		Extensive	Cautious	Minimal	Rapid	Incremental	Limited	Complete	Partial	Negligible	Sustained	Nonsustained
Stakeholder alignment	Strong	0.29	0.06	0.06	0.12	0.18	0.12	0.18	0.18	0.06	0.41	0
	Weak	0.41	0.06	0.12	0.12	0.29	0.18	0.18	0.18	0.24	0.41	0.18
Chi-squared test P-value		92%			93%			52%			11%	

Source: Based on Rethinking Power Sector Reform utility database 2015.
Notes: Fractions denote the share of the country sample in each category and sum to 1.00 within each of the boxes. The Chi-squared test examines whether it is possible to reject the null hypothesis of a statistically significant relationship between the severity of the crisis and (1) the extent of the reform announcement; (2) the boldness of the reform implementation; (3) the completeness of the reform delivery; and (4) the long-term sustainability of reform. A value below *X* percent denotes statistical significance at the *X* percent level.

TABLE 3B.7 Legislation and patterns of power sector reform

		Announcement			Implementation			Delivery			Sustainability	
		Extensive	Cautious	Minimal	Rapid	Incremental	Limited	Complete	Partial	Negligible	Sustained	Nonsustained
Legislation	Strong	0.47	0.12	0.06	0.18	0.29	0.18	0.29	0.24	0.12	0.59	0.06
	Weak	0.24	0	0.12	0.06	0.18	0.12	0.06	0.12	0.18	0.24	0.12
Chi-squared test P-value		30%			88%			32%			21%	

Source: Based on Rethinking Power Sector Reform utility database 2015.
Note: Fractions denote the share of the country sample in each category and sum to 1.00 within each of the boxes. The Chi-squared test examines whether it is possible to reject the null hypothesis of a statistically significant relationship between the severity of the crisis and (1) the extent of the reform announcement; (2) the boldness of the reform implementation; (3) the completeness of the reform delivery; and (4) the long-term sustainability of reform. A value below *X* percent denotes statistical significance at the *X* percent level.

ANNEX 3C. WORLD BANK SUPPORT FOR POWER SECTOR REFORM OBSERVATORY COUNTRIES AND STATES

Country/State	Major World Bank loans to power sector
Colombia	(1994) Regulatory reforms, US$11m (1995) Facilitating operation of wholesale market, US$249m (1997) Regulatory reforms, US$12.5m
Dominican Republic	(1988) Improve operating efficiency of CDE and build T&D infrastructure, US$105m (2005) Transmission and service expansion, US$150m (2008) Improve cash recovery and quality of supply, US$42m (2015) Financial viability of distribution companies, US$120m
Egypt, Arab Rep.	(2006–10) Natural gas power plants, US$900m (2010) Transmission for RE, US$70m (2016–20) Pricing and structural energy sector reforms, US$3,000m
India – Andhra Pradesh	(1999) Comprehensive power sector reforms, US$576m

(Annex continued next page)

Country/State	Major World Bank loans to power sector
India - Odisha	(1996) Comprehensive power sector reforms, US$350m
India - Rajasthan	(2001) Comprehensive power sector reforms, US$226m (2016) Financial health of distribution utilities, US$250m (2017) Financial health of distribution utilities, US$250m
Kenya	(1997) Sector reforms, US$125m (2000) Generation capacity, US$80m (2004) Sector recovery projects, US$80m (2010–11) Generation projects, US$500m (2012) Regional transmission interconnection, US$441m (2015) Modernization of sector, US$450m (2017) Off-grid access, US$150m
Morocco	(2005–07) Regulatory framework for RE, US$100m (2007–16) Renewable generation US$965m (2013) Improving quality of supply, US$40.5m (2013–15) Supporting renewable expansion, US$600m
Pakistan	(1994) Support for private power policy, WB: US$475m; IFC: US$378m; MIGA: US$31m (1995–2014) Hydro generation, US$2,688m (2006–11) Supporting private sector in energy projects, US$475m (2016–17) Transmission and distribution modernization, US$680m
Peru	(1994) Support power sector reform policy, US$150m (2006–11) Rural access, US$110m
Philippines	(1989, 1993, 1996) Rehabilitate, upgrade transmission and distribution systems, US$425m (1990) Improving generation, transmission, and distribution, US$390m (1992) Rural access, US$91m (1994) Geothermal, US$438m (2003) Rural access, US$40m (2008) Rehabilitating transmission in Bicol, US$13m
Senegal	(1998) Reforming power sector (private entry), US$100m (2004) Rural access, US$30m (2005) Rehabilitation of transmission and distribution, US$6.9m (2008, 2012, 2016) Improving Senelec financials and efficiency, US$235m (2014) Generation and transmission, US$99m
Tajikistan	(2005) Loss-reduction program, US$33m
Tanzania	(2001) Natural gas generation, US$205m (2004) Emergency rentals, US$46m (2007) Access expansion, US$134m (2010) Transmission expansion, US$60m (2013–14) Improve power and gas sector financials and performance, US$198m (2016) Rural access, US$209m
Uganda	(1991) Rehabilitating infrastructure, US$153m (2000) Support for UEB restructuring and privatization, US$8m (2001, 2009, 2015) Rural access, US$282m (2007) Generation, US$115m (2007) Supply until new generation comes online, US$306m (2011) Improve quality of supply, US$84m (2016) Grid expansion and reinforcement, US$100m
Ukraine	(1996) Sector reform (restructuring/privatization), US$76m (2001–03) Regulatory framework (economy wide), US$250m (2005, 2009) Hydro rehabilitation, US$197m (2007) Improving transmission, US$194m
Vietnam	(1990–99) Rural energy; power sector rehabilitation; power development; T&D, US$694m (2002) Improving efficiency, US$230m (2005, 2008) Rural energy (generation, transmission and distribution), US$375m (2006, 2011, 2013, 2015) Improving power grid, US$1,320m (2010, 2012, 2014) Power sector reform, US$712m

Source: World Bank.
Note: CDE = Corporación Dominicana de Electricidad; IFC = International Finance Corporation; m = million; MIGA = Multilateral Investment Guarantee Agency; RE = renewable energy; T&D = transmission and distribution; UEB = Uganda Electric Board; WB = World Bank.

NOTES

1. This chapter was informed in particular by Lee and Usman (2018). Further original research was done by a team lead by Ashish Khanna and comprising Anton Eberhard, Catrina Godinho, Alan David Lee, Brian Levy, Zainab Usman, and Jonathan Walters. The overall work program was coordinated by Vivien Foster and Anshul Rana.
2. In order to aggregate across the four dimensions of power sector reform considered in this study, a simple PSRI is constructed. The index gives each country a score of 0 to 100 on each dimension of power sector reform. The scores are based on giving equal weight to each step on each dimension of the reform continuum. The average of the four 0–100 scores is used to characterize a country's extent of reform. For more on the index, see annex 3A.
3. Overall reform announcement is the sum of all the reforms the country had committed to implementing from 1990 to 2015.
4. Pritchett, Woolcock, and Andrews (2010) describe isomorphic mimicry as the "adoption of the forms of other functional states and organizations which camouflages a persistent lack of function." It provides the "mechanism for avoiding needed reform or innovation while at the same time maintaining the appearance of legitimate engagement with developmental discourses."
5. *Acquis communautaire* is the accumulated laws and obligations of the European Union from 1958 to present. It includes all European Union treaties, laws, declarations, resolutions, and international agreements.
6. Significance is measured by the Chi-squared test; details can be found in annex 3B.
7. With a sample of 53 diverse countries, Erdogdu (2013) suggests that reforms went further when the minister in place at the outset of reforms had no previous experience in the power sector.

REFERENCES

APP (Africa Progress Panel). 2015. *Power, People, Planet: Seizing Africa's Energy and Climate Opportunities.* Geneva: APP.

Besant-Jones, John E. 2006. "Reforming Power Markets in Developing Countries: What Have We Learned?" Energy and Mining Sector Board Discussion Paper no. 19, World Bank, Washington, DC.

Cheng, Chaoyo, YuJung (Julia) Lee, Galen Murray, Yuree Noh, Joseph Van Horn, and Johannes Urpelainen. 2016. "Political Obstacles to Economic Reform: Comparative Evidence from the Power Sector in 20 Indian States." Paper presented at the Texas A&M Conference on Energy, College Station, Texas, September 26–28.

Dubash, Navroz K., ed. 2002. *Power Politics: Equity and Environment in Electricity Reform.* June. Washington, DC: World Resources Institute.

Erdogdu, Erkan. 2013. "A Cross-Country Analysis of Electricity Market Reforms: Potential Contribution of New Institutional Economics." *Energy Economics* 39 (5): 239–51.

———. 2014. "The Political Economy of Electricity Market Liberalization: A Cross-Country Approach." *Energy Journal* 35 (3): 91–128.

Foster, Vivien, Samantha Witte, Sudeshna Ghosh Banerjee, and Alejandro Moreno. 2017. "Charting the Diffusion of Power Sector Reforms across the Developing World." Policy Research Working Paper 8235, World Bank, Washington, DC.

Gore, Christopher D., Jennifer N. Brass, Elizabeth Baldwin, and Lauren M. MacLean. 2018. "Political Autonomy and Resistance in Electricity Sector Liberalization In Africa." *World Development* 120 (August): 193–209.

Henisz, Witold J., Bennet A. Zelner, and Mauro F. Guillén. 2005. "The Worldwide Diffusion of Market-Oriented Infrastructure Reform, 1977–1999." *American Sociological Review* 70 (6): 871–97.

Jayarajah, Carl, and William Branson. 1995. *Structural and Sectoral Adjustment: World Bank Experience, 1980–92.* A World Bank operations evaluation study. Washington, DC: World Bank Group.

Lee, Alan David, and Zainib Usman. 2018. "Taking Stock of the Political Economy of Power Sector Reform in Developing Countries: A Literature Review." Policy Research Working Paper 8518, Rethinking Power Sector Reform, World Bank, Washington, DC. http://documents.worldbank.org/curated/en/431981531320704737/Taking-stock-of-the-political-economy-of-power-sector-reforms-in-developing-countries-a-literature-review.

Levy, Brian. 2014. *Working with the Grain: Integrating Governance and Growth in Development Strategies.* New York: Oxford University Press.

Moerenhout, Tom S. H. 2018. "Reforming Egypt's Fossil Fuel Subsidies in the Context of a Changing Social Contract." In *The Politics of Fossil Fuel Subsidies and Their Reform*, edited by Jakob Skovgaard and Harro van Asselt. Cambridge: Cambridge University Press.

Nakhooda, Smita, Shantanu Dixit, and Navroz K. Dubash. 2007. *Empowering People: A Governance Analysis of Electricity—India, Indonesia, Philippines, Thailand.* June. Washington, DC: World Resources Institute.

Pritchett, Lant, Michael Woolcock, and Matt Andrews. 2010. "Capability Traps? The Mechanisms of Persistent Implementation Failure." CGD working paper 234, Center for Global Development, Washington, DC, December.

Sen, Anupama, Rabindra Nepal, and Tooraj Jamasb. 2016. "Reforming Electricity Reforms? Empirical Evidence from Asian Economies." Oxford Institute for Energy Studies (OIES), February.

Urpelainen, Johannes, and Joonseok Yang. 2017. "Policy Reform and the Problem of Private Investment: Evidence from the Power Sector." *Journal of Policy Analysis and Management* 36 (1): 38–64.

Williams, James H., and Reza Ghanadan. 2006. "Electricity Reform in Developing and Transition Countries: A Reappraisal." *Energy* 31 (6): 815–44.

Williamson, John. 2005. "The Washington Consensus as Policy Description for Development." A lecture in the series, "Practitioners of Development," delivered at the World Bank, January 13, 2004. Peterson Institute for International Economics (PIIE).

World Bank. 1993. *The World Bank's Role in the Electric Power Sector: Policies for Effective Institutional, Regulatory, and Financial Reform.* Washington, DC: World Bank.

Worley, Heather, Sara Bryan Pasquier, and Ezgi Canpolat. 2018. "Good Practice Note 10: Designing Communication Campaigns for Energy Subsidy Reform." World Bank, Washington, DC.

Zelner, Bennet A., Witold J. Henisz, and Guy L. F. Holburn. 2009. "Contentious Implementation and Retrenchment In Neoliberal Policy Reform: The Global Electric Power Industry, 1989–2001." *Administrative Science Quarterly* 54 (3): 379–412.

Building Blocks of Reform

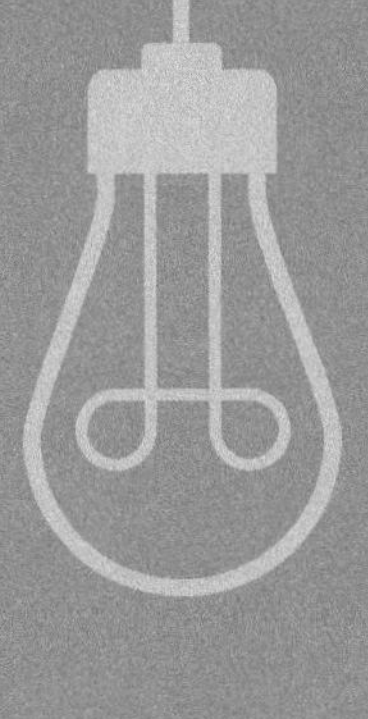

PART

What Has Been Done to Restructure Utilities and Improve Governance?

4

Guiding questions

- *How have countries gone about separating the main sectoral functions of policy making, regulation, and service provision?*
- *What measures have been adopted to improve the governance of corporatized power utilities?*
- *To what extent have countries pursued horizontal and vertical unbundling to pave the way for competition in the sector?*
- *How does the current wave of technological disruption in the power sector affect utilities' restructuring and management practices?*

Summary

- *Reform efforts began with the corporatization of power utilities, many of which had, until then, been operated as departments of energy ministries. This step was followed by the creation of a regulatory entity to provide arm's-length oversight.*
- *Amid efforts to reallocate responsibility for regulation and service provision, the need to strengthen core ministry functions was too often overlooked. In particular, scant attention was paid to developing capacity for sector planning, which is a critical omission given the vertiginous growth of electricity demand and the pressing need for new investments.*
- *A key aspect of the corporatization process was to instill sound management practices in the new utilities. Such practices aim to safeguard the autonomy and accountability of the company's board of directors, and responsibly manage human resources, financial discipline, and information technology. There is considerable variation in the management quality of state-owned utilities across countries, and by and large management practices fall short of those found in private utilities.*
- *Another major focus of reform efforts was the vertical and horizontal unbundling of corporatized national utilities. Such structural reforms were intended to be stepping-stones to greater private sector participation and competition in the power sector. In many countries, however, these goals have yet to be reached.*
- *Looking ahead, technological disruptions under way in the power sector will make planning increasingly complex; further sector restructuring may be needed to facilitate competition at the retail level.*

INTRODUCTION

This chapter examines efforts to restructure the power industry and improve utilities' governance.[1] At one level, reforms have called for a reengineering of the power sector's institutional architecture. This has often involved the breaking up of existing actors, the creation of new ones, and the reassignment of responsibilities across them. On a deeper level, reforms have sought to improve the internal functioning of all major actors, through encouraging the adoption of stronger governance and managerial practices. Thus, the guiding questions for this chapter are as follows: How did countries go about separating functions during the reform process? What structural models were used? What measures were adopted to assure improved utility governance? How does the current wave of technological disruptions in the power sector affect the restructuring and governance of utilities?

Under the 1990s reform model, restructuring the power sector was considered foundational for the subsequent implementation of deeper reforms meant to foster private sector participation and market liberalization. The process of restructuring the power sector entailed two key steps: the separating of key functions and the unbundling of utilities. Both were essentially directed toward reducing conflicts of interest.

The first step involved separating three key functions: policy making, regulation, and service provision. Under the arrangements typical up to 1990, all three functions fell under the mandate of a single national public utility charged with acting in the public interest, often embedded within the energy ministry and with no distinct institutional identity. Conflicts of interest were common. For instance, a utility that self-regulates is unlikely to hold itself accountable to the highest standards. Or, again, a utility whose economic clout and political influence are determined by the scale of its investment program may introduce biases into the planning process. In view of this, the 1990s reform model recommended clear separation of these functions under distinct institutions. It was envisaged that policy functions (including planning) should be undertaken by the line ministry, regulatory functions by a separate regulatory entity, and service provision by the utility.

Reform efforts sought to ensure that the service provider was corporatized and operating under sound governance and management arrangements. As of the early 1990s, some power utilities still functioned simply as departments of line ministries in the central government, subject to public sector arrangements for budgeting, employment, and decision making. A critical first step was to corporatize the utility, essentially converting it into a separate state-owned enterprise (SOE) distinct from the central government and operating under company law. Because of the greater managerial autonomy it provided, corporatization was expected to foster improved performance under public sector ownership and was also an indispensable precondition for private sector participation. Depending on the country, the resulting SOE might remain under the direct jurisdiction of the line ministry for energy or might find itself reporting to another ministry or entity charged with exercising public ownership and oversight functions for all SOEs across sectors. Corporatization was supposed to create incentives for the efficient management of the company, reduce scope for political interference in company decisions, and ensure financial discipline. However, the efficacy of the corporatization process depended on the quality of the governance framework and internal management arrangements that accompanied it.

The second step of the 1990s reform model was the sector's vertical and horizontal unbundling. Unbundling might have begun with separating accounting and managerial business unit segmentation but was supposed to eventually lead to full legal unbundling, and companies with distinct ownership. Unbundling was not seen as an end in itself but as an enabling measure that makes further reforms possible (Bacon 2018). The purpose of vertical unbundling was to separate out those elements of the supply chain deemed natural monopolies (notably transmission and distribution), from those considered to be competitive (such as generation and retail). This separation was to avoid the conflicts of interest that might arise when a generator operating in a competitive market also controls the transmission network. For example, such a generator might use its dominant position to restrict the participation of other generators. In some relatively developed jurisdictions, vertical unbundling was further extended to separate the retail (or commercial sale) function of power utilities from their physical distribution function. In principle, this makes electricity retailers compete for the same end consumers, even as distribution utilities continue to operate as local monopolies.

Once vertical unbundling was complete, reformers turned to horizontal unbundling. The purpose of horizontal unbundling was to create multiple competing entities in those segments of the supply chain where competition was possible. Horizontal unbundling has been particularly important in the generation sector, where it has been critical in limiting the market shares of individual generators to mitigate their abuse of market power. Horizontal unbundling has also been relevant in distribution, though here its effects are more indirect. For the functioning of a competitive wholesale power market, it is necessary to have multiple buyers as well as multiple sellers, with distribution utilities being among the major buyers in any market. Furthermore, the existence of multiple distribution entities, even if each serves a particular geographic area on a natural monopoly basis, provides comparative information that facilitates the task of regulation.

Depending on the depth of the structural reforms undertaken, a country might find itself with one of several possible power sector organizational models (box 4.1). At one end of the spectrum, in most countries, the starting point was a vertically integrated national monopoly power utility. At the other end of the spectrum, the endpoint of the 1990s sector reform model was at least a wholesale, if not a retail, power market. Because of the progressive and partial application of restructuring reforms, several intermediate structural models can be found (table 4.1). Some countries got no further than opening the market to independent power producers (IPPs) that competed for the right to build a new generation plant and supply the enduring national vertically integrated monopoly utility. Others went further, by vertically unbundling the entire generation tier, and allowing new IPPs and divested companies managing existing generation assets to compete for the right to supply a single buyer, which typically retained a monopoly in both transmission and distribution. An additional step was to complete the vertical unbundling process to create a separate transmission system operator, allowing third-party access to the grid so that generators could compete to provide power directly to large industrial customers, usually through long-term bilateral contracts. This situation could, relatively easily, evolve into a full-scale competitive power market through the creation of a market operator to provide a platform for short-term exchanges of electricity in a wholesale power market.

Although these intermediate models were originally intended as transitional phases in the pursuit of a wholesale power market, in many countries they became quasi-permanent (Gratwick and Eberhard 2008). This result occurred either because, in these countries, power systems were not suited to the implementation of a competitive market, or because political obstacles blocked further steps in the restructuring process (Besant-Jones 2006).

BOX 4.1 Selected power sector structures around the world

Depending on the extent of the reforms undertaken in recent decades, power sectors around the world may be classified by a variety of types of structures. A few examples are illustrated here.

Vertically integrated utility with independent power producers

Senegal, Tajikistan, and Tanzania. A vertically integrated utility (VIU) controls generation, transmission, and distribution. Several independent power producers (IPPs) have power purchase agreements with the utility and supply power to it (figure B4.1.1).

Variations: In Senegal and Tanzania, the regulator is autonomous; in Tajikistan, it is not.

FIGURE B4.1.1 Vertically integrated utility

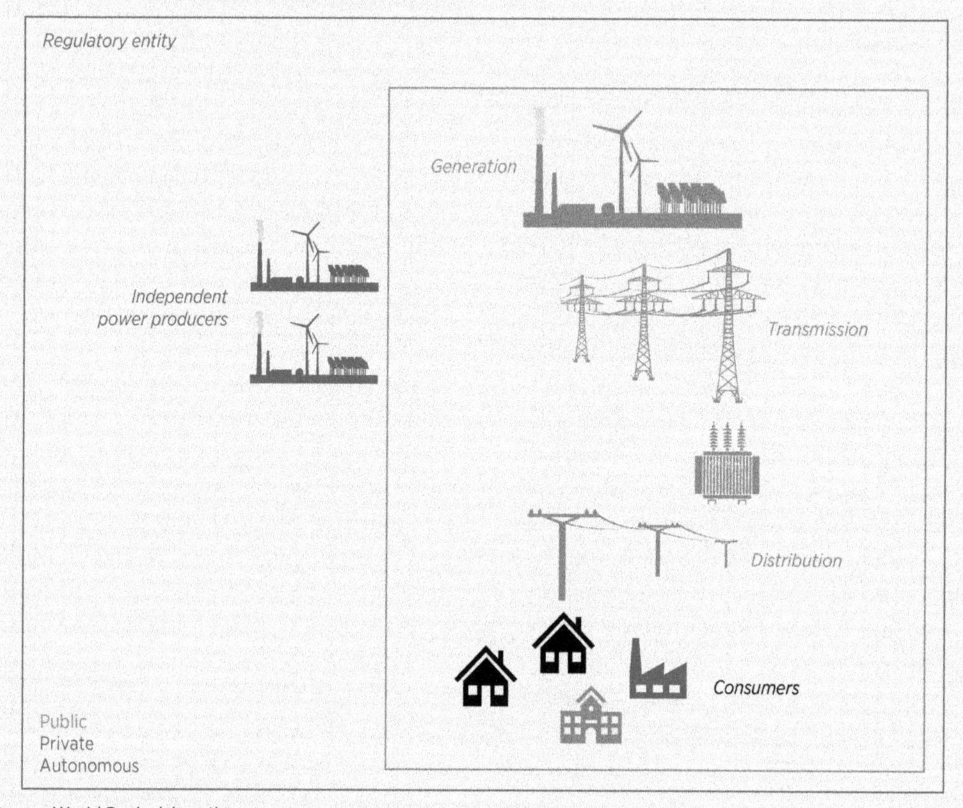

Source: World Bank elaboration.
Note: TSO = transmission system operator.

Single-buyer model

Arab Republic of Egypt, Kenya, Morocco, Pakistan, Uganda, Ukraine, and Vietnam. In this model, the VIU has been restructured. Generation is a mix of IPPs and public generation company(ies). Distribution and transmission are separate businesses but majority state owned (figure B4.1.2).

Variations: Ukraine—distribution companies are private and the transmission system operator is autonomous; Uganda—long-term private concessions exist for the generation and distribution companies; Pakistan—a privately owned VIU also exists in the system; Morocco—a VIU controls transmission and some distribution (as a single buyer); IPPs have a growing share; municipal-level private distributors have long-term concessions.

(Box continued next page)

BOX 4.1 Selected power sector structures around the world (*Continued*)

FIGURE B4.1.2 Single-buyer model

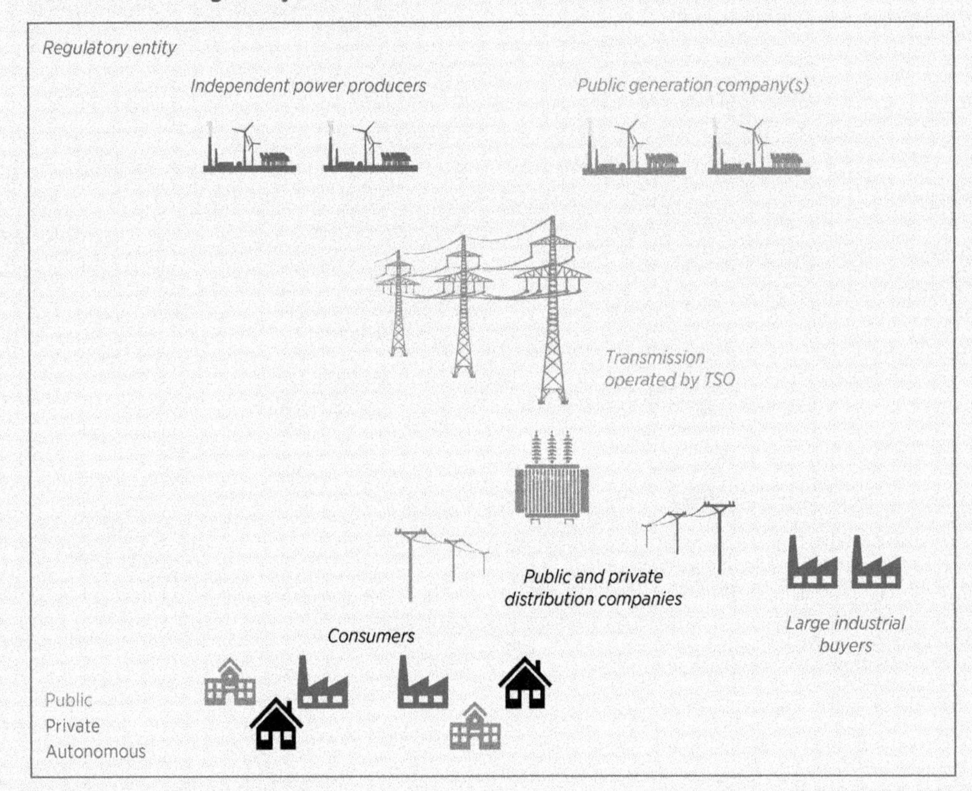

Source: World Bank elaboration.
Note: TSO = transmission system operator.

Wholesale-buyer model

Colombia, Dominican Republic, India, Peru, and the Philippines. A wholesale market exists for buyers (large industries and distribution companies) to purchase power from a mix of public and private generators. The wholesale buyers can also purchase directly from generators. Distribution is a mix of public and private ownership. The transmission network is operated by an independent transmission system operator (figure B4.1.3).

Variations: Colombia's power sector has some publicly owned vertically integrated utilities that also participate in the market.

(*Box continued next page*)

BOX 4.1 **Selected power sector structures around the world (*Continued*)**

FIGURE B4.1.3 **Wholesale-buyer model**

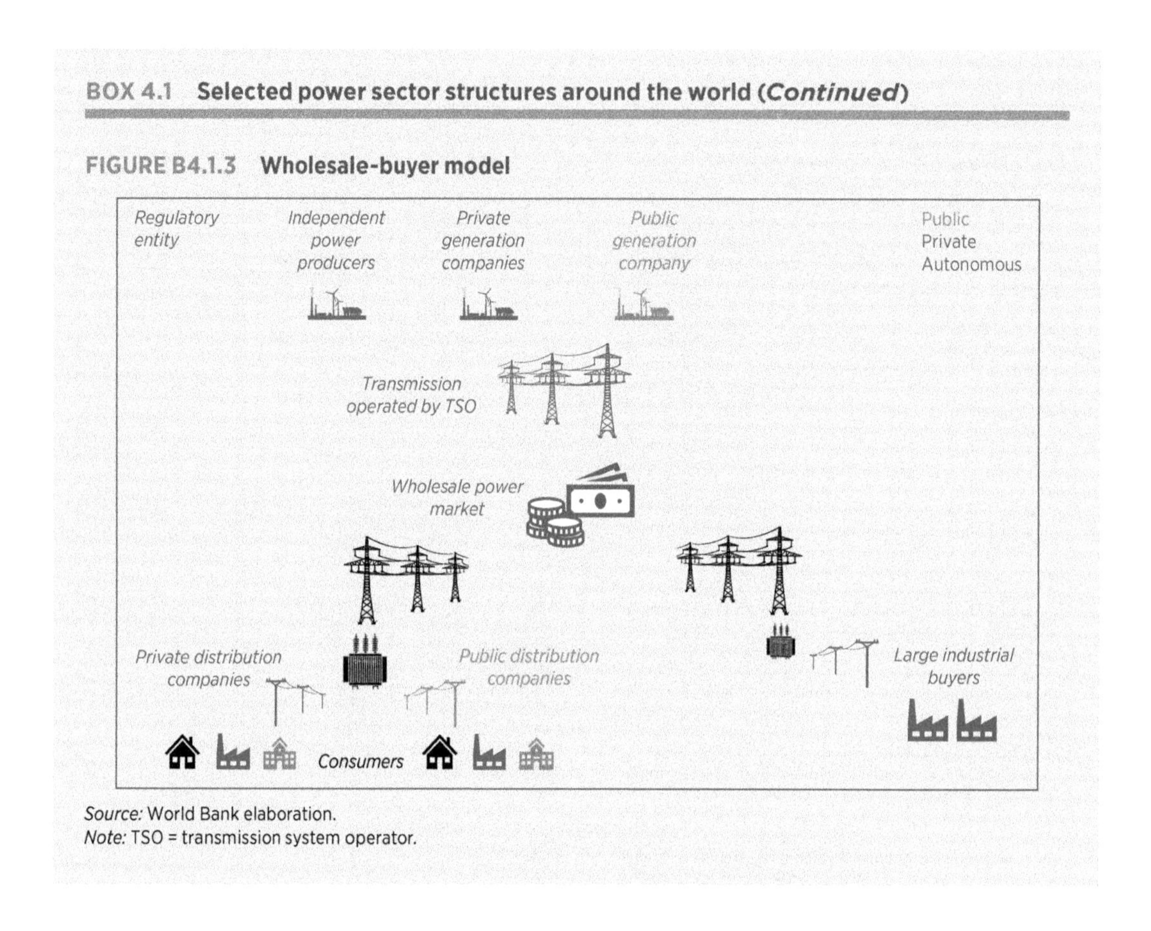

Source: World Bank elaboration.
Note: TSO = transmission system operator.

TABLE 4.1 **Unbundling is a means to remove conflict of interest and encourage competition**

Degrees of vertical and horizontal unbundling, and their effects on competition

	Vertical unbundling	Horizontal unbundling	Competition
Vertically integrated utility	None	None	None
Independent power producers	Incumbent remains vertically integrated	New entrants responsible for new generation plant	Companies compete for concessions to build new generation plant
Single-buyer model	Incumbent generation assets are separated out under a distinct company	New entrants responsible for new generation plant	All generators supply power under contract to single buyer
Third-party access to grid	Incumbent divided into distinct companies for generation, transmission, and distribution	New entrants responsible for new generation plant	All generation companies compete to serve large industrial customers under bilateral contracts
Wholesale power market	Incumbent divided into distinct companies for generation, transmission, and distribution	Incumbent generator and distributor divided into distinct companies	All generation competes to serve all large customers including distributors, with spot market
Retail power market	Incumbent divided into distinct companies for generation, transmission, distribution, and retail	Multiple retailers exist, allowing new entrants	All generation competes to serve all large customers with spot market, and retailers compete directly to serve residential customers

Source: World Bank elaboration.

KEY FINDINGS

Drawing on the experience of the 15 countries in the Power Sector Reform Observatory, the main findings of this chapter can be summarized as follows.

Finding #1: Although reform efforts prioritized the creation of regulatory entities, the critical function of sector planning was underemphasized

Achieving energy security remains a huge planning and procurement challenge for fast-growing developing economies. With electricity demand in most developing regions growing at rates of 6–7 percent per year since 1990 (Steinbuks and others 2017), achieving supply–demand balance and associated energy security calls for major investments that may entail a doubling of power system capacity every decade. Keeping up with this exacting pace requires countries to develop sound least-cost generation plans that identify the most cost-effective path of generation expansion, and to implement these in a timely fashion by creating a strong institutional link between planning and associated procurement of generation capacity.

The 1990s model had little to say on the issue of planning, with an implicit assumption that the advent of a wholesale power market would somehow circumvent the need for it. The end goal of the 1990s model was to create a competitive market. The supposition at that time was that private investments in power generation could be guided by price signals. The role of the state was seen primarily as the regulator of a privately owned and operated competitive sector, and great emphasis was placed on the creation of a capable regulatory institution and associated legal framework (Pardina and Schiro 2018) (which will be the subject of chapter 6). By contrast, central planning functions were overlooked or downplayed. Indeed, in some countries, the planning function traditionally housed in national power utilities or line ministries fell between the cracks as power sector reform processes worked toward the unbundling of the incumbent utilities and the creation of technical capacity in regulatory agencies outside of line ministries. In practice, power markets proved difficult to establish in all but a handful of developing countries, and even among those countries, price signals have not provided an adequate basis for investment decisions (Rudnick and Velasquez 2018) (a subject that will be taken up in chapter 7).

Good sector planning entails technically grounded plans for both generation and transmission that are fully integrated with other relevant plans. At a minimum, plans for generation and transmission need to be coordinated and mutually consistent. Ideally, generation planning should be aligned with the country's broader energy plan (for countries with primary energy resources) and be compatible with the overall national development plan. Consulting with stakeholders during the planning process helps to bring in these wider perspectives. In practice, deficiencies can be found in the planning framework for both generation and transmission (figure 4.1, panel a). Some countries lag further behind on transmission planning (such as the Arab Republic of Egypt, the Philippines, and Tajikistan), whereas others lag further behind on generation planning (figure 4.2).

The institutional responsibility for planning needs to be clearly assigned to an entity with adequate technical capacity. A wide range of institutional arrangements for power systems' planning can be observed across countries. Most prevalent are cases where the planning function is assigned either to the line ministry or to the power utility; in some cases, it is assigned to both. For sectors that are unbundled, it is the transmission utility that may retain the sector planning function (as in Egypt and Pakistan), and once competitive markets are introduced this responsibility may migrate

FIGURE 4.1 Plans must be mandatory in implementation and combined with transparent and competitive procurement

Best practices observed across 15 countries, by share of total countries

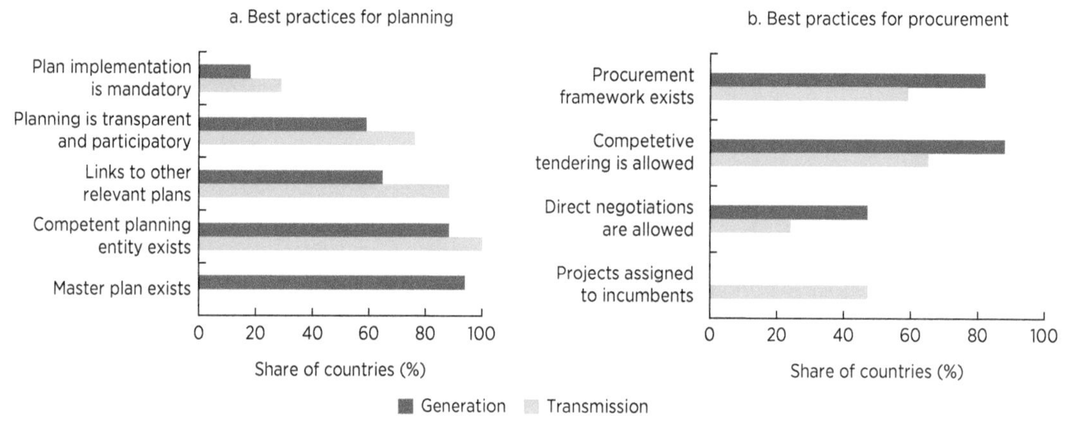

Source: Based on Rethinking Power Sector Reform utility database, 2015.

FIGURE 4.2 Sector planning remains necessary and calls for adequate institutional capacity and sound processes

Generation and transmission planning across 15 countries

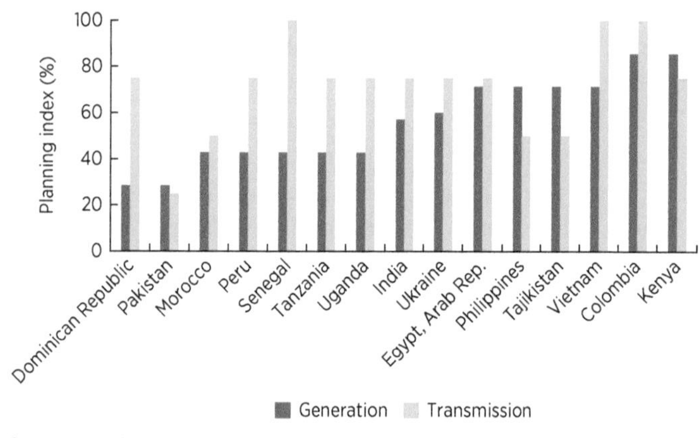

Source: Based on Rethinking Power Sector Reform utility database, 2015.

to the system operator (as in Ukraine). In a handful of relatively large countries, where a culture of planning is particularly ingrained, sector planning may be the responsibility of an independent entity, such as Brazil's Empresa de Pesquisa Energetica or the Central Electricity Authority of India. An interesting variation is found in Kenya and Uganda, where the regulator is given explicit responsibility for power system planning. Strictly speaking, planning and regulation are two distinct functions, one more strategic in nature and the other more supervisory. Nevertheless, unifying both functions at least takes advantage of the creation of a critical mass of technical expertise among sector regulators, which can also be brought to bear on planning. Even when planning and regulation are not merged in this way, it is still important for the regulator to review the legitimacy of investment plans put forward by the utility, which will have a major impact on tariff determinations.

Ultimately, the impact of good sector planning depends on its being linked to timely, competitive, and transparent procurement processes for new capacity. Critical investments in generation and transmission have significant lead times and must be initiated well ahead of when they are needed, which requires that plans be clearly time bound and that their implementation be mandatory. However, only a few of the 15 observatory countries make the implementation of plans mandatory (figure 4.1, panel a). It is also important to ensure that new generation and transmission projects are procured on a competitive basis to provide value for money. The existence of a procurement framework for generation and transmission projects is far from universal, and in many cases noncompetitive procurement methods—such as direct negotiation—continue to be allowed (figure 4.1, panel b).

In Tanzania, for example, few plans have been realized despite the country's planning capacity. The Ministry of Energy and Minerals has a well-defined responsibility for developing power generation and transmission expansion plans. The Tanzania Electric Supply Company (TANESCO), the state-owned utility, plays a central role in the development process, which includes experts from various ministries and government agencies. The resulting power system master plans take their cue from short-, medium-, and long-term national development plans. When it comes to implementation, however, experience shows that master plans are rarely followed, leaving the country without adequate reserves to withstand recurring drought situations, and leading to emergency procurement of rental plants and direct negotiation of oil-fired plants at exorbitant costs. To ensure that power system master plans are followed and new capacity is procured in a timely and transparent manner, the government plans to establish a multidisciplinary, interministerial committee that would coordinate the procurement of power projects.

The Dominican Republic has a long history of overlooking sector plans, despite institutional reforms of the sector's planning function. Expansion plans have been developed by the vertically integrated utility, Corporación Dominicana de Electricidad (CDE), since the 1960s. In 15 years during the 1970s and 1980s, more than 10 expansion plans were produced but none were actually followed, contributing to CDE's deterioration. Consequently, generation deficits were tackled by purchasing emergency oil-based turbines, which are both inefficient and costly to operate. Following power sector reforms, responsibility for long-term planning has fallen to the national energy commission. The latest plan covering the period 2011–25 aims at diversifying the generation mix and has been regularly updated every five years although always by a different international firm. Despite a reform process and changes in planning roles, expansion plans continue to be ignored.

Finding #2: The governance of corporatized public utilities leaves a lot of room for improvement, and still falls considerably short of governance practices in comparable private utilities

In many countries, the first step toward reform was the corporatization of the public utility. Before 1990, many public power utilities operated as administrative departments of their respective line ministries without any separate corporate existence. This situation left them subject to the vagaries of public administration and unable to adopt a proper commercial orientation. For this reason, the first step toward power sector reform in many countries was to separate out the operational functions associated with service provision into a distinct state-owned corporation, typically operating under company law. In doing so, countries made many important decisions regarding the governance of the company and the establishment of its management practices.

There is considerable agreement regarding the nature of good management and governance practices for utilities. All of the 15 observatory countries have corporatized their power utilities, in most cases during the reform process. (In some countries, such as the Philippines, the major utility was corporatized well before power sector reforms.) The effectiveness of the process, however, depends on the adoption of sound governance and management practices within the corporatized enterprises. Fortunately, well-established principles of good governance exist for SOEs such as power utilities (OECD 2015; World Bank 2014). On this basis, good governance practices can be ranked by several criteria, including the autonomy and accountability of the board of directors, the exercise of financial discipline within the company, and the prevalence of good management practices both for human resources and for information technology (figure 4.3).

A Utility Governance Index measures the extent to which specific utilities conform to good practices. It is difficult to say exactly when and how good governance and management practices have been adopted over time, because such measures are usually implemented within institutions and do not necessarily involve major legal or structural changes that can readily be tracked in the public record. Nevertheless, it is possible to measure the current rate of adoption of such practices. Using a sample of 19 state-owned and 9 privatized utilities from the 15 observatory countries, the Utility Governance Index measures the existence of best practices in utility rules and regulations. For example, a utility may, on paper, allow managers to hire and fire employees on the basis of performance—and the index captures this rule but is unable to tell whether the manager actually follows it.

The governance gap between corporatized public utilities and privatized ones is significant, and public utilities practice better governance when they coexist alongside private utilities. It is instructive to look at the variations in utility

FIGURE 4.3 Overview of utility governance performance indicators

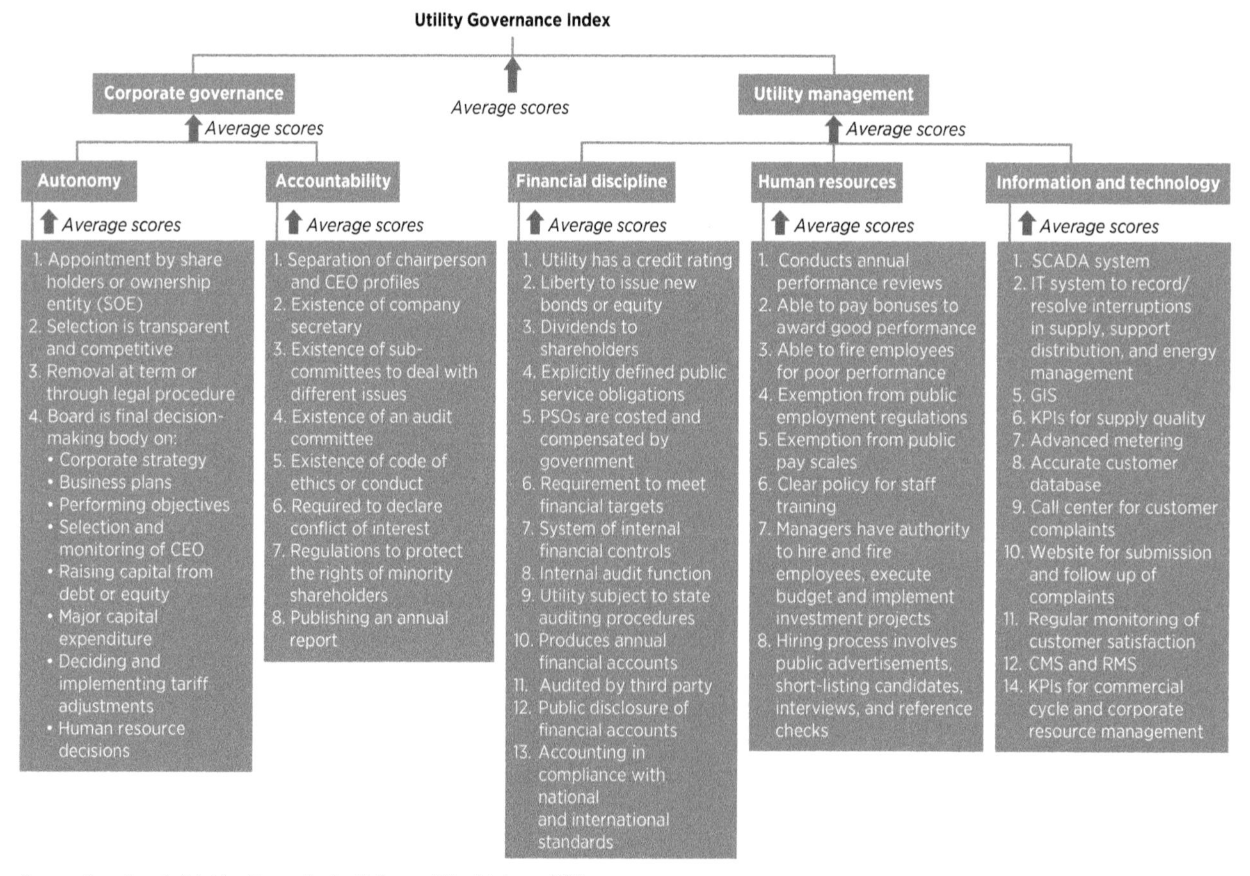

Source: Based on Rethinking Power Sector Reform utility database, 2015.
Note: CEO= chief executive officer; CMS = commercial management system; GIS = geographic information system; IT = information technology; KPI = key performance indicator; PSO = public service obligtion; RMS = resource management system; SCADA = Supervisory Control and Data Acquisition; SOE = state-owned enterprise.

governance scores across countries, and even more within countries for those cases where both public and private utilities exist side by side (figure 4.4). For those jurisdictions where companies are entirely state owned, utility governance scores tend to be systematically low with a median score of 55 percent. Those scores tend to be higher, in the 60–90 percent range, for the private utilities from the observatory group. The gap in scores is particularly striking, for example, between the public utility Lahore Electric Supply Company (LESCO) and the privatized Karachi Electric in Pakistan, and the provincial public utility Hidrandina and the metropolitan privatized utility Luz del Sur in Peru. In fact, the highest utility governance score for a public utility is for the Kenya Power and Lighting Company (KPLC), which has 49 percent of its capital floated on the Nairobi Stock Exchange, with the remainder held by the Ministry of Finance.

FIGURE 4.4 Private utilities implement more governance best practices, but government-owned utilities improve governance if there is some private competition

Good governance practices followed by public vs. private utilities across 17 jurisdictions, 2015

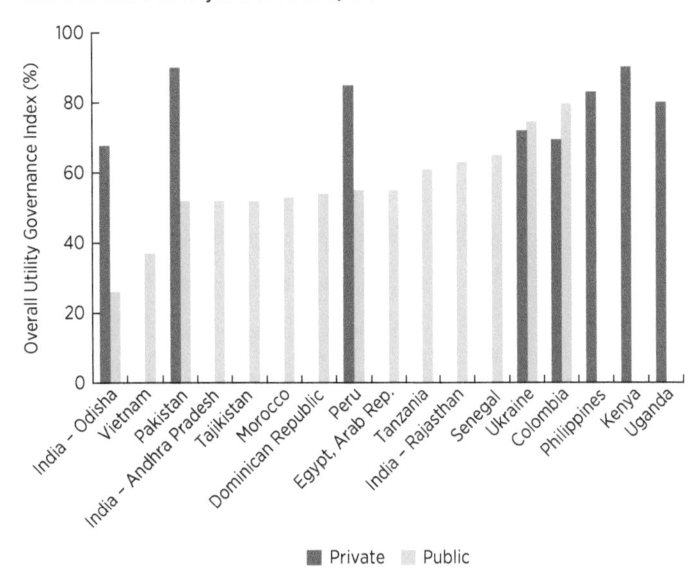

Source: Based on Rethinking Power Sector Reform utility database, 2015.
Note: The private utility in Kenya has a 49 percent stake floated on the Nairobi Stock Exchange and 51 percent is owned by government. The Western Electricity Supply Company of Orissa Ltd. (WESCO) in India - Odisha was a private utility during the study period, but the state took over the utility in 2016. The Central Electricity Supply Utility of Odisha (CESU) had gone back to the state in 2001 itself and as such is considered a public utility for the study.

Public utilities are more likely to lag behind with respect to certain areas of governance. In this section, the performance of public and private utilities across the sample is averaged out and disaggregated by components of their governance frameworks. The results serve to illustrate that both public and private utilities fall short of best practice in many areas, but that their strengths and weaknesses tend to differ.

Boards of private utilities enjoy almost complete decision-making autonomy, whereas public utility boards have limited freedom on critical matters of finance and human resources. The first area to consider relates to the autonomy and accountability of the company's board of directors. The board should in principle have the final say on all major business-related decisions, including the definition of strategies, plans, and performance objectives; important financial decisions on investment programs and related financing; and significant human resource decisions such as the appointment of the chief executive officer and the hiring and firing of staff. For the most part, the private utilities considered have full board autonomy on almost all of these decisions (figure 4.5). By contrast, the autonomy of the public utilities' boards is significantly constrained on all of these points, particularly the raising of finance and the appointment of the chief executive, which indicates that governments continue to be closely involved in the business decisions of state-owned utilities.

Public utilities also suffer considerable interference in the appointment and removal of board members. To be accountable, boards must have practices that enhance transparency and contain potential conflicts of interest. For example, it is important to have a transparent process for the appointment of suitably qualified board members, as well as clear and reasonable justification for their premature removal from office. To avoid conflicts of interest, it is important to separate out the roles of

FIGURE 4.5 **Private utilities far outpace their public counterparts when it comes to making decisions independently**

Practices to ensure board decision-making autonomy in public vs. private utilities in 17 jurisdictions

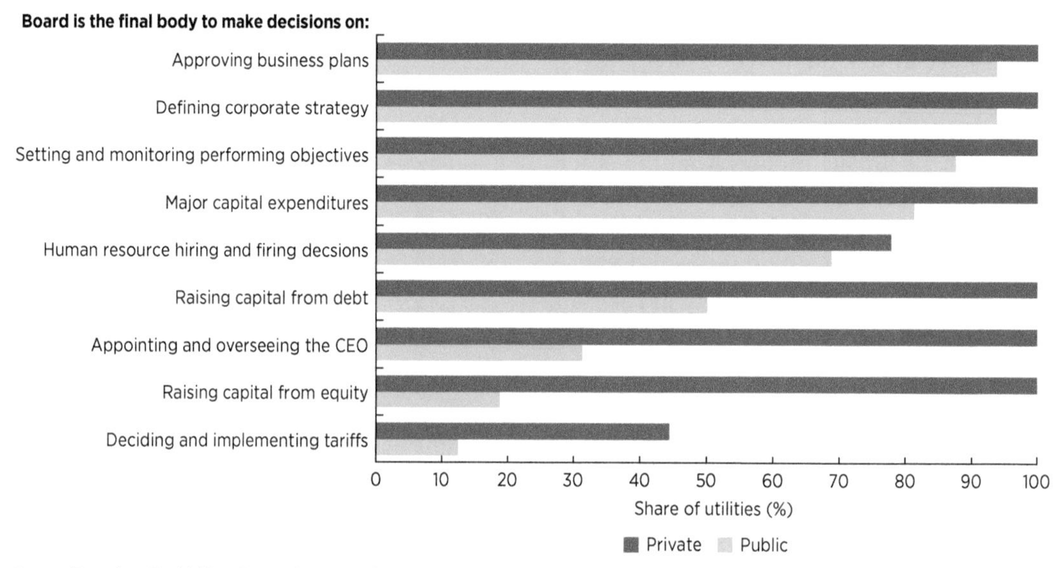

Source: Based on Rethinking Power Sector Reform utility database, 2015.
Note: CEO = chief executive officer.

FIGURE 4.6 **Boards in state-owned utilities are less accountable than their private counterparts, which must answer to various shareholders**

Practices to ensure board transparency and limit conflicts of interest in public vs. private utilities in 17 jurisdictions

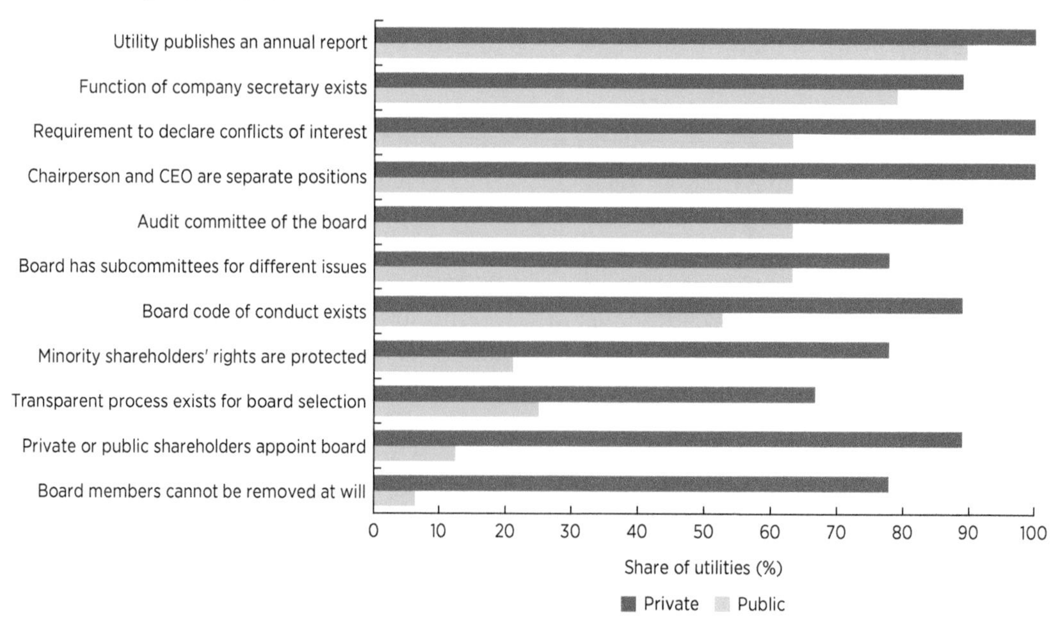

Source: Based on Rethinking Power Sector Reform utility database, 2015.
Note: CEO = chief executive officer.

the board chairperson, chief executive officer, and company secretary; to create board committees with clear responsibility for auditing and other functions; and to apply a code of conduct. Good accountability practices are not universal among the group of private utilities, and the gaps between the practices of public and private utilities are relatively small on some dimensions (figure 4.6). The most striking differences can be found with respect to board appointments and removals: about 75 percent of private utilities follow good practices (at least on paper), compared with less than 10 percent of public utilities.

Public utilities tend to fall short of basic good accounting practices, which are universal in the private sector (figure 4.7). Public utilities' financial discipline is related to the rigor of their accounting and auditing practices, as well as the extent of financial oversight by owners and investors. Among private utilities, financial accounts are universally produced and are externally audited. Almost all of the private utilities (90 percent) prepare accounts in accordance with international standards and disclose them publicly. Among public utilities, financial accounts are produced by 95 percent, are externally audited by 89 percent, are publicly disclosed by 74 percent, and meet international standards only in the case of 42 percent. In addition, public utilities are almost twice as likely to follow national accounting standards as international ones. Private utilities clearly have more freedom when it comes to raising various forms of finance and are exposed to the financial discipline that goes with that freedom, but less than half has a credit rating. For public and private utilities alike, public service

FIGURE 4.7 Public utilities have little independence when it comes to raising capital and tend to follow national rather than international accounting standards

Financial reporting practices of public vs. private utilities in 17 jurisdictions

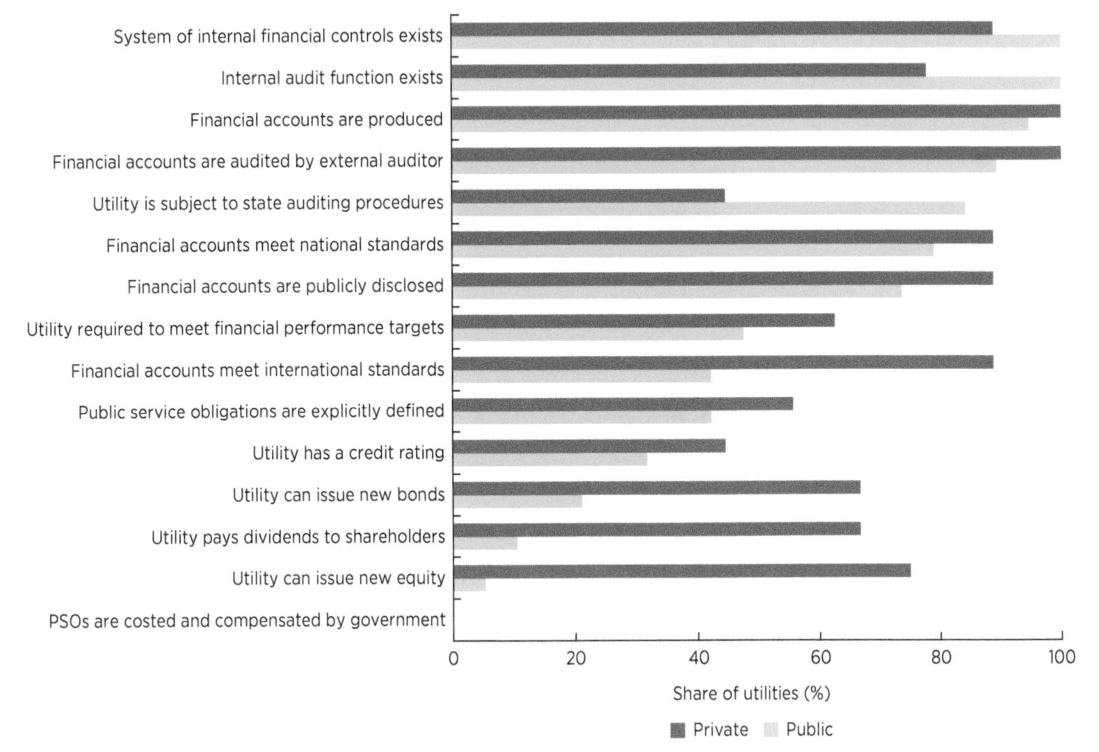

Source: Based on Rethinking Power Sector Reform utility database, 2015.
Note: PSO = public service obligation.

obligations are explicitly defined in less than half of cases, and are never properly costed and compensated for. Overall, public utilities outperform private ones only in internal auditing and financial controls.

Human resource practices tend to be more rigorous in private than in public utilities, which tend to have far less ability to reward good performers and fire bad ones. Good human resource practices entail objective hiring processes, adequate remuneration and performance-related pay, and the ability to fire poor performers. Once again, a comparison between private and public utilities has mixed outcomes (figure 4.8). The largest differences stem from the fact that private utilities are much less likely to be constrained by public sector wage scales and employment regulations. They also adopt a more rigorous approach to hiring and are much more likely to be able to fire employees. Although public and private utilities advertise only about 75 percent of available positions, recruitment in private utilities is more likely to involve short-listing, interviewing, and reference checks. Annual performance reviews are prevalent in over 80 percent of public utilities and are universal in their private counterparts, but the latter are more likely to offer performance-related bonuses. Finally, for private and public companies alike, decisions to hire and fire employees can rarely be taken at the management level but need to involve the board. Almost 90 percent of private utilities, however, can ultimately fire employees for poor performance as opposed to only about 70 percent of public ones.

Both public and private utilities are doing quite well at adopting information technology to improve internal management practices.

FIGURE 4.8 Public utilities have little freedom in making staffing decisions and have less transparency in hiring as compared to their private counterparts

Human resource practices of public vs. private utilities in 17 jurisdictions

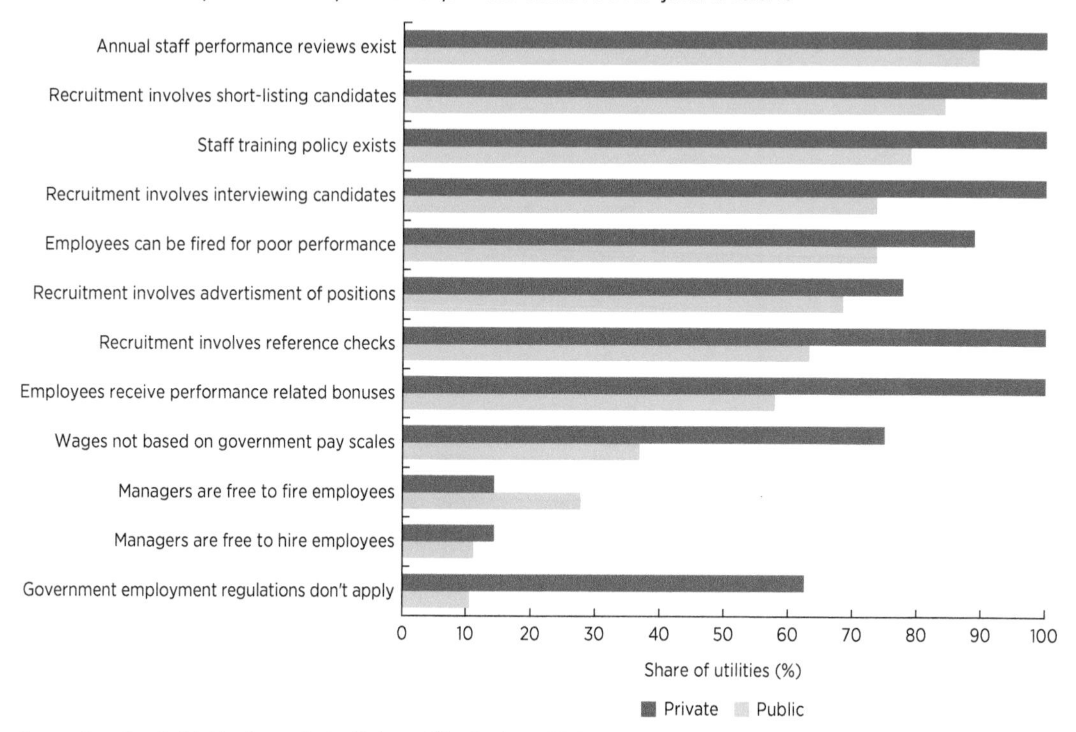

Source: Based on Rethinking Power Sector Reform utility database, 2015.

The continuous advance of digital technologies allows for greater automation and remote management of electricity networks. Information technologies can significantly enhance a utility's ability to deliver on many core areas such as network management and the commercial cycle, but uptake of information technology is far from being universal for either public or private utilities (figure 4.9). Among the most widespread applications of information technology are customer databases, call centers, and SCADA (Supervisory Control and Data Acquisition) systems, adopted by over 85 percent of both public and private utilities. Among the least prevalent information technologies are commercial management systems, resource management systems, and advanced metering infrastructure, which are adopted by close to half of the sample utilities. In certain areas, notably the use of online customer interfaces, public utilities seem to have the edge.

The example of the Indian state of Andhra Pradesh shows how governance reforms within the public sector can have a material effect on utility performance. Andhra Pradesh unbundled its state electricity board in 1999, creating four distribution utilities. Eschewing privatization of its distribution utilities, the state appointed a visionary managing director to the state transmission company, who was given unprecedented autonomy to oversee the distribution utilities that were subsidiaries of the transmission company in those days. This arrangement was a big change from a time when the state electricity board was run by the line ministry on a day-to-day basis. The new leadership went about implementing a

FIGURE 4.9 Both public and private utilities have adopted the latest information and technology solutions and are mostly at par

Adoption of information and technology solutions by public vs. private utilities in 17 jurisdictions

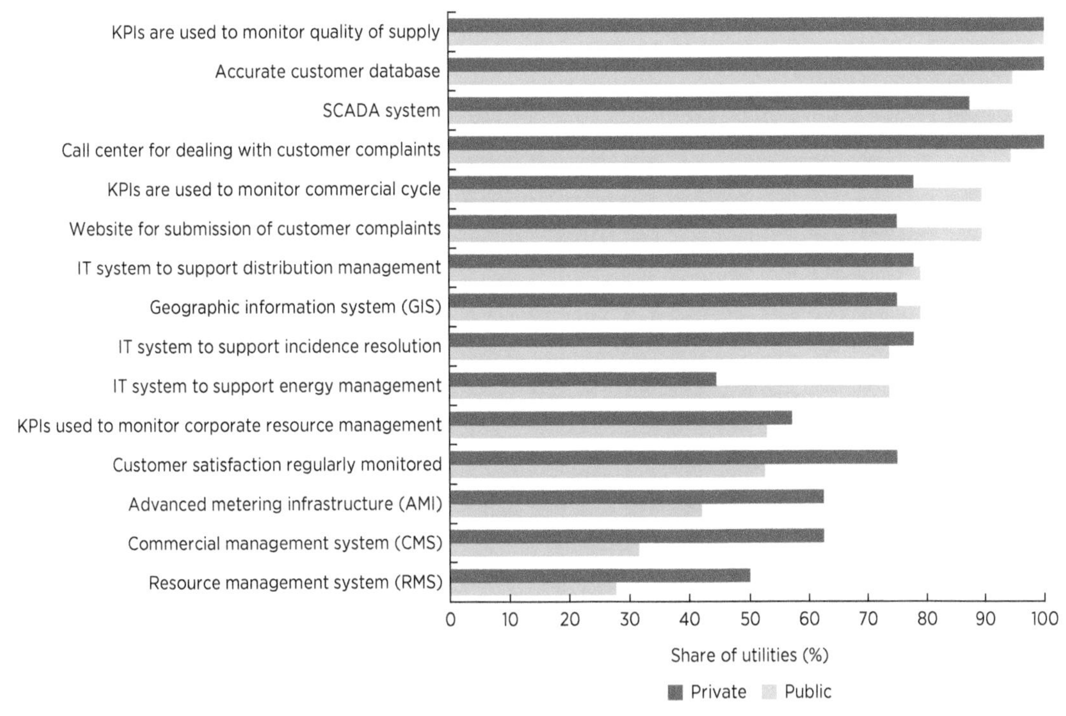

Source: Based on Rethinking Power Sector Reform utility database, 2015.
Note: IT = information and technology; KPI = key performance indicators ; SCADA = Supervisory Control and Data Acquisition.

culture change aimed at strengthening accountability by introducing monthly performance reviews for all officials from the top down. Simple and easily calculated key performance indicators were introduced, and staff at various levels reported their performance monthly. The most important of the key performance indicators was a simple reporting of cash collected per unit of energy that went into each administrative unit. (Before reform, the system had emphasized technical rather than commercial performance.) In addition, a significant enhancement of employee pay scales increased incentives. In parallel, significant investments were made in meter modernization, as well as in improving the quality of supply and expanding access. The state also enacted legislation that criminalized power theft, making it easier to prosecute felons, and setting an example for other customers. The combined effect of these measures was to double utility revenues in the space of five years (2002–07).

Finding #3: The unbundling of vertically integrated incumbent power utilities is unlikely to deliver major benefits unless it is accompanied by other reforms

Once utility corporatization is in place, the next stage of reform is to restructure the incumbent utility. As noted above, unbundling is a prerequisite for fostering competition in the power sector. It can also be helpful to separate out elements of the electricity supply chain that may be suitable for privatization from those that may not.

Full unbundling of the power sector has not been widely adopted across the developing world (figure 4.10). Few developing countries (less than 20 percent) have managed to implement full vertical and horizontal unbundling of their power sectors, and many (close to 60 percent) continue to operate with a vertically integrated national monopoly utility. The remaining 20 percent is at an intermediate stage of partial unbundling—that is, those

FIGURE 4.10 Close to 60 percent of developing countries still operate with a vertically integrated national monopoly utility

The power sector in developing countries, by structure and degree of unbundling, 1995–2015

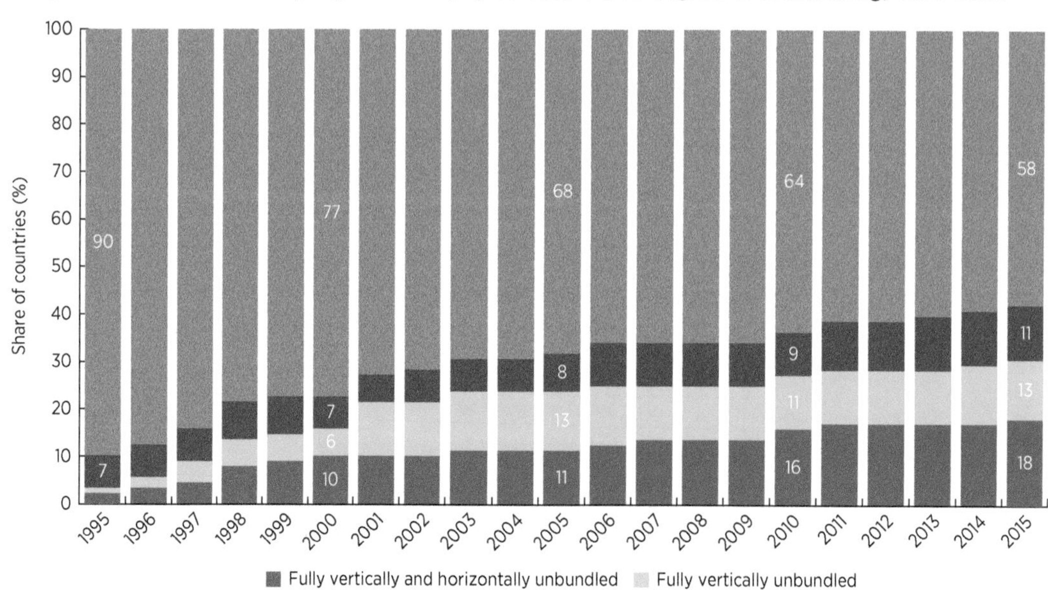

Source: Foster and others 2017.

companies have separated out either generation or distribution but not transmission, or have completed vertical unbundling without further horizontal unbundling of the generation or distribution sectors. Among the 15 observatory countries, some completed full unbundling as a basis for greater private sector participation and a wholesale power market (such as Peru and the Philippines), whereas others unbundled yet retained a single-buyer model largely dependent on the public sector (such as Egypt and Pakistan). Among African countries (such as Senegal, Tanzania, and Uganda), vertical integration of the sector remains widespread (table 4.2).

Some variation of the single-buyer model remains the most widely adopted organizational model for the power sector in the

TABLE 4.2 Overview of restructuring and competition reforms in 17 jurisdictions, 2015

	Vertical unbundling	Horizontal unbundling	Sector model
Colombia	Yes: but with some vertically integrated utilities remaining	Yes: multiple generators and distributors	Wholesale power market
Dominican Republic	Yes: generation, transmission, and distribution fully separated	Yes: 3 generators and 3 distributors	Wholesale power market
Egypt, Arab Rep.	Yes: generation, transmission, and distribution separated, but under a single holding company	Yes: 6 generators and 7 distributors under a single holding company	Single-buyer model
India - Andhra Pradesh	Yes: generation, transmission, and distribution fully separated	Yes: 4 distributors (after bifurcation, 2)[a]	Wholesale power market
India - Odisha	Yes: generation, transmission, and distribution fully separated	Yes: 2 generators and 4 distributors	Wholesale power market
India - Rajasthan	Yes: generation, transmission, and distribution fully separated	Yes: 3 distributors	Wholesale power market
Kenya	Yes: generation, transmission, and distribution fully separated	No: 5 different generation entities were all combined into KenGen	Single-buyer model
Morocco	No: vertically integrated national utility plus IPPs and 11 local distribution utilities	No: but IPPs are closing in on 50% of generation, and distribution in major cities is handled by city-level distribution companies	Single-buyer model
Pakistan	Yes: generation, transmission, and distribution fully separated with 1 vertically integrated utility	Yes: 4 generators and 8 distributors with 1 vertically integrated utility	Single-buyer model
Peru	Yes: generation, transmission, and distribution fully separated	Yes: Multiple generators and distributors	Wholesale power market
Philippines	Yes: generation, transmission, and distribution fully separated; utilities can own vertical business operations through subsidiaries	Yes: Multiple generators and distributors	Wholesale power market
Senegal	No: vertically integrated national utility plus IPPs	No	IPPs only; vertically integrated utility functions as single buyer
Tajikistan	No: vertically integrated national utility plus 2 IPPs and 1 regional distribution company	Partial unbundling of distribution in one region	IPPs only; vertically integrated utility functions as single buyer
Tanzania	No: vertically integrated national utility plus IPPs	No	IPPs only; vertically integrated utility functions as single buyer
Uganda	Yes: generation, transmission, and distribution fully separated	Partial unbundling in distribution, some rural concessions given out	Single-buyer model
Ukraine	Yes: generation, transmission, and distribution fully separated	Yes: 4 generators and 27 distributors	Single-buyer model, transitioning to market
Vietnam	Yes: generation, transmission, and distribution separated, but under a single holding company	Yes: 3 generators and 5 distributors under a single holding company	Single-buyer model, transitioning to market

Source: Based on Rethinking Power Sector Reform utility database, 2015.
Note: IPP = independent power producer.
a. In 2014 Andhra Pradesh was divided into two states and the four distribution companies were divided equally between the two.

FIGURE 4.11 **Most countries deploy some version of the single-buyer model in their power sectors**

The power sector in developing countries, by degree of competition, 1995–2015

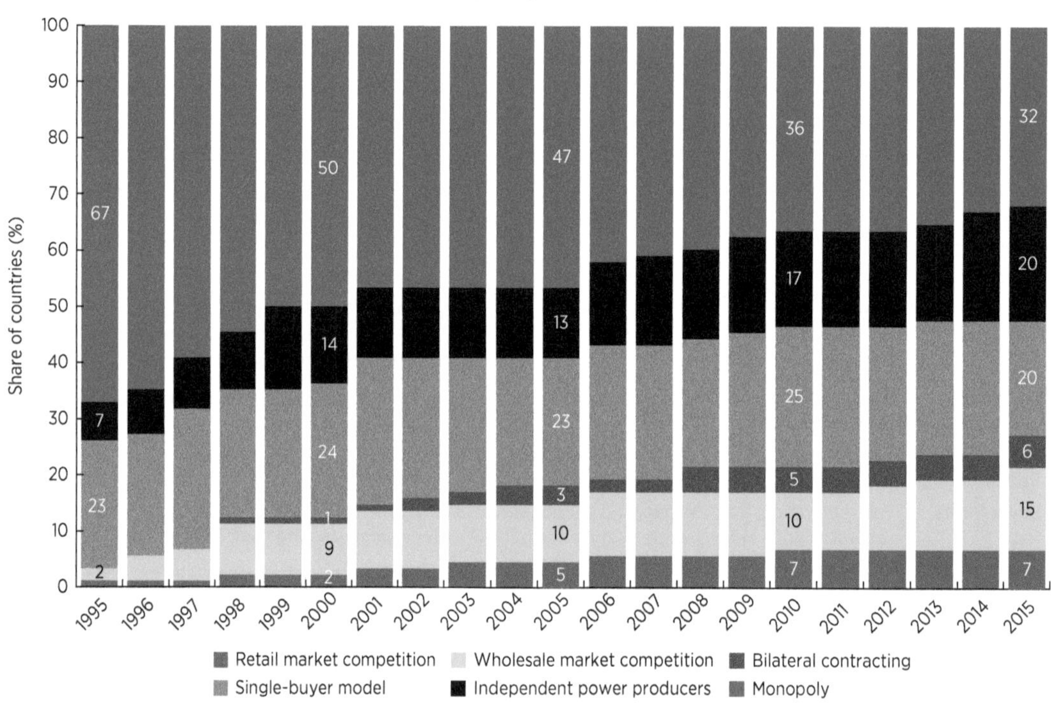

Source: Foster and others 2017.

developing world (figure 4.11). Overall, this represents about 44 percent of all developing countries. In half of these cases, the single-buyer model purely consists in an opening of generation to IPPs, whereas in the other half the incumbent's generation assets have been divested to reduce conflicts of interest. Among the 15 observatory countries, almost all operate some form of the single-buyer model, except for 4 with fully functioning wholesale power markets (Colombia, India, Peru, and the Philippines) (table 4.2). A few others are currently in transition from the single-buyer model to a wholesale power market (including Ukraine and Vietnam).

Although unbundling may bring some benefits of its own in terms of transparency and accountability, it is debatable whether those benefits are enough to drive performance improvements. An unbundling reform conducted in isolation simply creates several smaller state-owned companies, that absent other measures may suffer from similar performance problems as the original parent company. Thus unbundling cannot be counted on *in and of itself* to improve the performance of the system overall (Bacon 2018). The one benefit that may result from unbundling per se is an increase in transparency and managerial accountability. Whether this benefit has an impact on utility performance is likely to depend on the quality of governance and regulatory oversight, to ensure that the resulting information is put to good use.

In the Indian state of Andhra Pradesh, restructuring led to increased transparency and accountability, affecting utility performance. By the late 1990s, the power sector in the state was suffering from power shortages and declining quality of service, and contributing to a significant fiscal crisis. The state embarked on a comprehensive power sector reform plan, which began in 2000 with unbundling the Andhra Pradesh State Electricity Board (APSEB) into a generation company, Andhra Pradesh Power Generation Corporation (APGENCO); a transmission company, Transmission Corporation of Andhra Pradesh Limited (APTRANSCO); and four distribution companies. The restructuring allowed the unbundled entities to focus on their core business areas while the government provided the necessary backing through improved regulation and legislation to outlaw power theft. The restructuring had an immediate impact on reducing transmission and distribution losses and improving collections. Statewide losses fell from 38 percent in 1999 to 20 percent in 2004, whereas collections rose from 92 percent to 98 percent in the same period.

The costs of implementing unbundling are not insignificant (Pollitt 2008; Vagliasindi 2012). Utility restructuring is a complex process entailing significant transaction costs, potentially amounting to tens of millions of dollars, due to the need for an exhaustive inventory of human and physical assets; full revision of company accounts; installation of meters to monitor power flows across new company boundaries; and the legal work associated with the creation of independent governance structures for each of the subsidiary companies. Unbundling also creates higher fixed costs, associated with replicating board structures across different companies, that may prove challenging to implement in countries with a scarcity of managerial skills. Furthermore, in power systems that lack strong payment discipline, the unbundling of the sector risks creating a cascade of indebtedness across the various resulting entities.

The horizontal unbundling of distribution utilities, where no real prospects for competition exist, also risks the loss of economies of scale and the creation of unprofitable business units. The breakup of national distribution utilities into smaller business areas needs to be considered with caution, particularly in smaller countries where the loss may be material and the prospects of developing meaningful competition may be limited. Moreover, although having multiple distribution companies in a country may increase managerial transparency and accountability, the associated benefits will materialize only if sound governance and regulatory oversight is in place, and even then may not be large enough to compensate for the loss of economies of scale. Often the profitability of distribution utilities reflects the concentration of industrial and commercial customers—as well as relatively affluent residential customers—in metropolitan areas, whereas provincial utilities have a more meager customer base and sometimes higher costs due to lower population density. This situation is illustrated by data from Peru, where the profit margin per client in four of the regional distribution utilities (with customer bases of 200,000–400,000 connections) is just a fraction of the profit margin of metropolitan area utilities (with customer bases of 700,000–900,000 connections) (figure 4.12). Hence the breakup of a national utility in a small country risks creating one or two business units that are commercially attractive and several others that may be much less viable. The presence of larger customers with a stronger ability to pay within

FIGURE 4.12 Horizontal unbundling of the distribution sector in Peru created a couple of large profitable metropolitan utilities and a number of small regional utilities with limited scope for profits, 2002

Net profit per client across distribution utilities in Peru after horizontal unbundling, 2002

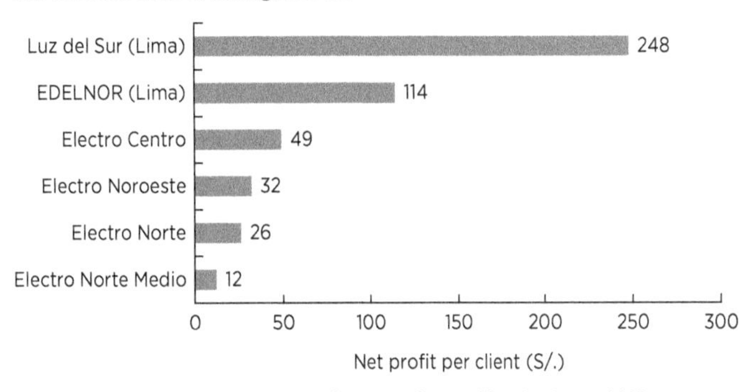

Source: Based on Rethinking Power Sector Reform utility database, 2015.
Note: S/. = Peruvian nuevos soles.

the context of a national utility also provides a basis for cross-subsidies across service areas, which often occur implicitly through national service areas. Although geographical cross-subsidies could also take place, in principle, across regional utilities through an explicit service levy of some kind, they may be more difficult to achieve politically because they involve a visible cross-jurisdictional financial transfer.

The balance between the costs and benefits of unbundling is unlikely to be favorable in smaller power systems, where transaction costs can weigh quite heavily. For this reason, unbundling is not typically recommended in power systems below 1 gigawatt (GW) in size (Besant-Jones 2006), and may not become a priority until significant scope for competition starts to materialize in systems of 3 GW and above. In larger power systems, the benefits of unbundling in terms of increased transparency and accountability are likely to outweigh the costs. Overall, although 70 percent of countries with power systems smaller than 1 GW have vertically integrated power utilities, that share falls to just over 20 percent for countries with power systems larger than 20 GW. Nevertheless, a significant number of larger countries have still not unbundled incumbent utilities. Indonesia is the world's fourth-most-populous country with over 60 GW of installed capacity (in 2017), and it still has a state-owned vertically integrated national utility, Perusahaan Listrik Negara (PLN), which has a monopoly on electricity distribution and transmission while having a majority share of generation assets as well.

The impact of unbundling, when all companies remain under a common state ownership structure, may be limited. Several countries—such as Egypt and Vietnam—have unbundled in the context of continued state ownership to create large state-owned holding company structures, as part of a transition toward a competitive market. When unbundling is envisaged as the foundation of a competitive market, common ownership of the unbundled companies may limit the extent to which the various subsidiary companies can operate independently of one another.

For example, the Egyptian power sector has seen various forms of unbundling and restructuring since the 1960s; however, decision-making power has remained centralized throughout. In the 1960s, private utilities were nationalized and the Egyptian Establishment for Electricity, under the Ministry of Energy, was created to oversee the generation, distribution, and transmission businesses. In 1976, the Egyptian Electricity Authority (EEA), also under the Ministry of Energy, was created to jointly manage both the generation and distribution businesses, and in 1978 seven regional distribution companies were established. This framework continued until 1993, when distribution utilities were moved from the EEA to the Ministry of Public Enterprise. In 1998, distribution companies were transferred back to the Ministry of Energy, where the generation and distribution businesses were bundled into seven vertically integrated SOEs under the EEA. In 2000, a further sector reorganization took place, whereby EEA was again unbundled and turned into a holding company—the Egyptian Electricity Holding Company (EEHC)—with five generation utilities, a transmission utility, and seven distribution utilities. Despite this repeated restructuring, the underlying governance and management arrangements of the distribution companies were not substantially altered.

In Pakistan, too, old decision-making structures continued to prevail after the original national utility had been unbundled. Before 1994, Pakistan's entire population was served by two vertically integrated companies: the Water and Power Development Authority (WAPDA) and the Karachi Electric Supply Company (KESC). A 1970–75 plan proposed the separation of generation and distribution, making the case that WAPDA had become too large and unwieldy to "shoulder the responsibility of retail distribution and generation of power" (Planning Commission 1970). By the late 1980s and early 1990s, WAPDA was experiencing severe governance issues and declining operational performance, which provided the final impetus for reform. Following the WAPDA Amendment Act, WAPDA's power wing was unbundled in 1997 to form 15 incorporated state-owned entities comprising 3 thermal generating companies, 1 national transmission and dispatch company, and 8 regional distribution companies. Hydropower generation and water management remained with WAPDA. However, in 2005, seven years after the passing of the law, the distribution companies were still not operating in a fully autonomous manner. According to Parish (2006), the finances of the utilities and their tariff applications were wholly handled by WAPDA, which even had some of the senior managers of the unbundled companies on its payroll and was closely involved in appointments to the board of directors. The resulting institutional confusion prejudiced the performance of the unbundled companies.

LOOKING AHEAD

Technological disruptions underway in the power sector will make planning increasingly complex and may carry implications for the sector's structure. The combination of increased uncertainty on both the supply side and the demand side complicates system planning and grid operations. When power is generated from fully dispatchable resources and assumed to flow only one way—from a generation source, across wires, to a consumer—both supply and demand are relatively predictable. Day-to-day grid operations and long-term system planning focus on meeting the system peak, and ensuring adequate reserve margins. The increased penetration of both variable renewable energy resources and distributed energy resources (DERs) however, introduces uncertainty simultaneously to the supply and demand sides of electricity, respectively, and this relatively simple formula begins to change.

On the demand side, net load rather than peak demand becomes the moving target for planning. Rather than a mostly consistent load forecast, growing reliance on DERs leads to greater uncertainty about how much consumers will reduce their total demand, or how much consumers might supply back to the grid. As a result, the whole profile of expected demand may shift. If, for example, during the typical peak[2] system load occurring during the day,[3] an increased amount of solar production is available from customer-sited rooftop solar photovoltaic (PV), then the total amount of grid-connected (or utility-scale) resources needed to meet consumer demand decreases. This difference between the typical, expected peak load and the total amount of customer-sited renewable resources[4] is known as the net load (IEA Wind Task 25 2013). This net load can change throughout the day, for instance, shifting the peak to the period when decentralized solar generation dips at the end of the day. Forecasting the net load—and the resources needed to quickly ramp up to the full output needed to meet this net load—are new challenges for system operators and system planners alike.

On the supply side, planners are increasingly concerned with building in greater system flexibility. Uncertainty about the output of renewable supply increases the need for fast-ramping resources (that is, resources that can adjust production up or down quickly) that can respond when the sun is not shining, or the wind is not blowing.[5] As renewable penetration increases, it becomes increasingly challenging to maintain supply/demand balance while ensuring sufficient voltage and frequency support to preserve reliability of supply. This challenge in turn creates the need to build out additional transmission capacity, both within and across national boundaries. Greater transmission capacity allows the system to fully absorb variable renewable generation when it is available, avoiding curtailment, and provides larger geographical areas across which to balance supply variability. With variable renewable energy, supply forecasting also becomes more complex, and requires better weather forecasts to ensure that the right mix of backup resources is available to meet system needs (including ramping, voltage control, frequency regulation, and black start), bearing in mind that these system needs could change across the day, over several hours, or even within the hour.

In addition, planning the distribution system is getting more challenging. The traditional system peak is declining, net load considerations are more relevant, and the concept of what constitutes peak system operations is changing altogether. Indeed, in some European distribution systems, peak system usage is already occurring during the times of greatest consumer *production*, not consumer *consumption* (Burger and others 2018).[6]

So much uncertainty brings a significant risk of stranded assets. Assets could get stranded if the customer demand that grid infrastructure, or generation plants, are designed to support is not realized. For example, large generation assets that are inflexible will be less helpful to system operators when flexibility is needed. Inflexible, typically thermal, generation assets are not designed to turn on and produce lots of electricity quickly. Instead, they are designed to turn on and operate within a particular range for long periods of time. If these units are constantly cycled (that is, turned on and off), thermal efficiency is reduced and the useful life of the plant shortened.

As storage technologies become commercially viable, they could provide a variety of grid services to facilitate the integration of renewable generation and distributed energy resources. How storage technologies are used, however, has implications for day-to-day grid operations and long-term system planning. Storage technologies, in general, enable generation supply to be stored and used for consumption later. Battery storage, in particular, is an example of a fast-ramping resource that can be discharged at any moment in response to system needs. This resource makes it easier to instantaneously match supply with demand, because there could always be stored supply available to meet demand at any time, if storage capacity is available at adequate levels and in appropriate locations. Unlike any other resource, storage can also be used for transmission. Transmission lines are designed to allow a certain amount of electricity to flow through them, and they must be operated within these limits to avoid damaging the transmission lines or damaging generation units that are connected to these lines.[7] By shifting when and how generation resources are used, storage helps to manage transmission congestion (that is, manage the desire to flow more electricity across transmission lines than is technically feasible) or even increase the total hosting capacity of distribution system lines (that is, accommodate customer-sited resources, like rooftop solar PV, on an existing system without requiring additional upgrades). Storage can also avoid the need for some transmission and distribution upgrades altogether. Reducing the overall system peak lessens the total system-wide build-out that grid operators need to plan in order to meet system peak demand.[8] Finally, storage can extend the life of existing transmission and distribution system assets by allowing for the optimal use of those assets.

How these different resources, on both the supply and demand sides, can be used to enable the transition to a low-carbon grid will be an ongoing challenge. Long-term planning is already complex even for traditional power systems based on one-way power flow from centralized grid resources. These challenges are compounded in emerging modern systems where consumer demand fluctuates and variable renewable energy introduces the need for different kinds of system resources. Going forward, the changing electricity paradigm will require decisions about the kinds of resources that need to be available to grid operators to integrate these variable resources, and system planning models will need to evolve to accommodate consumer-sited DERs, and the many possible uses of storage technologies.

Furthermore, the advent of low-cost DERs increasingly calls into question the validity of the traditional utility business model. This model reflects the historic cost structure of the power sector, specifically, the presence of economies of scale in power generation, transmission, and distribution, as well as large economies of scope in the tight coordination of energy flows along the electricity supply chain. These elements led to the creation of national-scale vertically integrated power utilities. As technological disruption shifts the underlying cost structure of the industry, the logic behind the institutional organization of the sector also changes. The interplay between shifting cost structures and regulatory responses is likely to affect the institutional evolution of the sector going forward.

Distribution system operators are moving from a passive role to an active one. Previously, they sent electricity from central generators directly to consumers, whereas now they increasingly manage various kinds of DERs that might be consuming, producing, or managing loads. Many of these activities are being undertaken by a range of private sector actors, including prosumers as well as new entrants that see business opportunities in aggregating demand or providing decentralized energy resources across the distribution grid. The choice of how to best integrate DERs often depends on existing market structures and existing regulatory models. Where vertically integrated utilities are dominant, utilities are likely to continue playing a role in integrating various DER technologies. Where competitive markets at the retail level exist, the ongoing role of the utility is less clear. In these markets, utilities do not own generation. DER integration instead focuses more on enabling third-party ownership of DER assets and transitioning the utility into an entirely different role.

Many unanswered questions remain regarding how DERs should be integrated into both transmission and distribution systems. In markets where the transmission and distribution segments are clearly separated, with different system operators managing each network, questions about how to manage the integration of DERs with separate system operators are largely unanswered. If DERs participate in both retail and wholesale markets, which system operators do these resources respond to? Can the different services DERs provide (for example, energy, regulation, reserves, and load management) be clearly distinguished in either system? If so, how should their market impact on the other system be considered? These are complicated questions, whose answers depend on different legal and jurisdictional realities that remain unclear in many of these markets.

Most regulators have focused instead on the future role of the distribution system operator, and what regulatory models might be needed to enable it. Generally, two approaches are being considered. Under the first approach, the distribution system operator retains its functions as owner and operator of the distribution system, amid changes to the regulatory model to facilitate the integration of DERs. Under the second approach, an independent entity manages the operation and planning of the distribution system to ensure that all market activities are separated from these functions. In other words, an independent distribution system operator stripped of retail functions exists and is similar to some markets at the transmission level.

The distribution utility becomes a market enabler in this platform/network model, whereby various sellers aggregating various DERs can trade or provide services to the distribution system through the distribution system platform. New York state is considering several utility cost-recovery models that facilitate platform development and data sharing with third parties to foster transparent and open markets. More advanced metering and more software and controls are needed to enable these kinds of structures. Regulators may allow utilities to recover the costs associated with infrastructure that enables such a transactive platform (for example, smart metering, or other grid modernization efforts), or enable utilities to recover new revenue streams associated with providing "value-added" services (for example, providing system data for third-party usage) to create a more platform-based structure.

These emerging regulatory models, however, require regulators to rethink where, when, and how utility capital investments are made. They also require regulators to consider whether utility ownership of various assets (for example, storage or electric vehicle charging infrastructure) that enable the greater integration of renewable generation could also enable (or hinder) a more transactive platform—or competitive distribution markets more generally. The challenge is to determine whether the utility or some other entity altogether is best positioned to manage the integration of these resources. In unbundled, competitive distribution markets, there is concern about utility ownership of distributed energy resources and the functions associated with the management of the distribution system that enable a more active distribution system operator. If regulators want to ensure that competitive distribution markets are maintained, then open and transparent access to data coming from consumer meters or electric vehicle charging stations—data that utilities typically do not share—will be required (Kufeoglu, Pollitt, and Anaya 2018).[9] If regulators want to ensure that competitive wholesale or retail markets are maintained, then a utility's ownership of distribution system assets (like energy storage or electric vehicle charging stations, that could also provide services to the wholesale market) may not be advisable.

The relevance of platform-based distribution models very much depends on the legacy market structure in the sector. In places where vertically integrated utilities remain dominant, or where no clear separation exists between the transmission system operator and the distribution system operator, questions about the utility ownership of DER assets, or about the participation of these resources in various systems, are less relevant. Moreover, a vertically integrated utility that still manages generation may find it easier to deploy DERs where they would provide the most system benefits, as well as capture the entire value that these resources can provide.[10] In places where utilities are largely unbundled and no longer own generation assets, and competitive retail markets already exist, it may be possible to consider a complicated new market design to fully integrate DERs. And, as with the challenges associated with integrating renewables into grid operations, if the price signal alone is intended to incentivize new resources, then markets and tariffs must be designed so as to accurately value the full services these resources provide to grid operations, including both their time and their locational value.

Where the development of a utility service remains incipient and regulatory capacity is weak, there is the real possibility of leapfrogging toward decentralized models of service provision. In many developing countries, neither of the above conditions holds. Utilities may still be incipient, serving only a fraction of the potential market, with electrification rates occasionally as low as 10–20 percent, and more typically in the 40–60 percent range. Even where service is available, it may be highly unreliable with frequent interruptions that leave consumers to rely on their own backup generation options, which have traditionally been diesel based. Although explicit or implicit subsidization of grid electricity is widespread across the developing world, a significant subset of countries may nonetheless face exceptionally high tariffs (and even higher costs) for grid electricity due to geographical disadvantages (as in small islands and landlocked countries) that leave them dependent on small-scale oil-fired generation. Added to this problem may be high levels of inefficiency in the distribution segment, and a marked tendency toward cross-subsidization—to the detriment of nonresidential customers. All of these factors conspire to make DERs increasingly attractive in many parts of the developing world, both as off-grid alternatives to grid electrification and as backup for inadequate grid supply. As storage technology further improves and becomes more cost-effective, the possibility of relying on self-generation, rather than the often unreliable electricity supply from the grid, will become increasingly attractive to individual consumers, irrespective of whether self-generation is economically efficient for the system as a whole. Mini/microgrids with innovative payment structures already play a key role in expanding energy access in rural areas and could increasingly provide alternative supply options to urban customers as well. Grid defection, however, will exacerbate the existing cost-recovery challenges of developing countries' utilities, and integration of (both centralized and) distributed renewable energy will exacerbate existing technical challenges in grid operation. The relatively weak capacity of regulators in such environments will make it challenging to guide this process in an optimal manner.

CONCLUSION

In summary, corporatization has led state-owned utilities to adopt many good governance and management practices, even if they still lag practice in the private sector. Good governance and management depend on the adoption of a range of widely accepted measures that relate to the autonomy and accountability of the board of directors, the extent of financial discipline, and the handling of human resources and information technology. Overall, only about 55 percent of such practices can be found (on paper at least) in entirely state-owned utilities, compared with about 80 percent in private utilities. In countries where public utilities coexist alongside private utilities, the share of good-governance practices among the public utilities rises to 57 percent. The governance practices seen the least in public utilities include the autonomy of board decision making, transparency of board appointment and removal processes, and the adoption of good financial accounting and staff recruitment practices.

Restructuring utilities was a central focus of reform efforts but proved difficult to implement and may have distracted attention from more fundamental issues. In the 1990s, as recommended by International financial institutions, many countries undertook major sector restructuring exercises. Barely 20 percent achieved the full vertical and horizontal unbundling envisaged. As a result, close to half of the world's developing countries operate under some variation of the single-buyer model. Because unbundling was primarily conceived as a means to deeper reform, the case for vertical and horizontal unbundling is questionable in smaller systems not able to progress toward a wholesale power market. The benefits of the process need to be balanced against the potential loss of economies of scale and scope. With their emphasis on regulation and restructuring, the sector reforms of the 1990s often underemphasized the critical function of sector planning, which remains inadequate in many countries.

Looking ahead, technological disruptions currently underway in the power sector will make planning increasingly complex and may carry implications for the sector's structure. The advent of prosumers and other third-party actors (such as demand aggregators) able to deploy decentralized generation, storage, or demand-response solutions complicates the task of planning, which affects both generation expansion and grid development and has prompted a shift toward concepts of net load. At the same time, the presence of these new players in markets that have already undergone extensive restructuring underscores the possibility of competition in the retail tier, with the utility acting as a distribution system operator that provides the platform across which other parties trade energy on the local grid. Where sectors remain vertically integrated, the issue is more one of incentivizing the utility to consider the deployment of decentralized off-grid alternatives to balance demand and supply. Finally, where utility networks remain underdeveloped, new technologies offer the option of leapfrogging to a decentralized service model.

ANNEX 4A. UTILITY RESTRUCTURING INDEX, 2015

Percent

Country/State	Utility restructuring (overall)	Vertical unbundling	Horizontal unbundling
Colombia	35	70	0
Dominican Republic	67	100	33
Egypt, Arab Rep.	37	40	33
India - Andhra Pradesh	57	80	33
India - Odisha	73	80	67
India - Rajasthan	57	80	33
Kenya	25	50	0
Morocco	0	0	0
Pakistan	73	80	67
Peru	73	80	67
Philippines	100	100	100
Senegal	0	0	0
Tajikistan	0	0	0
Tanzania	0	0	0
Uganda	55	60	50
Ukraine	63	60	67
Vietnam	47	60	33
International benchmark	**45**	**55**	**34**

Source: Based on Rethinking Power Sector Reform utility database, 2015.

ANNEX 4B. PLANNING AND PROCUREMENT INDEX, 2015

Percent

Country/State	Planning and procurement (overall)	Generation planning	Generation procurement	Transmission planning	Transmission procurement
Colombia	95	86	95	100	100
Dominican Republic	72	29	86	75	100
Egypt, Arab Rep.	82	71	100	75	83
India - Andhra Pradesh	78	57	95	75	83
India - Odisha	78	57	95	75	83
India - Rajasthan	78	57	95	75	83
Kenya	82	86	100	75	67
Morocco	61	43	100	50	50
Pakistan	63	29	100	25	100
Peru	77	43	90	75	100
Philippines	59	71	100	50	17
Senegal	59	43	50	100	42
Tajikistan	64	71	100	50	33
Tanzania	77	43	100	75	92
Uganda	76	43	95	75	92
Ukraine	38	60	0	75	17
Vietnam	59	71	50	100	17
International benchmark	**71**	**56**	**85**	**72**	**68**

Source: Based on Rethinking Power Sector Reform utility database, 2015.

ANNEX 4C. UTILITY GOVERNANCE INDEX, 2015

Percent

Country/ State	Utility	Overall utility governance	Corporate governance	Accountability	Autonomy	Utility management	Financial discipline	Human resource	Information and technology
Colombia	EPM	**80**	76	75	78	83	76	86	87
	CODENSA	**69**	96	92	100	43	69	60	0
Dominican Republic	EDESUR	**51**	50	33	67	52	29	50	79
	EDENORTE	**57**	63	58	67	52	21	50	86
Egypt, Arab Rep.	EEHC	**53**	44	33	56	61	53	71	60
	EEHC (Discos)	**55**	49	42	56	61	53	71	60
India – Andhra Pradesh	APSPDCL	**52**	47	50	44	56	53	43	73
	APEPDCL	**52**	47	50	44	56	53	43	73
India – Odisha	WESCO	**68**	86	83	89	50	43	70	36
	CESU	**26**	13	25	0	40	36	40	43
India – Rajasthan	JVVNL	**63**	67	67	67	60	64	36	80
	JDDVNL	**63**	67	67	67	60	64	36	79
Kenya	Kenya Power (KPLC)	**90**	100	100	100	80	76	64	100
Morocco	ONEE	**53**	35	25	44	70	57	79	73
Pakistan	LESCO	**52**	56	67	44	48	50	43	50
	KE	**90**	94	100	89	86	79	86	93
Peru	Luz del Sur	**85**	85	92	78	85	86	90	80
	Hidrandina	**55**	40	58	22	70	65	71	73
Philippines	MERALCO	**90**	100	100	100	81	71	79	93
	BENECO	**76**	83	67	100	68	53	86	67
Senegal	Senelec	**65**	74	58	89	56	57	57	53
Tajikistan	Barki Tojik	**52**	42	17	67	62	64	57	64
Tanzania	TANESCO	**61**	57	58	56	65	64	71	60
Uganda	UMEME	**80**	85	92	78	76	77	71	80
Ukraine	Khmelnitskoblenergo	**75**	82	75	89	67	64	57	80
	Dniproblenergo	**72**	69	50	89	74	65	79	80
Vietnam	NPC	**37**	8	17	0	65	53	50	93
	HPCMC	**37**	8	17	0	65	53	50	93
International benchmark		**63**	62	60	63	64	59	62	71

Source: Based on Rethinking Power Sector Reform utility database, 2015.

ANNEX 4D. UTILITY CLASSIFICATION, 2015

Utility	Country/State	Ownership
APEPDCL	India-Andhra Pradesh	Public
APSPDCL	India-Andhra Pradesh	Public
Barki Tojik	Tajikistan	Public
BENECO	Philippines	Private
CESU[a]	India-Odisha	Public
CODENSA	Colombia	Private
Dniproblenergo	Ukraine	Private
EDENORTE	Dominican Republic	Public
EDESUR	Dominican Republic	Public
EEHC	Egypt, Arab Rep.	Public
EEHC (Discos)	Egypt, Arab Rep.	Public
EPM	Colombia	Public
Hidrandina	Peru	Public
HPCMC	Vietnam	Public
JDDVNL	India-Rajasthan	Public
JVVNL	India-Rajasthan	Public
Karach Electric	Pakistan	Private
Kenya Power (KPLC)[b]	Kenya	Private
Khmelnitskoblenergo	Ukraine	Public
LESCO	Pakistan	Public
Luz del Sur	Peru	Private
MERALCO	Philippines	Private
NPC	Vietnam	Public
ONEE	Morocco	Public
Senelec	Senegal	Public
TANESCO	Tanzania	Public
UMEME	Uganda	Private
WESCO[a]	India-Odisha	Private

Source: Rethinking Power Sector Reform Observatory.

a. The Western Electricity Supply Company of Orissa Ltd. (Wesco) in India-Odisha was a private utility during the study period, but the state took over the utility in 2016. The Central Electricity Supply Utility of Odisha (CESU) had gone back to the state in 2001 itself and as such is considered a public utility for the study.

b. The utility in Kenya has a 49 percent stake floated on the Nairobi Stock Exchange and 51 percent is owned by government. However, day-to-day decision making is not in government hands for all intents and purposes. The utility is treated as private for this study.

ANNEX 4E. UTILITY RESTRUCTURING INDEX, 2015

Applicable only to the Rethinking Observatory countries

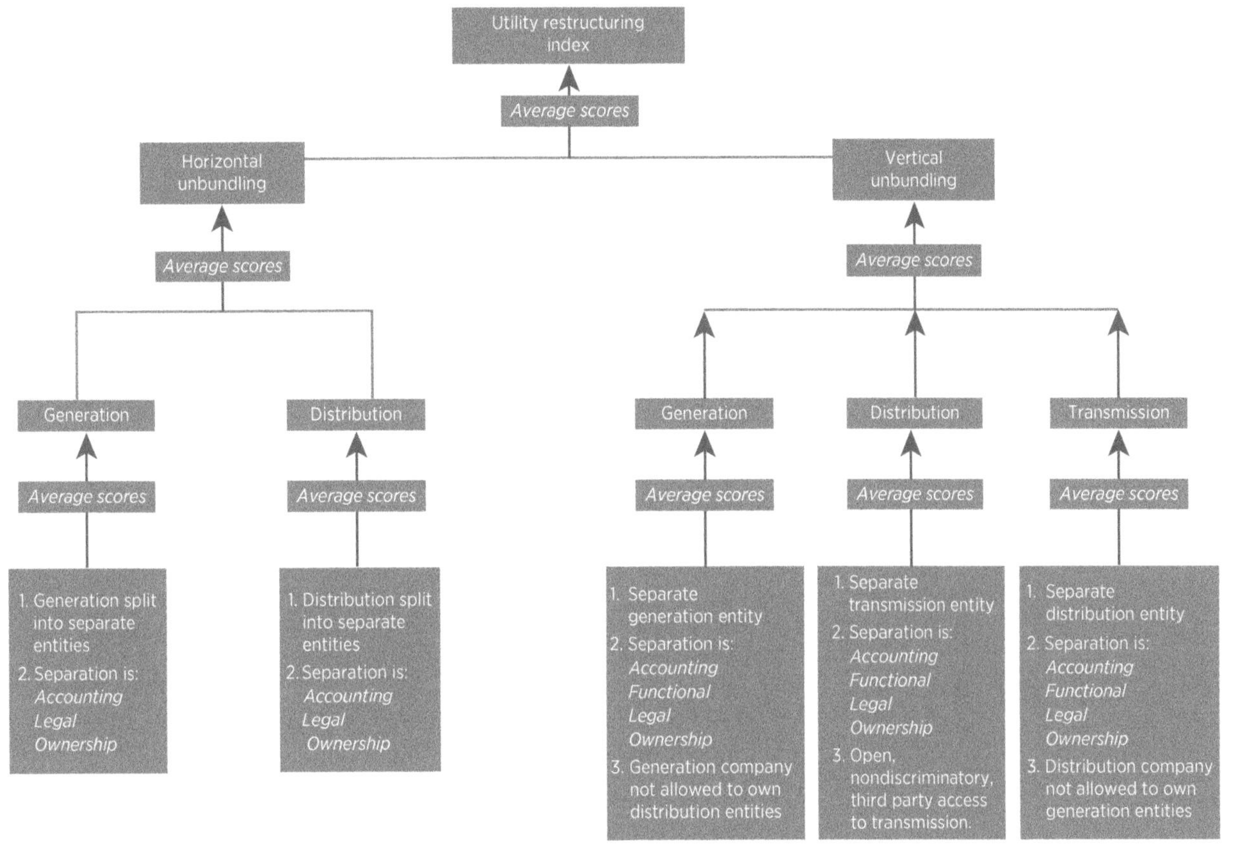

Source: Original figure for this publication.

NOTES

1. This chapter is based on original research conducted by Vivien Foster, Anshul Rana, Joeri de Wit, and Victor Loksha, supported by an advisory team comprising Pedro Antmann, Pedro Sanchez, Elvira Morella, and Mariano Salto.
2. The peak could change substantially, depending on the kind of DERs that are dominant. A significant amount of distributed rooftop solar PV could diminish the typical peak, whereas a significant number of electric vehicles or their charging stations could increase the overall system peak.
3. This example assumes a system load forecast with a daytime peak. Although daytime is a common peak period, as we will see, at the circuit level on distribution systems, the peak may differ.
4. The net load is typically considered in terms of the difference between the expected peak and the amount of renewable resources, but it is in fact the difference between *any* type of customer-sited DER and the expected peak load.
5. Renewable resources can also provide ramping capability to the grid, because they can quickly ramp up to full output or down to zero output (that is, by being curtailed), but only when the sun is shining or the wind is blowing. Thermal resources, by contrast, require output at some minimum level, and so can only be ramped down to a specific level, or nuclear units, which cannot provide any ramping capability. Although solar and wind resources can be curtailed, some common operating procedures can limit the full curtailment of wind resources.
6. On some distribution circuits, the peak usage is driven by consumer production (that is, from rooftop solar).
7. Transmission lines must be operated within particular thermal, voltage, and stability limits.
8. As we shall see, consumer-sited DERs of all types, including storage, could also play this role.
9. For a discussion of some of these challenges, as well as of existing distribution system operator models around the world, see Kufeoglu, Pollitt, and Anaya (2018).
10. However, there are still integration challenges, as well as challenges with cost recovery.

REFERENCES

Bacon, R. 2018. "Taking Stock of the Impact of Power Utility Reform in Developing Countries." Policy Research Working Paper 8460, World Bank, Washington, DC.

Besant-Jones, John E. 2006. "Reforming Power Markets in Developing Countries: What Have We Learned?" Energy and Mining Sector Board Discussion Paper No. 19, World Bank, Washington, DC.

Burger, Scott, Jesse Jenkins, Carlos Batlle, and Ignacio Perez-Arriaga. 2018. "Restructuring Revisited: Competition and Coordination in Electricity Distribution Systems." Working Paper 007, MIT Center for Energy and Environmental Policy Research, Cambridge, MA.

Foster, Vivien, Samantha Witte, Sudeshna Ghosh Banerjee, and Alejandro Moreno. 2017. "Charting the Diffusion of Power Sector Reforms across the Developing World." Policy Research Working Paper 8235, World Bank, Washington, DC.

Gratwick, K., and A. Eberhard. 2008. "Demise of the Standard Model of Power Sector Reform and the Emergence of Hybrid Power Markets." *Energy Policy* 36 (10): 3948–60.

IEA WIND Task 25. 2013. "Design and Operation of Power Systems with Large Amounts of Wind Power." Final summary report, phase three, VTT Technical Research Centre of Finland Ltd.

Kufeoglu, Sinan, Michael Pollitt, and Karim Anaya. 2018. "Electric Power Distribution in the World: Today and Tomorrow." Cambridge Working Paper in Economics 1846, University of Cambridge.

OECD (Organisation for Economic Co-operation and Development). 2015. *OECD Guidelines on Corporate Governance of State-Owned Enterprises.* Vienna: OECD.

Pardina, M. R., and J. Schiro. 2018. "Taking Stock of Economic Regulation of Power Utilities in the Developing World." Policy Research Working Paper 8461, World Bank, Washington, DC.

Parish, D. 2006. *Evaluation of the Power Sector Operations in Pakistan.* Manila: Asian Development Bank.

Planning Commission, Pakistan. 1970. *The Fourth Five Year Plan (1970–75).* Islamabad: Planning Commission.

Pollitt, M. 2008. "The Arguments for and against Ownership Unbundling of Energy Transmission." *Energy Policy* 36 (2): 704–71.

Rudnick, H., and C. Velasquez. 2018. "Taking Stock of Wholesale Power Markets in Developing Countries." Policy Research Working Paper 8519, World Bank, Washington, DC.

Steinbuks, J., J. de Wit, A. Kochnakyan, and V. Foster. 2017. Forecasting Electricity Demand: An Aid for Practitioners." Live Wire Knowledge Note Series No. 73/2017, World Bank, Washington, DC.

Vagliasindi, M. 2012. "Power Market Structure and Performance." Policy Research Working Paper 6123, World Bank, Washington, DC.

World Bank. 2014. *Corporate Governance of State-Owned Enterprises: A Toolkit.* Washington, DC: World Bank Group.

What Has the Private Sector Contributed?

5

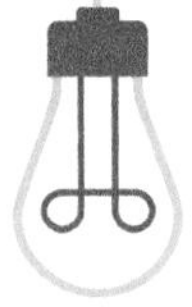

Guiding questions

- *How far did private sector participation (PSP) evolve in developing country power sectors?*
- *What form did PSP predominantly take? What were the main implementation challenges?*

Summary

- *A broad cross-section of countries introduced some degree of PSP in electricity, using various modalities and relying primarily on foreign investment.*
- *The private sector has contributed greatly to the expansion of power generation capacity in the developing world, even as public investment continues to play an important role. Nevertheless, governance issues have surfaced when capacity is not procured in a competitive fashion. Moreover, governments have struggled to strike the right balance in allocating public and private risk in contracts governing independent power projects.*
- *Although the reforms of the 1990s did not focus on privatizing transmission, Latin America and Asia had noteworthy and broadly positive experiences with PSP for power grids.*
- *PSP in power distribution was quite widely adopted among the early generation of reformers. Since the early 2000s, however, PSP in distribution has experienced notable setbacks and uptake has largely declined.*
- *Distribution privatization raises delicate challenges because the interests of key stakeholders are involved. Labor unions are anxious about potential layoffs, while customers are sensitive about the prospect of tariff hikes, as well as eager to see improvements in service quality and coverage.*

INTRODUCTION

This chapter evaluates the extent to which the private sector has played its envisaged role in power sector reform across the developing world.[1] It examines how countries went about incorporating the private sector into the electricity supply chain, particularly in generation and distribution. Was the introduction of private sector participation (PSP) feasible in developing country power sectors? What were the challenges? What form did PSP predominantly take? Did the experience of private sector ownership and operation function as originally envisaged?

PSP was a pillar of the 1990s power sector reform model (World Bank 1993). Many of its measures were designed to encourage PSP. Power utilities were unbundled in an effort to create an industry structure that isolated the natural monopolies (notably transmission and distribution [T&D]) in the supply chain, leaving intact the potentially competitive elements, notably generation and retail. PSP could then be introduced wherever relevant, thereby sidestepping the transfer of national monopolies from public to private hands. Similarly, regulatory entities were considered as important precursors for PSP because they can build on the commercial orientation of private operators. This commercial behavior works in favor of the public interest by ensuring adequate control over market power in monopoly segments—and by licensing market entry and encouraging competition. At the same time, the endpoint of the 1990s reform model—a wholesale competitive power market—was largely premised on the existence of multiple private sector players in the sector.

The private sector was expected to bring a more robust commercial orientation, which would bring a turnaround in sector performance. The state-owned enterprises (SOEs) prevalent in the 1990s power sector lacked clear commercial incentives (Bacon 1995). Enterprises tended to focus on social and political objectives, from affordable access to electricity for key electoral constituencies to furnishing patronage jobs. Remuneration to directors and managers was not linked to operational efficiency. Moreover, management seldom operated under budget constraints, accustomed as it was to periodic bailouts. The advent of private sector management and ownership greatly simplified matters. Utilities could focus on the bottom line, a focus that created strong incentives to cut costs and boost revenues. At the same time, oversight of output and service quality would safeguard the pursuit of legitimate public interests.

The private sector was also touted as a source of potential investment in the power sector. Investment by SOEs was often constrained by their weak balance sheets—constraints that prevented many of them from raising financing from commercial banks, capital markets, or internal cash generation. State-owned utilities were therefore reliant on loans and grants from the central government, often concessional in nature and originating with international donors. This dependence was particularly true in times of fiscal austerity, and it inhibited the timely expansion of the sector. Even where public utility finances were strong, a state guarantee was generally required to underwrite commercial borrowing, leaving the utility exposed to limits on public sector borrowing. Transfer of responsibility to the private sector was expected to ease these restrictions by allowing finance to be raised directly from the markets without fiscal checks, as long as private operators were able to improve the financial performance of utilities. Although the 1990s model was neutral about domestic or foreign investment in privatization, the advent of foreign participation opened up new reservoirs of international capital.

It was recommended that PSP be introduced first in distribution and then in generation. Power distribution and retailing are the cash cows of the electricity supply chain, capturing resources directly from customers that are then used to purchase transmission and generation services from upstream providers. The entire financial basis of the sector is undermined if these fundamental revenue-capture functions fail to function, which is why the power sector reform paradigm has reforms start at the distribution end. Once revenue capture was functioning efficiently under private management, the bankability of upstream investments in generation would be greatly enhanced.

TABLE 5.1 **Forms of private sector participation**

	Operations (operational risk)	Revenue (commercial risk)	Investment (investment risk)	Ownership (asset risk)
Service/management contract	Private sector	Public utility (management fee)	Government	Government
Lease contract/ distribution franchise	Private sector	Private sector (tariff revenues)	Government	Government
Concession/build, operate, transfer contract	Private sector	Private sector (tariff revenues)	Private sector	Government
Full privatization or divestiture	Private sector	Private sector (tariff revenues)	Private sector	Private sector

Source: World Bank elaboration.

A number of different PSP modalities can be placed along a spectrum according to the extent to which responsibilities are transferred to the private sector (table 5.1). At one extreme, a management contract temporarily delegates managerial responsibility for a utility from the public to the private sector for a period of two to three years. The management contractor is remunerated directly by the government according to a fixed fee that is sometimes performance related but does not depend on utility revenues. A further step would be a lease contract, where delegation of management is typically for a longer period of 5–10 years. The lease contractor takes full responsibility for operating the utility and depends on revenue collections for remuneration, paying a predetermined share of these revenues to the government as a lease to contribute to the financing of investments, which remain the responsibility of a state-owned holding company. A still-further step is a concession, whereby the private sector retains the entire sector revenue stream and assumes all responsibility for investment, entailing a much longer contract duration of at least 20 years. Whereas concessions are typically used to take over existing infrastructure, an important variation for the case of greenfield infrastructure is the use of build, operate, transfer (BOT) or equivalent contractual arrangements, whereby the private sector develops new infrastructure assets on a project-finance basis, on the basis of ring-fencing of the resulting revenues for an extended period of at least 20 years. The most complete and supposedly permanent form of privatization is asset sale or divestiture of either a minority or majority stake in the company. It may be done through a direct sale or auction, or via a stock market flotation.

In the case of the power sector, the deeper forms of PSP were believed to be the most relevant. For many areas of infrastructure, only some of the weaker forms of PSP proved to be feasible. Among water utilities, for instance, management, lease, and concession contracts are prevalent, whereas divestiture has been relatively unusual. In the transport sector, concessions (for ports, airports, and railways) and BOT contracts (for toll roads) are common. In the case of the power sector, the stronger characteristics of private electric service, combined with its potentially lucrative nature, meant that deeper forms of PSP, such as divestiture, appeared promising (Bacon 1999; Sen 2014).

KEY FINDINGS

Drawing on the World Bank's Private Participation in Infrastructure (PPI) database for broad global trends, and on the 15 countries of the Power Sector Reform Observatory for more detailed case experience, the main findings of this chapter can be summarized as follows.[2]

Finding #1: A broad cross-section of countries introduced a degree of PSP in electricity, using various modalities and relying primarily on foreign investment

Although prevalent in generation and distribution, PSP was comparatively rare in the transmission of power. At the global level, almost two out of every three developing countries implemented at least one transaction for PSP in power generation and electricity distribution (table 5.2). By contrast, private participation in the transmission segment occurred in only 15 percent of the observatory countries. Notwithstanding its prevalence in generation and distribution, private sector engagement varies greatly across the 15 observatory countries. In India, Pakistan, Senegal, and Tajikistan, PSP in distribution is far from the norm. The Philippines, Uganda, and Ukraine, by contrast, have privatized nearly all of their distribution utilities (table 5.3).

The preferred PSP modality varies along the electricity supply chain. In the generation segment, PSP comprised a mixture of divestiture of existing assets and independent power producers (IPPs) for development of new capacity. The majority of developing countries opened their markets to IPPs, making this among the most popular power sector reforms and often the first to be undertaken. IPPs for the construction of new generation plants were often, though not always, combined with divestiture of existing generation plants, although minority government stakes were often retained. In some of the more advanced markets of the observatory sample—such as Colombia, India, Peru, and the Philippines—there was also some entry of merchant plants into the generation market without the need for a public sector offtaker. In the distribution segment, PSP predominantly took the form of asset sales. For the most part, distribution privatizations were undertaken either in the same time frame as generation privatizations or (often) some years later.

In Sub-Saharan Africa, electricity divestitures were relatively rare, and PSP was skewed toward contract-based modalities. Although the overall percentage of countries adopting any form of PSP in Sub-Saharan Africa was similar to other developing regions, the modalities adopted were different. Divestitures were comparatively rare, with PSP in generation primarily taking the form of IPPs, and in

TABLE 5.2 Preferred private sector participation modality varies across regions/electricity supply chain (global)

Countries implementing (%) 1990–2016	Sub-Saharan Africa			Other developing countries		
	Generation	Transmission	Distribution	Generation	Transmission	Distribution
Management contract	3	0	34	5	1	10
Lease contract	0	0	3	4	7	3
Concession	20	0	26	29	5	16
BOT contract	46	3	3	59	8	5
Divestiture	14	0	9	38	7	51
Any of the above	63	3	66	75	16	64

Source: Based on the Private Participation in Infrastructure database 2018 (https://ppi.worldbank.org/).
Note: BOT = build, operate, transfer.

TABLE 5.3 **Private sector engagement across the 15 observatory countries**

	Generation			Distribution		
Country/State	Years	Market share (%)	Modality	Years	Market share (%)	Modality
Colombia	1993–2017	71	Extensive divestiture, numerous IPPs, and a few merchant plants	1996–2010	50	Mainly divestitures, plus one concession
Dominican Republic	1994–2017	61	Extensive divestiture, numerous IPPs, and a few merchant plants	1994–99	2	Partial divestitures that have been reversed
Egypt, Arab Rep.	1999–2017	6	Handful of IPPs	n.a.	0	n.a.
India - Andhra Pradesh	1995–2017	39	Numerous IPPs and a few merchant plants	n.a.	0	n.a.
India - Odisha	1998–2010	66	One partial divestiture and numerous IPPs	1999	0	Four instances of full divestiture all have been reversed
India - Rajasthan	2004–17	39	Numerous IPPs	2016	3	Partial divestiture as pilot projects in some areas
Kenya	1996–2014	31	One partial divestiture and numerous IPPs	2006	50	One partial divestiture with controlling government stake
Morocco	1997–2017	39	Numerous IPPs heading toward majority power generation share	1997–2001	33	Four urban concessions
Pakistan	1992–2017	47	Some partial divestiture and numerous IPPs	2005–16	16	One partial divestiture and one concession
Peru	1995–2017	83	Extensive partial and one full divestiture, numerous IPPs, and a few merchant plants	1994–2015	50	Several partial and full divestitures including some reversals, plus concessions for transmission
Philippines	1991–2017	90	Extensive full divestiture, numerous IPPs, and a few merchant plants	1990–2009	67	Extensive partial divestiture in distribution, and concession in transmission
Senegal	1997–2017	46	Several IPPs and a merchant plant	1999–2010	2	Reversed partial divestiture, and several rural concessions
Tajikistan	2006–08	17	Several IPPs	2002	5	One regional concession
Tanzania	1994–2011	20	Several IPPs	2002	0	n.a.
Uganda	2003–17	63	Numerous IPPs	2003–05	90	Several concession contracts
Ukraine	2002–17	14	Several partial and one full divestiture and numerous IPPs	1998–2010	74	Extensive full and partial divestiture in distribution and transmission
Vietnam	1996–2017	34	Several partial divestitures and numerous IPPs	n.a.	0	n.a.

Sources: Based on World Bank—PPIAF (Private Participation in Infrastructure database 2018, https://ppi.worldbank.org/) and Rethinking Power Sector Reform utility database 2015.
Note: IPP = independent power producer; n.a. = not applicable.

distribution focusing on concessions and management contracts.

In most regions of the world, foreign sponsors have been an important source of private investment in electricity generation and distribution.[3] Across Latin America and the Caribbean, East Asia and the Pacific, and Europe and Central Asia the relative contributions of domestic and foreign investors look to be evenly balanced (figure 5.1). South Asia stands out as having about three-quarters of private investment domestically sourced, which largely reflects the development of the domestic power sector and the

FIGURE 5.1 Private investment in electricity came predominantly from foreign sources and mostly in the generation sector

Private investment in power sector by region and source

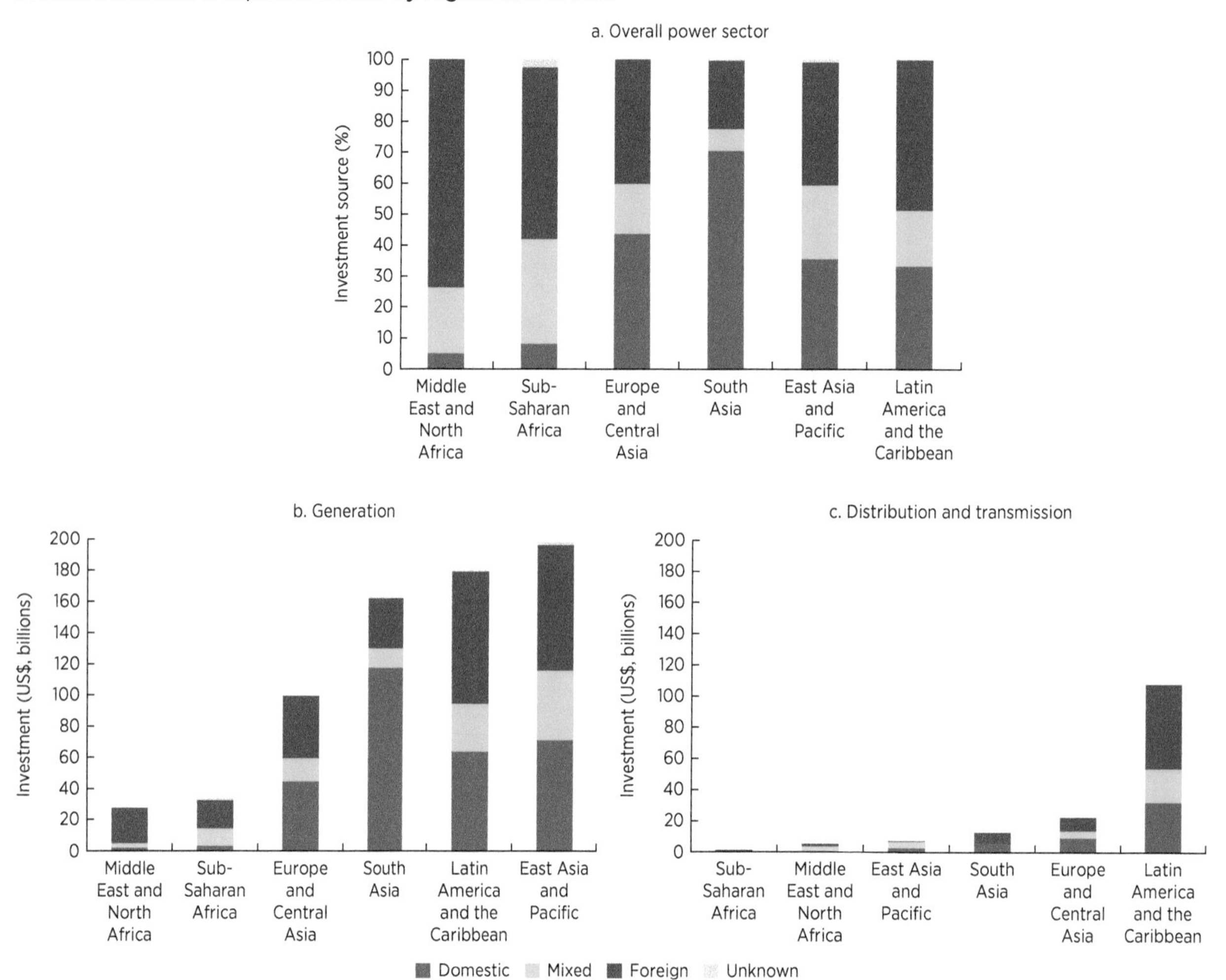

Source: Based on Private Participation in Infrastructure database 2018 (https://ppi.worldbank.org/).

depth of local capital markets in India. By contrast, in Sub-Saharan Africa, as well as the Middle East and North Africa, about three-quarters of private investment in electricity comes from foreign sources. These differences have significant implications in terms of the political economy of privatization, as well as sensitivity to foreign exchange risks. They may also affect government attitudes toward risk allocation with sponsors.

Finding #2: Working alongside public investment in capacity, the private sector has made substantial contributions toward the expansion of power generation capacity in the developing world

IPPs have had a big, albeit geographically concentrated, impact on developing country power systems. Since 1990, IPPs have attracted a cumulative total of almost

US$600 billion of investment to the developing world,[4] leading to the construction of almost 850 gigawatts (GW) of new generation capacity.[5] Over 80 percent of IPP investment was captured by just three developing regions—East Asia and the Pacific, Latin America and the Caribbean, and South Asia—with the top five countries (India, Brazil, China, Turkey, and Indonesia) accounting for nearly 60 percent of all investments in IPPs (figure 5.1).

IPP investment has been cyclical, reflecting movements in the largest markets. Following a relatively slow start over the decade from 1992 to 2002, IPPs underwent a particularly rapid expansion during the subsequent decade (2002–12), peaking at close to US$70 billion in 2012, and subsequently settling at a level of about US$30 billion per year (figure 5.2). The steep decline witnessed in 2013 is closely linked with the cessation of two of the world's largest IPP programs, first in India and later in Brazil. In India, regulatory setbacks were to blame, whereas Brazil was beset with political and economic crises.

IPPs typically coexist with continued public sector investment in power generation. Overall, IPPs contributed about 42 percent of the expansion in generation capacity in the developing world during the period 1990–2016,[6] with the balance continuing to be developed as public generation projects.[7] In just about half of the countries, the private sector contributed more than half of the expansion over the period (figure 5.3). Only in about a quarter of the countries did the private sector contribute more than 75 percent of new capacity additions in generation. The most salient examples were Cambodia, Georgia, Peru, and the Philippines, where the private sector contribution to capacity additions exceeded 90 percent.

The contribution of the private sector to new generation capacity shows some regional and income-group variations. Investment was highly skewed across income groups. Of the new capacity funded by the private sector during 1990–2016, almost 80 percent went to upper-middle-income countries; less than 1 percent went to low-income countries. Nevertheless, taking into account the varying sizes of power systems across these income groups, the contribution of the private sector to capacity expansions in each of these income

FIGURE 5.2 Private investment in independent power producers has been substantial though subject to fluctuations and concentrated in a few countries, 1990–2017

Private investment in independent power producers, by region and top five markets

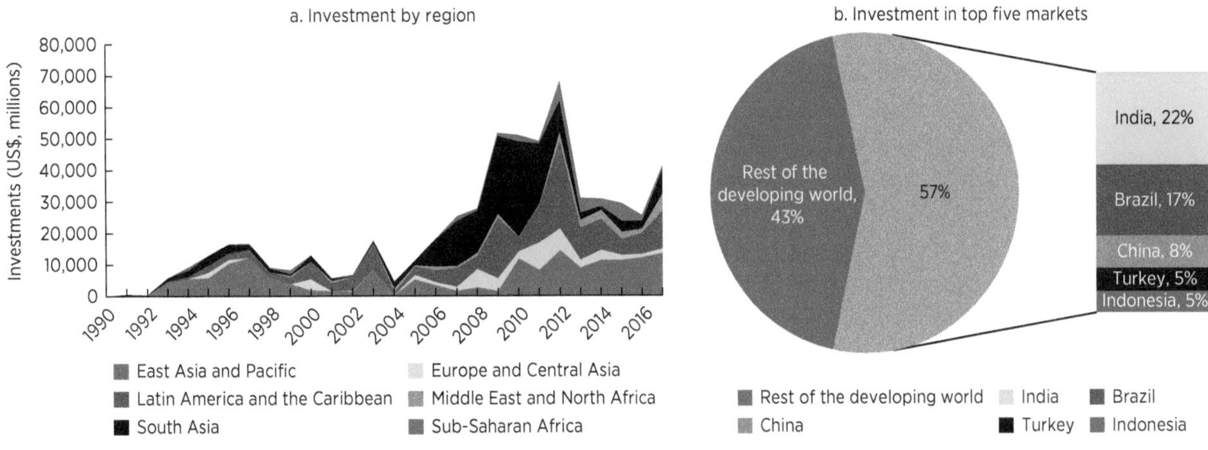

Source: Based on Private Participation in Infrastructure database 2018 (https://ppi.worldbank.org/).

FIGURE 5.3 Most countries still depend on a combination of public and private investment for the development of new power generation capacity

Countries with private projects as a percentage of total capacity additions

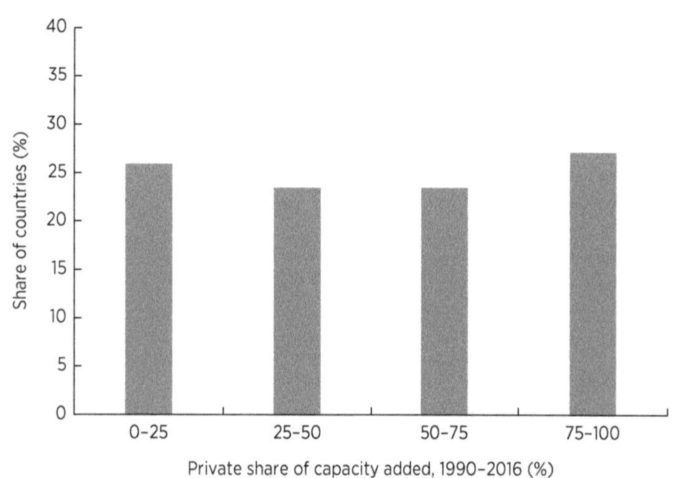

Source: Based on UDI World Electric Power Plants Database 2017.
Note: Considers only greenfield power projects and ignores any divestiture and consequent change of ownership.

groups was comparatively similar (figure 5.4, panel a). Differences were somewhat more pronounced by geography. Across the Middle East and North Africa, the private sector represented just over 20 percent of investment in new generation capacity. Interestingly, the share of private investment in new generation capacity in Sub-Saharan Africa was almost double that in the Middle East and North Africa. Although the absolute amount of private investment in generation capacity in Sub-Saharan Africa was small, at US$30 billion, it nonetheless represents a substantial share of the region's investment needs. For other developing regions, the private investment share was somewhat higher, in the 40–60 percent range (figure 5.4, panel b).[8]

Plant technology and country risk rating appeared to be much stronger drivers behind private sector investment in generation. Whereas coal was the dominant technology for IPP investments up to 2010, the balance has subsequently shifted strongly toward nonconventional renewable energy sources (figure 5.4, panel c, and figure 5.5). About 68 percent of the total wind and 78 percent of the solar capacity that was added in the period 1990–2016 was funded by the private sector compared to 45 percent of thermal capacity. Nonconventional renewable energy projects particularly lend themselves to private sector investment because of their relatively small scale and modular design, as well as their comparatively short and low-risk construction period. A second stronger driver of the share of private sector investment in new generation capacity is the country risk rating (figure 5.4, panel d). The share of new capacity contributed by the private sector is almost negligible in countries that are in default, speculative, or unrated. Countries that are rated Ba3[9] or above do significantly better at capturing private investment.

In some countries, public investments in generation are supported by large Chinese financing packages. For example, in Pakistan, an intergovernmental financing package as part of the China–Pakistan Economic Corridor project is providing approximately US$35 billion to support energy projects, including coal-fired, hydro, and solar plants totaling 17 GW of capacity. In East Africa, also, several projects are being developed with Chinese support, such as the Kinyerezi III and IV gas-to-power projects in Tanzania and the Karuma and Isimba hydropower projects in Uganda.

Finding #3: Governance challenges emerge when power generation capacity is not procured in a competitive fashion

IPPs have not always been procured by competitive means, raising concerns about transparency and value for money. IPP projects are fairly standard, as are the number of market participants. As a result, IPPs lend themselves to competitive procurement.

FIGURE 5.4 **Private investment shares reveal different drivers**

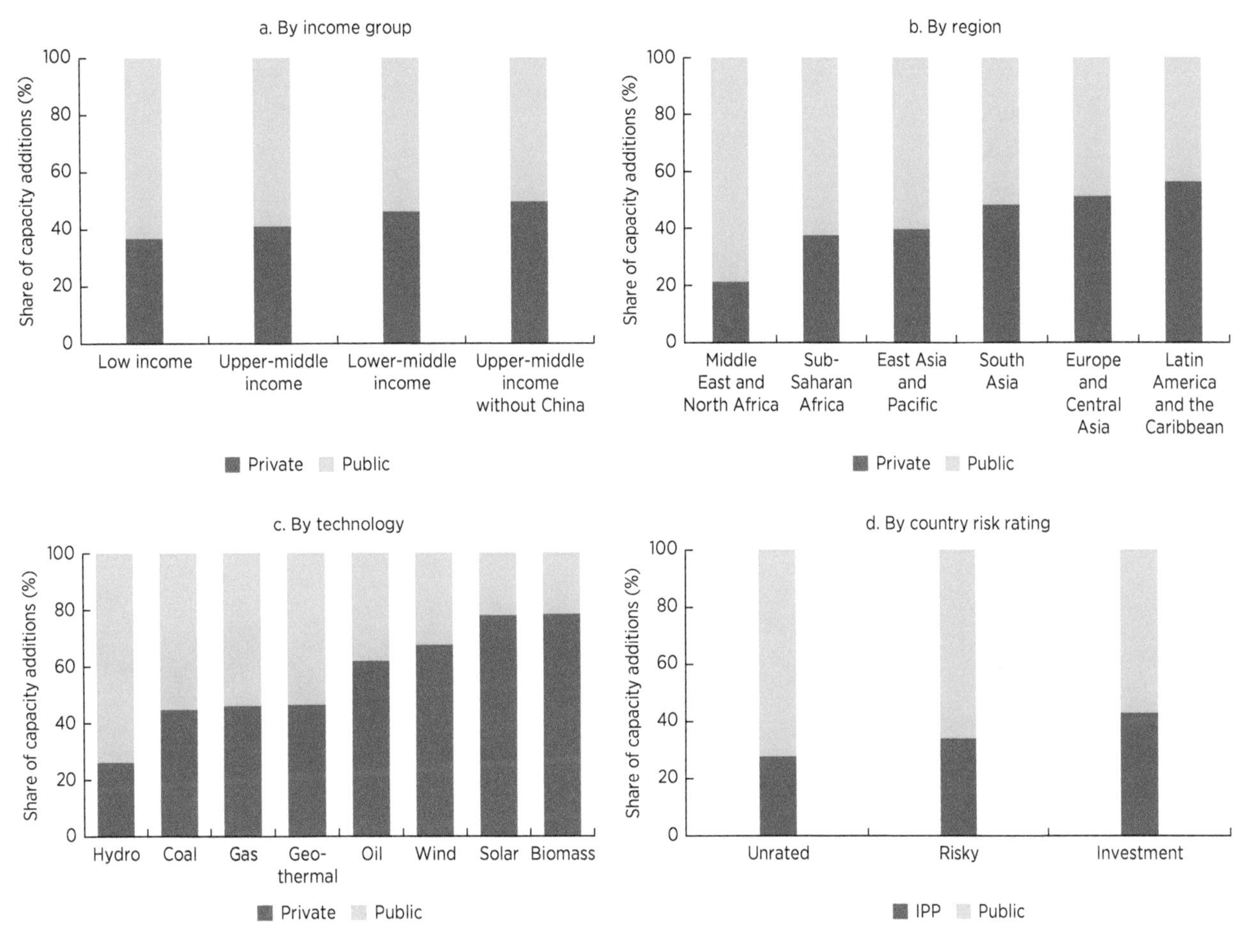

Sources: World Bank elaboration based on Moody's sovereign ratings and UDI World Electric Power Plants Database 2017.
Note: IPP = independent power producer.

Competition makes transactions relatively open and transparent, while competitive pressures help to keep costs down. Latin America has embraced competitive bidding, and several countries—including Brazil, Chile, and Peru—have established regular public auctions for contracting new generation capacity (figure 5.6). Nevertheless, across Africa and Asia, it remains commonplace for IPPs to arise as unsolicited bids or to be procured through direct negotiation. Exact figures are hard to come by because information on procurement methods is often missing from global databases; however, from 2005 to 2017 direct negotiation was as prevalent as competitive bidding in the developing world overall. For both South and East Asia, though, the number of reported transactions based on direct negotiation exceeds those reported as competitive procurements.

The case of Tanzania illustrates the difficulties that can arise when countries rely on direct negotiation of IPP contracts; especially

FIGURE 5.5 Independent power producer investment is moving toward cleaner sources of energy, although capacity expansion lags

IPP investment and capacity expansion, by technology

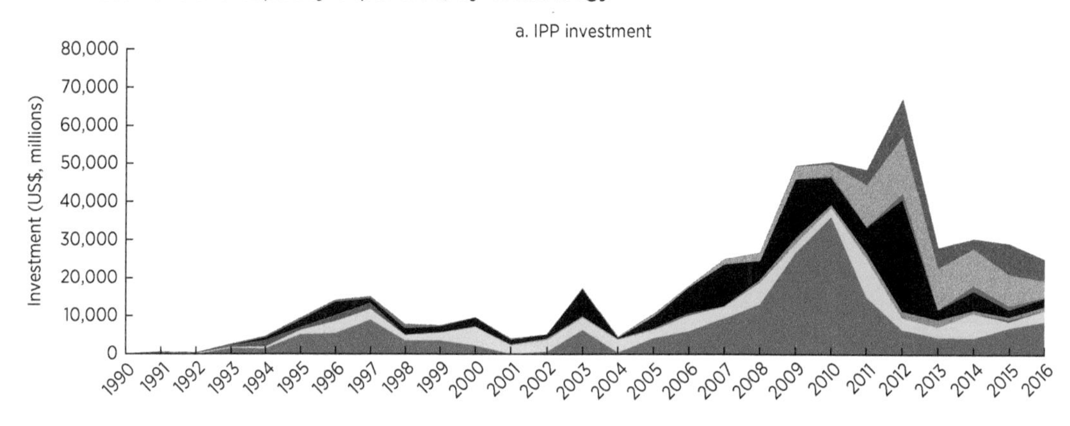

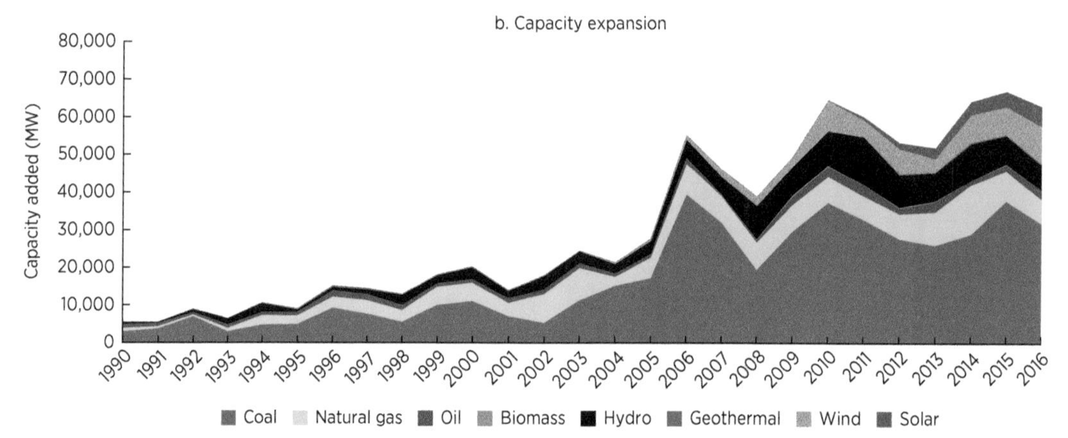

Sources: Panel a based on Private Participation in Infrastructure database 2018 (https://ppi.worldbank.org/); panel b based on UDI World Electric Power Plants Database 2017.
Note: IPP = independent power producer; MW = megawatt.

FIGURE 5.6 Competitive procurement of independent power producers varies widely between regions

Independent power producer procurement, by region, 2005–15

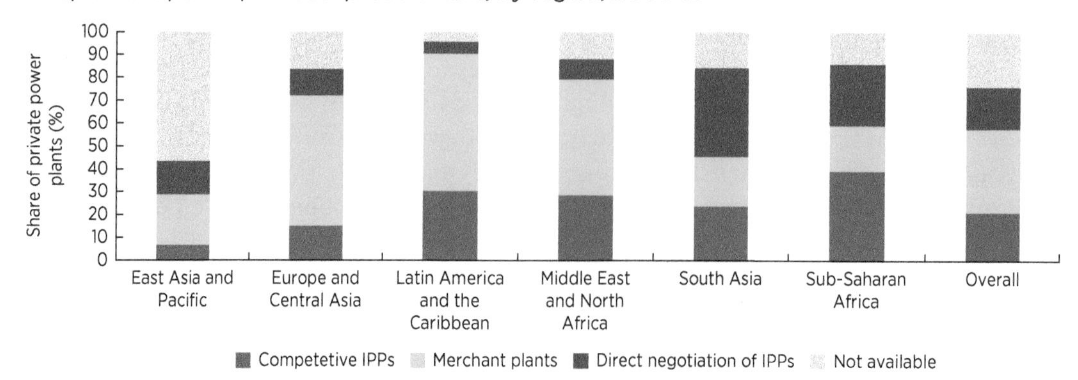

Source: Based on Private Participation in Infrastructure database 2018 (https://ppi.worldbank.org/).
Note: IPP = independent power producer.

when deals are struck during power shortages. In particular, the Independent Power Tanzania Limited (IPTL) diesel plant negotiated in 1994–95 was exceptionally costly. Negotiated just after a drought-related load shedding created a sense of emergency, the IPTL signed a power purchase agreement (PPA) with a tariff of US$0.31 per kilowatt-hour (kWh). Worse, it became embroiled in corruption charges. Lengthy court battles ensued, along with arbitration proceedings between the government and IPTL. During another drought-related crisis not long afterward, in 2006, a contract for another diesel plant was directly awarded to Richmond Development Company, an enterprise lacking any relevant power sector experience. Resulting delays meant the plant was not completed until the power shortage was over. Meanwhile, the investigation of corruption by a select committee appointed by the parliament caused the prime minister and the minister of energy and mines to resign. Tanzania adopted a new policy that allowed the public utility to play a leading role in the development of generation projects through a public–private partnership model.

Uganda faced repeated challenges while implementing the Bujagali hydropower project as an IPP, which eventually led to a policy of greater public sector involvement in power generation. In Uganda, the 1999 legal framework required all new generation to be provided by IPPs. The first of these IPPs was the 250 megawatt (MW) Bujagali I hydropower plant in 1998, seen as critical to relieving power shortages in the country. The associated direct procurement process lacked transparency. Allegations of corruption led to the cancellation of the project in 2003. The following year, the project was retendered on a competitive basis as Bujagali II. Nevertheless, the associated PPA tariff was relatively high at US$0.10/kwh. As a result end-user tariffs were hiked. Bujagali tariffs were indexed to debt repayment and given a backloaded payment structure, portending a substantial hike in the PPA tariff. This hike was avoided through a refinancing process that spread the repayments of the final 5 years over a period of 15. The challenges associated with both Bujagali I and II led the government to adopt a new policy on generation in 2010, requiring the public sector to participate in all generation projects exceeding 25 MW in capacity. Uganda awarded two large hydro projects—Karuma (600 MW) and Isimba (183 MW)— to Chinese developers through direct negotiation.

Finding #4: Governments have found it challenging to strike the right balance in allocating risk between the public and private sectors in contracts governing IPPs

The allocation of risk between the public and private sectors is a critical issue under IPP contracts. The key issue in the design of IPP contracts is finding the appropriate allocation of risk between the government and the private investor. IPPs face a plethora of risks, including risks regarding demand, fuel price, exchange rate, and termination. Such risks can weaken investor interest, particularly in untested markets, until a reliable track record has been established. In response, governments may provide contractual protections of various kinds. Oil price and currency fluctuations, for instance, may be passed through directly in the PPA tariff. "Take-or-pay" clauses may guarantee purchase of power even in the absence of demand, or capacity charges may at least ensure that fixed capital costs can be covered. Sovereign guarantees may be provided to compensate investors in case of premature termination.

At one end of the spectrum, IPP programs have sometimes stalled when private sector demands for risk mitigation are not matched by the willingness of governments to provide them. The public sector's retention of risks creates contingent liabilities for the state, which needs to be appropriately accounted for and monitored by the finance ministry. Where governments have been reluctant to retain risk, the private sector has tended to stay away. The experiences of the Arab Republic of Egypt and Vietnam are illustrative.

Following a financial crisis in the early 2000s, Egypt redesigned its IPP program to shift more risk to the private sector, a move that led to a loss of investor interest. Egypt's original IPP program, launched in 1996, had provided incentives that ranged from tax exemptions and full repatriation of profits to take-or-pay provisions and a sovereign guarantee for payments. The competitive procurement process attracted a lot of interest, and three dollar-denominated PPAs were signed, eliminating currency risk for the developer. The 2001–03 currency devaluation, however, created significant liabilities for the offtaker, doubling its bill as the Egyptian pound crashed. The experience prompted a major rethinking of the IPP program, followed by a decision to shift more risk to the private sector. The new IPP program required developers to source all foreign currency from foreign banks[10] and all local costs to be paid in the local currency. More important, the new framework favored bids that had larger shares of equity financing and more local investment (all earlier IPPs had no local partners). The proportion of the electricity output to be covered by the take-or-pay clause was also reduced. These changes inhibited further foreign investment, and the program was scrapped in 2003 as new funding streams from various donors became available.

In Vietnam, further expansion of the 2007–08 IPP program has been affected by disagreements over acceptable risk allocations. International investors argue that they require dollar convertibility of profits for repatriation, as well as sovereign guarantees to cover offtaker payment risk and the risk of early termination. The government has been reluctant to provide this level of protection, concerned about the contingent liabilities and their impact on the achievement of public debt limits set as part of macroeconomic policy. Furthermore, without a standardized international contract template that meets standards of bankability, terms must be negotiated on a case-by-case basis for each project, a process that can sometimes take several years to conclude. In contrast, Vietnam's program for private investment in smaller-scale renewable energy projects does not require government support, owing to lower concerns about risk among the domestic investors at whom it is targeted. Nevertheless, these factors have limited the government's capacity to support the larger-scale projects the country needs.

At the other end of the spectrum, when governments have assumed excessive risk, IPP programs have occasionally triggered financial crises. Where IPP programs are large and private investors are heavily protected against risks, governments may find themselves heavily exposed to unpredictable events such as exchange-rate devaluations, oil price hikes, recessions leading to declining demand, or civil strife resulting in contract termination. Given the economic weight of the sector in many developing countries, a power sector crisis can rapidly translate into a macrofiscal crisis, as happened in Pakistan and the Philippines.

Protections offered to IPP investors in Pakistan left the power sector highly exposed to an exchange-rate devaluation, combined with an oil price hike. The opening up of the market to IPPs in 1994 rapidly attracted 4,500 MW of additional power generation capacity. Private investors were offered an attractive package, with a guaranteed fixed rate of return and remuneration indexed to the U.S. dollar and the U.S. inflation rate, as well as numerous tax exemptions and an exoneration from fuel purchase responsibilities, which remained with the government. The policy fixed PPA tariffs on a "cost plus" basis and gave developers complete freedom to choose the fuel. As a result, most IPPs used heavy oil, which made Pakistan's fuel mix one of the most expensive in the region. Almost immediately the cost of oil started rising, increasing the cost of purchasing electricity for the utility offtaker. The major currency depreciation in 1998 led to another unaffordable hike in PPA tariffs, leading to nonpayment by the utility and the accumulation of US$1.6 billion of arrears to the IPPs by the end of 1998. The situation was further exacerbated as oil prices continued to rise, doubling by 2004, and increasing the financial burden on the government. Although the IPP program led to a rapid increase in power supply—even creating excess supply for a while—this financial crisis, triggered by default on IPP contracts, gave birth to the circular debt crisis that bedevils the sector to this day. The crisis is described as "circular debt" because customers fall behind on their power bills, and government on its subsidy payments, so distribution utilities cannot honor their power purchase bills with the single buyer, who cannot keep up with payments to generators, who in turn do not pay their fuel suppliers, almost all of whom are government entities. Hence, "circular debt."

The Philippines' IPP program initially succeeded in averting a power supply crisis, but it later amplified a macroeconomic crisis caused by a currency devaluation. Power shortages in 1993 prompted the government to sign 42 PPAs with IPPs, mostly through direct negotiation. In these contracts, the government provided generous protection to the private sector, assuming commercial risks (through take-or-pay clauses), fuel price and exchange-rate risks (through pass-through mechanisms), termination risk (through sovereign guarantees), and fiscal risks (by offering tax exemptions). Although the program expanded capacity and restored supply–demand balance in the country, the East Asian financial crisis of 1997 hit the Philippines hard. The currency devaluation and a slowdown of demand left the National Power Corporation with huge liabilities on the take-or-pay of surplus power as well as compensation for the currency devaluation. By 2001 the sector accounted for 25 percent of the national debt. The government responded by renegotiating 20 of the IPP contracts and canceling 7. Wholesale reform of the power sector followed.

Finding #5: Although privatizing transmission was not a focus of the 1990s' reforms, Latin America and Asia gained a lot of experience in the segment

The most widely used modality for PSP in transmission has been the BOT contract, analogous to those widely used in the generation sector.[11] Privatization efforts in the 1990s focused primarily on the generation and distribution sectors. Because it is a public good and has a role in the coordination of electricity dispatch, the transmission grid was considered less amenable to PSP. Nevertheless, a number of countries experimented with this approach under a variety of modalities (table 5.4). Of these, the most widely adopted has been the BOT contract, which is equivalent to an IPP arrangement for transmission. Popular in Latin America, these contracts entail the payment of a fee for the availability of capacity covering both operating and maintenance costs, as well as an annuity for capital investment. In other instances, the entire national transmission grid has been privatized

TABLE 5.4 Transmission: A few observatory countries have embraced private participation

Country	Year	Modality	Outcomes
Colombia	2000–07 2015–20	Partial divestiture BOOT	31% private shareholders; ISA, the new transmission company, owns 70% of national transmission system 16 projects under a 25-year BOOT system to be awarded with a total investment of US$1.5 billion
India	2002–17	BOT	Invested US$8.6 billion since 2002 and currently represents 15% of capacity expansion
Peru	1998–2015	30-year BOOTs 30-year concession	Invested US$2.6 billion; all new capacity since 1998 added through BOOTs
Philippines	2009	25-year grid concession	Invested in over 600 circuit kilometers of lines, and consistently met targets for availability and losses

Sources: Based on Rethinking Power Sector Reform utility database 2018, and Private Participation in Infrastructure database 2018 (https://ppi.worldbank.org/).
Note: BOOT = build, own, operate, transfer; BOT = build, operate, transfer.

under a concession modality, as in the Philippines. Somewhat rarer are permanent divestitures of the national grid, or merchant lines developed at the sole risk of the private investor.

Peru is one of the countries that pioneered PSP in transmission using the BOT mechanism. The transmission sector is almost entirely private and operated by 13 different companies, of which the two largest are Red de Energía del Peru (REP) and Consorcio Transmantaro (CTM) with 40 and 20 percent of the market, respectively. The process began in 1992, when the sector was unbundled into two public transmission companies, which went on to tender 30-year build, own, operate, transfer (BOOT) contracts to the private sector for extension of the grid, while retaining 15 percent public sector stakes. Once these projects had completed the national grid, the remaining public assets were privatized under 30-year concessions. The system was strengthened in 2006 with coordinated generation and transmission planning capability housed within the system operator. In addition, a clear regulatory framework allows the wheeling tariff to be determined through the tendering process with subsequent indexation but no periodic regulatory review. The tender price reflects a combination of operations and maintenance expenditure and a capital annuity set at a 12 percent rate of return. These improved arrangements led to an upswing of US$1.5 billion of private investment during the period 2006–15. Overall, some 6,000 kilometers (km) of transmission lines and associated transformer stations have been developed in Peru through this modality.

The Philippines took another route: an outright concession as part of the reform process initiated through the Electric Power Industry Reform Act (EPIRA) in 2001. But efforts to attract private investors for the transmission network did not begin well. Auctions in 2003 and early 2007 failed over regulatory uncertainty surrounding the transmission company's (TransCo) revenue stream. The Energy Regulatory Commission (ERC) managed to set up a performance-based regulatory framework by 2003 but took some time to fine-tune the revenue cap methodology that set rates for the transmission company. By December 2007, however, several private companies bid for the TransCo, with the National Grid Corporation of Philippines—a group of local and international companies—putting in the winning bid for a 25-year concession. The group invested about US$4.2 billion in the sector—US$1.9 billion of which was invested in physical assets (ESMAP 2015).

Finding #6: Among the first generation of reformers, PSP in distribution was widely adopted and gained considerable traction

Some of the early-reforming countries achieved high levels of private sector penetration in their power distribution sectors. The financial health and operational strength of distribution utilities is a key driver of overall power sector performance. A financially precarious distribution utility can undermine the entire payment chain, and operational weaknesses in the local grid can prevent power from reaching customers even when it is available. For precisely these reasons, the 1990s model prescribed PSP in the distribution tier as one of the first measures that needed to be taken to turn an ailing power sector around. This recommendation is reflected in the surge of PSP in distribution that took place during the 1990s, with an average of 88 transactions per year (1990–99), primarily concessions (and to a lesser extent) divestitures (figure 5.7). Divestiture of distribution utilities was prevalent among early reformers in Latin America and the Caribbean (such as Bolivia, Brazil, Colombia, Dominican Republic, El Salvador, Guatemala, Panama, and Peru) and in Europe and Central Asia (Georgia, the Russian Federation, and Ukraine). From Asia, Malaysia and the Philippines were comparatively unusual in divesting a large share of their distribution sector. Privatization of distribution was also undertaken in a handful of Indian states, although under a concession modality.

Colombia's macroeconomic crisis in the late 1980s was compounded by a blackout in 1994. These crises led to a complete transformation of the power system. Laws 142 and 143 aimed to increase competition and private investment and provide unrestricted access to the grid. The thrust was on restructuring the powerful vertically integrated utilities such as the ones in Bogotá (Empresa de Energía de Bogotá, EEB) and Medellín (Empresas Públicas de Medellín, EPM). The EEB was emblematic of all that ailed the Colombian power sector. The utility's

FIGURE 5.7 In the early years, most private sector participation in distribution was through concessions, primarily in Latin America and the Caribbean

Type of private sector participation in distribution, by region 1990–99

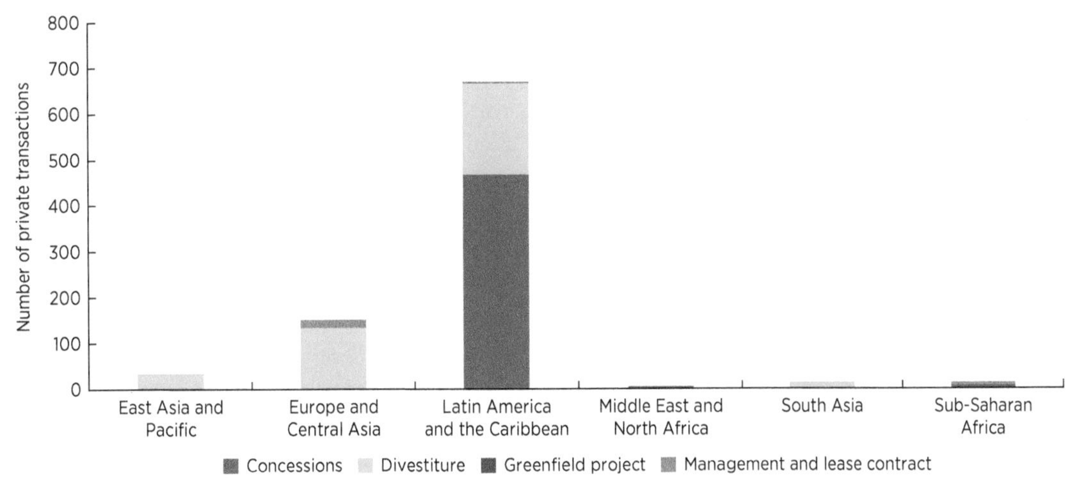

Source: Based on Private Participation in Infrastructure database 2018 (https://ppi.worldbank.org/).

system losses were at 27 percent (1995), and it owed US$900 million to Financiera Eléctrica Nacional (FEN—a national entity set up to finance the energy sector). With the utility unable to reduce its losses and its financial situation worsening, the government decided to restructure and privatize the utility. By 1997, EEB became a holding company (81 percent national government, 7 percent Bogotá government, and the rest owned by various companies) with a generating company (EMGESA), a transmission company (EEB), and a distribution company (Codensa). The Enel group owns 48.5 percent of the distribution company, whereas the rest is owned by Grupo Energía Bogotá (GEB—the government of Bogotá owns 76.28 percent of GEB). With management in private hands, Codensa's system losses fell, reaching a low of 8.9 percent in 2006. The utility's debt was restructured, and the company has maintained a healthy profit margin since 2008. It is considered to be one of the better-run utilities globally.

Ukraine began its power sector reform in 1995 by restructuring the eight vertically integrated utilities in the country. The process created four regional generation utilities and a single transmission company. Distribution assets, in contrast, were divided into 27 utilities (one for each region and one each for Kyiv and Sebastopol). Although a major change in the system, it was limited to legal and accounting separation; ownership remained with the state. The privatization process began in 1998, when shares in five distribution companies were sold by competitive tender to a group of local and international companies called Investment Pool. This process was followed by the sale of six more distribution companies for US$160 million. This early phase of privatization was largely successful; however, privatization of utilities undertaken between 2001 and 2004 was mired in allegations of corruption. At the same time, inadequate regulatory frameworks and governance structures ensured that the sector could not resolve its critical issues, and the privatization process came to a halt.

Finding #7: Since the early 2000s, PSP in distribution has experienced major setbacks and uptake has declined

Despite the strategic importance of PSP in electricity distribution, there has been relatively little interest in it since 2000 (figure 5.8). Further uptake of PSP for distribution utilities tailed off after 2000, with an average of only 15 transactions per year (2000–17) subsequently recorded, largely confined to five countries (Brazil, Bulgaria, India, Russia, and Turkey) that together account for 60 percent of the transactions.

Even among countries undertaking privatization of power distribution utilities, few have privatized the entire distribution sector. Cameroon, Côte d'Ivoire, and Uganda privatized their national distribution utility; and Argentina, Chile, Jordan, Nigeria, and Turkey privatized all of their regional distribution companies. More typically, public and private distribution utilities coexist within the same country, with private operators often serving capital cities or larger commercial centers. For example, in Pakistan, the privatization of Karachi Electric in 2006 was originally intended to be the first in a wider privatization program for distribution utilities, but the plan was later shelved because of opposition generated by the Karachi case.

The decision to privatize only some distribution utilities may reflect differences in the commercial viability of the service areas or variations in the local political authorizing environment. The fact that electricity distribution remains a subnational responsibility in many countries adds another layer of political complexity to the decision to privatize distribution. Moreover, with lucrative commercial and industrial demand often geographically concentrated in larger cities, the financial

FIGURE 5.8 **The rise and fall of private sector participation in electricity distribution, 1990–2017**

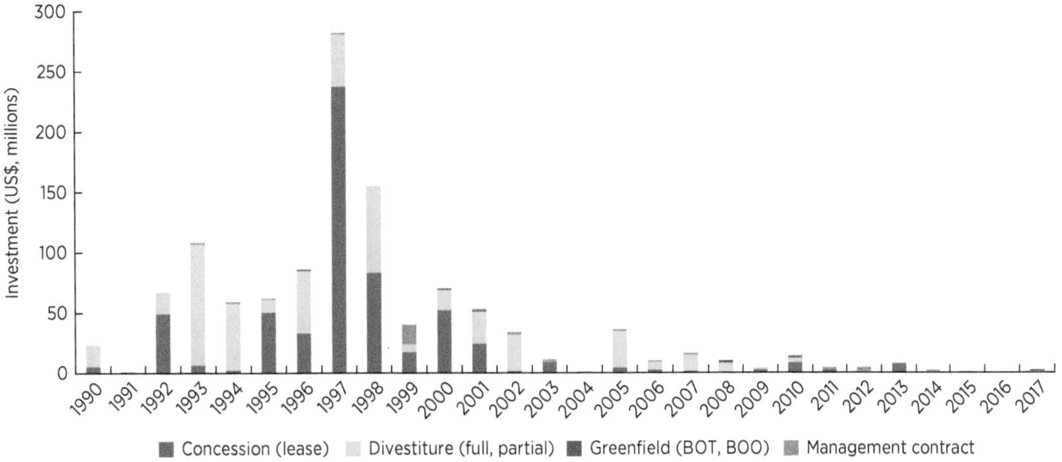

Source: Based on Private Participation in Infrastructure database 2018 (https://ppi.worldbank.org/).
Note: BOO = build, own, operate; BOT = build, operate, transfer.

viability of distribution utilities—and hence their attractiveness to the private sector—may vary significantly across jurisdictions. In Colombia, the strong municipal presence in the electricity sector led to a hybrid approach, whereby some major urban distribution utilities (in Bogotá and on the Caribbean coast) were privatized, whereas others (such as in Medellín) remained in municipal hands. In Morocco, private players in Casablanca, Rabat, Tangier, and Tetouan signed 30-year concession agreements to provide water and electric services under municipal jurisdiction in these cities, despite reluctance to unbundle or privatize the vertically integrated state-owned national utility, the Office National de l'Electricité et de l'Eau Potable, which still distributes electricity to the rest of the country. In the Philippines, where privatization of urban distribution utilities has been widespread, cooperatives continue to be the dominant model of provision in rural areas.

Peru provides a particularly interesting case of the difficulties encountered in attempting to extend distribution privatization beyond the capital city. In 1994, the government of Peru completed the privatization of Edelnor and Luz del Sur, two large distribution utilities serving different areas of metropolitan Lima. The success of this process prompted moves to privatize the eight state-owned provincial distribution utilities. In 1998, the Gloria group emerged as the successful bidder for a package of four of the regional companies (Electro Norte, Electro Norte Medio, Electro Noroeste, Electro Centro), paying a total of US$145 million for a 30 percent share, or almost twice the reference price for the assets. The privatization process envisaged the transfer of a second 30 percent tranche of shares after three years. In 2001, however, the regulator approved a new tariff based on the preprivatization reference price for the assets, rather than on what the company actually paid. The new tariff levels would not be able to cover the investment made by the Gloria group, so the deal collapsed and the government took over. In 2002, under a different administration, attempts were made to privatize the utility in the city of Arequipa. That plan was subsequently reversed because of popular opposition.

PSP in distribution has proved susceptible to reversals. Overall, 32 distribution transactions in 15 developing countries have been subjected to reversal (in the case of divestitures) or premature termination (in the case of concessions and other contractual instruments), particularly during the first decade of reform. Cancellations have typically taken place about five years after the privatization transaction. Most documented cancellations were instigated by the government, often in the form of renationalization of divested assets, although a significant number were also prompted by the voluntary departure of private operators dissatisfied with contractual conditions. About half of the recorded cancellations took place in Latin America and the Caribbean region, which was privatizing very rapidly during this period. These cancellations included a cluster of cases in Argentina following the peso crisis in 2003; Bolivia in the form of renationalization following a change of government in 2010; and Brazil, the Dominican Republic, and Peru, following disputes between the government and the private operators in the early 2000s. A significant number of cancellations is also reported for Europe and Central Asia—Albania, Kazakhstan, and Russia—which was also privatizing rapidly during this period. In Sub-Saharan Africa, many of the early PSPs saw contracts canceled (Comoros, Gambia, Guinea, Mali, Togo) and divestitures renationalized (Cabo Verde, Senegal). In South Asia, the earliest Indian PSP in the state of Odisha encountered difficulties and was renationalized. No cancellations are reported in East Asia and the Pacific or in the Middle East and North Africa.

The risk of reversal of PSP in electricity distribution is particularly high in Sub-Saharan Africa. Globally, premature cancellation of contracts with the private sector affects barely 1 percent of IPPs, but the rate has been about 3 percent for private sector arrangements in the distribution sector. Although the largest absolute number of cancellations took place in Latin America (17 cases, but only 2 percent of total transactions in the region), the highest rate of cancellation (affecting only 7 cases but representing 22 percent of transactions) occurred in Sub-Saharan Africa. Cancellation rates have been particularly high for divestitures (more than 60 percent) and management contracts (more than 30 percent) (figure 5.9).

From the private sector perspective, it can be challenging to achieve the required financial returns in uncertain operating environments. The financial equation for distribution utilities is highly sensitive to the condition of network assets and the level of distribution losses, factors that are often poorly understood and poorly documented in many publicly owned companies. At the time of privatization, prospective private operators and owners conduct their due diligence and make their bids on the basis of the limited information available. Once private operators take possession of companies and are able to see things at first hand, the situation may turn out to be significantly worse

FIGURE 5.9 Private sector participation in distribution suffers from premature contract cancellation

Private sector contract cancellations in distribution, 1990–2017

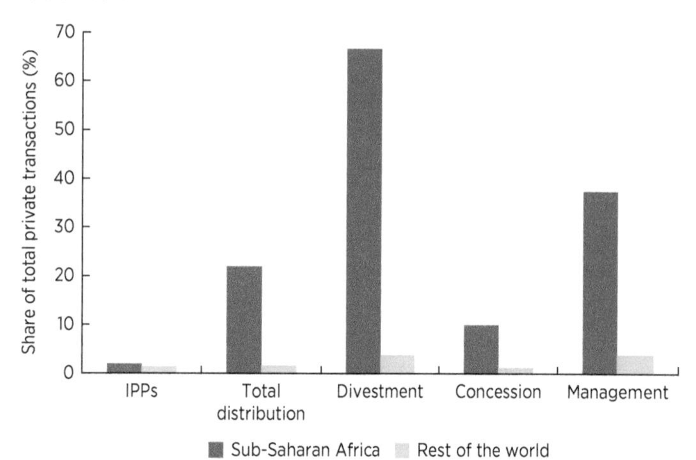

Source: Based on Private Participation in Infrastructure database 2018 (https://ppi.worldbank.org/).
Note: There was no case of cancellation in greenfield (BOT) distribution projects. BOT = build, operate, transfer; IPP = independent power producer.

than originally understood, triggering attempts to renegotiate contractual terms to restore financial returns to the required level. This situation often leads to renegotiation of the terms of privatization contracts, or in some cases, where the government may be unwilling to revise initial conditions, it may lead to some form of reversal or cancellation, as took place in the Dominican Republic and the Indian state of Odisha.

In the Dominican Republic, repeated oil price hikes, combined with an unwillingness to apply tariff regulation principles, made the financial position of the privatized distributors untenable. A 1997 law restructured the vertically integrated utility, Corporación Dominicana de Electricidad (CDE), into three distribution companies (EDENORTE, EDESUR, and EDEESTE), three generation companies, and a transmission company. The three distribution utilities were divested in 1998, despite the fact that all of them faced severe financial and operational conditions and reported billing only 60 percent of energy purchased. Almost as soon as the private utilities began operations, oil prices escalated, necessitating a tariff hike, which the government refused to allow because it violated its electoral promises. Instead, the government promised to pay the utilities the revenue difference between required and actual tariffs, a promise it never kept. Despite management efforts to improve commercial and operational performance, the financial situation worsened; PPAs with the generators were renegotiated. In 2002, the government finally eliminated subsidies and passed all operational costs on to the consumer, providing relief to utilities. The country soon had a banking crisis on its hands, followed by a currency devaluation. The resulting escalation of the costs of predominantly oil-based power generation once again led to a breakdown of the electricity payment chain, prompting the renationalization of EDENORTE and EDESUR in 2003, and of EDEESTE in 2009.

In the Indian state of Odisha, the privatization of the distribution companies unraveled after it was discovered that targets set by the regulator were based on faulty data. In 1996, the state electricity board began the restructuring process, which by 1999 succeeded in privatizing four distribution companies. The bids were obtained through a competitive international process. The state regulator (Odisha State Electricity Regulatory Commission, OERC) specified T&D loss targets for the privatized utilities on a baseline figure of 39.6 percent from 1996–97. In reality, however, T&D losses just prior to privatization had been much higher, at about 48.6 percent. The faulty baseline meant unrealistic loss-reduction targets had been set for the private distributors. Matters worsened when a cyclone later that year wreaked devastation on the T&D networks. Despite lobbying from the utilities, the government refused financial support to the privatized utilities. The regulator agreed to change the performance targets in 2001 only after a private player dropped out. With tariff revisions disallowed between 2000 and 2010, the utilities' financial crisis worsened. In addition, the investments needed to stem T&D losses could not be made, and the utilities struggled to meet their targets. By 2015 the regulator acted unilaterally and canceled the remaining three licenses for nonperformance.

Finding #8: The privatization of distribution raises delicate questions for key stakeholders such as labor unions

Privatization of distribution can create significant winners and losers among stakeholders. These include the government, private investors, utility employees, and utility customers. Distribution utility privatization typically focuses on operational efficiency and financial viability, with concomitant emphasis on raising tariffs, improving collections, containing distribution losses, and cutting costs. Such a package of measures can yield dramatic savings, reducing the need to subsidize the sector and bringing tangible fiscal benefits to the government while making operations more profitable for investors. The picture is different, however, for customers and labor unions. For them, this kind of reform agenda does not necessarily generate tangible benefits and may, in fact, inflict sizable costs. Stakeholder opposition in these cases becomes likely and can disrupt, if not overturn, privatization.

From the labor union perspective, the threat of layoffs is real. Overstaffing is common in distribution utilities in developing countries. In a study of 24 utilities in the Middle East and North Africa region, the median value for residential connections per full-time equivalent employee was found to be half that of the utilities outside the region (Camos and others 2018); 23 of the 24 utilities reported overstaffing, and another study found that 35 out of 39 countries in Sub-Saharan Africa had issues of overstaffing (Kojima and Trimble 2016). Nevertheless, both studies found that the costs associated with overstaffing are typically small compared with other types of inefficiencies. For example, in Sub-Saharan Africa, overstaffing costs represented only 10 percent of the hidden costs of inefficiency (or "quasi-fiscal deficit") in the sector, compared with 40 percent attributable to underpricing, 30 percent to T&D losses, and 20 percent to undercollection of revenues. Similarly, in the Middle East and North Africa region, overstaffing accounted for only 5 percent of the hidden costs of inefficiency in the sector, with underpricing accounting for at least 72 percent of inefficiency costs. Nevertheless, for a private operator, staff layoffs cut costs quickly and without requiring any major investment, so labor tensions are common during utility privatizations. If they are not adequately addressed, they can derail privatization.

In Senegal, for example, organized union opposition to the privatization of Senelec, the national utility, led to its renationalization. The main trade union, SUTELEC, was always opposed to Senelec's divestiture—opposition that intensified when the government retrenched some 15 percent of company staff through a voluntary severance package. In response, and against a backdrop of heavy load shedding in the country, SUTELEC organized a strike. The three-day-long countrywide blackout led to the arrests and imprisonment of union leaders. By way of compromise, the government altered its plans to sell a majority stake in Senelec, limiting privatization to a minority stake of 26 percent. The transaction was completed in 1999, and a consortium comprising Hydro-Quebec International (Canada) and Elyo (France) won the 25-year concession for Senelec; but the utility's infrastructure was much worse than determined during the bidding process and clearly required major investment. The consortium submitted a request for a tariff hike to the regulator, arguing that the agreed tariff level would no longer provide the anticipated rate of return. The request was not well received, given the lack of a clear investment plan or output targets that might hold the concessionaire accountable for improving the dire power supply situation in the country. To make matters worse, disputes arose within the company over the division of responsibilities between foreign management and local staff. The two foreign partners then clashed over the financing of investments. Tensions with SUTELEC continued. Over the course of 18 months, as the 2000 elections approached, privatization became drawn into electoral politics. Senelec was renationalized and the concession contract canceled following the change of administration.

Despite such tensions, interactions with labor unions can be constructive. Governments can make political choices about labor policy up front and build these into the terms of a privatization. Clear communication with unions throughout the process, carefully managed layoffs through attrition, and voluntary severance programs can all contribute to positive outcomes.

For example, in the Indian state of Odisha, union opposition to privatization was surmounted by binding the private investor to a program of gradual staff reductions. At the outset, in 1996, the unions had been vehemently opposed to privatization. In response, the state government built employment guarantees into the privatization contract—terms that assuaged labor and allowed the transaction to proceed. A careful government "staff rationalization" exercise identified an excess of 2,800 staff. Nevertheless, immediate reductions were rightly seen as both costly and disruptive. Instead, private bidders agreed to address overstaffing over a five-year period in line with the government's transition plan, which emphasized attrition, retraining, and redeployment. The surplus staff did not affect the sale price, and the deal went through without much disruption from the unions (Ray 2001).

In Tanzania, labor unions were mollified with clear, proactive communication and a well-designed severance package. in 2002, the private contractor negotiated a new management contract by engaging with staff and designing a severance package that allowed for amicable retrenchments of 20 percent of the staff. The government-approved labor agreement led to more than 1,200 workers leaving amicably—about 21 percent of the workforce. Tanzania Electric Supply Company Limited (TANESCO) financed the voluntary retirements from revenues totaling US$21 million (Ghanadan and Eberhard 2007). The managers continued to focus on maintaining good relations with the union. As a result, the union was in favor of an extension to the management contract in 2004, and again in 2006, although the second extension did not take place for other reasons.

The introduction of expatriate managers often spells friction with local staff, particularly in management contracts for African power utilities. Management contracts, with durations of just two to three years, have been introduced in some 17 African countries (table 5.5). They appear to offer a practical approach to boosting managerial capabilities with a view to reaping quick wins in operational performance while building local capacity. The experience has been checkered, however. Difficulties may arise even in attracting contractors with requisite skills, and thereafter in retaining qualified managers in-country. In Rwanda, for example, the sole bidder for the management contract tendered in the early 2000s was a company with no operational experience with distribution utilities. Even when good managers can be retained, cultural differences and resentment over wage differentials can lead to friction. In addition, the private operator may lack the motivation to transfer skills to a local leadership cadre—skills transfers being a key objective of a management contract. Most critically, in many cases, management contracts lack performance-improvement plans (with targets and incentives), making it difficult to channel the efforts of contractors whose remuneration is a fixed service fee.

Kenya's management contract, in force from 2006 to 2008, illustrates both the tensions and the benefits contracts can bring. A Canadian company, Manitoba Hydro International, won the management contract for the Kenya Power and Lighting Company (KPLC) in 2006. The reform champions in government saw a management contract as a way to shield KPLC from political interference—interference that had been hindering performance improvements even after structural reforms were introduced in the late 1990s. Designed to promote greater managerial autonomy, the reforms encouraged KPLC staff to use the time gained under the management contract to learn, develop, and empower themselves to run KPLC after the contract expired. Although some skills were transferred to the

TABLE 5.5 Management contracts for power utilities in Africa: A difficult history

Country/ State	Year	Renewed	Deepened	Issues
Guinea-Bissau	1991–97	No	No	—
São Tomé and Príncipe	1993–96	No	No	—
Ghana	1994–97	No	No	—
Mali[a]	1995–98	No	Yes	Differences between board and contractor over priorities; blackouts in third year of contract.
Namibia	1996–2002	No	No	National utility wanted to take over distribution business.
Togo	1997–2000	No	No	Contractor won court case regarding underdelivery.
Chad	2000	No (Canceled)	No	—
Malawi	2001–03	No	No	—
Tanzania	2002–06	Yes (once)	No	Political opposition coincided with drought period.
Lesotho	2002–Present	Yes	No	Outright privatization attempts failed and led to a management contract, which led to a turnaround.
Namibia	2002–present	Yes	No	After the previous contract (1996–2002) a new contract was given to new joint venture between the national utility, local, and regional governments.
Rwanda	2003	No	No	Contractor had no prior experience in electricity distribution.
Madagascar	2005–07	No	No	The management contractor, mired in corruption scandal elsewhere in Africa, was blacklisted.
Kenya	2006–08	No	Yes	Recurring tensions arose between foreign and local management.
Gambia, The	2006–11	No	No	—
Liberia	2010–16	Yes	No	—
Guinea[a]	2015–19	No	No	—

Sources: Based on Rethinking Power Sector Reform utility database 2015, and Private Participation in Infrastructure (PPI) database 2018 (https://ppi.worldbank.org/).
Note: — = not available.
a. Not in PPI database.

local cadre, tensions nevertheless mounted between locals and the foreign leadership team. Some staff felt that management was overcompensated and disrespectful to local staff.

Finding #9: Customers are sensitive to the tariff increases that come with privatization and will have legitimate expectations for improved and expanded services

From the customer perspective, distribution privatization generally entails higher service payments, with only meager prospects for service improvements. Because baseline tariffs typically fall well below the cost of service in many developing countries, tariff hikes help utilities recover costs so that private operators can run the business from revenues. To the extent that investment finance also switches from public and often concessional sources to private sources on commercial terms, the cost of capital may also increase. The result? More upward pressure on tariffs. At the same time, customers who pilfered electricity or failed to pay their bills will no longer be able to freeride after a private operator rolls out its revenue protection measures. All this adds up to more expensive electricity for customers, which is hard to justify unless the service improves. Distribution privatization contracts too often overlook the quality-of-service dimension by failing to include and enforce reliability standards. In some instances, unreliable service may have more to do with inadequate generation capacity than with deficiencies in the distribution network; in such cases, improvements are beyond the immediate control of the distribution operator.

The privatization of the national electric utility in Uganda brought immediate tariff increases. Meanwhile, output goals were deferred. In 2005, the national distribution utility was concessioned to a private operator, UMEME. As the sole bidder for the concession, the Eskom-Globeleq consortium was able to negotiate comprehensive contractual protections as well as a guaranteed fixed rate of return on investment of 20 percent. The privatization took place against the backdrop of serious load shedding. A drought had curtailed the supply of hydropower, and the government contracted costly emergency power plants to bridge the gap. A renegotiation of the contract took place in 2006, with a relaxation of regulatory performance targets and hikes in two retail tariffs (41 percent and 35 percent, respectively). Tariff hikes and unreliable supply led to public discontent. In time, opposition to the concession produced two public inquiries, in 2009 and 2011, each calling for the cancellation of the concession. Only in 2012, with the commissioning of new generation capacity, did the situation start to improve.

The regulator negotiated a new tighter suite of key performance indicators (KPIs) for the distribution concessionaire, which translated into substantial improvements in operational efficiency and service coverage. By 2016, revenue collection rates had climbed to 100 percent, and distribution losses had been halved to 17 percent. UMEME is now ranked as one of the better-performing utilities in Sub-Saharan Africa. Progress in increasing access was underwhelming, however, for two reasons. First, access had only recently been made a KPI. Second, the high guaranteed rate of return for the concession made the regulator hesitant to allow more access-related investment by the utility, because it would have implications for tariffs, already among the highest in Africa. Nonetheless, following the inclusion of access targets in 2012, electrification climbed from 16 percent in 2012 to 27 percent in 2016 (IEA and others 2018). Access improvements came at a cost to consumers, who have seen cumulative tariff hikes of more than 300 percent since the start of the reform process. At US$0.17/kWh, Uganda's average tariff is among the highest on the continent.

In Karachi, Pakistan, the privatization of K-Electric caused a hike in tariffs that, together with unreliable supply, led to consumer disaffection. K-Electric is a large, vertically integrated utility that serves nearly one-third of the country. It was divested in 2006 under martial law provisions that circumvented normal processes. Viewed as illegitimate, its privatization was quickly challenged in the courts. After company ownership changed hands in 2009, the new owners improved revenue collection and cut down on theft. Power sales increased 21 percent and 18 percent in the first two years. Reduction in nontechnical losses, usually associated with theft, brought losses down to 32 percent in 2011 from a high of 36 percent in 2009; but staff who tried collecting on outstanding bills faced violence in several Karachi neighborhoods. Meanwhile, the average tariffs increased by almost 90 percent in the years 2009 to 2011, without improvements in supply reliability. Massive public protests ensued when the utility began disconnections for nonpayment of outstanding bills. More than a decade later, the privatization of K-Electric remains a prominent political issue.

Special efforts are needed to ensure that private operators invest in the distribution network. Beyond first-order efficiency gains, service improvements require major investments in network infrastructure. Unless otherwise incentivized or obligated, private operators will make simple revenue enhancement and cost reduction their priorities. These goals involve modest investments and yield prompt paybacks. Hefty investments in long-lived distribution networks to expand electrification or improve service quality represent a big risk to the private operator. Such investment generally requires explicit contractual terms and reliability targets reinforced by regulatory entities and tariff-based remuneration. Expansion of access to new service areas is a thorny issue because the incremental costs of expansion tend to outstrip the incremental tariff revenues.

TANESCO's experience with management contracts between 2002 and 2006 illustrates the challenges of holding private operators accountable for output targets. In 2002 the NETGroup signed a management contract for TANESCO. The contract identified five KPIs—increasing revenue, reducing power losses, improving customer care, boosting reliability, and expanding electrification—all designed to turn the utility around. The new management grew revenues straightaway by upping the collection rate from a lowly 69 percent in 2001 to 89 percent in 2002. They accomplished this growth primarily by obliging public institutions to pay their bills. TANESCO's revenue grew by 140 percent over the four-year period (Ghanadan and Eberhard 2007). The other KPIs—power losses, customer care, reliability, and electrification—were not met. Power losses rose to 30 percent and then returned to precontract levels (22 percent) by 2006. Forced outages averaged 1,500 incidents and 2,500 hours per month with no improvement. New management met only half of the electrification target (Ghanadan and Eberhard 2007). While there were attenuating circumstances, such as drought and high generation costs, the contractual incentives to reduce distribution losses and increase service reliability were very weak, amounting to small fines that did not match the magnitude of the associated efforts. Despite these deficiencies, the contractor received 99 percent of its success fees from revenue improvements and paid only small fines for failing to improve reliability, power losses, and electrification.

CONCLUSION

Working alongside public sector investment, the private sector has become a major contributor to the expansion of power generation capacity in the developing world. The adoption of IPPs is one legacy of the standard reform model. IPPs have contributed greatly to power generation capacity since 1990 and are linked to the spread of renewable energy technologies. Nevertheless, private investment remains far from the norm—and far from providing the most funds—in the financing of new generation capacity in developing countries. Governance deficiencies made IPPs vulnerable to corruption through directly negotiated deals. The resulting scandals have sometimes lent them a bad name. Governments around the world struggle to balance investor protections with risk transfers.

The private sector has played an important role in the power distribution sector for some large middle-income countries. Its contributions elsewhere have proved more difficult. An early wave of distribution privatizations, primarily in Latin America and Eastern Europe, and particularly in larger urban centers, brought sustainable outcomes. Interest in the approach dropped off steeply after the early 2000s, and about 30 reversals of privatization have occurred. These reversals amounted to 3 percent of transactions globally but as much as 30 percent in Sub-Saharan Africa. In only one-third of developing countries does one find PSP in power distribution. A singular challenge has been to design privatization transactions so that they balance the benefits across key stakeholders, including consumers and labor unions.

New private entrants in the power sector should be able to compete alongside the incumbent utility, whether public or private.

ANNEX 5A. PRIVATE SECTOR PARTICIPATION INDEX

Specific to the 15 observatory countries

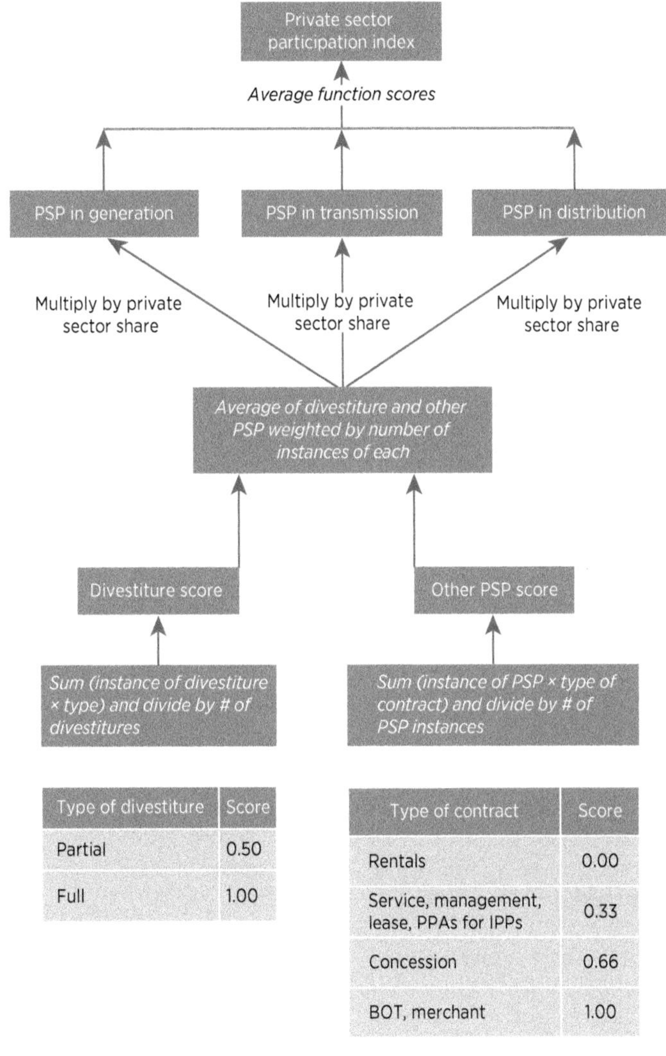

Type of divestiture	Score
Partial	0.50
Full	1.00

Type of contract	Score
Rentals	0.00
Service, management, lease, PPAs for IPPs	0.33
Concession	0.66
BOT, merchant	1.00

Source: Original figure for this publication.
Note: BOT = build, operate, transfer; IPP = independent power producer; PPA = power purchase agreement; PSP = private sector participation.

ANNEX 5B. PRIVATE SECTOR PARTICIPATION INDEX, 2015

Percent

Country/State	Private sector participation (PSP)	PSP in generation	PSP in distribution	PSP in transmission
Colombia	49	63	35	50
Dominican Republic	19	52	5	0
Egypt, Arab Rep.	2	7	0	0
India - Andhra Pradesh	13	39	0	0
India - Odisha	21	63	0	0
India - Rajasthan	14	39	1	2
Kenya	16	25	25	0
Morocco	25	53	22	0
Pakistan	19	40	9	8
Peru	61	78	19	88
Philippines	62	84	39	66
Senegal	11	34	1	0
Tajikistan	6	15	3	0
Tanzania	3	10	0	0
Uganda	38	53	63	0
Ukraine	30	13	52	25
Vietnam	10	31	0	0
International benchmark	24	41	16	14

Source: Based on Rethinking Power Sector Reform utility database 2015.

NOTES

1. This chapter is based on original research conducted by Vivien Foster, Anshul Rana, Joeri de Wit, and Victor Loksha, supported by an advisory team comprising Pedro Antmann, Pedro Sanchez, Elvira Morella, and Mariano Salto.
2. The Power Sector Reform Index (PSRI) explained in chapter 2 is used to calculate the extent of PSP globally. A more refined index is used to calculate PSP in the rethinking observatory countries. This index is detailed in annex 5A. Scores for the 15 countries from the observatory can be found in annex 5B.
3. Data on the geographical origin of private investment in the power sector are incomplete, because project records show the nationality of project sponsors but not the magnitude of their contribution. On this basis, it is possible to classify as foreign projects where all sponsors were foreign and as domestic projects where all sponsors were domestic, with the remainder falling into the mixed category.
4. Calculated using the PPI database, includes projects that have achieved financial closure.
5. Based on UDI World Electric Power Plants Database 2017.
6. Based on UDI World Electric Power Plants Database 2017.
7. This analysis is limited to capacity addition during 1990–2016, focusing only on the plant ownership at the time of commissioning. The figures cited do not consider existing plants that became private owing to divestiture and without adding to overall capacity. Nor do they take into account the change in ownership that may occur from divestiture after the plant is built.
8. A somewhat different picture emerges if the analysis is repeated to look at the composition of existing plants in 2016. In regions with notable divestitures of generation plant, the private ownership shares for generation are much higher than the incremental capacity additions: 62 percent for South Asia, 85 percent for Europe and Central Asia, and 87 percent for Latin America and the

Caribbean. By contrast, in regions marked by few divestitures of generation plant, the private ownership shares are much lower than the incremental capacity additions: 11 percent for the Middle East and North Africa, 15 percent for East Asia and the Pacific, and 22 percent for Sub-Saharan Africa.

9. Under the Moody's rating scale, "Ba" signifies obligations judged to have speculative elements and that are subject to substantial credit risk. The number 3 indicates a ranking in the lower end of that category.
10. This seems to be a reaction to the immediate aftermath of the financial crisis in 2001–03, when the country had a shortfall of dollars.
11. This section on finding #5 draws heavily on ESMAP (2015).

REFERENCES

Bacon, R. 1995. "Privatization and Reform in the Global Electricity Supply Industry." *Annual Review of Energy and the Environment* 20 (1): 119–43.

———. 1999. "A Scorecard for Energy Reform in Developing Countries." Viewpoint, Note No. 175, World Bank, Washington, DC.

———. 2018. "Taking Stock of the Impact of Power Utility Reform in Developing Countries." Policy Research Working Paper 8460, World Bank, Washington, DC.

Camos, D., R. Bacon, A. Estache, and M. M. Hamid. 2018. *Shedding Light on Electricity Utilities in the Middle East and North Africa: Insights from a Performance Diagnostic.* Washington, DC: World Bank.

ESMAP (Energy Sector Management Assistance Program). 2015. "Private Sector Participation in Electricity Transmission and Distribution: Experiences from Brazil, Peru, the Philippines and Turkey." ESMAP Knowledge Series 023/15, World Bank, Washington, DC.

Ghanadan, R., and A. Eberhard. 2007. "Electricity Utility Management Contracts in Africa: Lessons and Experiences from the TANESCO-NET Group Solutions Management Contract in Tanzania, 2002–2006." MIR Working Paper, Management Program in Infrastructure Reform and Regulation, University of Cape Town, South Africa.

IEA (International Energy Agency), International Renewable Energy Agency, United Nations Statistics Division, World Bank Group, and World Health Organization. 2018. *Tracking SDG7: The Energy Progress Report.* Washington, DC: World Bank Group.

Kojima, M., and C. Trimble. 2016. "Making Power Affordable for Africa and Viable for Its Utilities." World Bank, Washington, DC.

Ray, P. 2001. "HR Issues in Private Participation in Infrastructure: A Case Study of Orissa Power Reforms." World Bank, Washington, DC.

Sen, A. 2014. "Divergent Paths to a Common Goal? An Overview of Challenges to Electricity Sector Reform in Developing versus Developed Countries." Paper EL 10, Oxford Institute for Energy Studies, Oxford, UK.

World Bank. 1993. *The World Bank's Role in the Electric Power Sector: Policies for Effective Institutional, Regulatory, and Financial Reform.* Washington, DC: World Bank.

Did Countries Establish Meaningful Power Sector Regulation?

6

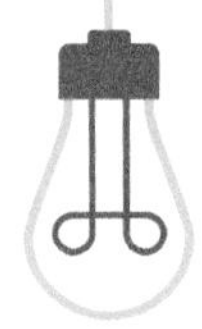

Guiding questions

- *To what extent were developing countries able to introduce regulatory regimes?*
- *What new challenges does technological disruption pose for regulators?*
- *Did the regulatory regimes operate the way they were originally designed?*

Summary

- *Creating a sector regulator was one of the most common power sector reforms, owing to the relative ease of implementation. However, there is a big difference between establishing a good regulatory framework on paper and building an effective regulatory system in practice. Quality of service regulation—in particular—too often exists on paper alone.*
- *Although tariff setting is the central function of regulatory agencies, their tariff recommendations are not necessarily respected or applied. Even in such cases, regulators can still play a valuable role in defining the magnitude of subsidies needed to ensure that the sector meets its revenue requirements.*
- *Although many countries have embraced incentive regulation, tariff regimes have often been poorly designed.*
- *While incentive regulation was originally designed with private utilities in mind—due to limited appetite for privatization—many regulators primarily oversee state-owned utilities.*
- *Regulators play a role in licensing market entry but are not always involved in the critical area of power purchase agreements.*
- *Looking ahead, price regulation and tariff structures will need to be significantly overhauled to provide adequate incentives for adoption of emerging technologies.*

INTRODUCTION

This chapter evaluates the quality of regulatory regimes adopted as part of power sector reforms in the developing world.[1] It considers the extent to which these regimes were able to meet the good practice standards for governance articulated in the sector reform literature. Considering both the design and quality of the regulatory systems put in place, the chapter probes the feasibility of introducing regulatory regimes in the power sectors of developing countries. It explores the modifications introduced during implementation and asks whether the regimes operated as designed.

The power sector reform paradigm of the 1990s emphasized the separation of powers. Policy making (setting strategic direction), service provision (implementing strategic direction), and regulation (monitoring and overseeing implementation) would be managed by separate entities. The previous institutional model for power utilities featured vertically integrated and state-owned monopolies, where all three functions resided in a single, reliable entity acting in the public interest. With the separation of powers, policy making would remain with the line ministry. An autonomous entity responsible for regulation would help to corporatize service provision. A privatized and independent company would be guided by a profit motive or take on a commercial orientation (chapter 4).

Regulation was believed to be particularly important in the context of intended private sector participation in electricity distribution. The replacement of a public monopoly with a private one raised consumer concerns about the potential abuse of market power either through overpricing or through erosion of quality. At the same time, investors worried that making large sunk investments with lengthy payback periods in a politically charged sector would leave them vulnerable to the risk that governments would rewrite the rules after irreversible investments had been made. Regulation was intended to balance the competing interests of investors, consumers, and governments, ensuring that investors achieved reliable and fair returns, consumers received value for money, and governments were constrained in their ability to exploit the sector for political purposes.

Practice in the United Kingdom and the United States envisioned autonomous regulatory entities that are largely separate from the line ministry. A variant of the approach, originating in the French-speaking world, held that regulation could be largely conducted through a legal contract (such as a concession or a lease) between the policy maker and the service provider. Contracts would be enforced by the courts, without the need for a specialized regulatory institution. In practice, "regulation by agency" proved to be more widespread than "regulation by contract"; however, in jurisdictions that adopted concessions or leases for power distribution, the associated contract was undoubtedly an important instrument of regulation.

Regulatory reforms in the power sector have focused on economics rather than on health and safety, or reforms of technical, financial, or environmental matters. This focus is understood as "the combination of institutions, laws, and processes that, taken together, enable a government to exercise formal and informal control over the operating and investment decisions of enterprises that supply infrastructure services" (Brown and others 2006,5). For analytical purposes, regulation has two dimensions: governance and substance (Brown and others 2006, 5).

Regulatory governance is defined by the laws, processes, and procedures that determine the enterprises, actions, and parameters that are regulated; the government entities that make the regulatory decisions; and the resources and information that are available to them

(Brown and others 2006). An effective and sustainable system of regulatory governance needs to have

- Credibility, so investors are confident that the system will honor its commitments;
- Legitimacy, so consumers trust that the system will protect them from the exercise of monopoly power whether through high prices, poor service, or both; and
- Transparency, so investors and consumers can see how and why decisions are made.

Above all, however, regulatory governance must balance autonomy with accountability (Rodriguez Pardina and Schiro 2018). Without autonomy, regulators cannot influence outcomes. Full autonomy arises from the exercise of a range of powers, including the freedom to appoint qualified and independent decision makers, to take and enforce decisions affecting stakeholders, to capture adequate budgetary resources, and to manage internal processes. Without accountability, however, regulators would have no obligation to the stakeholders they are meant to serve and protect. Accountability is secured through government oversight, the right of companies to legal appeal, and transparent regulatory processes and decisions. Autonomy and accountability are meant to work together. It is pointless to have a regulator that is highly accountable but lacking in autonomy. By the same token, a highly autonomous regulator that lacks accountability may abuse its power.

Regulatory substance refers to the content of regulation—the explicit or implicit decisions made by regulators, along with their rationale (Brown and others 2006). Substance includes the setting of tariffs. This complex process begins with applying a tariff regime (or methodology) to determine the average tariff level commensurate with efficient cost recovery. The process may also seek to determine the tariff structure—the suite of charges customers pay—which then defines the average tariff. In addition, regulators set minimum service standards to ensure that consumers receive adequate quality of service. A third aspect of regulatory substance is entry, meaning control over whether and how new operators may enter the regulated market (Rodriguez Pardina and Schiro 2018).

Many of the principles of tariff and quality regulation are premised on the commercial orientation of the power utility. The creation of a regulatory entity was intended to be just one element in a broader power sector reform package that emphasized utility governance reforms, private sector participation, and market liberalization. These wider reforms were intended to increase private sector participation in power utilities, or at least to induce in them a much stronger commercial orientation. For this reason, much of the substance of regulation is oriented toward the creation of financial incentives to invest, drive down operating expenditures, and meet quality standards. In each case, the utility's supposed profit motive is harnessed by the regulator to bring about socially desirable behavior. In the absence of these complementary reforms, the rationale for incentive-based regulation methods tends to break down.

These issues are addressed with reference to quantitative and qualitative evidence on the design and practice of regulatory systems in the 15-country Power Sector Reform Observatory (introduced in chapter 1). Quantitative evidence is based on a survey of the categorical data regarding the regulatory system. Using the conceptual framework outlined earlier, a Regulatory Performance Index was created. It scores countries according to their conformity with good regulatory practice (box 6.1 and figure 6.1). To capture any divergences between regulatory design and regulatory practice, two versions of the same index are calculated for each country. A first score portrays the country's de jure regulatory framework as it appears in laws, regulations, and

BOX 6.1 Introducing the Regulatory Performance Index

The survey instrument applied to the 15 Power Sector Reform Observatory countries included 355 categorical and quantitative questions on the regulatory system. The questions were both descriptive and normative.

Descriptive questions collected facts. How many regulatory commissioners are there? How frequently are tariffs revised? They are useful in understanding the nature of practice across the developing world and are summarized where appropriate throughout this chapter.

Normative questions aimed to capture regulatory best practices using the literature. To synthesize the normative data in a convenient and intelligible format, a Regulatory Performance Index was created. This index builds on earlier regulatory indexes (Pargal and Banerjee 2014; Vagliasindi and Nellis 2010), but encompasses as many as 74 features of the regulatory framework.

The structure of the Regulatory Performance Index is depicted in the flow chart (figure 6.1 in the main text) and closely follows a conceptual framework derived from the literature (Brown and others 2006; Rodriguez Pardina and Schiro 2018). The overall score is the product of two subindexes of regulatory governance and regulatory substance. All indexes and subindexes are scored on an interval of 0 to 100.

The subindex on regulatory governance is the product of two further subindexes capturing the degree of autonomy (based on 17 questions) and accountability (28 questions). Examples of good practices for autonomy include actual powers: if a regulator's decisions are legally binding; if the regulator can determine the agency's organizational structure and the use of its budget, and so on. Examples of good practices for accountability include the regulator's duty to report to another institution, to publicize its decisions and recommendations, and to involve nongovernment stakeholders in its decision-making processes.

The subindex on regulatory substance is the simple average of three subindexes capturing performance on tariff regulation (16 questions), quality regulation (12 questions), and entry regulation (7 questions). Some examples of good practices for tariff regulation are that tariff setting is based on a specified regulatory framework; that a written, publicly available formula prescribes how end-user tariff levels are to be set; that the regulator and utility are obliged to adhere to that formula; and that regulatory accounting guidelines are followed. Examples of good practices in regulation of service quality include mandatory and publicly available standards of quality and the existence of fines or penalties for noncompliance. Examples of good practices for entry regulation include the power to monitor compliance with the terms of licenses or permits and the authority to impose penalties for noncompliance.

Multiplication of subindexes, rather than averaging, is used where paired variables (autonomy and accountability, governance and substance) are meaningful only when they go together. Averaging of subindexes is adopted as the aggregation method when these are largely independent from each other (tariff and entry regulation).

Two versions of the same index were calculated for each country. First, a de jure index derives from the country's regulatory framework as captured on paper in laws, regulations, and administrative procedures. Second, a perception index determines whether the paper provisions are applied in practice. The local consultant in each country provided the perception index; his or her professional opinion was informed by some 20 interviews with key stakeholders in the reform process. The perception index was also reviewed by the World Bank country energy team knowledgeable about local context. Despite best efforts, this second index is more subjective than the first.

administrative processes. A second score assesses a perceived score—the country's regulatory framework *as it is practiced* according to local experts. This information draws on qualitative evidence from interviews with some 20 key stakeholders in each country. It forms a rich source of narratives that shed further light on regulatory experience.

KEY FINDINGS

The key findings of the analysis of power sector regulation are based on experiences in the 15 observatory countries. They are summarized in the following key findings.

Finding #1: Sector regulation was a popular measure in the package of power sector reforms, but the limited appetite for privatization meant that regulators were often overseeing state-owned utilities

The creation of a regulatory entity was a key component of the 1990s power sector reform paradigm. By 2015, more than 70 percent of developing countries had created a power sector regulator, making this the most widely adopted measure among all the reforms. Unlike the other reforms, momentum for the regulatory measure remained strong through

FIGURE 6.1 **Overview of Regulatory Performance Index**

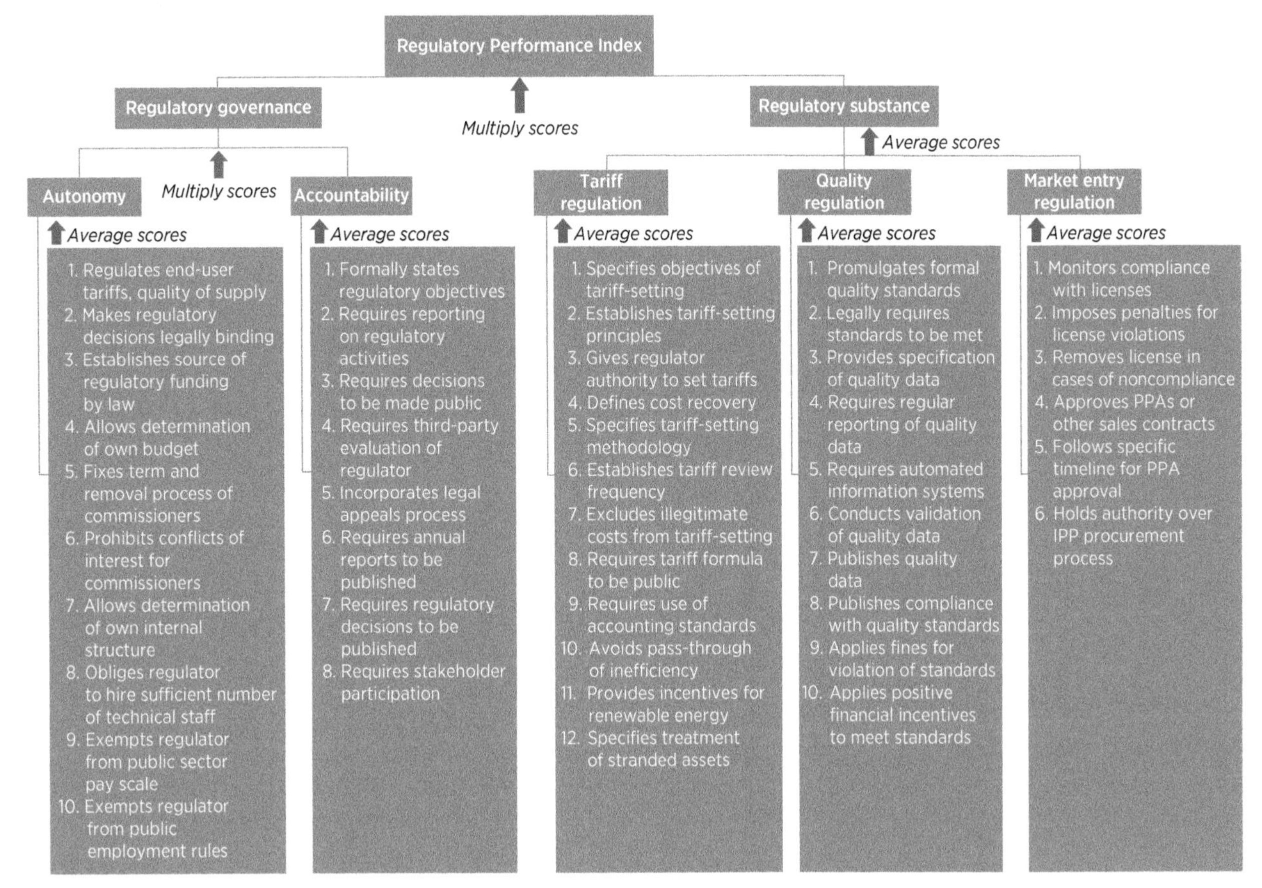

Source: Based on Rethinking Power Sector Reform utility database 2015.
Note: Detailed scores for the rethinking observatory countries are provided in annexes 6A and 6B. IPP = independent power producer; PPA = power purchase agreement.

the period 1995–2015; every year about 3 percent of countries established a regulator. In about one-third of developing countries, the introduction of a regulatory entity was the first reform to be implemented. For some it was the only reform.

Latin America initiated the trend in favor of power sector regulators. Adoption there was rapid—by 1998 nearly 80 percent of countries had established an entity (figure 6.2). In other developing regions, the process lagged. Nevertheless, by 2015, Sub-Saharan Africa and much of Asia had caught up with Latin America. These regions can claim regulatory agencies in 60 to 80 percent of their countries. In contrast, the diffusion of regulators in Europe and Central Asia as well as Middle East and North Africa plateaued in the early 2000s; the penetration level is about 40 percent.

Regulators may undergo governance changes over time—for good or ill. For example, in Kenya, regulation was initially entrusted to the Electricity Regulatory Board (ERB), which started operations in 1998. Almost 10 years later, the 2006 Energy Act led to further restructuring in the sector and the ERB was transformed into the Energy Regulatory Commission (ERC); it regulates the entire energy sector, including renewable energy and petroleum downstream activities. In Peru, the

FIGURE 6.2 Latin American countries were early adopters of regulatory reform, but countries in Africa and Asia caught up fast

Creation of regulatory entity for power sector, by region

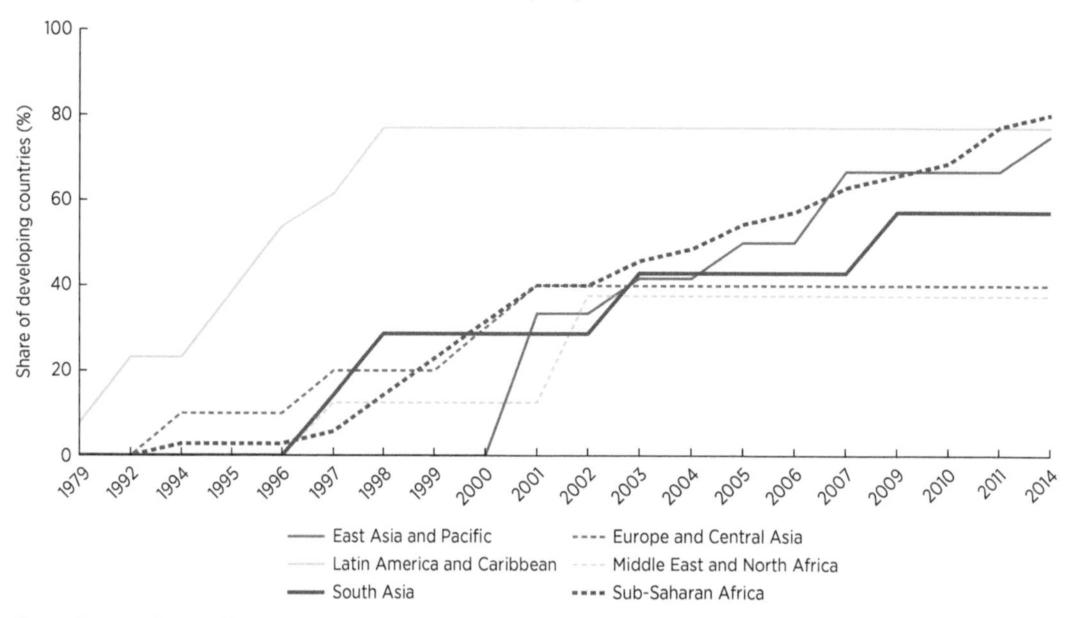

Source: Foster and others 2017.

FIGURE 6.3 Tariff and quality regulation are the core responsibilities of regulatory institutions

Regulatory mandate in the observatory countries, 2015

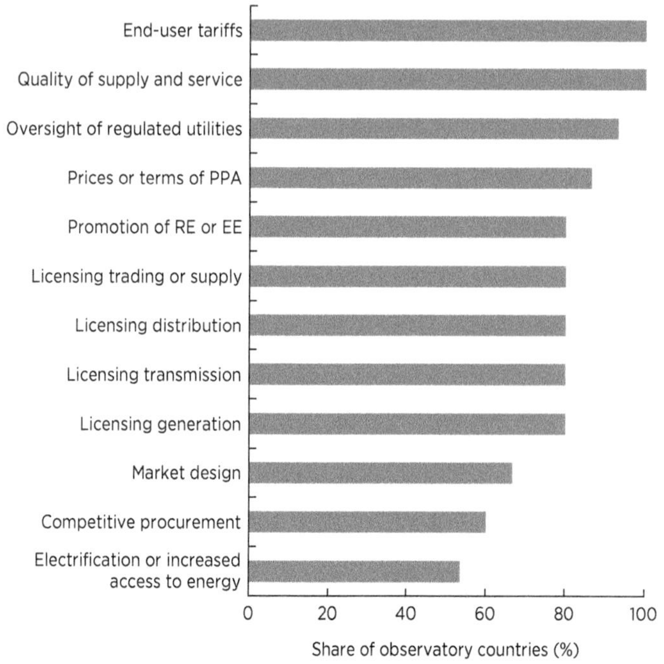

End-user tariffs
Quality of supply and service
Oversight of regulated utilities
Prices or terms of PPA
Promotion of RE or EE
Licensing trading or supply
Licensing distribution
Licensing transmission
Licensing generation
Market design
Competitive procurement
Electrification or increased access to energy
0 20 40 60 80 100
Share of observatory countries (%)

Source: Based on Rethinking Power Sector Reform utility database 2015.
Note: EE = energy efficiency; PPA = power purchase agreement; RE = renewable energy.

original energy regulator, OSINERG, created in 1996, had its scope of responsibilities expanded in 2007 to include the mining sector. It is now called Organismo Supervisor de la Inversión en Energía y Minería (OSINERGMIN). Pakistan's electricity regulator, the National Electric Power Regulatory Authority (NEPRA), was originally established as a separate institution in 1997. In 2016 it was subsumed under the line ministry in 2016, although the courts reversed that decision in 2017.

In addition to being widespread, regulatory entities have noteworthy commonalities in both governance and methodology, with similar features seen across at least 80 percent of the observatory countries.

The regulation of tariffs and quality is the core of the economic regulation function and forms part of the regulator's responsibilities in all observatory countries (figure 6.3). In almost all countries, these functions combine with oversight of the utilities to ensure compliance with tariff and quality regulation. The role of the regulator in market entry, though not universal, is also prevalent, with at least 80 percent of countries giving the regulator responsibilities in licensing entry and negotiating terms of power purchase agreements (PPAs). In addition to these core functions, regulators at times also play a role in enforcing sector policies on clean energy (80 percent), market design (65 percent), and electrification (55 percent). Despite their roles in negotiating the terms of PPAs with independent power producers (IPPs) (85 percent), regulators are less involved with competitive procurement (60 percent).

An interesting exception is Colombia, where regulatory and supervisory functions, combined in most countries, are separated into two distinct institutions. These are the Commission for the Regulation of Electricity and Gas (Comisión de Regulación de Energía y Gas, CREG) and the (multisector) Superintendence for Public Household Services (SSPD). SSPD provides oversight for the provision of public utility services, as well as protection of consumers' rights. This arrangement allows the SSPD to take over management of underperforming companies in extreme situations and imposes discipline on the market.

A significant minority of countries opted not to create an economic regulator for the sector. Among the 15 countries in the Power Sector Reform Observatory, Morocco and Tajikistan stand out. Their experience shows that regulatory functions remain relevant. In both countries, the national power utility itself plays a big role in self-regulating, raising concerns about conflict of interest.

In the case of Morocco, tariff adjustments are enacted by an interministerial committee (Commission Interministérielle des Prix). Led by the Ministry of General Affairs and Governance (MAGG) with representatives from the ministries of the interior, energy, finance and economy, trade and industry, and agriculture and fisheries. The commission also has members from the national power utility (Office National de l'Electricité et de l'Eau Potable, ONEE), municipal power distributors, and large private power distributors. Tariff adjustments are made on recommendation of a tariff study prepared by an independent consultant but supervised by ONEE. Quality of service is either regulated by contract—in case of private distributors—or self-regulated in the case of ONEE, as part of its mission statement.[2]

In Tajikistan, regulatory powers are spread across several government institutions. All of these have advisory roles, as opposed to decision-making authority, in their jurisdictions. Recommendations on tariffs are made by the general antitrust agency. The national standards agency is responsible for regulating quality of service. The Ministry of Energy and Water Resources issues licenses and is involved in regulating the terms of PPAs, market design, competitive procurement, and oversight of regulated utilities. Finally, the nation's state-owned, vertically integrated monopoly power utility, Barki Tojik, self-regulates, particularly with regard to tariffs. On the basis of submissions made by Barki Tojik, the Agency for Antimonopoly Services recommends electricity tariffs to the government of Tajikistan.

Many regulators continue to oversee state-owned power utilities that have limited managerial incentives to respond to regulatory instruments. As noted earlier, the creation of regulatory entities was intended to pave the way for private sector participation, particularly in distribution. Nevertheless, of the 70 percent of developing countries that created regulatory agencies since 1990, less than half of them also introduced private sector participation into power distribution. Most countries left the regulator to engage with largely state-owned distribution utilities (Foster and others 2017). Even countries that introduced some private sector participation in distribution left numerous state-owned enterprises (SOEs) coexisting alongside private ones.

Regulatory systems appear to function more effectively with a sizable private sector presence in the power sector. The presence of

FIGURE 6.4 Countries with higher levels of private sector participation in distribution have stronger perceived regulatory performance

Private sector participation vs. perceived regulation, 2015

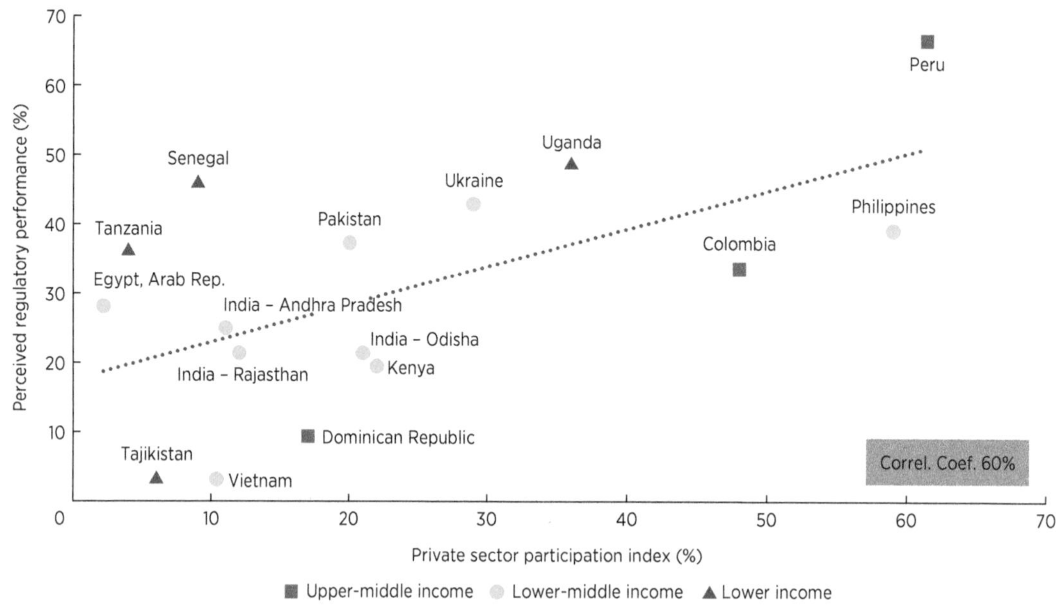

Source: Based on Rethinking Power Sector Reform utility database 2015.
Note: Correl. Coef. = correlation coefficient.

private sector players in power distribution seems to affect the performance of regulators (figure 6.4). One reason for this effect may be that, in the presence of private sector operators in the sector, it is more difficult for the government to circumvent its own regulatory framework because it can be held accountable by the private sector. The magnitude of the gap between de jure and de facto regulatory performance is smallest in countries where the private sector is extensively involved in power distribution.

Finding #2: There is a difference between establishing a good regulatory framework on paper and building an effective regulatory system in practice

Sound regulatory principles can be enshrined in the legal framework, but there is no guarantee that they will be implemented. The discrepancy between *formal*, or de jure, regulation (written in law and regulations) and *perceived* regulation (what actually happens) is well known in the literature and has been empirically documented both for Latin America (Andrés and others 2007; Correa and others 2006; Rodriguez Pardina, Schlirf Rapti, and Groom 2008) and for Western Europe (Gilardi and Maggetti 2011; Hanretty and Koop 2013), though the gap is believed to be larger in the case of developing countries. Explanations cite the dissonance between imported regulatory frameworks and the local legal environment (Rodriguez Pardina, Schlirf Rapti, and Groom 2008), or the difficulty of defining a complete contract between the political decision makers (the "principal") and the regulatory agency (the "agent") (Gilardi and Maggetti 2011).

De jure scores are systematically higher than perceived scores, and the latter are also more variable. The Regulatory Performance Index has been calculated both for the de jure and the perceived situation in each observatory

FIGURE 6.5 Regulatory frameworks are not always perceived to function as written on paper

Country performance on the Regulatory Performance Index, formal (de jure) vs. perceived practice

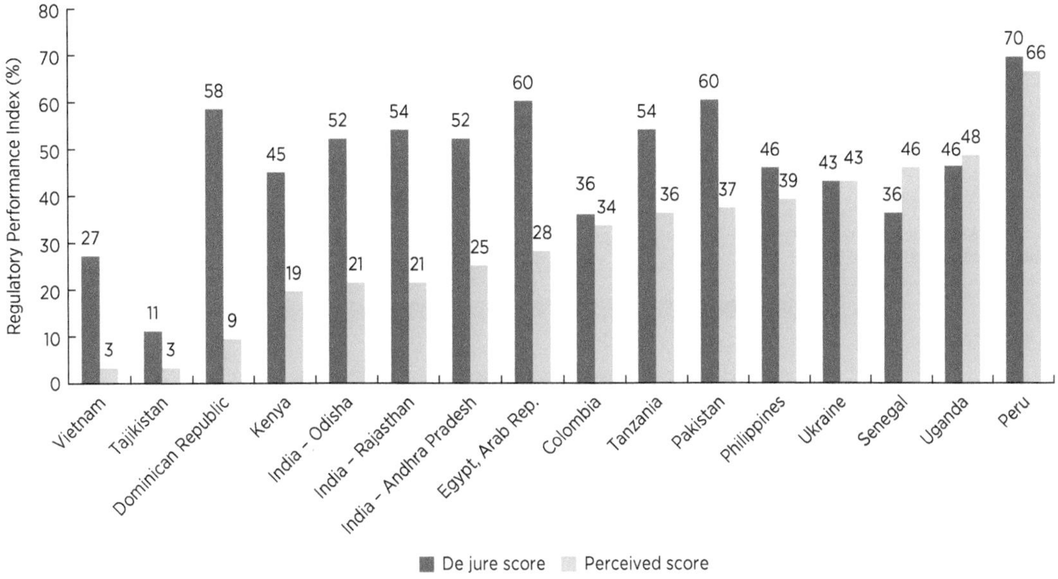

Source: Based on Rethinking Power Sector Reform utility database 2015.

country, making a comparison between the two highly informative (figure 6.5). For some countries, the difference between de jure and perceived scores is modest, with both being either consistently high (as in Peru) or consistently low (as in Tajikistan). For about half of the countries, the divergence can be more than 20 percentage points. The only area where perceived regulatory practice exceeded de jure regulatory requirements was on public consultations, with some regulators opting to make use of such channels even when not legally obligated to do so.

The Dominican Republic has the most divergent scores (de jure and perceived), with a gap of more than 50 percentage points. The country's regulatory entity, Superintendencia de Electricidad (SIE), was established in 1998 when the sector was restructured and the three distribution utilities were privatized. An impasse in parliament delayed passage of the new energy law, so SIE had to be created in the first instance by decree, weakening its legal standing. The regulator is responsible for determining tariffs in the sector under a price-cap mechanism that indexes tariffs to inflation, exchange rate, fuel prices, and so on. Almost as soon as the private utilities began operations, oil prices rose, so SIE ordered a tariff hike. The government rejected the hike, however, promising instead to pay the distribution companies the difference between cost-reflective and actual tariffs. The government then reneged on payment. The distribution companies' worsening finances led eventually to the renationalization of the three distribution utilities between 2003 (EDENORTE and EDESUR) and 2009 (EDEESTE). As just one government actor among many, SIE struggled to implement its decisions regarding tariffs, quality, and market entry. Successive governments used the utilities for political rather than commercial objectives. Thus, in the Dominican Republic, the regulator has no real authority over tariffs. Tariffs are not set according to a regulatory framework, nor are they adjusted in line with legal mandates.

FIGURE 6.6 **Gaps between paper and practice are particularly large on some aspects**

Formal (de jure) power vs. perceived practice, 2015

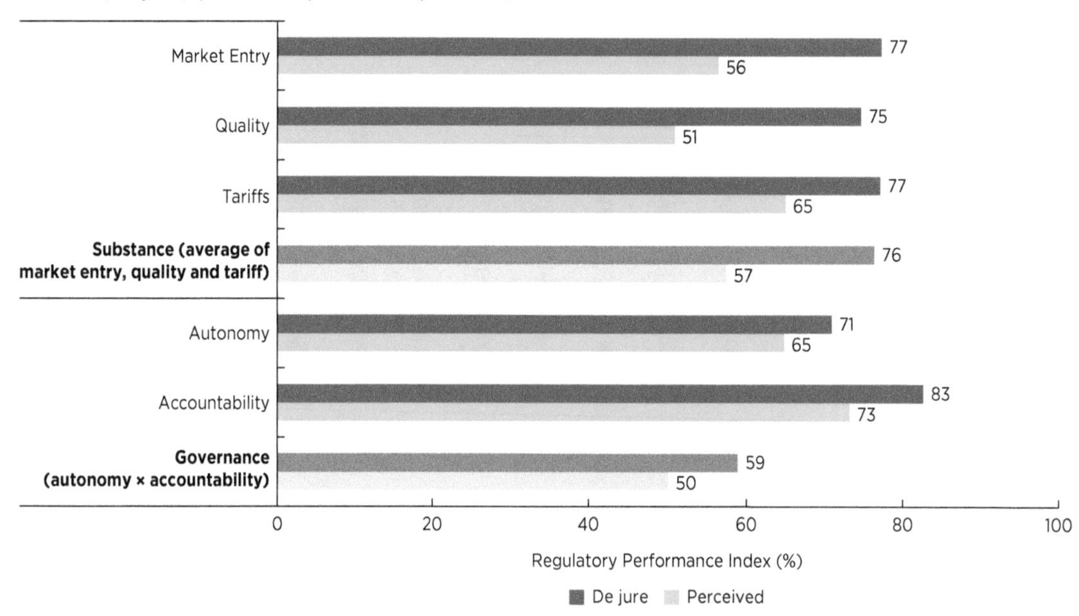

Source: Based on Rethinking Power Sector Reform utility database 2015.

There are also systematic differences in the gap between de jure and perceived regulatory scores for different aspects of regulation (figure 6.6). The average divergence is much greater in regulatory substance, where it reaches some 20 percentage points, than on regulatory governance, where the gap is only about 10 percentage points. Despite the fact that regulators have legally binding powers to set tariffs and regulate quality of service and market entry, in practice their decisions may often be delayed or overruled by other government entities (see the next section for further details). Quality-of-service regulation is even weaker, with many regulators reluctant to apply penalties for noncompliance with quality standards (as in the Arab Republic of Egypt, India, and Pakistan), and others not having established any penalties at all (Kenya, Tajikistan, Uganda, and Vietnam).

The performance of regulation reflects the institutional characteristics of the regulated companies. As noted earlier, countries with higher levels of private sector participation in distribution tend to have both stronger perceived and de jure regulatory performance. Interestingly, they also have smaller gaps between the two (figure 6.7). The quality of utility governance seems to be related to positive regulatory performance (figure 6.8). Because private sector participation and governance are strongly correlated, however, it is hard to disentangle the effects.

Finding #3: Most countries' regulatory regimes have gone further on accountability than they have on autonomy

Regulatory autonomy appears to pose greater political risks than does regulatory accountability. Because it involves the delegation of powers from politicians to regulators, autonomy is less prevalent than accountability, which places regulators under political scrutiny. As might be expected, the average

FIGURE 6.7 Countries with higher private sector participation have smaller gaps between de jure and *perceived* regulation

De jure vs. perceived regulation, 2015

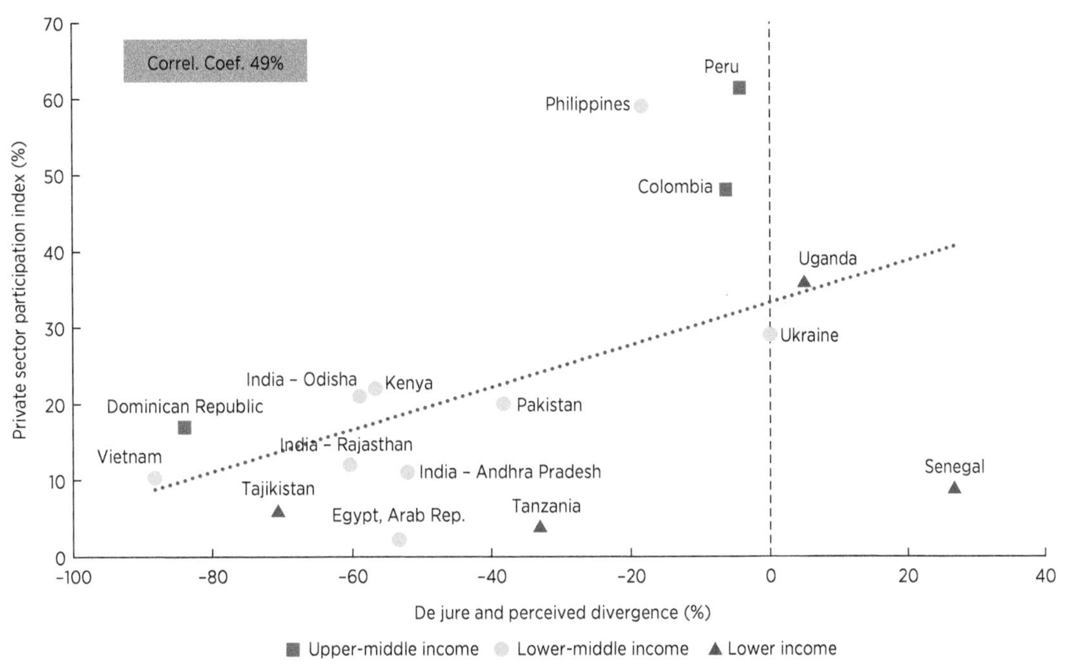

Source: Based on Rethinking Power Sector Reform utility database 2015.
Note: Income classification according to 2018 data from the World Bank. Correl. Coef. = correlation coefficient.

FIGURE 6.8 Quality of utility governance seems to be related to positive regulatory performance

Utility governance vs. perceived regulatory performance

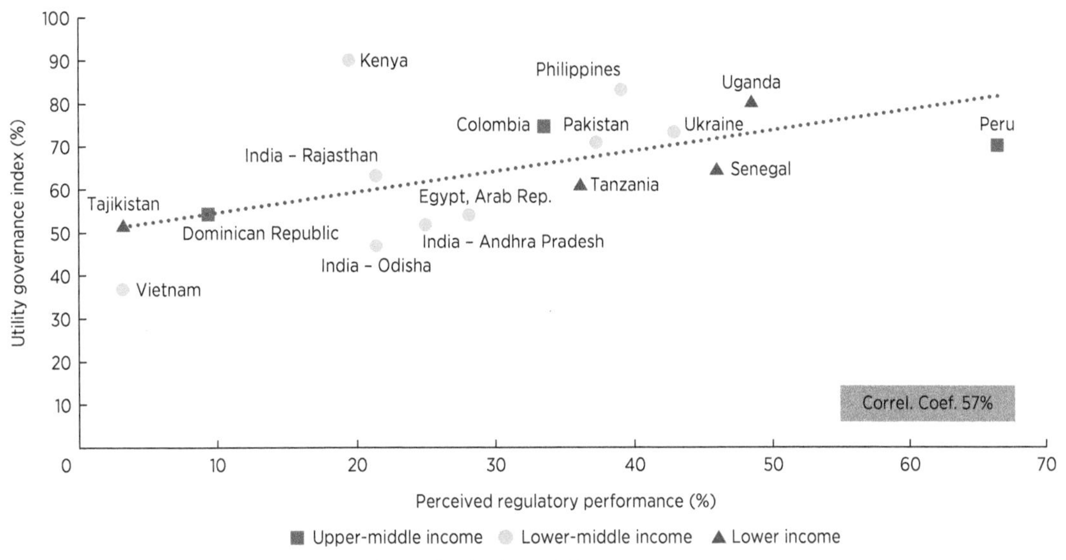

Source: Based on Rethinking Power Sector Reform utility database 2015.
Note: Income classification according to 2018 data from the World Bank. Correl. Coef. = correlation coefficient.

de jure score across countries for autonomy (at 71 percent) is well below the score for accountability (83 percent). More countries have accountability scores that exceed their autonomy scores (placing them below the 45-degree line) than the other way around (8), even if scores on autonomy and accountability are also correlated (figure 6.9).

Certain accountability measures are universal, whereas others are practiced in only about half of the countries (figure 6.10). Among the most ubiquitous practices on paper are legal appeals, publication of final decisions on tariffs, and production of annual reports. Much less widely adopted are measures such as regular performance evaluations for regulators and the participation of nongovernment stakeholders in regulatory decision making. Gaps between de jure and perceived performance in the area of regulatory accountability tend to be minor, except in certain areas. For instance, although all regulators are required to publish decisions on wholesale or PPA prices, this requirement is complied with in practice for only 73 percent of cases. Regulators are also much less likely to involve nongovernment stakeholders in decisions to license new generation than the legal framework suggests. Regulated entities are universally allowed to appeal decisions, but appeals are made in only 81 percent of cases. One area where regulatory practice on accountability exceeds the legal requirements is the publication of regulatory recommendations, which are legally required in only 33 percent of cases but practiced in 58 percent of cases, reflecting voluntary disclosure.

Peru's regulatory framework provides an outstanding example by incorporating many good-practice measures on accountability. Indeed, OSINERGMIN explicitly articulates the "Principle of Transparency" in the regulator's Action Principles (Decree No. 054-2001-PCM). Consultation drafts of proposed new regulations are required to be published for public

FIGURE 6.9 Achievement of autonomous regulators remains a significant challenge even as regulators are more accountable

Autonomy vs. accountability

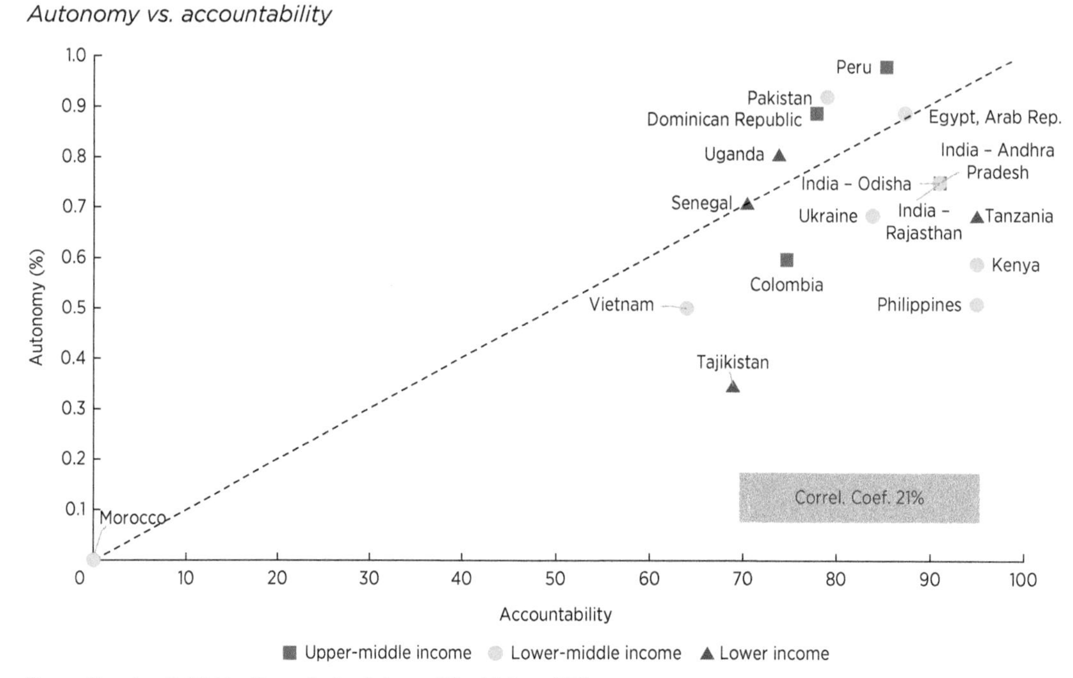

Source: Based on Rethinking Power Sector Reform utility database 2015.
Note: Income classification according to 2018 data from the World Bank. Correl. Coef. = correlation coefficient.

FIGURE 6.10 Certain accountability measures have been universally adopted, whereas others are practiced in only about half of the countries

Formal (de jure) power vs. perceived practice

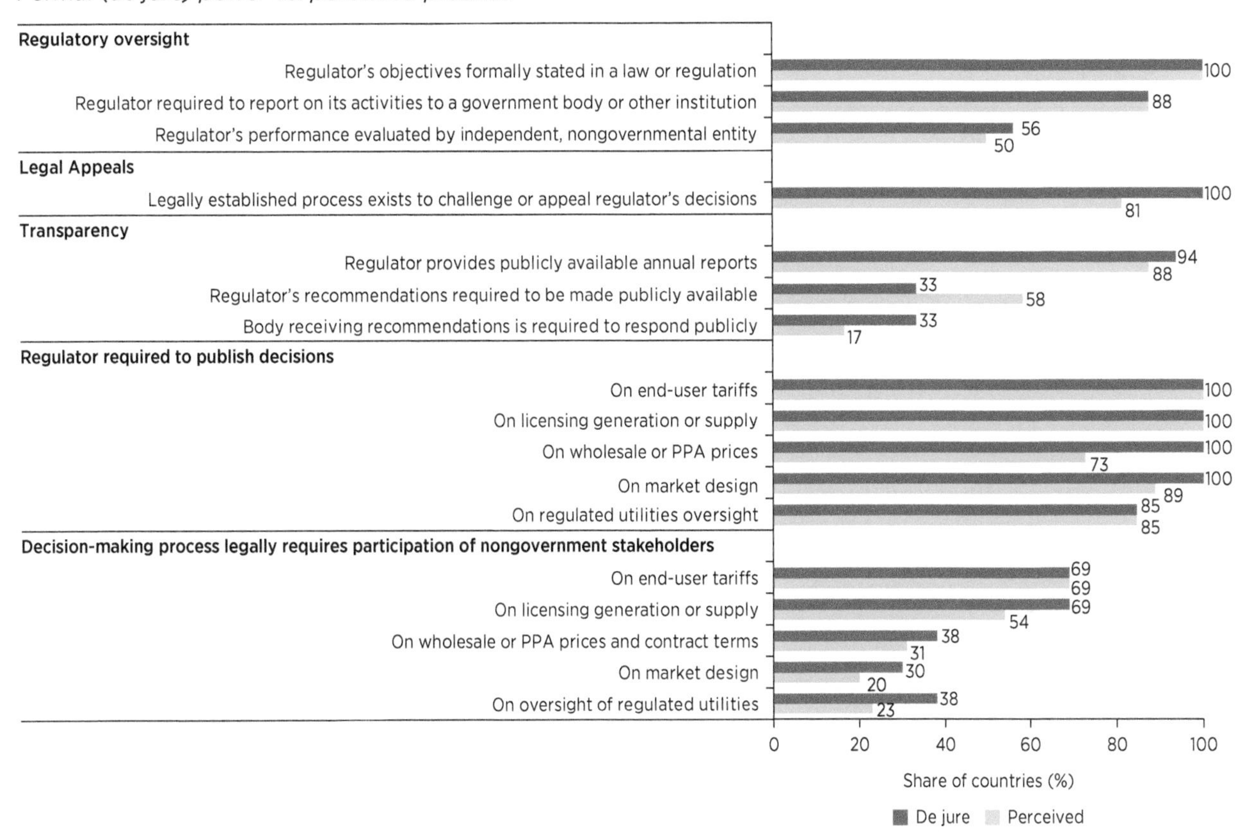

Source: Based on Rethinking Power Sector Reform utility database 2015.
Note: PPA = power purchase agreement.

comment, and all final decisions appear in the official government gazette. Stakeholder participation in this and all other regulatory decisions is mandatory and usually takes place through public hearings. All regulatory decisions must be justified through clearly articulated principles that help promote predictability. OSINERGMIN reports annually on its activities to the prime minister and also publishes quarterly updates on its website, allowing progress to be tracked against a range of performance targets and indicators. "Participatory regulatory impact assessments" are undertaken to evaluate the costs and benefits of any new regulatory decision for society as a whole.

In practice, the perceived autonomy regulators have over legally binding decisions is appreciably less than their de jure autonomy (figure 6.11). On paper, regulators have authority over tariffs, quality, licensing, and PPAs, but these powers appear weaker in practice. Whereas 53 percent of regulators have legally binding powers over end-user tariffs, this share drops as low as 20 percent in practice. Sizable gaps (on the order of 20 percentage points) also exist for the powers to approve grid access charges and PPA/wholesale prices. Although the funding basis for regulatory entities is almost always determined by law, coming mainly from levies from consumers or regulated companies, in only about half of cases are regulators free to determine their own budgetary allocations. Many regulators also lack the freedom to determine their own

FIGURE 6.11 Regulator's autonomy to make legally binding decisions on key issues is *perceived* to be significantly lower in practice than it looks de jure

Formal (de jure) power vs. perceived practice

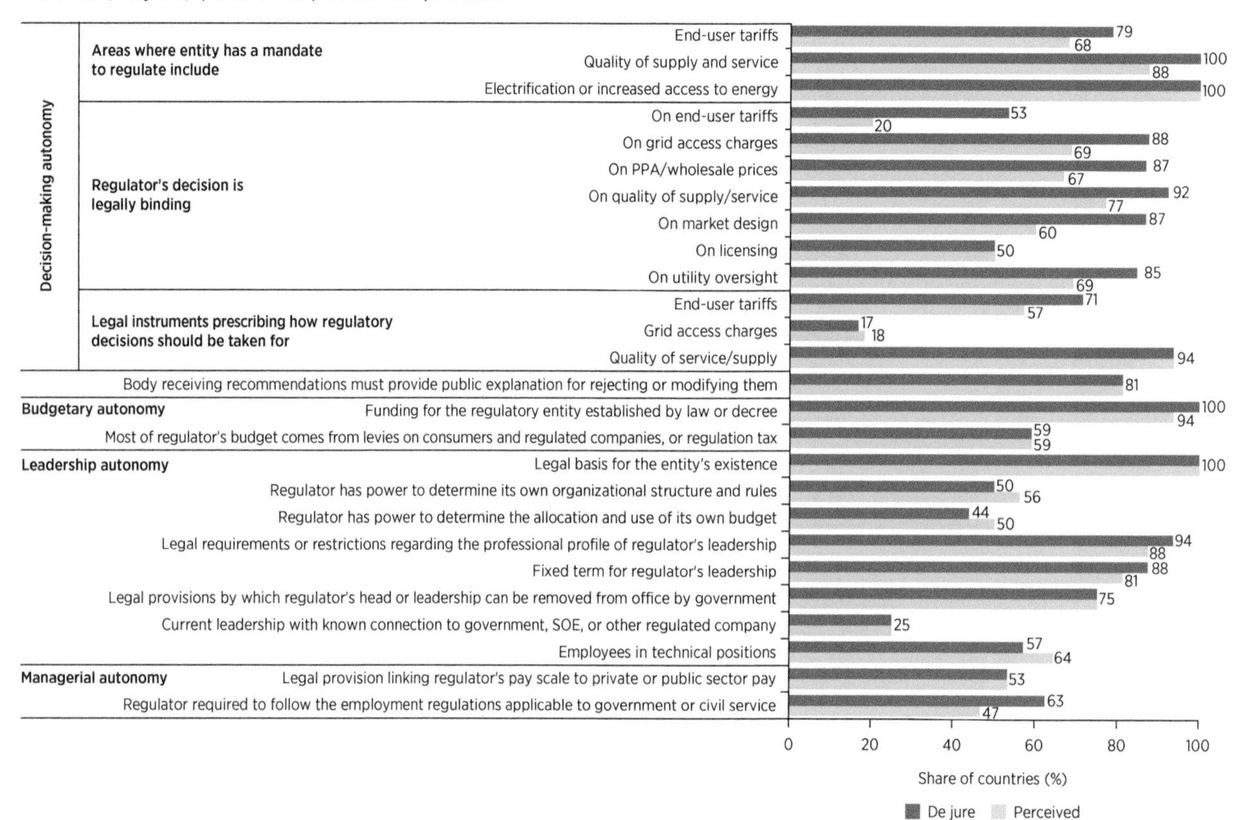

Source: Based on Rethinking Power Sector Reform utility database 2015.
Note: PPA = power purchase agreement; SOE = state-owned enterprise.

organizational structures and rules. Regarding leadership, multimember commissions are the norm, with such commissions having between 3 and 13 members. Commissioners are selected mostly by the executive, either the head of state or the ministry of energy, though in a few cases a special independent committee is charged with the selection process. In almost all cases, there are legal requirements for the professional profile of the leadership. The law specifies fixed terms for their mandates (anywhere between three and seven years), which can in most cases be renewed once or at most twice.

Despite these legal provisions, achieving autonomy for the leadership of regulatory entities can be challenging. Egypt provides a particularly striking example, inasmuch as the minister of energy chairs both the board of the regulatory agency, EgyptERA, and that of the national power utility holding company, the Egyptian Electricity Holding Company (EEHC). Furthermore, tariff decisions made by EgyptERA are subject to further approval by the cabinet. In India, government control is exerted through the appointment of commissioners to the federal- and state-level regulators—the Central Electricity Regulatory Commission (CERC) and the State Electricity Regulatory Commissions (SERCs), respectively. Elsewhere, autonomy of regulatory commissioners can be revoked through their removal from office before they complete their terms. Tanzania appoints its regulatory commissioners in a transparent and competitive fashion. Nevertheless, board members have been removed from office prior to the completion of their full term. Similarly, in the Dominican Republic, the president has been known to change the regulatory leadership, and in Senegal the head of the regulatory entity was reassigned to another public agency.

Finding #4: Many countries embraced incentive regulation, but tariff regimes tend to be poorly designed and the incentives questionable

At the heart of tariff regulation is the regulatory regime used to adjust tariffs over time. The two broad schools of thought on regulatory regimes are known as rate-of-return regulation and incentive-based regulation.

Rate-of-return regulation originated in the United States and seeks to create a stable environment for investment. According to Jamison (2007):

> Rate of return regulation adjusts overall price levels according to the operator's accounting costs and cost of capital. In most cases, the regulator reviews the operator's overall price level in response to a claim by the operator that the rate of return that it is receiving is less than its cost of capital, or in response to a suspicion of the regulator or claim by a consumer group that the actual rate of return is greater than the cost of capital.

Key features of pure rate-of-return regulation are that price adjustments can be as frequent as required to maintain the rate of return, and in practice are often annual; that price adjustments are made in such a way that all costs are passed through to the consumer, including any historic deviations from anticipated costs; and that regulators often get involved in setting tariff structures. The main concerns associated with rate-of-return regulation are pass-through of inefficient or imprudent costs to the consumer, and potential overinvestment in service quality as a response to the guaranteed rate of return on investment.

Incentive-based regulation originated in the United Kingdom and seeks to drive down operational inefficiencies. Following an earlier application to the telecommunications sector, this kind of regulation was first introduced to the power sector by the regulator Stephen Littlechild as part of the power sector reform of the 1990s. Incentive-based regulation normally sets a price or revenue cap that is then fixed for a period of several years (Joskow 2014). Under price cap regulation, the utility is required to keep the weighted average increase in its basket of prices beneath the increase in a specified price index, minus X percent. This means that prices should decline by X percent per year in real terms, where X is an efficiency factor based on anticipated productivity improvements in the sector. Because profits can be made only by outperforming the price cap over time, this type of regulation creates strong efficiency incentives. The four key features of pure price cap regulation are as follows: (1) infrequent price reviews give utilities time to respond to the inherent efficiency incentives—in practice, every four to five years; (2) prices come to be based on efficient market benchmarks rather than the actual costs of the utility; (3) there is no cost pass-through or retrospective adjustment for historic cost deviations; and (4) regulators are not involved in setting tariff structures. Incentive-based regulation can create the potential for utilities to underinvest as a means of increasing profits and thereby compromising quality of service; in addition, investment risks are higher under a regime that permits utilities to make supernormal losses or profits.

In practice, the pure forms of both types of regulation define extremes on a spectrum populated with many intermediate approaches.

The practice of either regime is seldom extreme, because regulators practicing price caps will refer to a utility's actual costs in setting the cap, and regulators practicing the rate-of-return model will do some form of benchmarking to ensure that imprudently incurred costs are not passed through to consumers. In general, it is possible to create a range of hybrid or intermediate regimes between the pure forms by adjusting the length of the review period, using efficiency benchmarking, and varying the degree of cost pass-through and the extent of retrospective carryover of cost deviations.

There has been considerable debate regarding the suitability of rate-of-return and incentive-based regulation for developing countries (Alexander 2014; Kessides 2012; Laffont 2005). Because it is supposedly less information intensive, incentive-based regulation is said to be more suited to environments where audited accounts are scarce. Nevertheless, incentive-based regulation was developed for mature markets that do not require major investments and where the focus is driving out inefficiency. Most developing countries have rapidly growing demand and a major investment backlog; the more stable investment incentives associated with rate-of-return regulation may be more relevant. Developing countries also suffer from inefficiencies, but incentive-based regulation reduces inefficiencies only when the utility operates under a strong profit motive. This motivation is not always the case in countries where utilities continue to be state-owned. Furthermore, developing countries may lack the institutional, legal, and financial acumen needed to apply incentive-based regulation.

The spread of developing countries adopting rate-of-return and incentive-based regulation appears even, although most lie somewhere in between. Among the 15 observatory countries, approximately one-third follow incentive-based regulation, whereas one-third follow rate-of-return regulation; the remainder have hybrid systems. This self-identified mechanism, coupled with the length of the regulatory period, gives a sense of the incentive power of the regimes in the analyzed countries:

FIGURE 6.12 Most regulatory regimes are closer to rate-of-return regulation, with some incentive-based elements

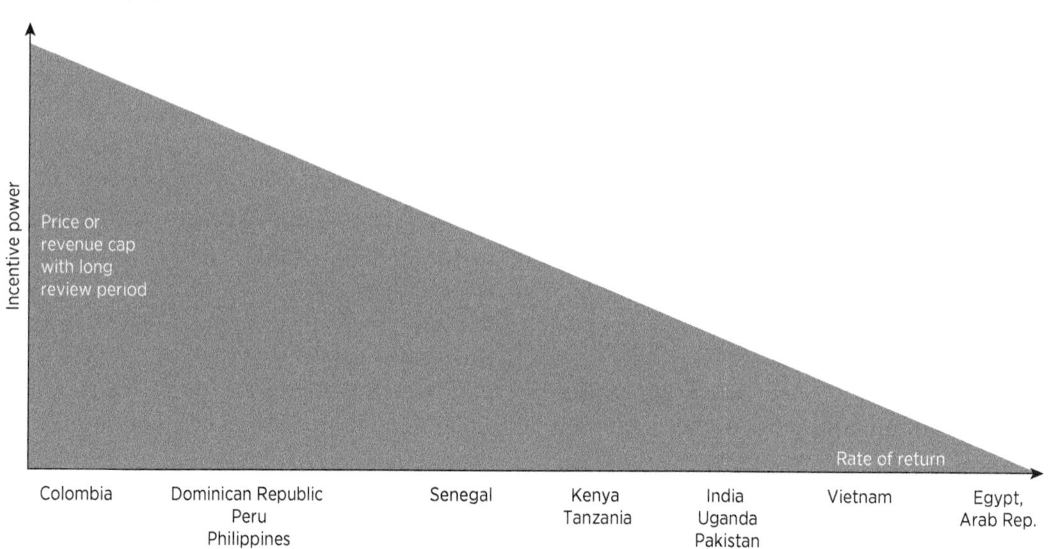

Source: Based on Rethinking Power Sector Reform utility database 2015.

incentive-based regimes with relatively long regulatory periods would give greater incentives to efficiency than cost-plus regimes (figure 6.12).

Nevertheless, the design details of the regulatory regime are often more informative than official descriptions. Countries practicing pure incentive-based regulation would tend to have long review periods, combined with little cost pass-through, and full write-off of historic cost deviations. Close analysis of the design of different regulatory regimes shows that almost all of them mix features, so that almost no country perfectly corresponds to either regime (table 6.1). Some countries approach the incentive-based end of the spectrum (such as Colombia, Peru, the Philippines, and Senegal), whereas others are closer to the rate-of-return end of the spectrum (such as India, Kenya, Pakistan, Tanzania, Uganda, and Vietnam). In only two cases (Peru and Senegal) have regulatory regimes fully written off the costs of over- or underrecovery from previous tariff-control periods rather than considering them in the next period. In all other cases, those costs are either partially or fully included. In most cases, end-user tariff-setting regulations incorporate incentives to improve efficiency by using benchmark efficiency or quality parameters to determine prices, which are usually related to transmission and distribution losses. This is the case in nine countries.[3] In Kenya, for example, the aggregate transmission, distribution, and nontechnical losses benchmark was 16 percent in the period 2015–16. For Pakistan utilities, transmission losses were set at 3 percent and distribution losses at 15 percent. In the Philippines, these targets vary by utility—in the case of MERALCO, for example, the system loss target is 9 percent. Finally, maximum technical and commercial losses are 12 percent in the Dominican Republic.

The tariff-setting methodology is remarkably consistent across countries, regardless of the regulatory regime. Where we have information on the tariff formula, it usually includes operating expenditure, depreciation, taxes, and a return on capital. Most countries base depreciation on the straight-line method applied to the historic cost–asset valuation. The allowed weighted average cost of capital is typically based on a sector-risk premium methodology (such as the Capital Asset Pricing Model) applied to the asset base. A handful of countries work with a preestablished fixed numeric value for the cost of capital. For the few cases where it is publicly available, this value lies in the 12–14 percent range. Interestingly, in the case of Colombia and Peru, investments are converted into annuities.

Nevertheless, regulatory accounting standards have yet to be developed in many countries. For regulators to be able to meaningfully interpret accounting information in determining the cost of service, it is important that this information be submitted according to regulatory accounting guidelines. Such guidelines have been developed in only about half of the countries considered. Good-practice examples include Colombia, Egypt, Pakistan, Peru, the Philippines, Tajikistan, Uganda, Ukraine, and Vietnam.

Automatic indexation mechanisms are prevalent and usually provide protection against oil price shocks, foreign-exchange movements, and domestic inflation. Two-thirds of the observatory countries practice automatic indexation of tariffs, which is important in the developing country context owing to the power sector's exposure to oil price and foreign-currency shocks that lie beyond the control of utilities. These adjustments are usually done quarterly, though in some cases (Dominican Republic, Kenya, and Pakistan) they are done every month. Semiannual inflationary adjustments are seen in Kenya and Tanzania. The most common elements for cost pass-through are oil prices, foreign exchange, and domestic

TABLE 6.1 Design characteristics of tariff regulatory regimes: An overview

Country	Regulatory regime	Tariff review period (years)	Use of efficiency benchmarks	Cost pass-through	Cost deviations		Basis of asset valuation	Basis for depreciation	Basis of rate of return
					OPEX	CAPEX			
Colombia	Incentive based	5	Yes	No	Partially carried over		Replacement cost	Straight line	Based on sector risk premium (13–14%)
Dominican Republic	Incentive based	4	Yes (AT&C 12%)	Yes (CPI, PPI, XR, oil price, PPA)	Wholly carried over		Market value / Replacement cost	Straight line	Fixed numeric value
Egypt, Arab Rep.	Rate of return	No fixed period	No	No	Written off	Partially carried over	Net historic cost	Straight line	Based on sector risk premium
India	Rate of return plus incentives	1	No	Yes (unspecified)	Partially/ wholly carried over		Net historic cost	Straight line	Other
Kenya	Rate of return	3	Yes (AT&C 15.9%)	Yes (CPI, XR, oil price, PPA)	Wholly carried over		Net historic cost	Straight line	Fixed numeric value
Pakistan	Rate of return plus incentives	1	Yes (AT&C 18.25%)	Yes (CPI, oil price)	Partially carried over	—	Net historic cost	Reducing balance method	Based on sector risk premium
Peru	Incentive based	4	Yes	Yes (CPI, PPI, XR, oil price, PPA)	Written off	—	Replacement cost	Straight line	Annuity computed using a 12% discount rate over 30 years
Philippines	Incentive based	4	Yes	Yes (CPI, XR)	Partially/ wholly carried over	Subject to regulatory discretion	—	Straight line	Based on sector risk premium (12%)
Senegal	Incentive based	3	No	Yes (unspecified)	Written off		Net historic cost	Straight line	Based on sector risk premium
Tajikistan	—	—	No	No	Written off	—	Net historic cost / Market value / Replacement cost	Sum of the years' digits	Fixed numeric value (12.5%) Based on sector risk premium
Tanzania	Rate of return	3	No	Yes (PPI, XR, oil price, PPA)	Wholly carried over	Partially carried over	Net historic cost	Straight line	Based on sector risk premium
Uganda	Rate of return plus incentives	1	Yes	Yes (CPI, PPI, XR, oil price)	Wholly carried over		Net historic cost	Straight line	Fixed numeric value
Vietnam	Rate of return	6 months	Yes	Yes (CPI, PPI, XR, oil price, PPA, generation mix)	Partially carried over		—	Straight line	—

Source: Based on Rethinking Power Sector Reform utility database 2015.
Note: Information unavailable for Ukraine and inapplicable for Morocco. AT&C = aggregate technical & commercial losses; CAPEX = capital expenditure; CPI = consumer price index; OPEX = operating expenditure; PPA = power purchase agreement; PPI = producer price index; XR= exchange rate; — = not available.

inflation, which are included in two-thirds of indexation formulas. Several countries also include other elements, such as force majeure (figure 6.13).

Uganda provides an example of a carefully developed automatic indexation mechanism. Base tariffs are adjusted quarterly to account for inflation, exchange rate, and fuel costs. These adjustments apply only to the energy charge (not to fixed monthly or maximum demand charges or to reconnection fees or the lifeline end-user tariff) and are capped so as not to increase end-user tariffs by more than 2.5 percent in any given quarter. Inflationary adjustments are applied only to the local currency portion of operating expenditure, and exchange-rate adjustments are applied only to the foreign currency portion of operating expenditure as well as to the return on capital (given that the utility operator is a foreign private sector investor).

FIGURE 6.13 Two-thirds of the observatory countries practice automatic indexation of tariffs

By index component

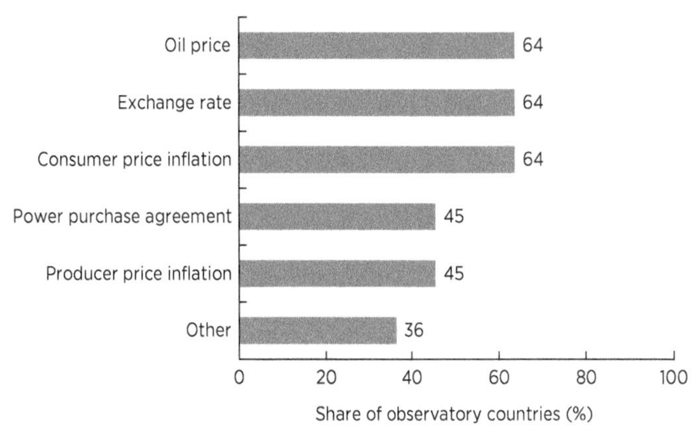

Source: Based on Rethinking Power Sector Reform utility database 2015.

Finding #5: Although tariff setting is the central function of regulatory agencies, their tariff recommendations are not necessarily respected or applied

Tariff regulation on paper diverges substantially from the practice of tariff regulation. The gap between de jure and perceived scores is as large as 20 percentage points for some critical elements of tariff regulation (figure 6.14). Particularly striking is the fact that, although most countries (nearly 90 percent) have a clear regulatory framework for tariff setting, barely 60 percent rely on this framework in practice. Even more telling, although 94 percent of countries give the regulatory entity authority over tariff setting, this authority is perceived not to prevail in some 35 percent of cases. Tajikistan and Vietnam are the only two countries where regulatory tariff setting is not legally binding, but several other countries have regulators that, in practice, play only an advisory role in tariff setting, including the Dominican Republic, India, Kenya, Senegal, and Tanzania. Discrepancies between de jure and perceived scores for tariff regulation at the country level can be as large as 20 percentage points—even 50 percentage points in the case of the Dominican Republic (figure 6.15).

In India, where a common regulatory framework for tariffs exists at the federal level, the practice of tariff regulation nonetheless varies substantially across states, reflecting local political dynamics. Across three Indian states (Andhra Pradesh, Odisha, and Rajasthan), actual tariff adjustments in local-currency terms have been on the order of 200 percent since 2010, and the corresponding regulators had mandated adjustments of 400–700 percent over the same period (figure 6.16). The politically sensitive nature of tariff decisions appears to produce the systematic rejection of regulatory decisions.

The state of Rajasthan provides a dramatic illustration of how the regulator's political authorizing environment shifts over time, with potentially dire financial consequences for the utility. The Rajasthan Electricity Regulatory

FIGURE 6.14 **Substantial divergence exists between tariff regulation as it appears on paper and as it is actually practiced**

Formal (de jure) power vs. perceived practice

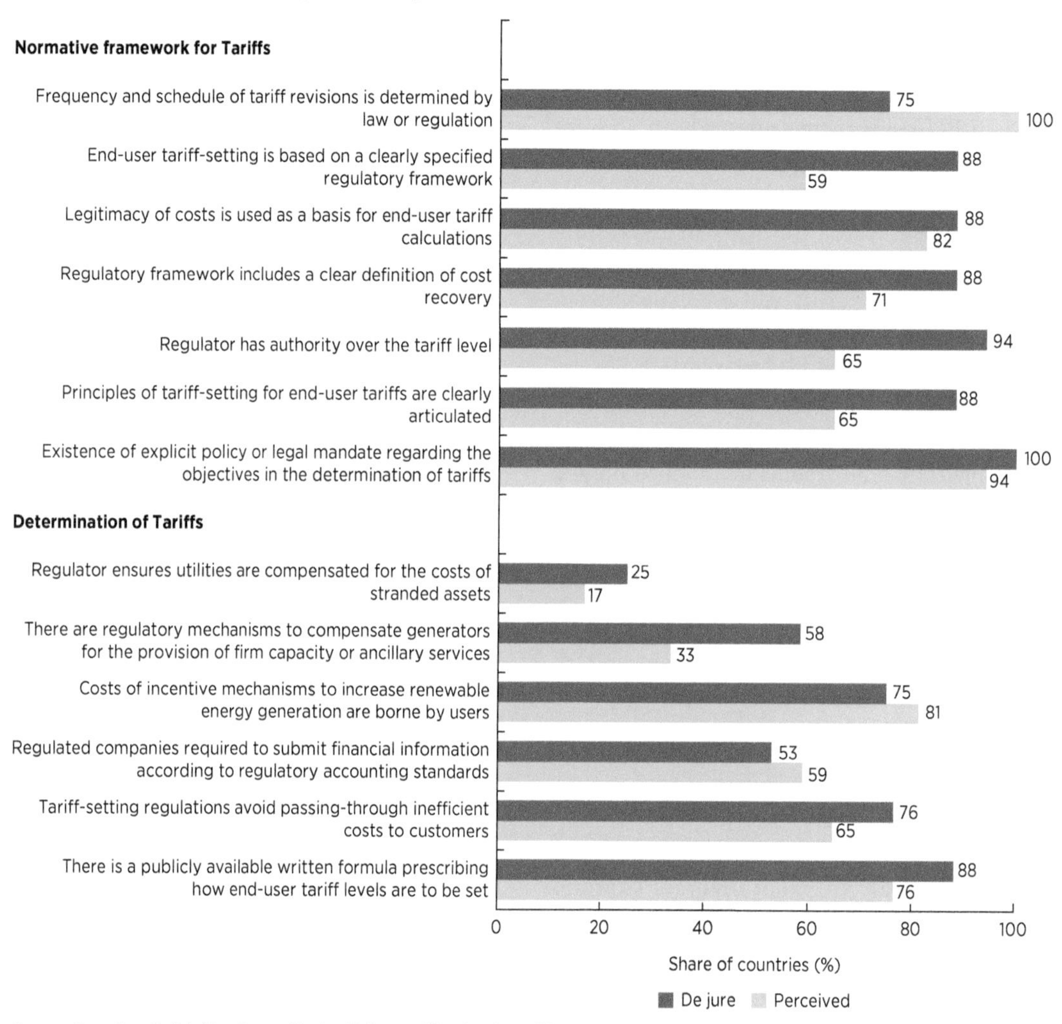

Source: Based on Rethinking Power Sector Reform utility database 2015.

Commission (RERC) was established in 2000. Despite strong legislation that laid out clear objectives for determining end-user tariffs, the regulator never really managed to take control of the tariff-setting process. Until 2004, the state government allowed regulatory tariff regulations to be partially implemented. Gradually, losses built up to US$50 million. The next government adopted a more radical position, refusing to consider any tariff revisions recommended by the regulator. A decade-long tariff freeze caused the utility's losses to reach a cumulative value of US$9 billion by the end of 2014, higher than in any other Indian state. Meanwhile, distribution losses and quality of supply remain problematic. The regulator is unable to implement its orders on the state-controlled utilities.

Countries across Sub-Saharan Africa have had a variety of experiences with regulatory tariff setting. In Uganda, for instance, tariff adjustments tracked those mandated by the

FIGURE 6.15 **Tariff recommendations made by regulatory entities are not necessarily respected or applied**

Formal (de jure) power vs. perceived practice

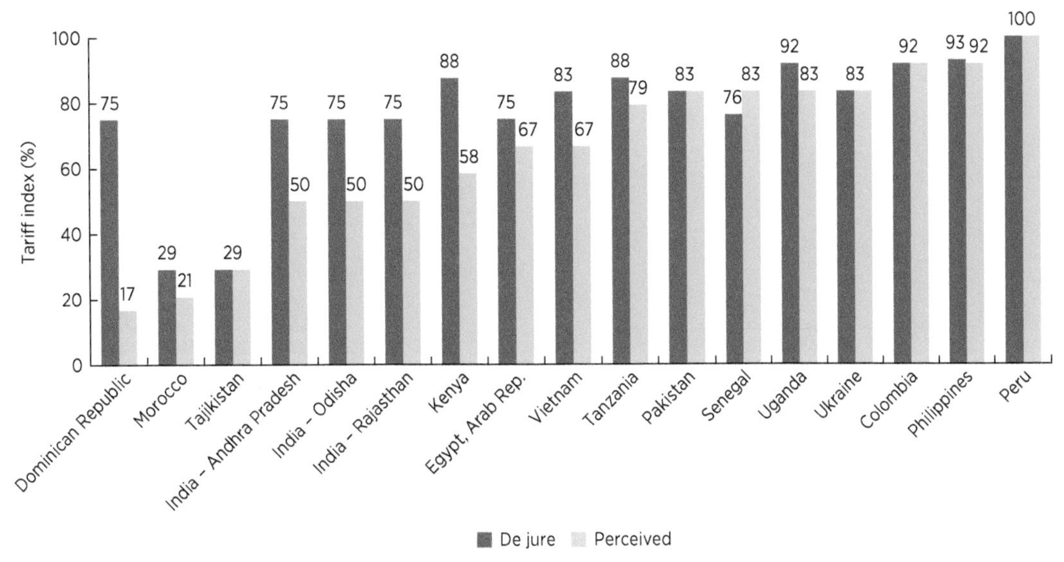

Source: Based on Rethinking Power Sector Reform utility database 2015.

FIGURE 6.16 **In India, authorized tariff increases fall well short of those approved by state regulators**

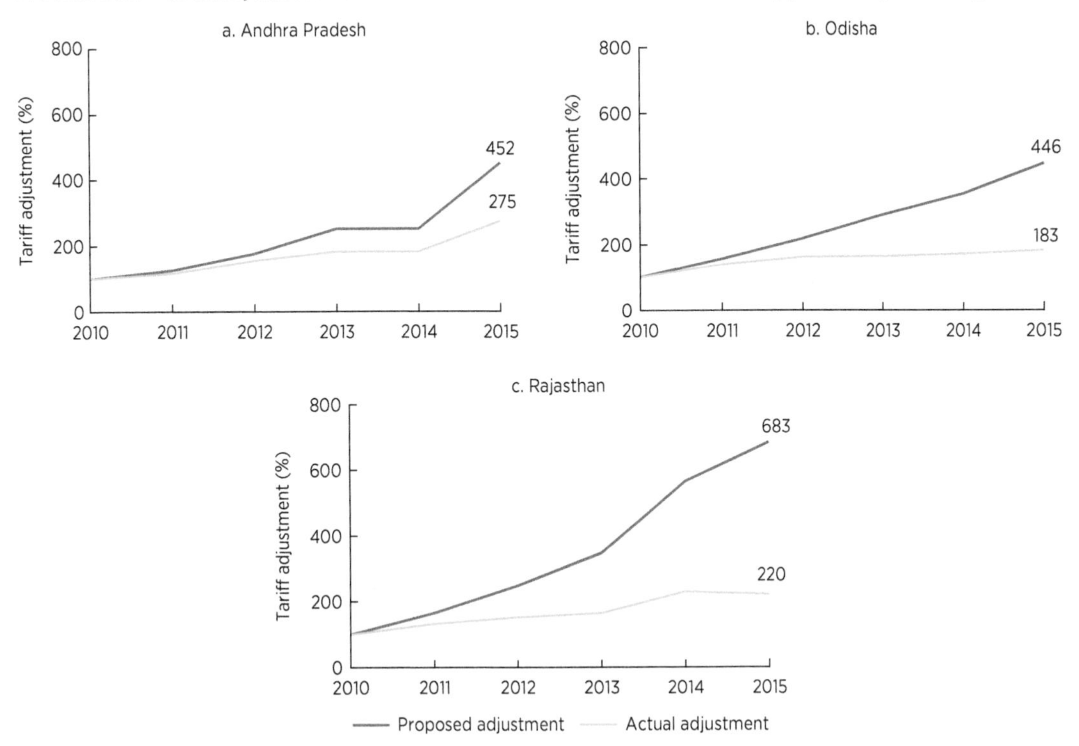

Source: Based on Rethinking Power Sector Reform utility database 2015.

FIGURE 6.17 **Tanzania and Uganda represent the extremes of the range of experiences of African countries**

Proposed vs. actual adjustments, 2008–17

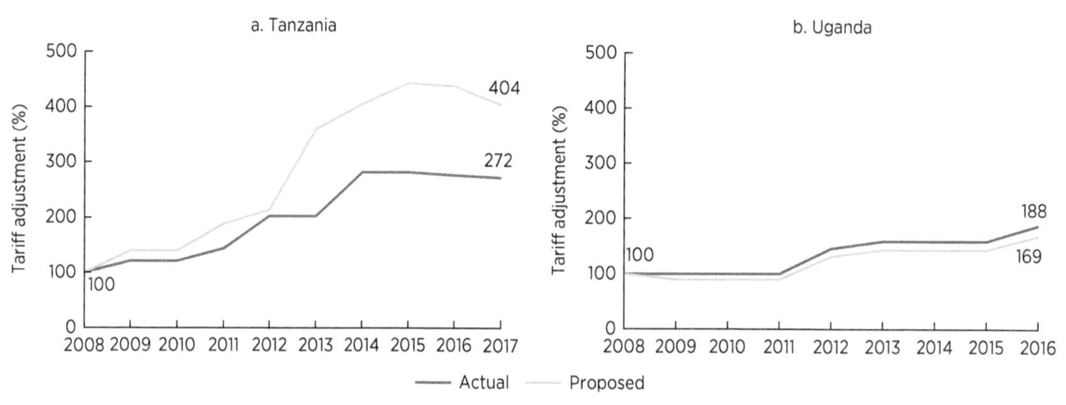

Source: Based on Rethinking Power Sector Reform utility database 2015.

regulator fairly closely from 2008 to 2016. That a private concessionaire was running the national utility partly explains why mandated tariff adjustments were largely honored. In Tanzania, regulatory tariff adjustments were largely implemented from 2008 to 2012, though the escalation of costs associated with the drought period meant that subsequent tariff adjustments were kept far below levels mandated by the regulator (figure 6.17). Not only were regulatory decisions overturned, but also the senior leadership and management of the regulatory agency were sacked. In Senegal, tariffs have been frozen since 2009 by government fiat, irrespective of regulatory advice.

Kenya's experience with tariff regulation illustrates the interaction with the electoral cycle, as well as the role that can be played by automatic indexation. The ERC replaced the ERB in 2007 and is considered one of the better-performing regulators in the region, having greatly improved its technical capacity over the years. Solid legislation has given the regulator a stronger voice in decision making. The ERC has moved tariffs closer to cost-recovery levels despite government disregard for the regulatory process. According to law, ERC is meant to conduct a tariff review every 3 years, yet over a period of about 20 years only three reviews were permitted: in 1999, 2008, and 2013. Political sensitivities have led successive governments to prevent full tariff reviews during electoral periods. Nevertheless, Kenya's retail tariffs incorporate a cost pass-through mechanism for the monthly indexation of foreign-exchange fluctuations and fuel costs, as well as semiannual indexation of domestic price inflation. Despite the prohibition of regular tariff reviews, this automatic mechanism has been allowed to function smoothly since its introduction in 1997; it partially compensates for the absence of regular tariff revisions. This changed in 2017, an election year, when drought conditions would have led to a large indexation adjustment owing to higher fuel costs, but the indexation was permitted to pass after the elections were over.

Finding #6: In countries where cost-recovery tariffs recommended by the regulator are not implemented, the regulator may play a role in ensuring that sector revenue requirements are met through subsidies

Ideally, regulators should be able to set tariffs at cost-recovery levels and enforce their application, but this is not always politically feasible. Most regulators have the formal legal authority to set tariffs. This authority can be overridden, however, by concerns about the social impact of higher tariffs during electoral periods. As a result, tariff adjustments in many jurisdictions have been only a fraction of what the regulator considered necessary, leaving utilities with a hefty financial gap. In principle, the financial viability of the utility can be safeguarded if below-cost tariffs are offset by fiscal transfers large enough to meet the shortfall. Where cost-recovery tariffs cannot be fully implemented, regulators can counsel a combination of tariff and subsidies to meet the utility's revenue requirements. The choice of the exact tariff and subsidy combination is therefore left to the political authorities.

Regulators have sometimes counseled on the magnitude of subsidies needed to compensate for below cost-recovery tariffs, but this approach is risky. Both Egypt and Senegal used a similar approach, and it met with some temporary success. The regulator does not really have the power to hold the ministry of finance accountable for paying the requisite subsidy, leaving the approach vulnerable during periods of fiscal duress. Moreover, once established, such arrangements may be difficult to reverse and in some specific cases may even lead to sustained inefficiency in utilities as seen in Pakistan.

In Egypt, regulations stipulate that the regulator must inform the Cabinet of Ministers of the compensating subsidy owed to the utility if cost-recovery tariffs are not approved. The regulatory agency, EgyptERA, has authority over the tariff level and structure as well as the frequency of tariff revisions; but every tariff order requires cabinet approval, and the cabinet does not always endorse the adjustments recommended by EgyptERA. Following the enactment of a new energy law in 2015, the associated regulations stipulate that the cabinet may choose to set tariffs below the level recommended by EgyptERA only if the government provides a compensating subsidy to the relevant utilities. EgyptERA informs the cabinet of the tariff-subsidy combinations that are compatible with financial equilibrium for the sector. The minister of finance is involved in these decisions, because the level of subsidy has to be accounted for in the national budget. This approach has been successfully applied since 2014.

In Senegal, the regulator played a key role in calculating the magnitude of fiscal subsidies needed to maintain the financial equilibrium of the utility, at least for a time. According to the 1998 law, regulation of electricity tariffs in Senegal is based on a revenue-cap system that determines the revenue required for the economic and financial viability of the utility, National Electricity Company of Senegal (Senelec). In 2008, when the last 18 percent increase in tariffs occurred, the Senegalese government decided to freeze tariffs, promising compensation to Senelec. No payment schedule was specified, however, delaying compensation and forcing Senelec to take on expensive commercial bank debt to continue operations. A legal amendment introduced in 2011 aimed to tackle this issue by having the regulator calculate Senelec's maximum authorized revenue on a quarterly basis and requiring the ministry of finance to pay compensation accordingly. If compensation cannot be paid, the government must provide Senelec with a "letter of comfort," allowing the company to borrow from banks while the government commits to guarantee all financial fees and principal debt repayment. The introduction of a more formal, legally based system for compensation payments resulted in more disciplined fiscal transfers for several years, until Senegal faced fiscal challenges in 2017 and compensation payments once again ceased.

Pakistan's experience with so-called tariff differential subsidies illustrates the perverse efficiency incentives that can arise from this approach. In Pakistan, for social reasons, the retail pricing approach involves both a tariff determined by the regulator and a tariff notified by government; the latter is the one actually applied. The regulatory tariff is computed to afford the utility adequate revenues to cover its costs after meeting certain efficiency parameters for revenue collection, system losses, and fuel purchases. The difference between the government's notified tariff and the regulator's cost-based tariff, known as the tariff differential subsidy, is paid by the government to the utilities. In principle, this subsidy should assure the financial viability of the sector. In practice, however, it does not; because utility performance falls short of the efficiency benchmarks the regulator uses to determine the cost-based tariff, losses are incurred despite the subsidy. These losses become circular debt when companies with insufficient revenues fail to pay their suppliers in full. Because the government is ultimately responsible for this debt through its ownership of state-owned companies, and because such circular debts are periodically paid off by the government, managers of the loss-prone utilities have little incentive to improve the efficiency of their operations.

Finding #7: Quality-of-service regulation leaves much to be desired

Quality-of-service regulation is an important complement to tariff regulation, particularly under incentive-based regulatory regimes. Regulation of service quality entails first establishing a suitable quality standard and then creating a system of incentives to induce utilities to meet that standard (Adam 2011). Quality standards should be industrywide, clearly defined, and long term to provide regulatory certainty and foster investment. The desired quality level should be informed by a balancing of costs to the industry against benefits to the consumer, because standards can sometimes be set too high. Incentive mechanisms for meeting quality standards will likely require a combination of financial penalties for failing to meet the standard, with adequate capital expenditure allowances in the regulatory revenue base so that the investments needed to meet the standard can be funded. Financial penalties need to be set at a level high enough to affect behavior and should ideally reflect the cost of outages to consumers. For example, in the European Union, the economic costs suffered by customers as a result of power outages (known as Value of Lost Load) are estimated to range from €5 to €10 per kilowatt-hour. Quality-of-service penalties are critical in price cap regulatory regimes, where the utility may face incentives to cut costs by reducing quality of service. Beyond financial penalties, regular, transparent reports on the achievement of quality standards can also affect a utility's public reputation and improve its performance.

Quality-of-service regulation is close to universal in developing countries, but enforcement is weak, even on paper. Near-universal formal quality standards exist for

- Product quality (such as frequency or voltage variations),
- Service quality (interruptions), and
- Customer service (for example, response to complaints).

Utilities are legally required to meet these standards and must submit data periodically regarding their compliance. Despite the standards, however, enforcement remains weak. Only about one-third of countries in the sample attaches positive or negative financial incentives to the achievement of quality standards, and only half of the countries publishes information on compliance.

Moreover, the discrepancy between the official quality-of-service regulation and actual regulatory practice is particularly large in some cases. Whereas the average score for formal (de jure) quality-of-service regulation across countries is 75 percent, it drops to 51 percent in perceived terms. The situation varies drastically across countries (figure 6.18). In middle-income countries of Latin America, East Asia,

FIGURE 6.18 Quality of service regulation has some of the largest gaps between rules on paper and in practice

Formal (de jure) power vs. perceived practice

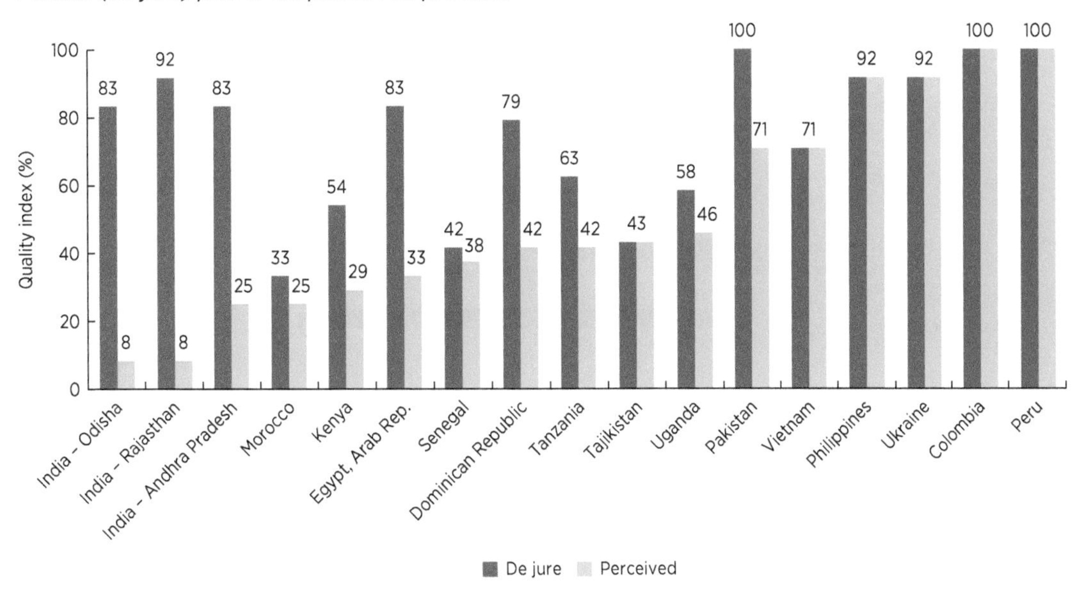

Source: Based on Rethinking Power Sector Reform utility database 2015.

and Eastern Europe—such as Colombia, Peru, the Philippines, Ukraine, and Vietnam—a good system appears to be regulating quality of service, and it also functions effectively in practice. For the countries of Sub-Saharan Africa—Kenya, Senegal, Tanzania, and Uganda—the de jure scores for quality of regulation are much lower, at about 40–60 percent, and even these partially developed regulatory frameworks are not fully observed. Particularly striking is the case of the Indian states, where the de jure framework for quality-of-service scores are relatively high at 80–90 percent, but the perceived scores range from 8 to 25 percent. The situation is similar for the Dominican Republic and Egypt.

Many fundamental components of a quality-of-service regulation system fail to be applied in practice. A closer examination reveals numerous areas where formal features of the regulatory framework for quality of service are overlooked in practice. Although all countries have legal requirements for quality-of-service regulation and standards, only about two-thirds actually publishes such standards. Fines for noncompliance are a legal requirement in about 60 percent of cases but have been defined in only 24 percent. Whereas all utilities are required to report quality-of-service data to the regulator, only about 70 percent does so, and little more than half operates an automated information system. In about 65 percent of countries, quality performance is supposed to be made public, but this happens in only about 40 percent of cases. The lack of compliance with quality-of-service regulation can be attributed not only to the inefficiency of the utilities but also in some cases to standards being set at unrealistically high levels. Many countries report that utilities try to observe standards but fail because of the technical challenges (figure 6.19).

Colombia provides a good counterexample of a country where quality-of-service regulation is working effectively. Prior to the power sector reforms of 1994, Colombia barely considered quality-of-service regulations. The new regulator, CREG, published

FIGURE 6.19 **Quality-of-service regulations are not widely implemented, often for lack of technical capability within utilities**

Formal (de jure) power vs. perceived practice

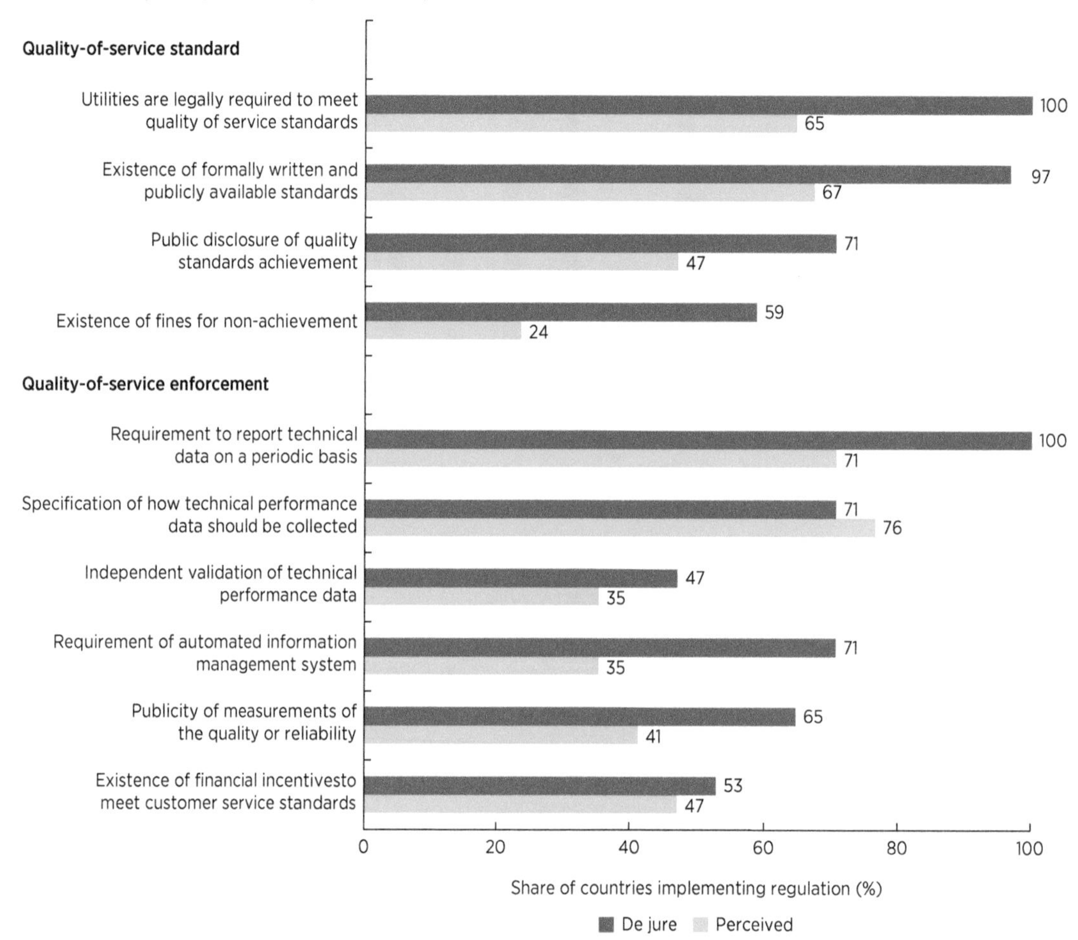

Source: Based on Rethinking Power Sector Reform utility database 2015.

quality-of-service regulations for power distribution in 1998, adopting the international standards, SAIFI (System Average Interruption Frequency Index) and SAIDI (System Average Interruption Duration Index), as the two main indicators. A third indicator capturing the percentage of lost load was added in 2008. Each of Colombia's four geographical regions had a different standard to meet—variations that acknowledged the varying local distribution systems and the magnitude of the associated technical challenges. Under the current system, network operators record duration and frequency of outages for their respective users on each circuit and voltage transformer. Figures are reported monthly to the regulator, which then compiles a quarterly review of compliance against maximum allowed values for each of the four regions. Utilities take quality-of-service standards seriously and generally comply, inasmuch as shortfalls result in compensation payments to users. The quality standards, together with the performance data reported by the utilities and information on any resulting penalties paid, are all available to the public. Formal measurements of overall customer satisfaction are also required and undertaken both by utilities and the regulator, both through customer surveys and website comment forms.

Finding #8: Regulators play a role in licensing market entry but are not always involved in the power purchase agreements, a critical area

With most developing countries opening power generation to IPPs, they need a reliable way to admit new players into their markets. A licensing process for new entrants, one that is operated by the regulator, is a good way to go given the sensitive nature of these activities. In some cases a government agency, such as the ministry of energy, could oversee licensing new entrants. Regulators are sometimes involved with procuring IPPs, or they review the terms of PPAs prior to their signature to ensure that the PPAs represent value for money, because costs will be passed on directly to consumers through retail tariffs.

Market-entry regulations either are underdeveloped or fail to be properly implemented. Overall, countries scored 77 percent on the de jure regulatory framework for market-entry regulation, dropping to 56 percent for perceived regulation. Most countries with well-developed formal regulations for market entry lag on practice (figure 6.20). This finding is true in the Dominican Republic, Kenya, the Philippines, Tanzania, and Vietnam. In Egypt, India, and Pakistan, even the formal framework is not well defined. Colombia and Peru have advanced regulatory frameworks, but their scores on practice are not especially high.

The most serious deficiencies in the regulatory framework for market entry relate to the revocation of licenses and the award of IPPs. The evidence suggests that, once licenses are awarded, regulators fail to monitor compliance with license conditions. Worse, they do not impose the legally stipulated fines, making it difficult to force nonperforming companies to relinquish their licenses (figure 6.21). Particularly striking is that, whereas half the countries empower regulators to conduct IPP procurement, only about a quarter of them

FIGURE 6.20 Most countries with well-developed formal regulations for market entry lag far behind in terms of practice

Formal (de jure) power vs. perceived practice

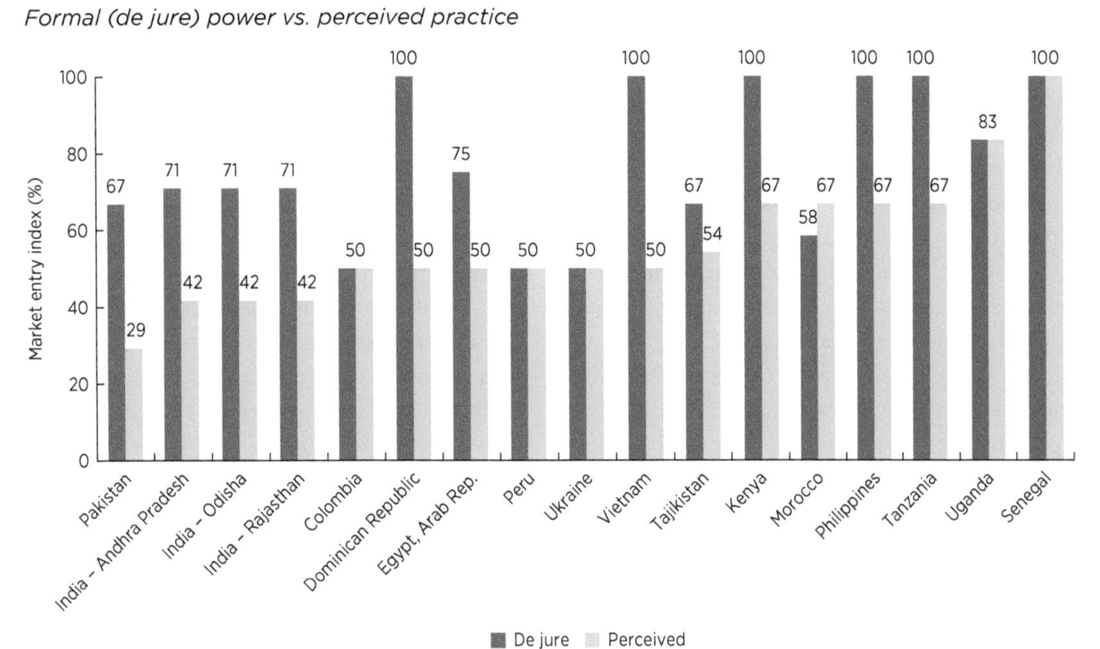

Source: Based on Rethinking Power Sector Reform utility database 2015.

FIGURE 6.21 Serious deficiencies relate to the revocation of licenses and the award of IPPs
Formal (de jure) power vs. perceived practice

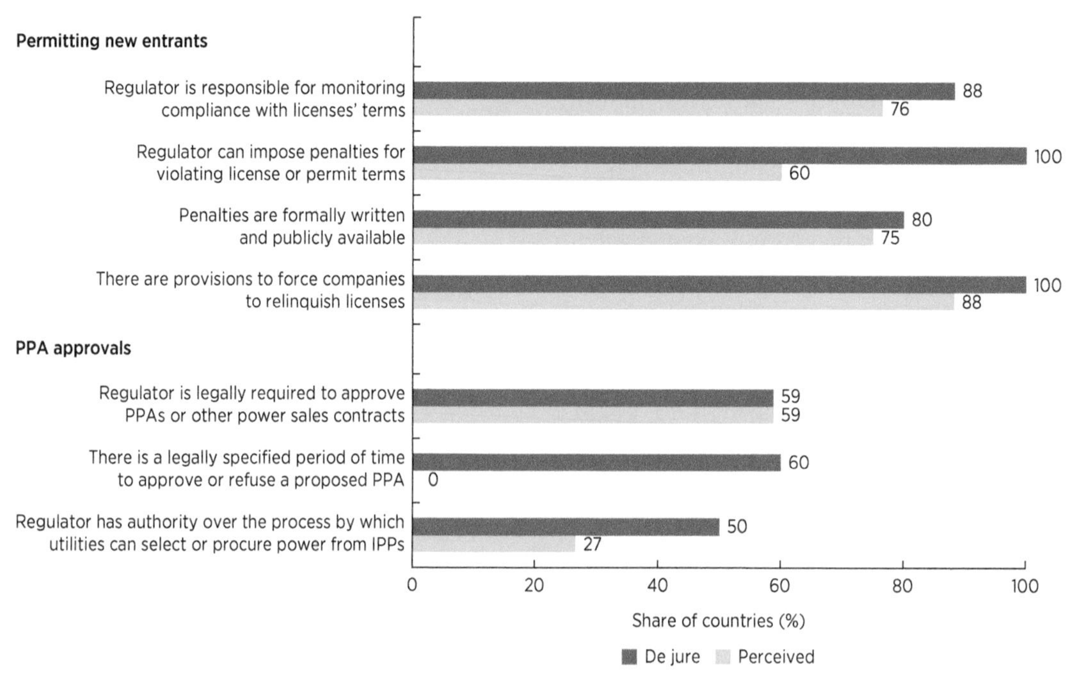

Source: Based on Rethinking Power Sector Reform utility database 2015.
Note: IPPs = independent power producers; PPA = power purchase agreement.

actually allow the regulator to do so in practice. Regulators generally review the terms of PPAs, but they also, invariably, exceed legal time limits for doing so.

Pakistan has an ineffective system of market-entry regulation, with damaging results. Pakistan has allowed IPPs in generation since 1994 under a clear framework for market-entry regulation. In practice, however, new entrants (particularly for solar) struggle through the initial stages of approval. The major roadblocks involve land allocation. First, for both wind and solar technologies, the number of letters of intent issued with associated land-allocation rights is greater than the land available and the interconnection potential of the grid, which means that only private developers with government connections will have land allocated and be able to move on to the next stage (feasibility studies). Second, the government issues interconnection permits based on the findings of a required interconnection study. Thirty days are allowed for government comments; in practice, delays can last a year or more, sometimes well past the expiry of the incentive program that originally encouraged the projects.

LOOKING AHEAD

Disruptive technologies will require new regulatory frameworks. The historical practice of regulation reviewed thus far does not consider the implications of new technologies, such as distributed energy resources, battery storage, and smart grids. Emerging experience from frontier markets suggests that such technological disruption is posing new challenges for regulators. Traditional practices of tariff regulation may thus need to be overhauled. The traditional cost-of-service model of utility regulation was well suited to advancing the policy objective of expanding the centralized grid (Graffy and Kihm 2014; Kihm and others 2015). Policy goals are now shifting, however, toward encouraging consumers to adopt distributed energy resources that serve to accelerate decarbonization, while also strengthening distribution networks (Graffy and Kihm 2014; Kihm and others 2015). As distributed energy resources become more readily available, the concern becomes how to adopt a regulatory model that ensures adequate cost recovery for the needed wires and poles of the distribution network infrastructure, while ensuring that there is also an incentive to embrace emission reduction, resiliency, and energy efficiency goals through increased reliance on decentralized resources.

Traditional cost-of-service approaches to rate making introduce incentives that run counter to the efficient integration of distributed energy resources. Perhaps the most widely adopted regulatory model for determining the revenue requirement of the distribution utilities is the cost-of-service approach, also known as rate-of-return regulation. According to this methodology, utility tariffs are set to earn a guaranteed return on a specified regulatory rate base of allowed assets plus an allowance for legitimate operating expenditures. This framework embeds a particular set of incentives, so utilities benefit financially either from increasing their investments or from decreasing their operating expenditures.[4] Moreover, because the focus of regulation is on the unit price, utilities also benefit financially from expansion in the volume of sales. All of these incentives run counter to the integration of distributed energy resources, which can be expected to reduce the need for capital investments by utilities, potentially increase the need for operating expenditures, and reduce the net flow of power from the utility to the customer. Utilities operating under traditional cost-of-service regulation have little incentive to embrace distributed energy resources.

Performance-based regulatory approaches provide utilities with more targeted incentives. The long-standing alternative to rate-of-return regulation is incentive- or performance-based regulation, as discussed earlier. Existing applications of performance-based regulation do not fundamentally change the overall cost-of-service model. Instead, they create mechanisms by which utilities can earn more revenue by improving operational efficiency, rather than by increasing capital expenditures. This has typically been achieved by lengthening the regulatory review period from one year to several years. With a longer review period, utilities that spend below the preapproved envelope for operating expenditures can hold on to the savings over a longer time period, strengthening their incentive to cut costs in the first place.[5]

Regulators are already adapting performance-based regulation to incentivize utilities to meet energy-efficiency goals and achieve specific quality-of-service outcomes. Newer forms of performance-based regulation introduce specific goals for peak reduction, resilience, emissions reduction, and customer satisfaction. Under performance-based regulation, utilities have the freedom to determine how best to reach those goals while meeting their targets for safety, reliability, affordability, and accessibility. Revenue caps, rate freezes, multiyear rate plans, or earnings adjustments based on achieving specific energy-efficiency targets (that is, revenue decoupling) are all examples of performance-based incentives.

The new regulatory challenge is to incentivize utilities to consider the "nonwires alternatives" alongside traditional grid investment. The challenges posed by incentivizing the adoption of distributed energy resources go beyond the examples cited above. The central issue is to encourage utilities to consider nonwires alternatives—or consumer-sited distributed energy resources—alongside (or even instead of) traditional grid investment. Customer-sited generation, demand response, energy efficiency, and battery storage could, in combination, serve as an alternative to a traditional distribution-grid upgrade, while producing the same result at lower cost. Where adopted, such nonwires alternatives could create more resilient distribution systems while avoiding investments that may prove difficult to recover.

Experiments underway in the United Kingdom and the U.S. state of New York illustrate how performance-based regulation can be adapted to encourage such technological innovation. The United Kingdom has introduced the "Revenue using incentives to deliver innovation and outputs" (RIIO) model. In the United States, New York is implementing the "Reforming the Energy Vision" (REV) model.

The British RIIO approach builds up the utility's revenue using a series of incentive-based components and further lengthens the regulatory review period. Under such a "total expenditure" (TOTEX) approach, capital expenditures and operating expenses are considered together rather than as separate costs, as in the past (Cave 2016). This approach does not directly tackle the uncertainty problem, but it dispenses with the need for a detailed review of cost forecasts for individual projects and may reduce the bias toward capital solutions. Additionally, TOTEX may, in some instances, be more stable over time and more comparable between companies than capital expenditure alone. The overall regulatory revenue requirement has three components (figure 6.22): incentives, innovation, and output (Ofgem 2015).

The incentives component focuses on the efficient delivery of outputs. The innovation component provides specific incentives for adoption of new technologies and permits third-party delivery of services. The output component focuses on service dimensions of concern to consumers. This approach departs from traditional cash-flow analysis, giving more weight to flexibility and optionality, and incorporating elements of output-based regulation. In order to ensure sufficient incentive for innovation under conditions of uncertainty, tariff review periods are lengthened beyond the traditional three-to-five-year cycle to a much longer eight-year cycle. According to Ofgem (2015), the first implementation of RIIO is proving successful, fundamentally altering behavior and board discussions at companies and encouraging stakeholders to present well-thought-out, detailed, and better-justified business plans.

FIGURE 6.22 **RIIO framework**

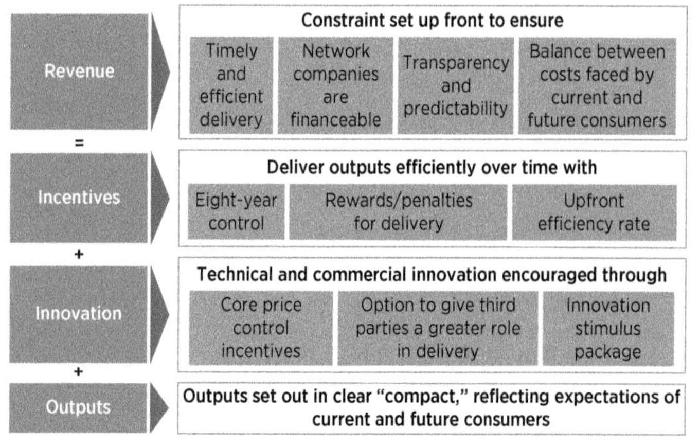

Source: Nixon 2013.
Note: RIIO = revenue using incentives to deliver innovation and outputs.

The New York REV model focuses on creating new earnings mechanisms and broadening the definition of efficiency gains. Under New York's REV procedures, utilities will provide distributed system platform services that enable a market for distributed energy resource providers. REV's goal is to integrate distributed energy resources into utility system planning and utility operations. By relying on third-party capital for increased deployment of distributed energy resources, utilities become less reliant on utility capital investment; however, by relying on a platform-based model, utilities may increase operating expenses associated with providing services for the platform marketplace. To offset this loss of revenue and account for higher operating costs, REV introduces new revenue sources so utilities can shift to a platform-based market system. The model introduces market-based earnings (in addition to earnings-impact and earnings-sharing mechanisms) that encourage gains based on performance and outcomes, in line with regulation based on performance instead of on capital investments alone. These new mechanisms are designed around the value-added services utilities can provide in support of a platform-based marketplace. Examples include data analysis or creating an online portal. The idea is to enable earnings for utilities that decrease reliance on utility capital investment.

Furthermore, traditional approaches to setting consumer-facing tariff structures are characterized by static volumetric charges driven more by social than economic considerations. Once regulators set the overall revenue requirement for the utility, the requirement must be met through tariff structures that allocate charges to particular consumer groups and consumption brackets. Traditionally, tariff structures have involved averaging costs over entire customer classes (such as residential, commercial, and industrial) without considering the temporal or spatial profiles of consumption, which may in practice be significant cost drivers. Although a high share of the costs of electricity production is fixed, at least in the short run, utilities have preferred to recover the bulk of their costs through volumetric charges. For residential customers, these charges are often structured as increasing block tariff structures where the marginal tariff rate is higher for higher blocks of consumption, under the premise of protecting low-income consumers believed to have lower energy demand. On the same pretext, cross-subsidies between customer classes are not unusual and often entail higher variable charges for commercial than residential customers.

Existing rate structures, especially for residential or small commercial consumers, assume that consumers receive power and generate none of their own. This fundamental assumption is no longer true, so the challenge going forward is understanding how to set rates in a world where consumers may have various options for electricity consumption and production. Because tariff structures fundamentally affect the consumption and production choices of consumers, traditional approaches to pricing often fail to give the appropriate incentives.

The widespread adoption of net metering as a mechanism for incentivizing distributed generation illustrates the challenges of creating appropriate incentives. Net energy metering allows residential and small commercial customers to receive a credit on their bills for any electricity generated by their rooftop solar panels that they feed back into the system, essentially by subtracting it from the demand used to calculate the customer's bill.[6] This approach, designed to incentivize the adoption of rooftop solar, essentially buys back distributed generation from prosumers at the marginal variable retail charge that they face in the tariff structure. Given that almost all the fixed costs of the grid are being recovered through the variable charge, prosumers are being exonerated from contributing to recovering the fixed costs of the grid while benefiting from the grid as a backup source when the electricity they generate falls short. From the utility's perspective, the consumer-supplied energy provides only variable supply but is being remunerated at full production costs. In that sense, the utility is likely to be financially harmed by distributed generation under net metering arrangements. As a recent U.S. Department of Energy study observes, "After a century of utility concerns over whether rate increases will be high enough to allow full cost recovery, the emergence of elastic demand for electricity will shift the focus to whether utility costs are simply too high to be recoverable" (Corneli and Kihm 2015).

The potential perverse incentives created by net metering arrangements are further exacerbated in the presence of various kinds of cross-subsidies. The effects described above are further accentuated where cross-subsidies exist between large and small consumers via increasing block tariff structures. Large consumers, on which utilities rely disproportionately for revenues, will face strong incentives to become prosumers, because their production is remunerated at the higher block tariffs. Doing so might enable them to reduce their use of grid electricity to a level compatible with the lower-cost consumption blocks. (This incentive may widen as consumers acquire electric vehicles, necessitating higher volumes of domestic electricity consumption.) In countries where tariff structures embody cross-subsidies from commercial to residential consumers, the incentive for commercial entities to become prosumers will be particularly strong because own-generation may enable them to escape the cross-subsidy. Considering that the cost of rooftop solar at the larger commercial scale of production is more attractive than for smaller residential loads, net metering may present utilities with a significant risk of revenue loss from the commercial segment.

A simple fix to the problem of net metering is to introduce a separate fixed charge for prosumers, although this solution creates problems of its own. The fixed charge would represent prosumers' contribution to the fixed cost of the network. It is difficult to set the fixed charge at the right level, however; if set incorrectly, it may discourage the adoption of distributed energy resources that might otherwise benefit distribution system operators.

A more sophisticated alternative is to introduce a separate charging structure for electricity sold into the grid, ideally based on time of use. Such a rate could reflect the energy, capacity, environmental, and locational benefits prosumers provide to the system as opposed to crediting them at the average retail rate. New York has a rate design for distributed resources, which is based on a "value stack." This design considers the locational value of distributed resources in addition to crediting them for their supply value. Most of these rate designs wrestle with quantifying locational or time-based values that adequately compensate consumers for their energy production. Existing rate designs do not take these considerations into account. There is growing interest in time-of-use rates that vary according to the time band in which energy consumption or production takes place, with a view to incentivizing peak-shifting behaviors. Critical peak pricing is a similar concept. Load charges that capture the extent to which customers contribute to the system peak may also be helpful. Many of these structures have already been in place for some time for commercial and industrial customers, particularly in developed countries, and the novelty lies in extending them to the residential segment.

CONCLUSION

In conclusion, despite widespread creation of regulatory entities and the adoption of solid legal frameworks, the practice of independent regulation remains elusive. The establishment of regulatory frameworks was one of the most popular reforms from the 1990s model, and many countries succeeded in enacting technically sound regulatory methodologies. Nevertheless, implementation has often fallen short. Several countries in the observatory show sizable discrepancies between the quality of formal (de jure) regulatory frameworks and the extent to which those frameworks are perceived to operate in practice. The discrepancy is particularly wide when it comes to regulating quality of service and market entry. Moreover, although almost all countries grant regulators legal authority over tariff setting, this authority is respected in only about two-thirds of cases. There is substantial evidence that tariff adjustments systematically lag behind regulatory recommendations; in some cases, regulators are being used primarily to determine the subsidy requirement for the sector over and above politically constrained tariffs. Although many countries have espoused incentive regulation, the incentive regimes are weak, and the relevance of incentives may be limited given that many regulators continue to oversee primarily SOEs with limited commercial orientation.

The likely impact of technological disruptions of the power sector suggests that the regulator's task will become not only more complex but also critical in driving the pace of innovation. Incentive-based regulation will grow in relevance as regulators struggle to encourage utilities to take on decentralized solutions and more sophisticated demand management. Regulatory involvement in utility tariff structures will also become more critical to ensure that rooftop solar generation by prosumers is not overincentivized and that all customers contribute their fair share to the fixed costs of maintaining the power grid.

ANNEX 6A. FORMAL (DE JURE) SCORES ON THE REGULATORY PERFORMANCE INDEX

Percent

Indicators	Colombia	Dominican Republic	Egypt, Arab Rep.	India – Andhra Pradesh	India – Odisha	India – Rajasthan	Kenya	Morocco	Pakistan	Peru	Philippines
Overall de jure	**36**	**58**	**60**	**52**	**52**	**54**	**45**	**n.a.**	**60**	**70**	**46**
Regulatory governance	**45**	**69**	**77**	**68**	**68**	**68**	**56**	**n.a.**	**72**	**83**	**48**
Accountability	**75**	**78**	**87**	**91**	**91**	**91**	**95**	**n.a.**	**79**	**85**	**95**
Regulatory oversight	*67*	*67*	*100*	*100*	*100*	*100*	*100*	*n.a.*	*67*	*67*	*100*
Legal appeals	*100*	*100*	*100*	*100*	*100*	*100*	*100*	*n.a.*	*100*	*100*	*100*
Transparency	*57*	*67*	*62*	*73*	*73*	*73*	*85*	*n.a.*	*70*	*89*	*85*
Autonomy	**60**	**89**	**89**	**75**	**75**	**75**	**59**	**n.a.**	**92**	**98**	**51**
Decision-making autonomy	*64*	*83*	*86*	*100*	*100*	*100*	*79*	*n.a.*	*92*	*92*	*79*
Budgetary autonomy	*88*	*97*	*94*	*50*	*50*	*50*	*94*	*n.a.*	*100*	*100*	*50*
Leadership autonomy	*88*	*75*	*75*	*50*	*50*	*50*	*63*	*n.a.*	*75*	*100*	*75*
Managerial autonomy	*0*	*100*	*100*	*100*	*100*	*100*	*0*	*n.a.*	*100*	*100*	*0*
Regulatory substance	**81**	**85**	**78**	**76**	**76**	**79**	**81**	**40**	**83**	**83**	**95**
Tariff regulation	**92**	**75**	**75**	**75**	**75**	**75**	**88**	**29**	**83**	**100**	**93**
Regulatory framework for tariffs	*100*	*100*	*100*	*100*	*100*	*100*	*100*	*33*	*100*	*100*	*86*
Determination of tariffs	*83*	*50*	*50*	*50*	*50*	*50*	*75*	*25*	*67*	*100*	*100*
Quality regulation	**100**	**79**	**83**	**83**	**83**	**92**	**54**	**33**	**100**	**100**	**92**
Quality of service standards	*100*	*75*	*100*	*100*	*100*	*100*	*75*	*50*	*100*	*100*	*100*
Quality of service enforcement	*100*	*83*	*67*	*67*	*67*	*83*	*33*	*17*	*100*	*100*	*83*
Market entry regulation	**50**	**100**	**75**	**71**	**71**	**71**	**100**	**58**	**67**	**50**	**100**
Permitting new entrants	*50*	*100*	*100*	*75*	*75*	*75*	*100*	*50*	*100*	*100*	*100*
PPA approvals	*50*	*n.a.*	*50*	*67*	*67*	*67*	*100*	*67*	*33*	*0*	*100*

Source: Based on Rethinking Power Sector Reform utility database 2015.
Note: n.a. = not applicable; PPA = power purchase agreement.

ANNEX 6B. PERCEIVED SCORES ON THE REGULATORY PERFORMANCE INDEX

Percent

Indicators	Colombia	Dominican Republic	Egypt, Arab Rep.	India – Andhra Pradesh	India – Odisha	India – Rajasthan	Kenya	Morocco	Pakistan	Peru	Philippines
Overall perceived	**34**	**9**	**28**	**25**	**21**	**21**	**19**	**n.a.**	**37**	**66**	**39**
Regulatory governance	**42**	**26**	**56**	**64**	**64**	**64**	**38**	**n.a.**	**61**	**80**	**47**
Accountability	**70**	**37**	**77**	**88**	**88**	**88**	**71**	**n.a.**	**85**	**81**	**86**
Regulatory oversight	*67*	*67*	*100*	*100*	*100*	*100*	*67*	*n.a.*	*67*	*67*	*67*
Legal appeals	*100*	*0*	*100*	*100*	*100*	*100*	*100*	*n.a.*	*100*	*100*	*100*
Transparency	*43*	*44*	*31*	*64*	*64*	*64*	*46*	*n.a.*	*89*	*78*	*92*
Autonomy	**60**	**70**	**73**	**73**	**73**	**73**	**53**	**n.a.**	**72**	**98**	**54**
Decision-making autonomy	*64*	*33*	*86*	*92*	*92*	*92*	*57*	*n.a.*	*62*	*92*	*92*
Budgetary autonomy	*88*	*97*	*94*	*50*	*50*	*50*	*94*	*n.a.*	*100*	*100*	*50*
Leadership autonomy	*88*	*50*	*63*	*50*	*50*	*50*	*63*	*n.a.*	*75*	*100*	*75*
Managerial autonomy	*0*	*100*	*50*	*100*	*100*	*100*	*0*	*n.a.*	*50*	*100*	*0*
Regulatory substance	**81**	**36**	**50**	**39**	**33**	**33**	**51**	**38**	**61**	**83**	**83**
Tariff regulation	**92**	**17**	**67**	**50**	**50**	**50**	**58**	**21**	**83**	**100**	**92**
Regulatory framework for tariffs	*100*	*17*	*83*	*50*	*50*	*50*	*67*	*17*	*100*	*100*	*100*
Determination of tariffs	*83*	*17*	*50*	*50*	*50*	*50*	*50*	*25*	*67*	*100*	*83*
Quality regulation	**100**	**42**	**33**	**25**	**8**	**8**	**29**	**25**	**71**	**100**	**92**
Quality of service standards	*100*	*50*	*50*	*0*	*0*	*0*	*25*	*50*	*75*	*100*	*100*
Quality of service enforcement	*100*	*33*	*17*	*50*	*17*	*17*	*33*	*0*	*67*	*100*	*83*
Market entry regulation	**50**	**50**	**50**	**42**	**42**	**42**	**67**	**67**	**29**	**50**	**67**
Permitting new entrants	*50*	*50*	*100*	*50*	*50*	*50*	*100*	*67*	*25*	*100*	*100*
PPA approvals	*50*	*n.a.*	*0*	*33*	*33*	*33*	*33*	*67*	*33*	*0*	*33*

Source: Based on Rethinking Power Sector Reform utility database 2015.
Note: n.a. = not applicable; PPA = power purchase agreement.

NOTES

1. This chapter draws on the background paper by Rodriguez Pardina and Schiro (2018) and original research from a team led by Katharina Gassner and Joseph Kapika. Martin Rodriguez Pardina, Julieta Schiro, and Kagaba Paul Mukibi were members of the team. The work program was coordinated by Vivien Foster and Anshul Rana.
2. Given this particularity, Morocco's broad regulatory and governance scores were not computed, nor was Morocco taken into account when computing average scores for broad regulatory performance and governance (and its subareas), or when correlating coefficients involving governance or any of their subareas.
3. These countries are Colombia, the Dominican Republic, Kenya, Pakistan, Peru, the Philippines, Uganda, Ukraine, and Vietnam.
4. Various quality-of-service outcomes are measured to ensure that utilities do not decrease operating expenses by lowering their quality of service.
5. Most of these savings are refunded to customers, though utilities can keep some of the savings, which is a key incentive.
6. Note that net metering is one compensation mechanism for rooftop solar. Other mechanisms exist where consumers are not credited at the retail rate at all, or are not credited for any self-generation at all, but are instead providing the entire output of rooftop solar resources directly to the utility.

REFERENCES

Adam, R. 2011. "Establishing Regulatory Incentives to Raise Service Quality in Electricity Networks." Point of View, Cisco Internet Business Solutions Group.

Alexander, I. 2014. "Developing Countries Experience and Outlook: Getting the Framework Right." *Utilities Policy* 31 (C): 184–87.

Andrés, L., J. Guasch, M. Diop, and S. Lopez Azumendi. 2007. *Assessing the Governance of Electricity Regulatory Agencies in the Latin American and Caribbean Region: A Benchmarking Analysis*. Washington, DC: World Bank.

Brown, A., J. Stern, B. Tenenbaum, and D. Gencer. 2006. *Handbook for Evaluating Infrastructure Regulatory Systems*. Washington, DC: World Bank.

Cave, M. 2016. "Thoughts on UK Economic Regulation, 2016." *Oxera*, March. https://www.oxera.com/agenda/thoughts-on-uk-economic-regulation-2016/.

Corneli, S., and S. Kihm. 2015. "Electric Industry Structure and Regulatory Responses in a High Distributed Energy Resources Future." Future Electric Utility Regulation Report No. 1, Lawrence Berkeley National Laboratory, University of California.

Correa, P., C. Pereira, B. Mueller, and M. Melo. 2006. *Regulatory Governance in Infrastructure Industries: Assessment and Measurement of Brazilian Operators*. Trends and Policy Options No. 3. Washington, DC: World Bank.

Foster, Vivien, Samantha Witte, Sudeshna Ghosh Banerjee, and Alejandro Moreno. 2017. "Charting the Diffusion of Power Sector Reforms across the Developing World." Policy Research Working Paper 8235, World Bank, Washington, DC.

Gilardi, F., and M. Maggetti. 2011. "The Independence of Regulatory Authorities." Chapter 14 in *Handbook on the Politics of Regulation*, edited by D. Levi-Faur. Cheltenham, UK: Edward Elgar Publishing.

Graffy, E., and S. Kihm. 2014. "Does Disruptive Competetion Mean a Death Spiral for Electric Utilities?" *Energy Law Journal* 35 (1): 1–44.

Hanretty, C., and C. Koop. 2013. "Shall the Law Set Them Free? The Formal and Actual Independence of Regulatory Agencies." *Regulation and Governance* 7 (2): 195–214.

Jamison, M. 2007. "Regulation: Rate of Return." In *Encyclopedia of Energy Engineering and Technology*, Vol. 3, edited by B. Capehart, 1252–57. New York: CRC Press, Taylor and Francis.

Joskow, P. 2014. "Incentive Regulation in Theory and Practice: Electricity Distribution and Transmission Networks." In *Economic Regulation and Its Reform: What Have We Learned?* edited by N. L. Rose, 291–344. University of Chicago Press.

Kessides, I. 2012. "Electricity Reforms—What Some Countries Did Right and Others Can Do Better." *Viewpoint*, Note No. 332, World Bank, Washington, DC.

Kihm, S., R. Lehr, S. Aggarwal, and E. Burgess. 2015. "You Get What You Pay for: Moving towards Value in Utility Compensation." https://www.seventhwave.org/sites/default/files/you-get-what-you-pay-for-part-one-2015.pdf.

Laffont, J.-J. 2005. *Regulation and Development*. Cambridge, UK: Cambridge University Press.

Nixon, H. 2013. "RIIO Incentive Framework." Presentation at the 10th EU-US Energy Regulators Roundtable, The Hague, April 8.

Ofgem. 2015. *RIIO-GD1 Annual Report 2013-14*. London: Ofgem.

Pargal, S., and S. Banerjee. 2014. *More Power to India: The Challenge of Electricity Distribution*. Directions in Development Series. Washington, DC: World Bank.

Rodriguez Pardina, M. A., and J. Schiro. 2018. "Taking Stock of Economic Regulation of Power Utilities in the Developing World: A Literature Review." Policy Research Working Paper 8461, World Bank, Washington, DC.

Rodriguez Pardina, M., R. Schlirf Rapti, and E. Groom. 2008. *Accounting for Infrastructure Regulation: An Introduction*. Washington, DC: World Bank.

Vagliasindi, M., and J. Nellis. 2010. "Building Sound Institutions." In *Africa's Infrastructure: A time for Transformation*, edited by Vivien Foster and Cecilia Briceño-Garmendia. Africa Development Forum. Washington DC: World Bank.

What Progress Has Been Made with Wholesale Power Markets?

7

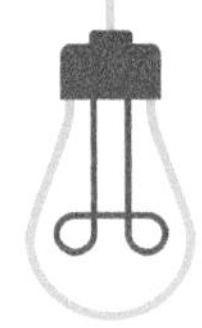

Guiding questions

- *What conditions must countries meet before attempting to create a wholesale power market?*
- *What transition challenges have countries faced in introducing and fine-tuning a competitive market?*
- *What are the emerging best-practice design features for wholesale power markets in the developing world?*
- *What are the effects of the current wave of technological disruption on wholesale power markets?*

Summary

- *The rarity of wholesale power markets in the developing world reflects a demanding list of preconditions that must be met before competition becomes viable. Until those preconditions are fully met, countries can benefit from wholesale competition by participating in regional power markets.*
- *There are significant risks of getting stuck in the transition to a competitive market. Once in place, constant monitoring is needed so that the market design can be fine-tuned as challenges arise.*
- *An independent and adaptable institutional and regulatory framework is a great aid to market efficiency. The role of system operator can be undertaken by the transmission company or by an independent entity; it may or may not be combined with the market operator role.*
- *Adequate system governance and open access are essential for operational efficiency and for attracting new entrants. Accurate short-term prices will follow from efficient and secure real-time dispatch. It is risky, however, to rely on short-term prices alone to incentivize investment in new capacity. Increasingly, incentives for new investment are being provided through the auctioning of long-term power supply contracts.*
- *Looking forward, there is a need to modify market designs to accommodate and incentivize variable renewable energy, battery storage, and demand-side participation.*

INTRODUCTION

This chapter reviews the experience of the few developing countries that have managed to implement wholesale power markets.[1] After asking why some countries have made more progress than others in introducing competition, the chapter goes on to examine the complex process of designing and introducing a power market. The guiding questions are as follows. What are the minimum conditions that countries must meet before attempting to introduce a wholesale power market? What challenges have countries faced in introducing and fine-tuning a competitive market? What are the emerging best practices for wholesale power markets in the developing world? Finally, how are wholesale power markets being affected by the current wave of technological disruption?

The creation of a wholesale power market was the endpoint envisaged by the model of power sector reform that prevailed in the 1990s. Most of the other reform measures—including restructuring, privatization, and regulation—were considered stepping-stones toward full competition in the sector. According to the model, competitive forces would efficiently balance supply and demand by driving investment, operation, and consumption decisions in both the short run and the long run. In the short run, power markets would improve system operations by promoting efficient scheduling and dispatch, ensuring reliability, and providing appropriate price signals for operation and investment. In the long run, adequately regulated power markets would incentivize optimal investments at the right locations and times, attaining desired levels of supply security through an efficient mix of generation technologies. In contrast to a vertically integrated electricity industry, power markets shift risks of technology choice, construction cost, and operating "mistakes" to suppliers and away from consumers. Strong profit-maximizing incentives would thus work to increase efficiency of the power sector (Joskow 2008b).

The model notwithstanding, mixed experience tells us that power markets need careful assessment, design, and implementation. Over the past three decades, power markets have demonstrated their ability to improve performance through an evolving mix of competition and regulation—for example, in Australia, Chile, Norway, the United Kingdom, and the United States. However, California's power market crisis in 2000 revealed the significant risks of establishing power markets without careful design and implementation, even in jurisdictions with ample resources and supportive conditions (Besant-Jones and Tenenbaum 2001). The core principles of market design, widely agreed to be essential, include open access to the grid, demand-side participation, coordination for short-term efficiency and reliability, and a workable framework for supply adequacy (Hogan 2002; Hunt 2002; Joskow 2008b; Rudnick and Velasquez 2018).

This chapter focuses primarily on wholesale, rather than retail, power markets. Wholesale power markets, in which competition is found in the generation segment, have gained significant traction in the developing world owing to their ability to deliver gains in efficiency and reliability. Retail competition, entailing the introduction of competition in the commercial segment, has evolved slowly, reaching fewer jurisdictions and leading to mixed outcomes (Defeuilley 2009; Littlechild 2009). For these reasons, the discussion will center on wholesale markets for generation, even if some of the issues may carry over to transmission and retail markets.

KEY FINDINGS

Relatively little is known about power markets in developing countries. The literature on power markets in developed countries and on the wider reform process in developing countries is extensive. The same cannot be said, however, for the specific experiences of developing countries with the introduction of power markets. Those experiences are sparsely documented (Rudnick and Velasquez 2018). As part of the Power Sector Reform Observatory undertaken for this report, in-depth case studies were produced for the four countries with relatively mature wholesale power markets in place: Colombia, India, Peru, and the Philippines (Rudnick and Velasquez 2019a, 2019b, 2019c, and forthcoming). Drawing on this new body of knowledge, the developing country experience with power markets can be conveyed through the following key findings.

Finding #1: The rarity of wholesale power markets in the developing world reflects the demanding preconditions for viable competition

Only one in five developing countries has established a wholesale power market. The share rose gradually from 11 percent in 2000 to just over 20 percent by 2015 (figure 7.1). Out of the 22 percent, only 7 percent has also introduced retail competition. This share compares with about 80 percent of Organisation for Economic Co-operation and Development (OECD) countries with established wholesale power markets, out of which 66 percent has also introduced retail competition. Most of the wholesale power markets in the developing world are in two regions—Latin America and the Caribbean and Europe and Central Asia—where about half of the countries have introduced them. By contrast, not one wholesale power market can be found in Africa and the Middle East. Uptake of power markets in the developing world is strongly related to system size. Just 5 percent of countries with systems under 1 gigawatt (GW) have them, compared with 25 percent of countries with systems above 20 GW. Nevertheless, many of the largest power systems in the developing world—such as those of the Arab Republic of Egypt, Indonesia, Pakistan, and South Africa—have not yet introduced wholesale power markets.

FIGURE 7.1 Only one in five developing countries has established a wholesale market
Competition in the power sector, 1995–2015

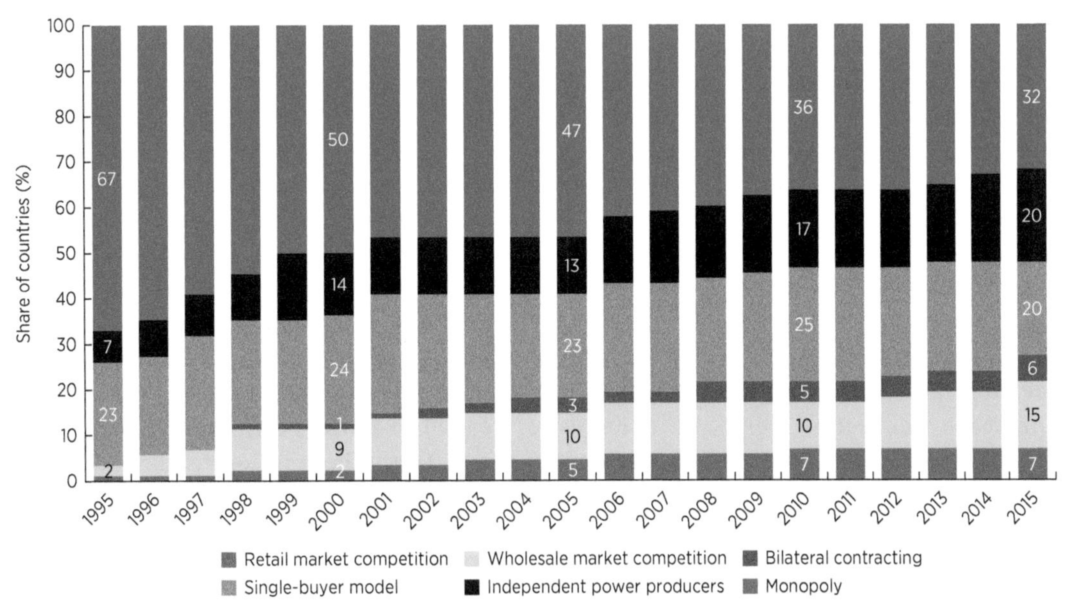

Source: Foster and others 2017.

The preconditions needed to establish a wholesale power market are relatively demanding and not met in much of the developing world. Starting conditions matter for power markets. Those conditions span a wide range, including market structure, transmission constraints, financial health, and the wider enabling environment. A few suboptimal conditions can be addressed by market design in some very specific situations, and the market must certainly be carefully tailored to local limitations and imperfect starting conditions. Nevertheless, power markets are easier to implement and more likely to succeed if starting conditions are supportive of competition. For countries where the risks of developing a market significantly exceed those of the status quo, markets are better introduced gradually and only after appropriate conditions are attained.

First and foremost, the market structure of the generation segment needs to be conducive to competition. The benefits of competition are unlikely to pass through to end-users if there is a significant concentration of market power in generation, or if certain actors control both generation and transmission assets and can strategically manipulate access to the grid. Accommodating competition under such circumstances entails not only vertical unbundling of generation and transmission but also horizontal unbundling of generation to make certain that the market includes enough players to ensure competitive pressure. (Five generators of roughly equivalent size are often considered to be a minimum.) Creation of a competitive generation segment also requires careful attention to how the assets are broken up across players, so as to ensure that each generation company controls roughly equal shares of the total capacity, particularly during the critical price-setting peak period, and that none of them can exercise market power in the relevant upstream fuel markets. Although some successful power markets (such as those of Chile and the United Kingdom) started out with highly concentrated generation segments; decreasing concentration before the market opens can help deter undesirable exertions of market power and the resulting public backlash. For example, Argentina dealt with market power by breaking up the generation sector at the outset of reforms, to a point where none of the many generators controlled more than a single plant. Finally, if a country is short of generation capacity, then all available plants will need to be deployed to meet the system peak. Without some plant redundancy, competition is meaningless, and market power will swell during the peak period.

Second, transmission constraints should be removed, because they can create temporary concentrations of market power. Adequate transmission infrastructure is needed to avoid congestion, which can create market power on the other side of the bottleneck and prevent optimization of available energy resources across large areas.[2] Conversely, connecting previously independent power grids can dramatically increase competition. For example, the previously independent wholesale markets of Chile began to integrate in 2017, boosting competition in supply auctions and improving system operations. Interconnection of power grids was also successful in the Philippines, where the Visayas island and its power grid have been integrated with the Luzon market since 2010, thereby lessening the horizontal concentration of the generation segment in the combined power grid. The Indian power system also went through a major transformation starting in the early 1990s, when interconnection among the states and regions was very weak. The process concluded in late 2013, when the southern region synchronized with the rest of the nation's grids.[3]

Third, the power sector must be financially healthy and underpinned by a reliable payment chain. Financial ill-health anywhere in the electricity supply chain lessens investors' trust in the market, preventing it from delivering investment and competition among generators. Market participants must have confidence that counterparties will pay for the power they sell. This confidence hinges on the integrity of payments across the power supply chain, the presence of creditworthy buyers of electricity, and cost-recovery tariffs that ensure the financial viability of the industry as a whole. The case of Ukraine highlights the challenges of establishing a power market under weak financial conditions. Despite considerable scale (25 GW peak demand), attempts to introduce a wholesale market in Ukraine during the 1990s failed partly because of poor collection from end-users and low levels of cash payments among companies (Besant-Jones 2006; Krishnaswamy 1999).

Fourth, wider country conditions also matter for power markets, including institutions and the macroeconomic, political, and social environment. A lack of government commitment to reform and signals of unwarranted regulatory intervention or instability can also limit investors' interest in the market. Protection of property rights was crucial to the successful Chilean reform. Argentina, by contrast, illustrates the negative effects of macroeconomic crises and political interference in electricity pricing (Pollitt 2004, 2008). The experience of Eastern European countries indicates that it is extremely difficult to carry out structural reforms of the sector during economic turmoil (Krishnaswamy and Stuggins 2003). Finally, permitting, siting, and social approval are persistent obstacles to investment in new generation and transmission infrastructure across Asia and Latin America.

Nonetheless, power markets have failed even under very advantageous conditions, as in the U.S. state of California. The state's electricity crisis in the summer of 2000 illustrates how power markets can fail even in a jurisdiction with plenty of resources and initial conditions supportive of competition. California's power market emerged from a political reform process that brought many distortions to the market, such as obliging distribution utilities to supply their customers directly from the wholesale market and preventing them from hedging against wholesale spot price volatility. (The reform required distributors to sell their generation assets and did not allow them to sign bilateral contracts.) These and other market design aspects, along with increased demand and diminished hydropower availability in the summer of 2000, resulted in acute price spikes (reflecting some abuse of market power) and mandated rolling blackouts, severely harming California's economy and end-users (Besant-Jones and Tenenbaum 2001; Wolak 2003b).

Other power markets have flourished even under challenging conditions, such as those that prevail in India. India's power exchanges deliver benefits of competition at least to a portion of suppliers and customers, despite underlying conditions that are far from ideal. Indeed, inadequate generation falling short of demand persisted in India until 2017, and the transmission grid remains weak, constraining the regional power system. Moreover, the Indian power sector has many financial problems, with continuing cross-subsidies, tariffs too low to recover costs, and even bankrupt utilities. Nevertheless, large open-access customers and other market participants have benefitted from increased participation in power exchanges. Trade in Indian power exchanges surged from 1,735 gigawatt-hours (GWh) in fiscal year 2008 (FY2008) to 41,120 GWh in FY2017, while prices fell 83 percent in real terms over the same period as generation capacity grew to exceed demand. Delivering the benefits of competition to a wider set of customers, however, requires tackling the fundamental problems of the sector, which include inadequate infrastructure and financial nonviability.

Finding #2: Small countries can benefit from wholesale competition by participating in regional power markets

For small power systems, a competitive market for generation may be neither feasible, owing to size constraints, nor desirable, owing to loss of scale economies. Caution should be exercised in moving toward a wholesale power market if the system falls below certain critical size thresholds (box 7.1). Underdeveloped power systems in countries with rapidly growing populations and economies—such as many in Sub-Saharan Africa—may soon reach the scale where a domestic power market becomes relevant. In other cases, small scale may be a permanent situation dictated by geographic constraints.

Countries that are too small to develop wholesale power markets often have the option of participating in regional power markets. Small countries for which a domestic wholesale power market is not viable may be close to larger countries or regional power markets that could allow them to reap the benefits of competition and increased security of supply. Working examples of regional power markets exist in Africa (Southern Africa Power Pool and West African Power Pool) and Central America (Mercado Eléctrico Regional, MER); the benefits of electricity integration have also been realized in South Asia through India's interconnections with neighboring markets in Bangladesh and Nepal (Andrews-Speed 2016; Asian Development Bank 2015; Oseni and Pollitt 2016).

The economic and supply-security benefits of regional integration and cross-border trade can be enormous, for countries large and small. Regional markets can improve resource optimization across countries by pooling generation resources, sharing reserves, and harnessing the seasonal and hourly complementarity of generation availability and of demand patterns across broad areas. For example, demand peaks in the countries of the South Asia region do not coincide over the year, implying huge potential savings on the costs of supply and of unserved demand—savings derived in both cases from resource pooling. These benefits have been estimated at more than US$9 billion per year (Timilsina and others 2015). Moreover, regional power markets can help reduce carbon emissions, both by displacing domestic thermal generation with imported renewable generation and by creating larger balancing areas that facilitate the integration of variable renewable energy resources (Chattopadhyay and Fernando 2011; Raineri and others 2013; Wijayatunga, Chattopadhyay, and Fernando 2015.

To deliver these benefits, however, successful regional power markets require significant up-front investments to build infrastructure and an institutional framework capable of delivering efficient outcomes (Oseni and Pollitt 2016). Connecting independent power grids presents challenges ranging from the purely technical to the financial and institutional. The challenges are worth addressing, because the market resulting from integration can rapidly intensify competition. Sufficient transmission capacity, both in interconnectors and in the domestic grid of each country, must be financed and developed to enable cross-border trading. Institutional challenges include merging independent operators to form a single regional one, regulating and overseeing cross-border trade, and developing efficient regional trade arrangements beyond bilateral long-term contracts. Such trade arrangements require a regional market operator to support short-term trading and settlement, complemented by financial contracts that allow for risk management while preserving dispatch efficiency (Rose, Stoner, and Pérez-Arriaga 2016).

BOX 7.1 How big must a power system be to support a wholesale power market?

The answer to this important policy question depends on both physical and financial considerations. The thresholds indicated here are notional and intended solely as a guide. They are not intended to suggest that power markets are impossible in smaller systems, but the balance between the costs and the benefits of introducing a wholesale power market varies with scale.

First, for competition to be meaningful, at least five generation companies should be present in the market. Each of them must be large enough to achieve scale economies. If the number of companies is too low, there is a risk of collusion. If their size is too small, there is a loss of efficiency in production. Given a minimum efficient scale of 400–1,000 megawatts for thermal generation, and the need for at least five companies, this suggests a minimum system size of 3 gigawatts of peak demand or 20 terawatt-hours of annual energy demand to support a competitive market. The reality may be more nuanced than this simple rule of thumb, because the threshold will also be affected by the extent to which competition exists specifically at the system peak, the conditions of access to fuel supply, and the extent of vertical integration between generation and distribution.

Second, the investment and operation costs of a wholesale power market are substantial, and the efficiency gains from competition are proportional to the size of the market. Although the evidence on costs is limited, some figures may be illustrative. Even in Singapore's small market, the costs of establishing the wholesale power market exceeded US$75 million, and annual running costs are US$15 million–20 million (Ching 2014). For a larger market, such as the United Kingdom, the cost of switching from the power pool market design to the New Electricity Trading Arrangements was estimated to be more than US$1 billion (spread over a five-year period), with annual operating costs approaching US$50 million (Newbery 2005). Ofgem (1999) notes that the up-front investment cost would be justified if the change caused prices to drop by 10 percent. The levelized cost of operation of a typical market operator has been estimated at US$0.20/megawatt-hour for a large market like the Pennsylvania–New Jersey–Maryland interconnection or PJM (in the eastern United States) to as much as US$1.00/megawatt-hour for smaller jurisdictions like Ontario, Canada.[a] Taken in conjunction, this evidence suggests that a country would need to exceed a potential market trading value of US$1 billion before the investment in a market trading platform could be justified; otherwise the value of the potential efficiency gains would be unlikely to surpass the investment and operating costs of the market.

Of the 15 countries in the Power Sector Reform Observatory, all of those with functioning wholesale power markets comfortably meet the size thresholds, except the Dominican Republic. Several countries meet the size thresholds but do not yet have a functioning wholesale power market. Of this group, several are in transition toward such a market: the Arab Republic of Egypt by 2023, Ukraine by 2025, and Vietnam by 2024 (table B7.1.1).

TABLE B7.1.1 Classification of Observatory countries, by size threshold

	Meets thresholds	Does not meet thresholds
Wholesale power market	Colombia, India, Peru, Philippines	Dominican Republic
No wholesale power market	Arab Republic of Egypt, Morocco, Pakistan, Ukraine, Vietnam	Kenya, Senegal, Tajikistan, Tanzania, Uganda

a. For Canada, see, for example, http://www.ieso.ca/corporate-ieso/regulatory-accountability/usage-fees; for PJM, http://www.pjm.com/-/media/committees-groups/committees/fc/postings/first-quarter-2018-schedule-9-rates.ashx?la=en.

Finding #3: A transition toward a competitive market is required, but there are significant risks of becoming stuck in the process

The transition to a competitive market has taken a variety of forms. Colombia and Peru opted for immediate implementation of power markets without a transitional phase (table 7.1). By contrast, in India and the Philippines, five years elapsed between the passage of legislation for a power market and its entry into operation. In the case of India, the delay was caused by the absence of a clear road map for implementation. In the Philippines, the transition period was deliberately aimed at easing in the new arrangements and giving market actors time to adapt.

TABLE 7.1 **Immediate or transitional implementation of a power market is possible; each path has benefits and challenges**

	Peru	Colombia	India	Philippines
Year transition began[a]	1992	1994	2003	2001
Year market began	1992	1995	2008	2006
Transition process	None	None	Law established principles of reform, with no clear transition roadmap	Requirement for distribution utilities to source at least 10 percent of their power supply from the spot market for first five years
Target type of market	Wholesale (cost-based) competition; regulated capacity payments	Wholesale (bid-based) competition; regulated capacity payments	Bilateral, with power exchanges	Wholesale (bid-based) and retail competition

Source: World Bank elaboration based on data provided by local consultants and independent research.
a. The transition is considered to begin with the enactment of the law mandating the future development of a competitive wholesale power market.

There is no doubt that significant time and resources are required to develop and establish the market. Once the legal framework governing the power sector reform is enacted, detailed bylaws, market rules, and procedures must be developed; discussed with key stakeholders; and approved. This process can easily take two years or longer if the tasks are not prioritized and managed properly. Moreover, the market requires development of information systems; optimization models for planning, scheduling, dispatch, and pricing; and billing and settlement procedures. Even when clear principles have been agreed upon, the minutiae of models and processes can affect the incomes and costs of each power company and can therefore be contentious. Moreover, many practical details of system operation are not formally codified through procedures, but rather left to common practice (for example, criteria for real-time redispatch by the system operator). Therefore, appropriate time and resources should be allocated when laying out the timetable for market implementation, considering the potential for conflicts over the details of the market during its definition and implementation.

Transition processes can create a risk that reforms may never be completed. Several countries have opted for a trial or pilot period of a few months to allow market participants to prepare for actual market operations. The danger of such periods is that they may slow the momentum of reform or divert it from its original objectives. For example, the California electricity market crisis of 2000 was used as a pretext to stall market reforms not only in the rest of the United States but also in other parts of the world, including Malaysia, which settled on a single-buyer market instead.

The single-buyer model is a risky transition measure because rigid contracts with independent power producers can deter participation in a subsequent competitive market. In a pure single-buyer model only the integrated monopoly is permitted to buy power (including energy, capacity, and ancillary services) from competing generators or from independent power producers at regulated prices (Arizu, Gencer, and Maurer 2006; Hunt 2002). It is considered by some to be a second-best alternative to comprehensive restructuring, providing time for a smooth transition toward fully competitive wholesale markets (Vagliasindi and Besant-Jones 2013). The single-buyer model can help alleviate supply shortages by introducing independent power producers to an industry with limited financing capabilities. However, the inflexible long-term power purchase agreements associated with the model, usually built around take-or-pay clauses, deter the evolution of competition, because plants operating under the agreements may have no incentive to participate in a competitive market and cannot be dispatched on a merit-order basis to minimize short-run production costs. Moreover, high-priced contracts could be undercut by competition in the wholesale market, becoming "stranded costs" that require a recovery mechanism. Although careful design can mitigate some of its problems (Arizu, Gencer, and Maurer 2006), the single-buyer model remains problematic because it transfers risks from generators to end-users, thus removing incentives for generators to manage those risks (Castalia 2013) and possibly leading to inefficient investment decisions (Thomas 2012).

Incipient power markets remain vulnerable to abuses of market power even when structural precautions have been taken. The physical and economic properties of electricity make its markets prone to unilateral exercise—and abuse—of market power. Such abuse can severely harm customers, as well as the reliability and efficiency of the overall market. The initial stages of a new market can be particularly vulnerable to abuse because of concentration in the ownership of the marginal generation resources that set the market price and because of residual congestion in transmission or a tight balance of supply and demand.

Vesting contracts are one device for containing market power during the transition period. Vesting contracts are hedge contracts assigned to incumbent retailers and generators when the electricity industry is disaggregated, but before asset divestiture or privatization (Kee 2001). They reduce the incentive of generators to bid strategically (and so exercise market power) by hedging revenues from the spot price. Moreover, vesting contracts can help market development during the initial phases, when the spot market may not provide enough revenue (or enough revenue certainty) for generators or retailers.

Cost-based power pools can also be adopted as a transitional measure to deter the exercise of market power. Cost-based pools are based on the "audited" variable costs of each power plant, whereas more sophisticated markets allow participants to submit bids. Bid-based markets thus allow generation firms to reveal their full opportunity costs, while also leaving room for gaming directly through the submitted bids. Compared to bid-based markets, cost-based pools are exposed to fewer opportunities for market power abuse and the dramatic price spikes that may result from gaming the market, especially in the presence of transmission congestion or tight supply–demand conditions. For example, the Pennsylvania-New Jersey-Maryland (PJM) market in the United States was administered as a transitional cost-based pool during its first year of operation, and even now the independent system operator requires bids to be cost-based in cases where transmission congestion creates risks of localized market power (Wolak 2003a).

Using cost-based pools as a transitional structure risks leaving the country with incomplete reforms, unless the target market is also cost-based. Although cost-based pools can be useful transitional measures (as in PJM), they are not exempt from short and long-term inefficiencies (Munoz and others 2017). In the short run, generators can game the parameters of the cost-based pool (such as the startup times or minimum load levels) to maximize profits. Also, the audited fuel cost may be very different from the true opportunity cost, which would reflect take-or-pay clauses in natural gas contracts, as well as the availability of gas storage and a secondary gas market. In the long run, the generation technology can be strategically selected by firms to maximize profits. Hence, initially adopting a cost-based market instead of a bid-based one poses the risk of miring the country in an inefficient transition. The Republic of Korea, for example, has been stuck with a transitional cost-based pool for nearly two decades, failing to evolve toward the envisioned bid-based market (Kim, Kim, and Shin 2013).

Nevertheless, in some situations cost-based pools may be the preferred long-term market design—particularly where market power remains concentrated or where the system depends heavily on hydropower. This was the case in several hydro-dependent Latin American countries (Brazil, Chile, and Peru), where cost-based pools provided a means of mitigating short-term market power and hydro-thermal coordination (Rudnick, Varela, and Hogan 1997). Cost-based power pools in Brazil and Chile have been effective in managing multiyear cascading hydrological reservoirs, realizing economies from coordination of hydropower resources across broad areas and of hydro and thermal generation plants. These coordination economies could be difficult to attain under a more decentralized trading arrangement where each company independently managed its own reservoir or power plant.

Finding #4: An independent and adaptable institutional and regulatory framework is a great aid to market efficiency

Power markets require regulation and oversight by independent and effective institutions (Jamasb, Nepal, and Timilsina 2015; Jamasb, Newbery, and Pollitt 2005; Nepal and Jamasb 2012). A strong, independent, and effective regulator is an important actor in any wholesale power market (Joskow and Schmalensee 1983). A tailored process for monitoring and oversight is needed from the start of reforms; it must be able to account for the technical and economic complexities of the power sector, including the physical laws governing power flows in the transmission system and the lack of economic large-scale storage (Bushnell, Mansur, and Saravia 2008).

The monitoring and oversight process should be capable of assessing market outcomes, proposing enhancements to market design, and providing a base for detection of abuse of market power. The oversight process should assess the performance of submarkets for energy, capacity, and ancillary services. Large sets of market indicators are calculated and published regularly in some countries (Chile, India, and Peru); other countries analyze the behavior of market participants (Colombia and the Philippines). Given the complexities of the power sector, indicators are not enough by themselves, but should be used in more detailed analyses to provide meaningful conclusions for regulators, policy makers, antitrust authorities, and market participants (Stoft 2002). Market monitoring in developing countries often falls short in assessing performance in terms of outcomes such as security of supply and competitiveness (table 7.2).

Adequate monitoring and oversight will occur only if they are formally required and purposefully organized. The definition and allocation of monitoring and oversight functions must be clearly established from the outset of reforms. Depending on the country, those functions might involve the regulator, the ministry, the antitrust authorities, and the system operator. The system or market operator can be very effective in compiling and publishing performance indicators. Because these indicators require deeper analysis before they can be meaningfully interpreted, some

TABLE 7.2 **A country-specific process for monitoring and overseeing the electricity market is required from the start of reforms**

	India	Philippines	Colombia	Peru	Chile
Market monitoring approach	Structural monitoring by regulator (CERC)	Structural and behavioral analysis by regulator, and independent entity within LTSO	Structural and behavioral monitoring by regulator (CREG)	Market oversight by regulator (OSINERGMIN)	No dedicated entity for market monitoring until 2016 law introduced a dedicated monitoring entity within system operator
Major reform adaptations	2014: Improvements in mechanism for balancing the grid (deviation settlement)	2015: Regulations on competitive selection procedure for distribution utilities	2006: Regulated capacity mechanism is replaced by the Firm Energy Market 2009: government begins driving technology-specific development of generation (hydro and gas)	2006: Centralized auctions to supply regulated customers; improved independence of system operator	2004–05: Centralized auctions for supplying regulated customers, and centralized transmission expansion planning

Source: World Bank elaboration based on data provided by local consultants and independent research.
Note: CERC = Central Electricity Regulatory Commission; CREG = Commission for the Regulation of Electricity and Gas; LTSO = legally unbundled transmission system operator; OSINERGMIN = Supervisory Agency for Investment in Mining.

jurisdictions have created an independent monitoring entity reporting to the board of the system operator. This approach has been successfully applied in PJM and other power markets in the United States; it has also been adopted in the Philippines. In the European Union, the market monitoring function is performed by the Agency for the Cooperation of Energy Regulators (ACER), a regulatory body for the entire union that collects electricity trading data and refers cases of market manipulation back to the national regulatory agencies. Elsewhere, simple reporting is conducted by the system operator or the regulator (Chile, India, and Peru), along with case-by-case analysis of specific incidents of abuse of market power and of potential market reforms (as in Colombia).

A persistent challenge for electricity markets is preventing abuses of market power. Several developing countries have introduced power markets despite concentration in generation (for example, Chile and the Philippines). In these situations, simple limits on ownership and concentration have proven ineffective at deterring the exercise of market power in wholesale electricity markets. The entry of new competitors, coupled with wider market integration, has been more effective at lowering concentration and increasing competition. As revealed by the energy concentration index (Herfhindahl-Hirschmann Index, HHI),[4] the entry of new competitors gradually drove down concentration in Peru, and the interconnection of the two major power grids in the Philippines brought about a sharp decline in concentration in 2010 (figure 7.2). In general, the abuse of market power is easier to detect in bid-based markets, where it manifests itself in the form of sudden price spikes, than in more tightly regulated cost-based markets; however, cost-based pricing can be selectively introduced into bid-based markets as a strategy for managing market power in the presence of localized transmission constraints.

A balance must be sought between managing the risks of sustained exertions of market power and restrictive regulations that could cripple the market (Wolak 2005). Market intervention (and even suspension) may be justified, for example, if capacity is withheld from the market, leading to excessive reliability risks.

FIGURE 7.2 Addressing market power is a critical component of establishing efficient power markets

Evolution of market concentration index for generation

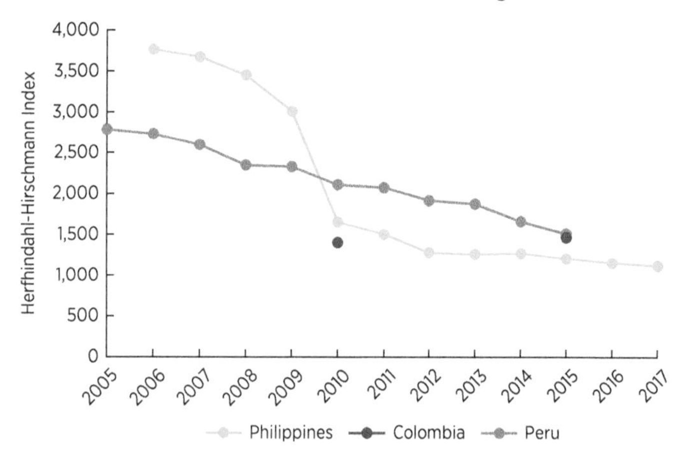

Source: World Bank elaboration based on Rethinking Power Sector Reform utility database 2015.

The intervention must follow a rational and predictable process, however, ideally specified in the form of rules and procedures. There is always a danger that provisions for market intervention will be used for political purposes, such as reducing the volatility of spot prices if they are perceived to be too high to be passed through to households. Intervention and suspension of power markets have been limited to very specific circumstances in the observatory countries; they have generally worked well but are not exempt from controversy.

Two conflicts in the Philippines' bid-based power market illustrate the difficulties of dealing with market power under tight supply conditions. The first stemmed from a concentration of market power in the early days of the market. The market was introduced in 2001, with the generation sector still highly concentrated. At the time, the government corporation created to manage privatization of generation assets, Power Sector Assets and Liabilities Management Corporation (PSALM), managed 54 percent of installed capacity through four trading teams that were required to act independently. However, spot prices surged during the first months of market operation; initial analyses found that prices had been altered by collusion among PSALM's trading teams. A series of exchanges and appeals then ensued, and a case is still pending in the Supreme Court of the Philippines (Abrenica 2009; Roxas and Santiago 2010). This conflict illustrates the difficulties of introducing a power market where generation remains highly concentrated, as well as the practical ineffectiveness of many simple mitigation measures.

The second conflict occurred when exogenous factors led to a period of tight supply. Several large plants were affected by maintenance activities and forced outages in late 2013, creating tight supply conditions that caused prices to surge (DOE 2014). To prevent further price spikes, the regulator lowered the cap on the spot market price and later imposed an even lower cap to be triggered in the event of any sustained period of high prices. In response, the major distribution utility, Meralco, attempted to pass these price spikes on to its retail customers to offset its own exposure. The Supreme Court temporarily restrained Meralco's price increase while the regulator conducted a probe, which found that generators had wielded market power by withholding capacity. The probe led to a suit that is also still pending in the Supreme Court. This case illustrates that, although regulatory interventions (such as price caps) may be effective in the short-term, they are no substitute for tackling underlying causes—in this case, insufficient competition in the spot market, plus lack of contracting to supply regulated customers in the retail market. The moral is that, although emergency interventions may occasionally be necessary, more permanent market adaptations are required to deal with the underlying issues that trigger price spikes and reliability problems.

Power markets require regulation to be flexible enough to adapt to market changes and public policy objectives. The market must correct problems and adapt to evolving supply and demand conditions by creating new markets or enhancing existing ones, while maintaining the confidence of investors by avoiding ex post adjustments of market rules for purely political reasons. The observatory countries provide plenty of examples of how market designs have gradually been improved through market monitoring efforts (see table 7.2). Several Latin American countries introduced supply auctions for regulated customers as a means of attracting investment in baseload generating capacity. This market-based approach has been successful in preventing costly rationing during dry hydrological periods, thus reducing the appetite for direct intervention in the market. Evolving public policy objectives also require market adaptations, but these should harness the benefits of competition as a means of efficiently attaining the desired outcomes. As an example, India introduced the policy objective of increasing renewable energy investment, and this was initially incentivized through feed-in-tariffs. Although effective, these have been gradually replaced by market-based instruments such as renewable auctions and renewable energy credits that can be traded through power exchanges. Peru's energy policy objective was to diversify hydrological risk through greater investment in gas-fired power generation. The wholesale power market was pushed in the desired direction through the use of technology-specific auctions and socialized market levies to finance the necessary natural gas infrastructure.

Finding #5: Adequate system governance and effective open access are key to ensuring operational efficiency and attracting new entrants

Open access to the transmission grid is a precondition for competition in the electricity industry. Because of the unique properties of electricity, centralized, real-time coordination is required to keep supply and demand in balance; the necessary coordination is made possible through the planned and real-time operation of the transmission system (Hogan 1998a). Open and nondiscriminatory access to the transmission network by wholesale sellers and buyers is required for efficient production and exchange. Moreover, because transmission is key for effective wholesale competition, transmission operations should be effectively independent from market participants such as generation companies, retailers, and distribution utilities (ESMAP 2013). Making transmission independent from the rest of market participants requires restrictions on cross-ownership (Arizu, Dunn, and Tenenbaum 2001). Once structural measures are in place, open access should be enshrined in law rather than left to bilateral private negotiations. Separate contracting should be allowed for energy and network services (to enable multibuyer, multiseller competition), and independent system operators should be established (Joskow 2008a; Newbery 2002; Rudnick and Velasquez 2018).

Successful power markets also require an adequate governance structure, especially for the system operator. The governance of the market defines how decisions are made and enforced. *Effective* governance leads to smooth and continuous improvements in market rules and procedures, and in their practical implementation, enhancing the benefits of competition. System and market operators play a central role in any power market, coordinating dispatch, determining pricing, and conducting settlement arrangements in the wholesale market. Good governance arrangements for the operators enable active involvement by all market participants, while ensuring that no single interest group dominates decision making. Broader elements of market governance include expedited dispute resolution, as well as regulatory review and approval of procedures (Barker, Tenenbaum, and Woolf 1997).

To ensure open access and efficient market operation, different approaches to transmission ownership and operation have been adopted in the industrialized countries of OECD (Pollitt 2012). On the one hand, several markets in the United States have been organized around an independent system operator,[5] which operates the transmission system and the wholesale markets without owning any transmission assets. Markets in Europe, on the other hand, have allowed transmission system operators to combine both ownership and operation of transmission. Incentives for cost control have been cited as one of the advantages of transmission system operators, which have been relatively successful in Europe, and particularly in the United Kingdom (Pollitt 2012). Overall, however, both integrated and separated models appear to function effectively in the OECD context.

In the developing country context, some evidence suggests that integrated transmission ownership and operation can pose institutional challenges. In principle, unbundling transmission ownership from operation by establishing an independent system and market operator with no ownership of transmission assets is not strictly necessary, as long as adequate governance and oversight arrangements are in place. For example, Colombia has a successful system and market operator owned by the major transmission company, Interconexion Electrica (ISA) (table 7.3). The operator's functional independence is ensured by appropriate arrangements; however, effective transmission system operators have been more difficult to implement in other emerging markets. In the case of India, the authority of state utilities that are often vertically integrated and financially distressed hinders open access and regional market integration. State operators have denied open access to large customers, probably to avoid financial losses for the state's distribution utility. These experiences highlight the importance of both the general structure of the power sector and the detailed rules governing open access in enabling the success of power markets; the experiences suggest that separation may be preferable, because it is less prone to concerns that the operator is discriminating against or in favor of third parties through opaque and complex operational criteria (Arizu, Dunn, and Tenenbaum 2001).

TABLE 7.3 Overview of power market governance across developing countries

Design element	India	Philippines	Colombia	Peru
Coordination of operations	1 national and 5 regional SOs coordinate state SOs	Centralized merit-order, though many contracts are physical	Fully centralized merit-order	Fully centralized merit-order
System operator	Government-owned TSOs at the regional and state level. Often integrated utility at the state level	LTSO: Market operator not independent (it is chaired by the Department of Energy).	LTSO: System/market operator is functionally independent, and subsidiary of major government-owned Transco ISA.	ISO: Private not-for-profit organization, independent from market participants (owns no transmission assets)
Open access (regulated / negotiated)	Regulated, administered by SOs. Some problems due to conflicts of interest with state utilities.	Regulated. Distributors may have some conflicts of interest due to vertical integration.	Regulated.	Regulated, established by law.

Source: World Bank elaboration based on Rethinking Power Sector Reform utility database 2015.
Note: ISO = independent system operator; LTSO = Legally unbundled transmission system operator; SO = system operator; TSO = transmission system operator.

Beyond unbundling of transmission ownership from operation, further separation of system and market operation may be organized depending on the chosen market design. In a centralized model, the functions of market and system operation might be assigned to different coordinated entities. For example, in the Philippines, the Wholesale Electricity Spot Market (WESM) is the independent market operator, and Transco the transmission system operator. Other countries with centralized markets, such as Colombia and Peru, established an integrated system and market operator. On the basis of their experience, the strength of governance arrangements seems more important than the separation of system and market operation per se. For less centralized markets, such as India's, power exchanges can be organized to enable voluntary exchange of standardized products; however, voluntary participation probably requires a large system to ensure sufficient liquidity in the power exchange. Overall, surveyed experience does not suggest a preponderance of either benefits or risks from separating market and system operations. It does, however, reinforce the general principle that centralized coordination is inevitable (whether among multiple operators and balancing authorities, or between the system and market operator) and that ultimate operational authority should be given to the (transmission) system operator to preserve supply reliability.

Finding #6: Efficient, security-oriented real-time dispatch is critical to getting short-term prices right

Short-term markets should be designed to fulfill their primary function of facilitating efficient, liquid, and transparent decisions about dispatch and adjustment. Short-term markets must also produce the right signals to inform and incentivize investment, with minimum resort to distorting interventions. The role of short-term markets in providing these signals is important, even though new-generation capacity in developing countries is largely hedged through long-term contracts or other mechanisms. Thus, pricing, dispatch, transmission capacity allocation, and congestion management are among key components for efficient short-term wholesale markets. Additionally, ancillary services and price volatility have become the focus of many market design discussions, given increasing penetration of variable renewable energy such as wind and solar.

The ideal pricing mechanism for wholesale short-term power markets involves high spatial and temporal resolution. A temporal resolution of one hour or less (for example, 5–15 minutes) is required. The greater the temporal resolution, the easier it becomes to integrate variable renewable energy and other new technologies (IEA 2016). Spatially, a price should be defined for each transmission node (so-called locational marginal prices), reflecting the physical properties of the transmission network, including losses and congestion. Nodal pricing enhances efficiency and transparency of the market by identifying and managing transmission bottlenecks (Hogan 1998b); it has been implemented in Peru and the Philippines (table 7.4). Simpler pricing options, such as zonal prices (for example, power exchanges with market-splitting in Europe and India) or even systemwide prices (for example, Colombia) are also a common choice in both developed and developing countries. Given the pervasiveness of transmission congestion in these markets, the potential benefits of nodal pricing and better congestion management are significant (McRae and Wolak 2016; Neuhoff and others 2013; Ryan 2014). Moreover, simplicity is not necessarily an advantage of zonal or system prices. Indeed, the system spot price of the Colombian market is arguably more complex and less transparent. The Colombian market is based on a uniform-price, bid-based auction, with a single

TABLE 7.4 Comparison of short-term power market design across developing countries

Design element	India	Philippines	Colombia	Peru
Centralized or bilateral market	Mostly bilateral; with centralized power exchanges	Centralized; partially bilateral	Centralized	Centralized
Cost or bid- based dispatch	Bid in power exchanges	Bid-based	Bid-based	Cost-based
Market for reserves and ancillary services	Nascent regional ancillary services market aimed at restoring frequency, relieving congestion	Reserves are cooptimized with energy scheduling, at prices agreed by the SO.	Regulation reserves optimized before energy, based on energy-bids by Gencos.	—
Mechanism for pricing and congestion management	Priority stack for allocating transmission capacity (with priority for long-term contracts). Market-splitting in power exchanges	Nodal spot prices paid to generators, while customers pay a zonal price.	Single-node price with up-lift for startup and shutdown costs. Congestion constraints settle against real dispatch.	Nodal spot prices and centralized system operation (with power and gas constraints suppressed from spot prices 2009–16).
Settlement approach	Multisettlement in power exchanges	Single-settlement	Single-settlement	Single-settlement

Source: Based on Rethinking Power Sector Reform utility database 2015.
Note: Gencos = generation companies; SO = system operator; — = not applicable.

system price that ignores transmission congestion and an ideal schedule that disregards the transmission system. A parallel, technically feasible schedule is also determined. Differences between the ideal schedule and the technically feasible are settled financially using a reconciliation price for each generator, different from the spot price, thus resulting in a more complex dispatch and pricing process than the nodal pricing alternative.

Distortions in pricing and dispatch should be avoided to the extent possible. Direct regulatory interventions to deal with undesired market outcomes should be used as an emergency measure only. Expedients that significantly distort economics, such as artificially low price caps and disregard for the transmission system, are best avoided. For example, price caps should be high enough to encourage investment. The transmission system should be expanded to reduce congestion and accommodate a market in liquid contracts (whether bilateral or organized futures) as a means of hedging risks and stabilizing retail prices.

Scheduling and dispatch can be centralized or decentralized. The scheduling and dispatch process should support competition and reliability, and market outcomes should be as transparent as possible. Centralized coordination of system operations is ultimately necessary for reliability, owing to the complexities of the physical interactions that take place in the transmission grid. The degree of centralization of system decisions varies, however, across power markets. In the fully centralized pool model, bids are submitted to a market run by a market operator that, through an optimization process known as security-constrained economic dispatch, determines the market-clearing and technically feasible schedule of production for each power plant to reliably and economically supply demand. In the power exchange model, by contrast, market agents can sign bilateral contracts and then declare their production or consumption schedules directly to the system operator before gate closure (such as one hour before real-time operation) or can submit bids to the trading platform of the power exchange. After gate closure, the system operator takes control to perform balancing of real-time operations (Batlle 2013). In either case, procedures and outcomes should be publicly available in a timely manner to all participants and interested participants, including available transmission capacities, prices, and schedules—all without disclosure of commercially sensitive proprietary information.

Centralized markets based on security-constrained economic dispatch have proven successful in delivering transparent market outcomes reflective of underlying demand and supply. That record of success supports the view that centralized security-constrained economic dispatch is preferable for wholesale electricity markets (Hogan 2002; Newbery 2005; Rudnick, Varela, and Hogan 1997). Among key advantages of centralized power pools is the tight integration of operational and economic considerations across energy, transmission, and ancillary services markets to determine the scheduling and dispatch of power plants. In theory, such integration enables higher productive efficiency (Sioshansi, Oren, and O'Neill 2008). These advantages are borne out by the successful experiences of Colombia, Peru, and the Philippines, which have all successfully employed some form of centralized short-term markets for dispatch (table 7.4).

Although decentralized bilateral markets with power exchanges offer some practical advantages over centralized pools, they also present significant challenges. The main advantages of decentralized markets are that they are easy to set up and they allow for simpler trading (resembling market-clearing in other commodity markets). This simplicity comes at the expense of reduced integration of energy trading and transmission management (Wilson 1998). Bilateral markets with power exchanges have worked well in Germany and the Nordic countries. They have also allowed for some limited efficiency gains in India, which has a bilateral market with a regulated mechanism for regional balancing complemented by power exchanges functioning as organized futures markets. India's decentralized short-term power markets have struggled, however, to deliver economic dispatch across wide geographic areas.

To improve performance of the power sector, the wholesale market should be complete and liquid, with competition covering a significant amount of supply. Competition requires completeness, that is, a full set of forward and spot markets, in addition to risk-management tools (Hunt 2002). Liquidity is particularly important in voluntary bilateral markets, where not enough customers or suppliers may be willing to participate. Centralized mandatory markets such as power pools, by contrast, are inherently liquid. Organized futures markets have been established only in India, where power exchanges emerged in 2008 with day-ahead markets, which later incorporated week-ahead and renewable energy credit markets. However, inflexible power purchase agreements in India relegate the role of power exchanges to a bare minimum, thus limiting the efficiency gains that competition can provide in both operation and investment. In other developing countries, only single-settlement centralized markets have been organized, primarily with day-ahead scheduling and monthly settlement.

Over the past decade, wholesale prices in the observatory countries have come to reflect market conditions, regulation, and central planning. There is significant interannual price variation across power markets, with variations of three to four times between high-price and low-price years (figure 7.3). The highest sustained annual prices observed are approaching US$200 per megawatt-hour (MWh), whereas the lowest are less than US$20/MWh. Among the key market conditions driving prices are the entry of new generation capacity (apparent in the downward price trend in India), fuel availability and prices in international and domestic markets, and hydrological conditions (especially for Colombia's hydro-dominated system). Evolving regulation is also

FIGURE 7.3 Significant interannual price variations across power markets in the observatory countries

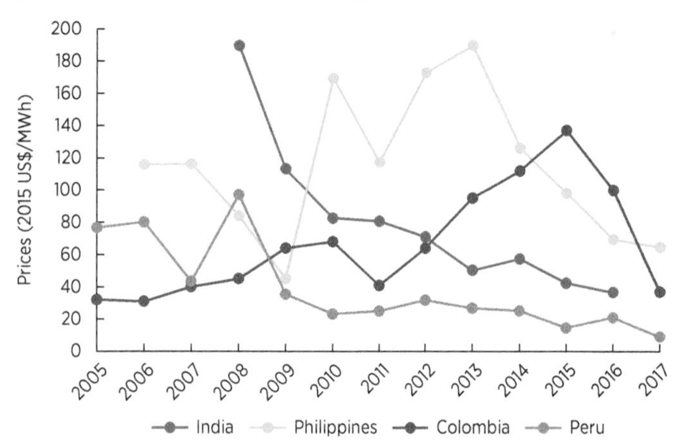

Source: World Bank elaboration based on Rethinking Power Sector Reform utility database 2015.
Note: MWh = megawatt-hour.

visible in Colombian prices, owing to the introduction of a firm energy market, which contributed to containing price spikes during the 2009–10 El Niño drought but failed to do so during 2015–16 because of design weaknesses and unexpected events. Central planning and government direction of markets are reflected in price trends for India and Peru, where prices fell consistently, partly because of surplus capacity that resulted from aggressive government-led investments in generation. These trends underscore the close interaction of short-term efficiency in pricing and dispatch with the longer-term development of the market.

Finding #7: Drawing new entrants into the market requires effective risk management, including the use of contracts

Incentivizing the optimal generation mix and attracting adequate levels of investment is one of the most important roles of power markets in fast-growing developing countries. Whereas developed countries introduced reforms primarily under conditions of surplus capacity and high reliability, many developing countries pursue power sector reforms to address shortages of generation capacity or to lower the financial burden of large investments in generation.

Developing countries must not fail to implement mechanisms to ensure the stable revenues that investors require before financing infrastructure. There have been few cases of merchant power plants built spontaneously by the private sector at the owner's risk and with full exposure to spot prices. Rather, the state has played an important role, ranging from direct investment in India to indicative planning in Colombia. New capacity additions across developing countries have almost always been driven by medium- to long-term contracts (with maturities ranging from a few years to 20 or more years). These contracts, freely agreed among market participants, can be physical (entailing physical and cash delivery) or purely financial (entailing only cash delivery) (Batlle 2013). In many cases, they are accompanied by special mechanisms designed to incentivize capacity additions (table 7.5). The design of such mechanisms is challenging and can make a material difference to investment decisions. In addition, investment outcomes also depend on the quality of institutional governance in the power sector, and on the security of fuel supplies and the efficiency of short-term markets.

Regulated capacity payments have proven effective in eliciting investment in generation—sometimes to the point of creating excess capacity. Regulated capacity payments are in place in Chile and Peru; they were also part of the initial approach to ensure adequate capacity that was adopted and later abandoned in Colombia. These payments, based on the theory of peak-load pricing (Boiteux 1960), provide power plants an income based on a regulated price that reflects the regulator's view of the marginal cost of efficient generating technology to provide peak power (often

TABLE 7.5 Comparison of mechanisms for generation investment across developing countries

Design element	India	Philippines	Colombia	Peru
Contracts and bilateral markets	Physical bilateral market for long and medium term, traders for short term	Physical bilateral contracts (financial accounting in WESM)	Financial bilateral contracts	Financial bilateral contracts
Capacity markets	None	None	Firm-energy payments (for generation available during poor hydrology)	Regulated capacity payments
Auctions	Renewable auctions	Auctions for privatization of generation assets and power purchase agreements	Centralized auctions (by transmission system operator) for firm energy	Distribution utility supply auctions for regulated customers (including technology-specific auctions)
Investment responsibility and risk allocation	Primarily driven by government planning and private investment. States hold a competitive bidding process when a demand-supply deficit is projected.	Responsibility and risks left entirely to private participants (government is forbidden from investing or underwriting new generation capacity).	Generators bear full risks and responsibility of investment, incentivized by short-term power market and firm-energy payments. (However, costs were passed through to customers during the 2015–16 El Niño event.)	Generators bear full investment risk in the core market design. Market has been adjusted to incentivize gas, hydro, and other renewable generation (extra costs levied directly on end-users).
Renewables	Renewable energy credits traded in power exchanges Renewable purchase obligations, feed-in-tariffs, and renewable auctions	Renewable portfolio standards, plus feed-in-tariffs	No specific mechanism in place	Renewable auctions, with price premium if spot price falls below auction price

Source: World Bank elaboration based on Rethinking Power Sector Reform utility database 2015.
Note: WESM = Wholesale Electricity Spot Market.

diesel-fired plants), without necessarily establishing a physical target for security of supply. Regulated capacity mechanisms, which are reflective of the supply side only, are effective in attracting peaking capacity, but they often elicit capacity that is perceived as excessive. For example, the regulated capacity payment in Chile has been called into question by some market participants, given continuous capacity additions in oil-fired peaking power plants, despite an apparent shortage of generating capacity. In this context, some aspects of the administrative procedure for calculating the capacity payments will be revisited by the regulator in the course of 2019.

Capacity markets are designed specifically to ensure that enough generating capacity is available to meet demand (adequacy). Although in theory an energy-only market could provide adequate generating capacity, in practice—at least in Europe and the United States—a series of market failures (such as low price caps on the spot price) have motivated the development of capacity markets (Newbery 2015). The Philippines has an energy-only market that has suffered from tight capacity margins and spot price volatility, leading to concerns about the adequacy of the system. Colombia has developed a parallel market that focuses on firm energy rather than capacity per se. Colombia is highly dependent on hydropower generation, with variable availability that can greatly diminish during dry seasons. Therefore, meeting peak demand is not the primary issue, as it is in systems based on thermal generation. Instead, the challenge is to ensure adequate energy supplies during periods of low hydrological availability. The Colombian firm-energy market procures firm energy through auctions, to ensure availability during such periods of scarcity (defined as periods when the spot price exceeds a regulated strike price called the scarcity price). Generators receive a monthly payment for firm energy according to the results of the auctions, in exchange for the obligation to supply energy under conditions of scarcity. Unfortunately, during the 2015–16 drought, the efficacy of the firm-energy market was undermined by unexpected events, including forced generation outages, reduced fuel availability, and falling international oil prices. These events left many generators with operating costs that exceeded the regulated scarcity price, leaving them unwilling to provide the contracted supplies and necessitating regulatory intervention of the scarcity price to prevent supply shortfalls.

Auctions for awarding supply contracts have been successful in fostering competition in contracts markets and in attracting investment. Various auction mechanisms have been successfully introduced across the developing world. For developing countries in which financial markets are underdeveloped and electricity futures markets are absent or illiquid, auctions can provide a workable means of harnessing competitive forces (Maurer and Barroso 2011). Peru accomplished this in 2004, followed by India in 2006; and the Philippines is currently considering auction arrangements. These auctions can be required by law, for example, for the supply of regulated customers by distribution utilities years in advance of supply to allow participation of new power plants. That is the case in Brazil and Chile, where supply auctions have been of great interest to international investors looking to enter these markets. Auctions are often technology-neutral, encompassing hydropower, gas, coal, oil, sugarcane biomass, solar, and wind, as well as cross-border imports (Rudnick and Velasquez 2018). They can also be purposely targeted at specific technologies, as is often the case when they are used to accelerate uptake of renewable energy. Even without legal mandates, auctions may still be encouraged as a means of competitive procurement, for example, by facilitating aggregation of smaller customers (for example, in Chile and the Philippines).

Market outlooks and indicative planning of capacity expansion have a role in guiding private investment and market participants. The use of capacity markets or auction mechanisms requires the government to determine desired volumes of investment based on indicative generation planning. This is the case of the firm-energy market in Colombia and the capacity markets in the PJM area of the eastern United States. Centralized planning of transmission capacity has also been chosen in some markets (for example, Chile) as the mandatory planning process. Other countries provide more comprehensive indicative planning for the power and energy sector (for example, Colombia). Such outlooks can also help to orient potential new entrants seeking investment opportunities in complex markets.

The evolution of the reserve margin over time provides some indication of success in attracting adequate new investment into generation (figure 7.4). The observatory countries faced different situations with regard to reserve margins. India's achievement was to reverse its chronic power deficit to reach overall balance, albeit without any significant reserve margin. The Philippines began its wholesale market with high reserve margins resulting from capacity procured under the independent power producer program of the 1990s, but those margins have been shrinking, in part because of greater efficiency, but also reflecting the weaker investment incentives associated with an energy-only market, compounded by the application of price caps, insufficient competition in the contracts market, and barriers to entry. For Colombia and Peru, reserve margins are much higher and growing, reflecting the need to diversify hydro-based systems toward natural gas so as to ensure firm energy during droughts.

It is important to strike a balance between the revenue stability provided by long-term contracts and the need to retain flexibility to adapt to emerging market conditions and opportunities. Contracts of very long maturity may be needed to cope with serious capacity shortages, making a wholesale market infeasible, or to incentivize investment in rapidly growing economies with country-specific risks. A large share of inflexible contracts of very long duration (10 or 20 years), however, can curb the ability of the market to adapt to changing conditions, such as lower-than-expected demand growth or technological disruptions like cheaper generation technologies (undercutting higher-cost contracts). In some developing country markets, a large share of power is traded under rigid long-term contracts that predate the establishment of a wholesale power market (figure 7.5). When these contracts expire, power plants can enter into more-competitive contracts or participate in competitive capacity markets. Moreover, uncontracted capacity can participate in power exchanges (as in India) or directly in the wholesale market (as in the Philippines), depending on the risk appetite of individual market participants. Competition in contracts is also important for centralized markets such as Peru's, where dispatch and spot pricing are independent from contractual positions (which by design fully cover actual demand).

Variable renewable energy can make a valuable contribution to the overall flexibility of the system, if allowed to compete alongside other technologies. In achieving flexibility, wind and solar power plants present a unique opportunity for developing countries, given their short lead times and lower scales compared with large coal-fired and hydropower plants (whether fed by pumped water or

FIGURE 7.4 Power markets have helped countries attract adequate investment into generation

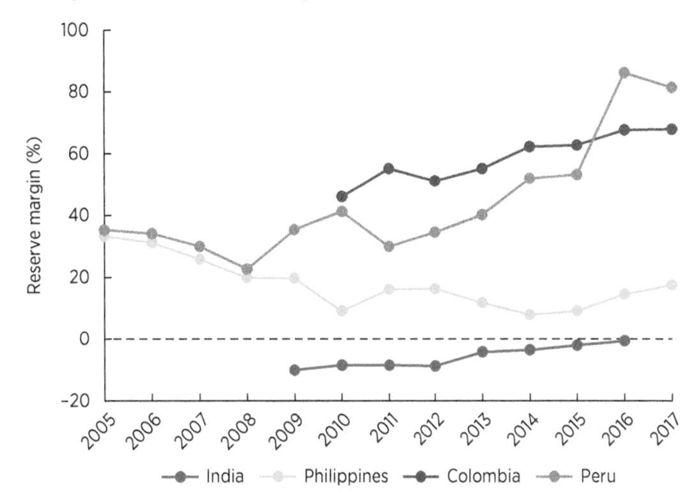

Source: World Bank elaboration based on Rethinking Power Sector Reform utility database 2015.
Note: The reserve margin is calculated as available generating capacity divided by the system peak demand (maximum hourly metered value for each year), without correcting for low hydropower availability. The reserve margin for the Philippines is for the Luzon grid through 2009; from 2010, the Luzon and Visayas grids are considered.

FIGURE 7.5 Inflexible contracts of long duration can limit a market's ability to adapt to changing conditions

Relative size of contract market as share of traded energy

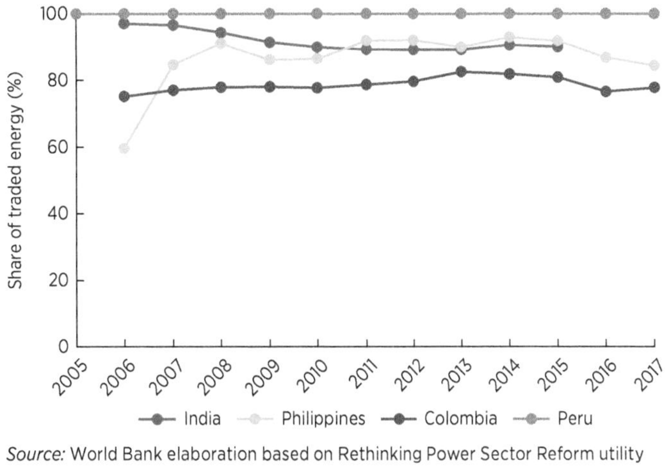

Source: World Bank elaboration based on Rethinking Power Sector Reform utility database 2015.

large reservoirs). Beyond benefiting from their contributions to the achievement of policy targets for low-carbon investments (for example, renewable auctions or renewable portfolio standards), the market can also allow renewables to compete alongside other technologies for the supply of regular energy and capacity markets, including bilateral contracts and centralized supply auctions. In Peru, regulation effectively bars renewables from participating in auctions for regulated customers, though discussions are underway to adapt the procedure for calculating capacity credits to recognize renewables and so allow them to compete for contracts.

Renewable generation in developing countries has been growing faster over the past decade. Growth has been driven by targeted incentives, support mechanisms, and declining investment costs for wind and solar photovoltaic power. In India, the plunge in solar investment costs has motivated more competitive market-based forces such as renewable auctions, instead of the previously established feed-in-tariffs. In Chile, renewables were initially incentivized through a quota law. The required share of renewable generation has been exceeded because of falling investment costs.

Finding #8: Demand-side participation, especially of large customers, is critical to several aspects of market efficiency and system reliability

Demand-side participation in power markets can greatly improve their performance. So-called demand participation has altered a previously one-sided market focused on generation and supply, providing benefits in terms of efficiency and reliability. Key benefits include (1) gradual replacement of price regulation with market-based forces; (2) limitations on the extent of market power in short-term markets, especially during tight conditions; and (3) greater flexibility in the power grid's operation. Organizing demand participation requires that distribution utilities and large customers participate in all key sector processes, such as discussions of market rules and procedures, regulatory processes, and legal reforms. Representatives of major consumers also need to be formally incorporated into the governance structure of system and market operators, which has not always been the case to date.

Deepening demand participation is often a gradual process that begins with the largest industrial customers. The usual starting point is to liberate the largest industrial consumers to seek their own power supply directly from generators, which creates an important market feedback loop and can provide an attractive commercial opportunity for new market entrants into generation. In many countries, the size threshold above which customers are freed from the captive or regulated market has gradually dropped. In the Philippines, customers above 1 megawatt (MW) have been eligible since 2013. In Colombia and Peru, where reform processes date further back, the threshold has progressively fallen to the current values of 0.1 MW and 0.2 MW, respectively. Deepening the retail market can increase contestability by lowering barriers to entry, enabling more customers to contract with new generators. It is therefore important to allow regulatory flexibility to revise downward the thresholds for customer choice according to evolving market conditions.

Demand aggregation is an emerging institutional model that allows smaller customers to harness the benefits of competition. For example, the Philippines is implementing a competitive supply procedure after successful aggregation experiences with smaller cooperatives that resulted in lower prices. In Brazil and Chile, a few load aggregators have also succeeded by organizing private auctions for power supply from load reductions across a cluster of customers. This experience highlights the need for regulation that allows such innovations and mechanisms wherever market conditions make them relevant and feasible.

Demand participation can also be an important mechanism for efficiently achieving supply reliability, and the operational flexibility required to integrate high levels of variable renewable energy. Growing shares of wind and solar energy emphasize the need for resources and mechanisms capable of accommodating sudden shifts in production (Strbac 2008). Demand response can potentially play this role. For example, Philippine's largest distribution utility developed a scheme of interruptible load demand to be used under conditions of tight supply as declared by the system operator. Such demand response programs help to avoid the high economic and political costs associated with rationing, as long as they are voluntary and properly compensated. Colombia implemented an emergency demand penalization program for household customers among the packet of measures adopted in the wake of the 2015–16 El Niño crisis. These cases illustrate the large potential benefits of demand participation, especially for developing countries such as Colombia and the Philippines, which are periodically subject to tight supply conditions.

LOOKING AHEAD

The growing penetration of variable renewable energy is challenging the established paradigm for price-setting in wholesale power markets. Competitive wholesale markets are designed to send a price signal that reflects the marginal cost of the last unit brought on to meet consumer demand (that is, the short-run marginal cost of generation). In these markets, a complex model ensures that generators bidding to supply the market are selected on the basis of least cost, while meeting all system constraints. The last unit selected to meet the system demand sets the price. This model works well when all units, including demand response, are dispatchable and have positive marginal cost; variable renewables, however, are not always dispatchable and typically have zero marginal costs.[6] This creates problems in terms of providing adequate incentives for investment in new plants and may even push existing (mainly thermal) resources out of the market, raising concerns over the potential for excessive decommissioning of thermal power plants and associated reliability risks, as well as the financial challenges associated with the stranding of assets.

These challenges associated with variable renewable energy further exacerbate longstanding concerns regarding the adequacy of investment incentives arising from short-run marginal price signals. Longstanding experience with traditional power markets dominated by thermal resources has shown that relying on purely market-based incentives for investment in new capacity entails allowing spot prices to spike at high enough levels for long enough periods. The very high price levels that result can be challenging to sustain for wider political and economic reasons, to the point that in some competitive markets regulators have set caps that stop prices from going above some maximum acceptable level, but at the same time prevent investors from receiving an adequate level of remuneration. Even when complex mechanisms exist to allow prices to escalate, grid operators must sometimes take actions outside of the market context in order to protect system reliability. In both cases, such measures mute the price signals needed in the short-term to ensure longer-term fixed cost recovery.[7] This has led to the development of a variety of parallel mechanisms to guide investment decisions, such as capacity charges, capacity markets, and supply auctions.

In addition, the rise of variable renewable energy creates the need for a different composition of investment geared toward flexible generation. Grid operators trying to integrate renewables need fast-ramping, flexible resources to come online when the sun is not shining or the wind is not blowing. Whereas flexible resources such as flexible generators, storage, interconnections to neighboring grids, and demand response can accommodate these variations to some extent, inflexible generation resources, such as nuclear and older coal power plants, are less helpful in responding rapidly to changes in net load. If the price is increasingly set by renewables, then there is no consistent energy price signal to incentivize the development of flexible resources. For this reason, a key current concern is to ensure that ancillary services (such as operating reserves, frequency regulation, and so on) are sufficiently priced to incentivize fast-ramping, flexible resources.[8] These essential reliability services would have to be priced very high, however, to make up for the zero (or very low) energy price seen for much of the day. There is growing concern that, in power systems made up mostly of renewable resources with zero marginal prices, a market framework that incentivizes investment decisions through price signals based on short-run marginal costs is unlikely to ensure the correct mix of resources needed to keep the lights on (see, for example, Sen 2014; Tierney 2018).

For this reason, grid systems are beginning to consider the introduction of new markets in ancillary services. Some are considering the introduction of specific ramping products or holding reserves that match the energy limitations of resources (for example, when the sun or wind is suddenly unavailable.) In addition, in centralized power markets, reserves are integrated to some extent with the scheduling and dispatch processes. These optimization-based processes performed by the system operator can directly accommodate complex bids (including start-up costs in addition to variable costs) and technical operating parameters such as ramping capabilities (IEA 2016). Although tight pool models have worked reasonably well in practice, there is room for improvement in developing countries, particularly through cooptimization of reserves and energy, because the values of energy and reserves are mutually dependent (MIT 2016). Chile, for example, is expected to introduce auctions for the provision of some ancillary services by 2020. Bilateral markets, conversely, rely on real-time balancing services coordinated by the system operator, which penalize participants for imbalances in their scheduled generation and demand. India, for example, is developing markets for centralized scheduling of regulation services at the regional level, although the outcomes of these reforms have yet to be seen (Mukhopadhyay and others 2016).

In any case, both centralized and bilateral markets can benefit from integration of nongeneration resources to provide flexibility and other ancillary services. Demand response is by far the most underused flexibility resource across developing countries. Ancillary services markets or mechanisms should allow market-based participation of large customers and aggregators of smaller customers, because even moderate amounts of customer participation can help to balance the grid, facilitating the integration of variable generation resources. Furthermore, distributed resources such as air conditioners, refrigerators, electric vehicles, rooftop solar photovoltaic, and household batteries can and should participate with larger generation resources in providing the ancillary services that the system operator requires (MIT 2016).[9] Along with sound institutions, solid regulatory frameworks, and adequate economic incentives, demand participation could flourish and help avoid politically costly rationing. Developing countries, however, have been slow to adopt the new technologies that enable demand response and related adaptations by wholesale markets.

The current wave of technological advances has changed the relative costs of generation technologies and of energy storage. Modern forms of renewable energy, notably wind and solar, have seen a dramatic decline in costs driven by economies of scale in equipment manufacturing, technological breakthroughs, and learning; they now compete with—and in some cases outcompete—conventional power, even without subsidies and before environmental costs are factored in. At the same time, a variety of storage options is becoming available to compensate for the variable nature of modern renewables. Beyond the longstanding location-specific option of pumped storage, thermal storage with molten salts is becoming available in areas with adequate solar radiation for concentrated solar power, to name just one new storage technology. Although electrochemical storage in batteries is still economic only for small-scale or short-term storage, costs are falling rapidly, and further technological breakthroughs are anticipated. In the future, the batteries of electric vehicles may also become storage sites that can supply grids when the vehicle is not needed. Compressed air and flywheels could provide additional forms of energy storage. Finally, there is the storage potential of "power-to-gas," for example, converting renewable electricity to be stored as hydrogen or other gases.

Given such rapid technological change, markets can be useful to enable price discovery of the most efficient resources to meet specific system needs. Even in contexts where competitive wholesale generation markets may not be applicable, market-like approaches could still bring considerable value. In small systems, this might take the form of a single buyer conducting cost-based dispatch and publishing information on the system's marginal costs. Competitive auctions to procure a variety of resources could be just as helpful. Auctions could be designed to procure not a specific technology but a service requirement defined by a set of characteristics including point of delivery, load curve, power quality, reliability, and carbon-intensity. For example, the integration of renewable energy could be facilitated by contracts with providers of fast-ramping resources or other essential reliability services, such as operating reserves, regulation, voltage support, and black start capability. Contract designs will matter even more than previously, because inflexible take-or-pay contracts will make it more challenging to integrate variable renewable energy. In addition, contracts should be designed to pay for ancillary services.

One of the challenges with planning for the least-cost mix of resources, given all of these technological changes, is that traditional cost-comparison mechanisms may no longer be relevant. For example, the concept of levelized cost of energy (LCOE)—the expected costs over the lifetime of a particular generation asset per unit of energy produced—assumes that resources are dispatchable and able to produce as needed during peak hours. Neither of these assumptions holds for solar photovoltaic or wind. Power produced mostly at night (for example, wind) is not as valuable when the greatest demand occurs during the day.[10] Solar resources are likely to produce during the day but may not operate consistently, and both solar and wind require the availability of flexible resources. In addition, location matters for solar and wind more than for other technologies. Some sites are better; if wind and solar are not sited at those locations, their overall output will be reduced or curtailed. Ignoring these key aspects of renewables makes the LCOE comparison, on its own, insufficient. Unless output restrictions are considered, renewable resources could be overvalued when compared to dispatchable resources (Joskow 2011). The levelized avoided cost of electricity (LACE) attempts to consider the costs of integrating new technologies into an existing resource mix, given their expected utilization. It is a much more complex analysis, because it requires an assessment of how a given project would operate within an existing grid system. When combined with the LCOE concept, it may offer a better indication of how benefits exceed costs.[11] If other resources are displaced when new resources are added to the grid system, the LACE concept can capture some of the potential costs associated with this displacement.[12]

Ultimately, the goal is to enable a power system that integrates various disruptive technologies and ensures that an appropriate mix of resources will be available to grid operators. Markets may be particularly helpful in the transition to a low-carbon electricity sector. Competition can also help protect against the power of vested interests seeking to protect conventional fuel supply and generation technologies that may no longer be cost-effective. Certainly, markets can be used to incentivize generation or load-management solutions that meet system needs, but how those markets will be created and enabled is an ongoing discussion.

CONCLUSIONS

Developing country power markets have the potential to improve the efficiency and reliability of electricity generation, once countries are ready to move to this stage; however, significant preconditions must be met before countries are ready to move toward a wholesale market. Those preconditions may include structural reforms, financial health, a supportive regulatory environment, and resolution of major generation and transmission constraints. Full vertical and horizontal unbundling of the sector is required, together with significant investments in a market platform, none of which is likely to make much sense until the power system reaches a threshold equivalent to 3 GW of peak demand, 20 terawatt-hours of annual energy demand, and wholesale power sales of at least US$1 billion. Nevertheless, there are growing opportunities for countries with smaller power systems to harness some of the benefits of trade by participating in regional power markets.

Creation of a wholesale power market is a complex process that may take some time and requires careful design choices. Countries have adopted various transitional strategies, such as single-buyer markets, vesting contracts, and cost-based pools. These strategies may play useful roles, but they introduce the risk of bogging down the transition. Once a market is operational, careful monitoring and regulatory oversight are needed to detect and act on signs of abuse of market power, as well as to adapt the market design to emerging conditions. System operators, which play a critical role in coordinating actions across the market, may be autonomous or housed within the transmission utility—either way, they must ensure broad-based stakeholder representation. Efficient security-constrained real-time dispatch is a critical requirement for correct short-term prices. Energy-only markets have struggled to attract adequate investment in future generation, and the investment deficit needs to be separately handled through long-term contracts complemented by some form of capacity mechanism or supply auctions. Enabling demand-side participation in the power market significantly adds to its efficiency and flexibility.

Experience shows that adhering to general principles of market design tends to deliver improvements in efficiency and reliability. Conversely, deviations from sound design principles imperil future performance. For example, various degrees of centralized planning in India and Peru have proven successful in attaining public policy objectives, but at the expense of higher costs being passed through to end-users. Looking ahead, power systems are bound to be transformed by disruptive technologies such as solar- and wind-fueled generation, electric vehicles, and widespread distributed generation. In this context, power markets can be an effective enabler for the transition of developing countries toward more sustainable, efficient, and reliable power systems.

The wave of technological disruption currently affecting the electricity sector only underscores the need for sophisticated market design. The rapidly changing landscape of electricity services, with technological change driving down the costs of variable renewable energy and battery storage, increases the value of a wholesale power market that can provide continuous price information. Market rules will also need to be adapted to remunerate the ancillary services that are so critical to the integration of variable renewable energy and to provide appropriate economic incentives for the use of batteries. Moreover, the need to design wholesale power markets to fully accommodate demand-side participation is becoming increasingly important as the evolution of smart grids facilitates ever-more-sophisticated ways of aggregating responses to demand.

NOTES

1. This chapter draws on Rudnick and Velasquez (2018, 2019a, 2019b, 2019c, and forthcoming). Further original research was conducted by a team led by Debabrata Chattopadhyay and comprising Hugh Rudnick, Constantin Velasquez, Martin Schroder, and Tatyana Kramskaya. The work program was coordinated by Vivien Foster and Anshul Rana.
2. Given the physical laws governing power flows, transmission constraints can occur between subsystems owing to internal constraints in the transmission network of a subsystem, even with adequate transmission links between subsystems.
3. See, for example, One-Nation-One-Grid announcement by Powergrid India (https://www.powergridindia.com/one-nation-one-grid) and also Ryan (2017).
4. The Herfhindahl-Hirschmann Index (HHI) presented here measures concentration based on the market share (in terms of generated energy) of each participant. Although the average energy HHI inadequately reflects the extent of market power and its potential impact in prices (which could be much higher during outages of generating facilities), it does provide a useful signal for the overall evolution of competition in the power market.
5. Although in the United States the model of the independent system operator has been widely adopted, operators are also known as regional transmission operators. However, in practice they follow the same general principles as independent system operators.
6. In addition, many renewable resources are often compensated outside of the wholesale market model through various subsidies. This makes their already low bids even lower, because they would be willing to produce up to the point where they lost money, and is part of the reason why negative pricing has begun to appear in competitive wholesale markets in the United States. A lack of transmission is also a reason for the negative pricing, because when transmission is insufficient, only generation within a particular area can meet demand in the same area, effectively bottling renewables.
7. Essentially, these actions mute the scarcity price signal often cited as the key mechanism to incentivize flexible resources.
8. In countries that have them, there is also a discussion about whether capacity markets should focus on procuring resources with specific characteristics, or whether the energy and ancillary services markets alone can produce price signals consistent with system needs.
9. This is a complicated regulatory challenge in many of the advanced markets, because the transmission (wholesale) market system is separated from the distribution (retail) market system.
10. This could change, for example, with the advent of widespread night-time charging of electric vehicles.
11. However, neither LCOE or LACE consider environmental regulations or policy decisions that also influence project development.
12. Or, benefits, especially if environmental or policy goals are taken into account.

REFERENCES

Abrenica, M. J. V. 2009. "Detecting and Measuring Market Power in the Philippine Wholesale Electricity Market." *The Philippine Review of Economics* XLVI (2): 5–46.

Andrews-Speed, P. 2016. "Energy Security and Energy Connectivity in the Context of ASEAN Energy Market Integration." ASEAN Studies Center, Chulalongkorn University; Energy Studies Institute, University of Singapore; Norwegian Studies Institute, February. http://www.asean-aemi.org/wp-content/uploads/2016/03/AEMI-Forum-November-2015-Andrews-Speed-Feb2016.pdf.

Arizu, B., W. H. Dunn, and B. Tenenbaum. 2001. "Regulating Transmission: Why System Operators Must Be Truly Independent." Public Policy for the Private Sector No. 226, World Bank, Washington, DC. http://siteresources.worldbank.org/EXTFINANCIALSECTOR/Resources/282884-1303327122200/226Ariz-1221.pdf.

Arizu, B., D. Gencer, and L. Maurer. 2006. "Centralized Purchasing Arrangements: International Practices and Lessons Learned on Variations to the Single Buyer Model." Energy and Mining Sector Board Discussion Paper No. 16, World Bank, Washington, DC.

Asian Development Bank. 2015. "Knowledge and Power: Lessons from ADB Energy Projects." Mandaluyong City, Philippines.

Barker, J., B. Tenenbaum, and F. Woolf. 1997. "Governance and Regulation of Power Pools and System Operators: An International Comparison." World Bank Technical Paper No. 382, World Bank, Washington, DC.

Batlle, C. 2013. "Electricity Generation and Wholesale Markets." In *Regulation of the Power Sector*, edited by I. J. Pérez-Arriaga, 341–95. London: Springer-Verlag. https://doi.org/10.1007/978-1-4471-5034-3.

Besant-Jones, John E. 2006. "Reforming Power Markets in Developing Countries: What Have We Learned?" Energy and Mining Sector Board Discussion Paper no. 19, World Bank, Washington, DC.

Besant-Jones, J., and B. Tenenbaum. 2001. "The California Power Crisis: Lessons for Developing Countries." ESMAP Discussion Paper No. 1, World Bank, Washington, DC. http://www-wds.worldbank.org/external/default/WDSContentServer/WDSP/IB/2004/03/25/000090341_20040325143555/Rendered/PDF/280840California0Power0crisis0EMS0no.01.pdf.

Boiteux, M. 1960. "Peak-Load Pricing." *The Journal of Business* 33 (2): 157–79. http://www.jstor.org/stable/2351015.

Bushnell, J. B., E. T. Mansur, and C. Saravia. 2008. "Vertical Arrangements, Market Structure, and Competition: An Analysis of Restructured US Electricity Markets." *American Economic Review* 98 (1): 237–66. https://doi.org/10.1257/aer.98.1.237.

Castalia. 2013. "International Experience with Single Buyer Models for Electricity." Report for Contact Energy, New Zealand. https://www.yumpu.com/en/document/read/28356228/castalia-review-international-experience-single-buyer-models.

Chattopadhyay, D., and P. N. Fernando. 2011. "Cross-Border Power Trading in South Asia: It's Time to Raise the Game." *Electricity Journal* 24 (9): 41–50. https://doi.org/10.1016/j.tej.2011.10.009.

Ching, T. L. 2014. "Electricity Spot Market: The Singapore Experience." Presentation at the GCCIA 3rd Regional Power Trade Forum, Abu Dhabi, September 29. http://www.gccia.com.sa/Data/PressRelease/Press_39.pdf.

Defeuilley, C. 2009. "Retail Competition in Electricity Markets." *Energy Policy* 37 (2): 377–86.

DOE (Department of Energy, Philippines). 2014. "Department of Energy Power Situation." Report, Department of Energy, Manila. http://www.doe.gov.ph/sites/default/files/pdf/electric_power/power_situationer/2014_power_situationer.pdf.

ESMAP (Energy Sector Management Assistance Program). 2013. "International Experience with Open Access to Power Grids: Synthesis Report." Knowledge Series 016/13, World Bank, Washington, DC. http://documents.worldbank.org/curated/en/2013/11/18699545/international-experience-open-access-power-grids-synthesis-report.

Foster, Vivien, Samantha Helen Witte, Sudeshna Ghosh Banerjee, and Alejandro Vega Moreno. 2017. "Charting the Diffusion of Power Sector Reforms across the Developing World." Policy Research Working Paper 8235, Rethinking Power Sector Reform, World Bank, Washington, DC.

Hogan, W. W. 1998a. "Competitive Electricity Markets: A Wholesale Primer." Center for Business and Government, John F. Kennedy School of Government, Harvard University, Cambridge, MA. http://www.hks.harvard.edu/fs/whogan/empr1298.pdf.

Hogan, W. W. 1998b. "Nodes and Zones in Electricity Markets: Seeking Simplified Congestion Pricing." In *Designing Competitive Electricity Markets*, edited by H. Chao and H. G. Huntington, 33–62. New York: Springer Science+Business Media.

Hogan, W. W. 2002. "Electricity Market Restructuring: Reforms of Reforms." *Journal of Regulatory Economics* 21 (1): 103–32. https://doi.org/10.1023/A:1013682825693.

Holttinen, H., A. Tuohy, M. Milligan, E. Lannoye, V. Silva, S. Muller, and L. Soder. 2013. "The Flexibility Workout: Managing Variable Resources and Assessing the Need for Power System Modification." *IEEE Power and Energy Magazine* 11 (6): 53–62. https://doi.org/10.1109/MPE.2013.2278000.

Hunt, S. 2002. *Making Competition Work in Electricity*. New York: John Wiley and Sons, Inc.

IEA (International Energy Agency). 2016. *Re-powering Markets: Market Design and Regulation during the Transition to Low-Carbon Power Systems*. Paris: IEA.

Jamasb, T., R. Nepal, and G. R. Timilsina. 2015. "A Quarter Century Effort Yet to Come of Age: A Survey of Power Sector Reforms in Developing Countries." Policy Research Working Paper 7330, World Bank, Washington, D.C.

Jamasb, T., D. Newbery, and M. Pollitt. 2005. "Core Indicators for Determinants and Performance of the Electricity Sector in Developing Countries." Policy Research Working Paper 3599, World Bank, Washington, DC.

Joskow, P. L. 2008a. "Capacity Payments in Imperfect Electricity Markets: Need and Design." *Utilities Policy* 16 (3): 159–70. https://doi.org/10.1016/j.jup.2007.10.003.

———. 2008b. "Lessons Learned from Electricity Market Liberalization." *The Energy Journal, Special Issue on the Future of Electricity, Papers in Honor of David Newbery*, 9–42. http://economics.mit.edu/files/2093.

———. 2011. "Comparing the Costs of Intermittent and Dispatachable Electricty Generating Technologies." *American Economic Review* 101 (3): 238–41.

Joskow, P. L., and R. Schmalensee. 1983. *Markets for Power: An Analysis of Electric Utility Deregulation*. Cambridge, MA: MIT Press.

Kee, E. D. 2001. "Vesting Contracts: A Tool for Electricity Market Transition." *The Electricity Journal* 14 (6): 11–22. https://doi.org/10.1016/S1040-6190(01)00211-1.

Kim, S., Y. Kim, and J. S. Shin. 2013. "The Korean Electricity Market: Stuck in Transition." In *Evolution of Global Electricity Markets*, edited by F. P. Sioshansi, 679–713. Academic Press.

Krishnaswamy, V. 1999. "Non-payment in the Electricity Sector in Eastern Europe and the Former Soviet Union." Technical Paper No. 423, World Bank, Washington, DC. http://documents.worldbank.org/curated/en/1999/06/440475/non-payment-electricity-sector-eastern-europe-former-soviet-union.

Krishnaswamy, V., and G. Stuggins. 2003. *Private Participation in the Power sector in Europe and Central Asia: Lessons from the Last Decade*. Working Paper No. 8. Washington, DC: World Bank. http://documents.worldbank.org/curated/en/2003/07/2477725/private-participation-power-sector-europe-central-asia-lessons-last-decade.

Larsen, E. R., I. Dyner, V. L. Bedoya, and C. J. Franco. 2004. "Lessons from Deregulation in Colombia: Successes, Failures and the Way Ahead." *Energy Policy* 32 (15): 1767–80. https://doi.org/10.1016/S0301-4215(03)00167-8.

Littlechild, S. 2009. "Retail Competition in Electricity Markets: Expectations, Outcomes and Economics." *Energy Policy* 37 (2): 759–63.

Lovei, L. 2000. "The Single-Buyer Model: A Dangerous Path toward Competitive Electricity Markets." Viewpoint Note 225, World Bank, Washington, DC.

Maurer, L., and L. Barroso. 2011. *Electricity Auctions: An Overview of Efficient Practices*. Washington, DC: World Bank. https://doi.org/10.1596/978-0-8213-8822-8.

McRae, S. D., and F. A. Wolak. 2016. *Diagnosing the Causes of the Recent El Niño Event and Recommendations for Reform*. Program on Energy and Sustainable Development, Stanford University. https://web.stanford.edu/group/fwolak/cgi-bin/sites/default/files/diagnosing-el-nino_mcrae_wolak.pdf.

MIT Energy Initiative. 2016. *Utility of the Future*. Cambridge, MA: Massachusetts Institute of Technology. http://energy.mit.edu/wp-content/uploads/2016/12/Utility-of-the-Future-Full-Report.pdf.

Mukhopadhyay, S., V. K. Agrawal, S. K. Soonee, P. K. Agarwal, R. Anumasula, and C. Kumar. 2016. "Detecting Low Frequency Oscillations through PMU-Based Measurements for Indian National Grid." In *Power Systems Computation Conference* (PSCC), 1–7, Genoa.

Munoz, F. D., S. Wogrin, S. S. Oren, and B. F. Hobbs. 2017. "Economic Inefficiencies of Cost-Based Electricity Market Designs." USAEE Working Paper No. 17-313, U.S. Association for Energy Economics, Cleveland, OH. https://doi.org/10.2139/ssrn.2974353.

Nepal, R., and T. Jamasb. 2012. "Reforming the Power Sector in Transition: Do Institutions Matter?" *Energy Economics* 34 (5): 1675–82. https://doi.org/10.1016/j.eneco.2012.02.002.

Neuhoff, K., J. Barquin, J. W. Bialek, R. Boyd, C. J. Dent, F. Echavarren, T. Grau, C. von Hirschhausen, B. F. Hobbs, F. Kunz, C. Nabe, G. Papaefthymiou, C.Weber, and H. Weigt. 2013. "Renewable Electric Energy Integration: Quantifying the Value of Design of Markets for International Transmission Capacity." *Energy Economics* 40 (November): 760–72.

Newbery, D. 2002. "Issues and Options in Restructuring Electricity Supply Industries." Cambridge Working Papers in Economics No. 0210, Department of Applied Economics, University of Cambridge. https://pdfs.semanticscholar.org/2e6e/c9ddece0f664bbc72ec6a6c85d28ddee59ba.pdf.

Newbery, D. 2005. "Electricity Liberalisation in Britain: The Quest for a Satisfactory Wholesale Market Design." *The Energy Journal* 26 (Special Issue): 43–70. https://www.researchgate.net/publication/227356675_Electricity_Liberalisation_in_Britain_The_Quest_for_a_Satisfactory_Wholesale_Market_Design.

Newbery, D. 2015. "Missing Money and Missing Markets: Reliability, Capacity Auctions and Interconnectors." *Energy Policy* 94 (July): 401–10. https://doi.org/10.1016/j.enpol.2015.10.028.

Ofgem (Office of Gas and Electricity Markets). 1999. *The New Electricity Trading Arrangements.* Ofgem, London. https://archive.uea.ac.uk/~e680/energy/energy_links/electricity/neta1.pdf.

Oseni, M. O., and M. G. Pollitt. 2016. "The Promotion of Regional Integration of Electricity Markets: Lessons for Developing Countries." *Energy Policy* 88 (July): 628–38. https://doi.org/10.1016/j.enpol.2015.09.007.

Pollitt, M. 2004. "Electricity Reform in Chile: Lessons for Developing Countries." *Journal of Network Industries* 5 (3–4): 221–62. http://heinonline.org/HOL/Page?handle=hein.journals/netwin5&id=221&div=&collection=.

Pollitt, M. 2008. Electricity Reform in Argentina: Lessons for Developing Countries." *Energy Economics* 30 (4): 1536–67. https://doi.org/10.1016/j.eneco.2007.12.012.

Pollitt, M. G. 2012. "Lessons from the History of Independent System Operators in the Energy Sector." *Energy Policy* 47 (August) 32–48.

Raineri, R., I. Dyner, J. Goñi, N. Castro, Y. Olaya, and C. Franco. 2013. "Latin America Energy Integration: An Outstanding Dilemma." Chapter 14 in *Evolution of Global Electricity Markets*. Academic Press. https://doi.org/10.1016/B978-0-12-397891-2.00014-6.

Rose, A., R. Stoner, and I. Pérez-Arriaga. 2016. "Integrating Market and Bilateral Power Trading in the South African Power Pool." WIDER Working Paper No. 2016/132, United Nations University World Institute for Development Economic Research (UNU-WIDER). https://www.econstor.eu/bitstream/10419/161515/1/873997999.pdf.

Roxas, F., and A. Santiago. 2010. "Broken Dreams: Unmet Expectations of Investors in the Philippine Electricity Restructuring and Privatization." *Energy Policy* 38 (11): 7269–77. https://doi.org/10.1016/j.enpol.2010.08.003.

Rudnick, H., R. Varela, and W. Hogan. 1997. "Evaluation of Alternatives for Power System Coordination and Pooling in a Competitive Environment." *IEEE Transactions on Power Systems* 12 (2): 605–13. https://doi.org/10.1109/59.589615.

Rudnick, H., and C. Velasquez. 2018. "Taking Stock of Wholesale Power Markets in Developing Countries: A Literature Review." Policy Research Working Paper 8519, World Bank, Washington, D.C. http://documents.worldbank.org/curated/en/992171531321846513/Taking-stock-of-wholesale-power-markets-in-developing-countries-a-literature-review.

———. 2019a. "Learning from Developing Country Power Market Experiences: The Case of the Philippines." Policy Research Working Paper 8721, World Bank, Washington, DC. https://openknowledge.worldbank.org/bitstream/handle/10986/31189/WPS8721.pdf?sequence=1&isAllowed=y.

———. 2019b. "Learning from Developing Country Power Market Experiences: The Case of Colombia." Policy Research Working Paper 8771, World Bank, Washington, DC.

———. 2019c. "Learning from Developing Country Power Market Experiences: The Case of Peru." Policy Research Working Paper 8772, World Bank, Washington, DC.

———. Forthcoming. "Learning from Developing Country Power Market Experiences: The Case of India." Policy Research Working Paper, World Bank, Washington, D.C.

Ryan, N. 2014. "The Competitive Effects of Transmission Infrastructure in the Indian Electricity Market." http://campuspress.yale.edu/nicholasryan/files/2014/10/ryan_competition_india_electricity_2014oct23-1gdti64.pdf.

———. 2017. "The Competitive Effects of Transmission Infrastructure in the Indian Electricity Market." NBER Working Paper 23106. National Bureau of Economic Research, Cambridge, MA. https://www.nber.org/papers/w23106.pdf.

Sen, A. 2014. *Divergent Paths to a Common Goal? An Overview of Challenges to Electricity Sector Reform in Developing versus Developed Countries*. OIES Paper EL 10. Oxford: Oxford Institute for Energy Studies.

Sioshansi, R., S. Oren, and R. O'Neill. 2008. "The Cost of Anarchy in Self-Commitment-Based Electricity Markets." In *Competitive Electricity Markets: Design, Implementation, Performance*, edited by F. P. Sioshansi. Oxford, U.K.: Elsevier Ltd.

Stoft, S. 2002. *Power System Economics: Designing Markets for Electricity*. Piscataway, NJ: Wiley-IEEE Press.

Strbac, G. 2008. "Demand Side Management: Benefits and Challenges." *Energy Policy* 36 (12): 4419–26. https://doi.org/10.1016/j.enpol.2008.09.030.

Sweeney, J. L. 2002. *The California Electricity Crisis*. Stanford University: Hoover Institution Press.

Thomas, M. 2012. "Does Your Single Buyer Make the Very Best Bad Decision?" *Lantau Pique*, July. http://www.lantaugroup.com/files/pique_single_buyer.pdf.

Tierney, S. F. 2018. "Resource Adequacy and Wholesale Market Structure for a Future Low Carbon Power System in California." White Paper, Analysis Group.

Timilsina, G. R., M. Toman, J. Karacsonyi, and L. De Tena Diego. 2015. "How Much Could South Asia Benefit from Regional Electricity Cooperation and Trade?" Policy Research Working Paper 7341, World Bank, Washington, DC. https://openknowledge.worldbank.org/bitstream/handle/10986/22224/How0much0could0operation0and0trade00.pdf?sequence=1&isAllowed=y.

Vagliasindi, M., and J. Besant-Jones. 2013. *Power Market Structure: Revisiting Policy Options*. Directions in Development. Washington, DC: World Bank. https://doi.org/10.1596/978-0-8213-9556-1.

Wijayatunga, P., D. Chattopadhyay, and P. N. Fernando. 2015. "Cross-Border Power Trading in South Asia: A Techno Economic Rationale." ADB South Asia Working Paper No. 38, Asian Development Bank, Manila. https://www.adb.org/sites/default/files/publication/173198/south-asia-wp-038.pdf.

Wilson, R. 1998. "Design Principles." In *Designing Competitive Electricity Markets*, edited by H. Chao and H. G. Huntington, 212. Kluwer Academic Publishers.

Wolak, F. A. 2003a. "Designing Competitive Wholesale Electricity Markets for Latin American Countries." Paper prepared for the First Meeting of the Latin American Competition Forum, Paris, April 7–8. http://www.oecd.org/daf/competition/prosecutionandlawenforcement/2490209.pdf

———. 2003b. "Diagnosing the California Electricity Crisis." *The Electricity Journal* 16 (7): 11–37. https://doi.org/10.1016/S1040-6190(03)00099-X.

———. 2005. "Managing Unilateral Market Power in Electricity." Policy Research Working Paper 3691, World Bank, Washington, DC. http://documents.worldbank.org/curated/en/2005/09/6246687/managing-unilateral-market-power-electricity.

Gauging Impact

PART III

Did Power Sector Reforms Improve Efficiency and Cost Recovery?

8

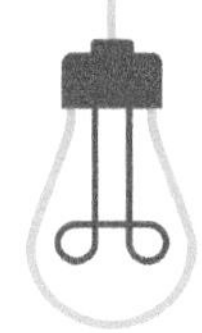

Guiding questions

- *Did countries make material progress toward improving cost recovery in the power sector?*
- *To what extent was the evolution in cost recovery driven by increasing tariffs versus reducing costs?*
- *How did cross-subsidies affect the achievement of cost recovery?*
- *To what extent were any persistent shortfalls in cost recovery absorbed through fiscal transfers or simply allowed to undermine the financial viability of the utilities?*

Summary

- *Electricity tariffs are rarely high enough to cover the full costs of service delivery. This underrecovery of costs is common both for countries facing high costs of electricity and for those facing low costs of electricity.*
- *The underlying reasons for lack of cost recovery can be found not only in relatively low tariffs but also in costs that are inflated because of inefficiency.*
- *Costs in the power sector fluctuate greatly over time because of exposure to foreign exchange and oil price fluctuations and because of hydrological risk. As a result, even countries that achieve cost recovery struggle to sustain this achievement over time.*
- *Power utilities often impose a substantial fiscal burden and contingent liabilities on government budgets. Nevertheless, financial losses due to the underrecovery of costs are seldom fully compensated by government subsidies. Rather, they are absorbed by the utility through the accumulation of arrears with suppliers and short-term loans from commercial banks.*
- *Tariff levels are highly differentiated across customer groups often with a view to preserving affordability for residential consumers; however, such large cross-subsidies further undermine the achievement of cost recovery.*
- *Overall, the cost recovery ratio for power utilities has improved somewhat over the last 25 years. Progress has been very uneven, however, with cost recovery improving in about half the countries, and deteriorating in the rest.*
- *In most cases, tariff erosion due to inflation outweighed the impact of tariff increases on cost recovery; however, system losses declined more consistently across the period and have been a major contributor to improvements in cost recovery.*

INTRODUCTION

This chapter evaluates the extent to which developing countries made progress over the period 1990–2015 on cost recovery and operational efficiency for power utilities.[1] The theory of change underlying power sector reform envisioned improved sector performance by way of two intermediate outcomes. First, it was thought that improved cost recovery and financial viability would attract needed investment in service quality and security of supply. Second, greater operational efficiency for the utilities would ease cost recovery, also thought to be critical to better service quality. Thus, the guiding questions for the chapter are as follows: Did countries increase cost recovery in the power sector? As cost recovery measures evolved, were they driven by increasing tariffs or by reducing costs? How did cross-subsidies affect cost recovery? To what extent were persistent shortfalls absorbed through fiscal transfers? Or were they allowed to undermine the financial viability of the utilities?

Cost recovery, financial viability, and fiscal sustainability are closely related concepts sometimes used interchangeably. Underpricing of electricity often leads to poor financial viability of a utility, which, in turn, results in explicit or implicit government subsidies. The terms are not identical, however. The following paragraphs define each term and explains how they are interrelated.

Cost recovery is an attribute of electricity tariffs—the average tariff corresponding to the average cost of service. It is usually measured as the ratio between the average effective electricity tariff and the reference cost, often expressed as a percentage.[2] Average effective tariffs are calculated by dividing total revenue from electricity sales by the volume of electricity sales in kilowatt-hours (kWh). Various definitions of costs exist, however, and all are useful in different circumstances. Therefore, although cost recovery is sometimes reported in binary terms, the reality—particularly in the developing world—is that degrees of cost recovery exist on a continuum, and that cost recovery can be analyzed from different perspectives. The World Bank has developed a conceptual framework to define different levels of cost recovery (table 8.1) The definitions include three levels—operating cost recovery, limited capital cost

TABLE 8.1 The cost recovery ladder from financial, fiscal, and economic perspectives

	A. Financial perspective	B. Fiscal perspective	C. Economic perspective
Level 1: Operating cost recovery	**Level A1:** Only those operating costs that are covered by the utility/sector (excluding various reserves, such as depreciation, bad debt allowance, and revaluation of assets)	**B1:** Operating costs that are covered on behalf of the utility/sector by the government through budgetary transfers and provision of subsidized goods and services	**C1:** Operating costs (excluding various reserves, such as depreciation, bad debt allowance, and revaluation of assets), irrespective of who bears them, required to adequately run the utility
Level 2: Operating and limited capital cost recovery	**A2:** A1 plus any financing costs (to the utility) for existing capital expenditure, such as debt service (interest and principal), required equity payments, and internally funded investments	**B2:** B1 plus the financing costs (assessed at the cost of existing capital incurred by the government) for the capital expenditure covered through sovereign funding/guarantee	**C2:** C1 plus existing capital expenditure (incorporated using the weighted average cost of commercial capital assessed at the opportunity cost of debt and equity)
Level 3: Full cost recovery of current and future costs	**A3:** A2 plus financing costs (to the utility) and the associated O&M costs for new capital investments (based on an adequate investment prioritization framework) required to meet future demand	**B3:** B2 plus financing costs (to the government) for new capital investments (based on an adequate investment prioritization framework) required to meet future demand	**C3:** C2 plus new capital investments required to meet future demand (incorporated using weighted average cost of commercial capital assessed at the opportunity cost of debt and equity capital)[a]

Source: Kojima 2017.
Note: O&M = operation and maintenance.
a. The full definition of C3 also includes externalities, but these could not be assessed appropriately in this study due to a lack of data and are therefore omitted from this definition.

recovery, and full cost recovery from financial, fiscal, and economic perspectives—to yield a total of nine different cost recovery definitions (A1–A3, B1–B3, and C1–C3). Over the past few decades the literature on cost recovery has moved from a financial to an economic perspective on cost of service. This shift reflects the literature's increased focus on the macroeconomic and environmental implications of the underrecovery of costs (Huenteler and others 2017). In line with these findings, this chapter starts off with a comprehensive economic definition of the cost of electricity service (levels C1 and C3). This analysis is then complemented by an analysis of cost recovery from a financial perspective (A1, A2, and A3). See box 8.1 for the details of the methodology for cost recovery analysis.

BOX 8.1 Methodology used in cost recovery analysis

The analysis presented in this chapter distinguishes itself from the existing literature on cost recovery by comparing pre- and postreform performance on cost recovery levels over a 20- to 30-year period. This goes well beyond the existing literature (annex 8A). The chapter analyzes power sector reform from the late 1980s and 1990s through the period 2010–17 of 17 jurisdictions[a] (14 countries and 3 Indian states). These jurisdictions represent diverse geographies, income levels, and approaches to reform.

Postreform cost recovery analysis is based on detailed cost and tariff data collected for the period 2010–17. The analysis collected data on tariff levels from 2010 to 17 (or a subset of these years) for all 17 jurisdictions on the country and utility levels and compared those data with estimates of six cost recovery levels (A1–A3 and C1–C3),[b] using average cost metrics. The total sample of the utility-level analysis includes 18 majority publicly owned and 7 privately owned utilities. Where possible, the analysis included cash collected along with revenue billed. Compiled from the financial statements of all utilities in the sector, the data helped us obtain a full picture of cost and revenues in the jurisdictions, complemented with information on indirect government support (for example, in the form of subsidized fuels). Exceptions were made for independent power producers and small power producers, the cost of which were approximated by the electricity purchase cost of the off-taking utility. The only two cases where such a holistic picture of costs and revenues in the sector was not possible were the Dominican Republic, where the analysis focused only on the utility Edesur, and Morocco, where the analysis relied on a previous study by Camos and others (2018) for information on full cost recovery levels in the sector. All data presented in the chapter are expressed in 2017 U.S. dollars to adjust for inflation. An overview of data compiled for the case studies, the years covered, and the data sources is provided in annex 8b.

Cost recovery analysis of the prereform period is based on historical studies of the 1980s–1990s that compare tariff levels with long-run marginal cost (LRMC). Data on tariff levels for this period were collected for all 17 jurisdictions and compared with LRMC estimates, which reflect the long-run marginal cost of supply that would need to be covered to expand the system, accounting for shadow prices for fuels, labor, and capital. Comparable LRMC estimates are available for most countries for the 1980s and 1990s, when the World Bank financed a number of LRMC studies. For 12 jurisdictions, the tariff and cost data are for 1987 and based on a World Bank study (1990) that uses a strictly comparable methodology. For two other jurisdictions, the data are based on the same LRMC methodology but for different years: 1991 (Tanzania) and 1993 (Vietnam) (World Bank 1993, 1995). For the remaining three cases, the assessment compares tariffs and current cost-of-service estimates (as opposed to LRMC) for 1991 (Senegal), 1994 (Ukraine), and 2003 (Tajikistan) (World Bank 1994, 1998, 2004). The exact years used for prereform analysis in each country are reported in annex 8B. The results of this second analytical step provide a baseline for assessments of reform impacts and allow assessment of whether countries were able to raise tariffs or reduce costs compared to the prereform period. The methodological differences between the historic analysis based on LRMC and the recent analysis based on average cost mean that the analysis cannot provide conclusions about whether or not actual costs have risen or fallen compared to prereform estimates. Because LRMC can legitimately be interpreted as the best historic estimate of *future* costs in fast-growing power systems,[c] however, the analysis can help us understand if countries' actual costs from 2010 to 2015 were either higher or lower than long-term average cost in the 1980s.

Cost recovery is assessed with respect to a utility's actual costs without passing judgment on its efficiency. Over time, cost recovery may improve through higher tariffs or reduced costs—that is, improving operational efficiency. The analysis of utility efficiency focuses on (1) excessive transmission and distribution losses above a suitable norm (ranging between 7–13 percent, see annex 8D) taking into account the operating context of the utility; and (2) the undercollection of bills. Utility inefficiencies are measured as a percentage of revenue the utility foregoes because of these two factors.

a. The cases are Colombia, Dominican Republic, Arab Republic of Egypt, India (Andhra Pradesh, Odisha, and Rajasthan), Kenya, Morocco, Pakistan, Peru, Philippines, Senegal, Tajikistan, Tanzania, Uganda, Ukraine, and Vietnam.

b. Externalities were not assessed as part of C3 because data were unavailable.

c. For Ukraine and Tajikistan, the prereform cost estimates are based on actual cost rather than LRMC to account for the fact that installed capacity did not grow by very much compared to the prereform period.

Financial viability is an attribute of utility companies[3] and is fulfilled when tariff revenues and other sources of income cover the cost of service. Cost recovery of tariffs is obviously a key determinant, but financial viability also depends on accessible government transfers, ready cash inflows (taking into account collection losses and timely allocation of government transfers), and proper spending priorities (for example, debts and payables are settled in a timely manner, or utilities pay out large dividends or lend to other state-owned enterprises [SOEs]). Therefore, although the two mutually reinforce each other and the literature sometimes uses them interchangeably, a utility can be financially viable even when cost recovery is below 100 percent—for example, if tariffs are set below cost recovery level but reliable fiscal transfers are made to compensate for the shortfall. Similarly, a utility may not be financially viable when tariffs are at cost recovery level, for example, when the utility uses its cash flows to finance new investments while accumulating arrears to its suppliers and financiers. Further, analyses of financial viability usually do not differentiate between revenues from electricity sales and revenue not related to the sale of electricity (for example, government transfers); take input cost at invoiced value (for example, fuels, capital, land, or labor at subsidized prices); and count SOEs' contribution to the government's revenues—for example, in the form of taxes, duties, and, for SOEs and any mandatory allocations from profits—as costs. When discussing the financial viability in sectors with multiple utility companies, the term can apply to each company separately or as an aggregate of all companies in the sector.

Electricity subsidies are understood as an attribute of the sector or the economy. Electricity subsidies can be defined as deliberate government policy actions targeting electricity services that (1) reduce the net cost of electricity or fuels purchased; (2) reduce the cost of electricity production or service delivery; or (3) increase the revenues retained by the electricity producer or service provider (Kojima 2017). This means that electricity can be subsidized whether or not the utility incurs a visible cash shortfall, and whether or not visible cash waterfall is covered by fiscal transfers from the budget (as opposed to commercial borrowing, deferred depreciation and so forth.).

The quasi-fiscal deficit (hidden costs in the case of private utilities, QFD) is a measure of implicit fiscal costs of the power sector (or hidden losses in the case of private utilities). The QFD is measured as the difference between the cash collected by the existing utility and the revenues that would be collected without bill collection losses by a utility applying cost recovery tariffs (in this analysis, using cost benchmark level C3) and achieving commercial and operational efficiency. In general, power utilities in most developing countries are state-owned and can be considered quasi-fiscal entities. Typically, these utilities display poor financial performance in part because they channel various transfers to consumers through underpricing, uncollected bills, and unmetered consumption. The total cost of such transfers, however, is not reflected in the public budget because it is implicit or involuntary (for example, theft). The resulting financial gap in the public utility has been called in the literature QFD, typically expressed as percentage of gross domestic product (GDP), or hidden cost, expressed in absolute terms.[4] The QFD can usefully be disaggregated to clarify how much is attributable to three main factors: (1) system losses, or the cost of electricity injected into the transmission system but not metered/billed, minus the cost of electricity lost for technical reasons within the normative level of 10 percent; (2) collection losses, or the value of electricity billed but not collected from customers; and (3) underpricing, or the difference between the amount billed to customers and the cost of the corresponding amount of electricity.[5]

KEY FINDINGS

The most salient results from the analysis of efficiency and cost recovery across the 25 utilities are summarized in the following key findings. These findings evaluate the current performance of utilities against the intermediate outcomes of efficiency and cost recovery and go on to examine historical trends to see to what extent performance has improved over time and why.

Finding #1: Electricity tariffs rarely cover the full costs of service delivery

Cost recovery remains an elusive goal for many power sectors. Governments made noteworthy efforts on cost recovery, largely along the lines of the standard model. Consistent with the methodological framework described above (table 8.1), cost recovery is analyzed from a full economic cost perspective (C3), while providing for comparison measures A1 and A2, which are the most immediate determinants of the financial viability of the electricity service. Cost recovery can be achieved through cost reduction or tariff adjustments or a combination. Cost reduction takes time, however, because it requires major changes in generation sources and efficiency improvement, whereas tariff adjustments may encounter political and social resistance. Subsidized electricity is so entrenched that it is seen as a social right in some countries (the Arab Republic of Egypt, India, Pakistan, and Senegal). This view is reflected in a continued pattern for underrecovered costs in most cases.

On the country level, the analysis of postreform cost recovery levels suggests that tariffs in only 2 out of 17 case studies cover the full cost of service delivery (C3). Tariffs in 7 out of 17 case studies do not even recover operating costs of electricity service. Ukraine (109 percent on average in 2010–17) and Colombia (104 percent) are the top performers on full economic cost recovery in this sample (figure 8.1). Egypt and the Indian states of Andhra Pradesh and Rajasthan are the lowest performers. In addition, in several cases, bill collection losses introduce additional financial burden on utilities. Bill collection losses are particularly high in Egypt and Tajikistan. These findings are broadly in line with recent studies (see annex 8A), suggesting that the sample of cases is representative of developing countries more broadly.

On the utility level, the analysis of postreform cost recovery suggests that tariffs in only 3 out of 24 case studies cover the full cost of service delivery (C3). Tariffs in 10 out of 23 case studies do not even recover operating costs of electricity service. Ukraine (109 percent on average in 2010–17) and Colombia (104 percent) are the top performers in this sample. Egypt and the Indian states of Andhra Pradesh (both utilities) and Rajasthan (all three utilities) are the lowest performers.

From a financial perspective, which excludes many implicit and "hidden" costs, the picture looks slightly better. Eleven out of 16 countries or states and 14 utilities out of 23 utilities are recovering their financial operating costs (A1; see figure 8.2, panels a and c). Eight of 16 countries or states and 8 of 23 utilities are meeting their operating and existing-capital expenditures from a financial perspective (see figure 8.2, panels b and d).

Finding #2: The underrecovery of costs is common among countries with both relatively high and relatively low costs of electricity

It is notable that cost recovery appears unrelated to the absolute level of costs. The cost of electricity service differs by an order of magnitude between the case studies, but underrecovery of costs is an issue at both ends of the cost spectrum (see figure 8.3). When costs are high, either because of the legacy of expensive generation sources and high-cost take or

FIGURE 8.1 Although most achieved operating cost recovery, few attained full cost recovery

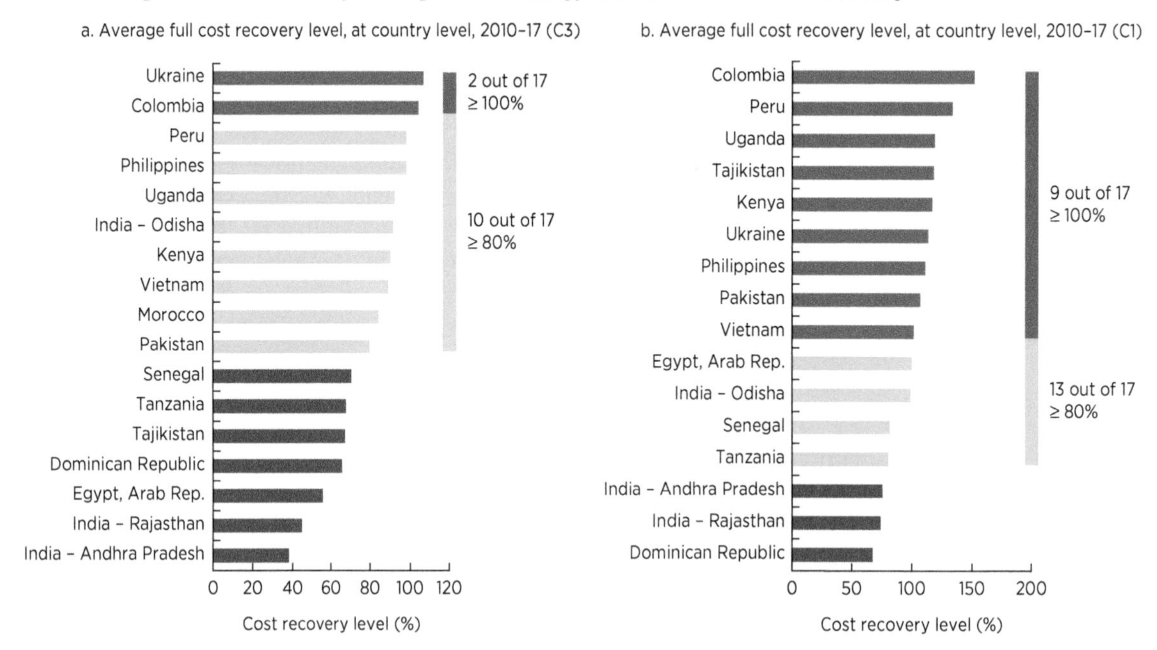

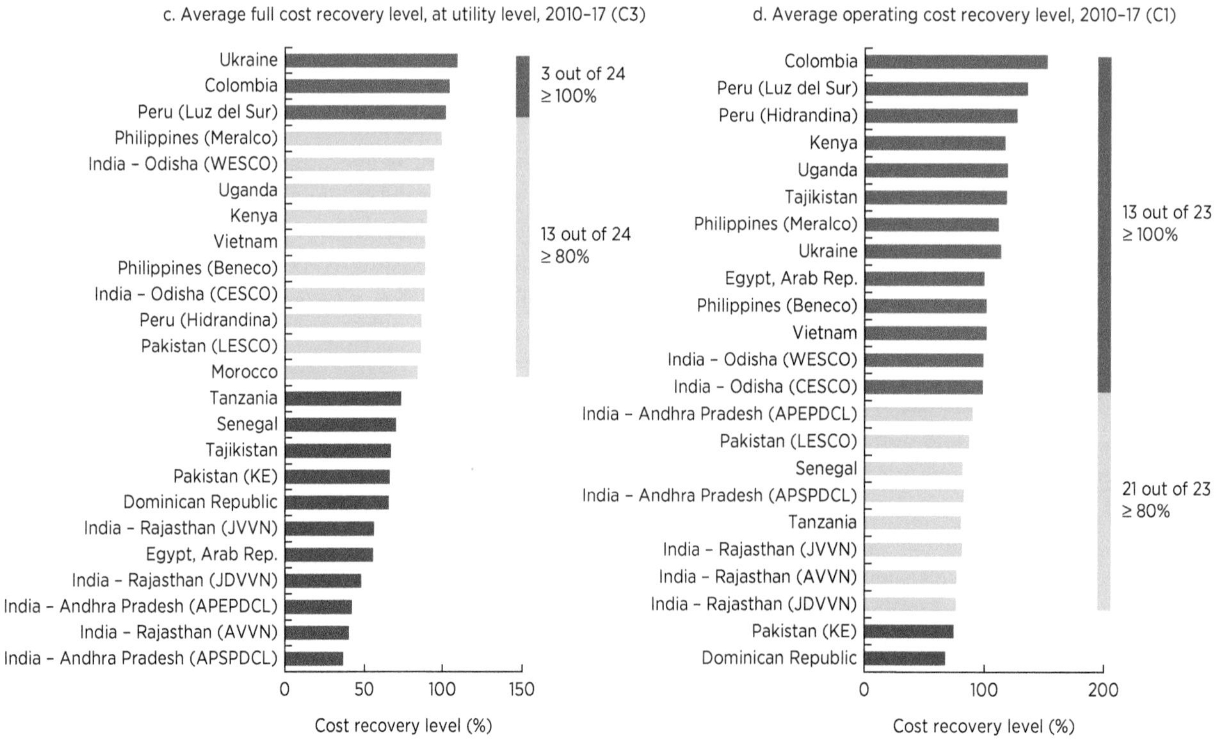

Source: World Bank elaboration based on Rethinking Power Sector Reform utility database 2015.

FIGURE 8.2 About half of cases cover cash needs for limited capital costs

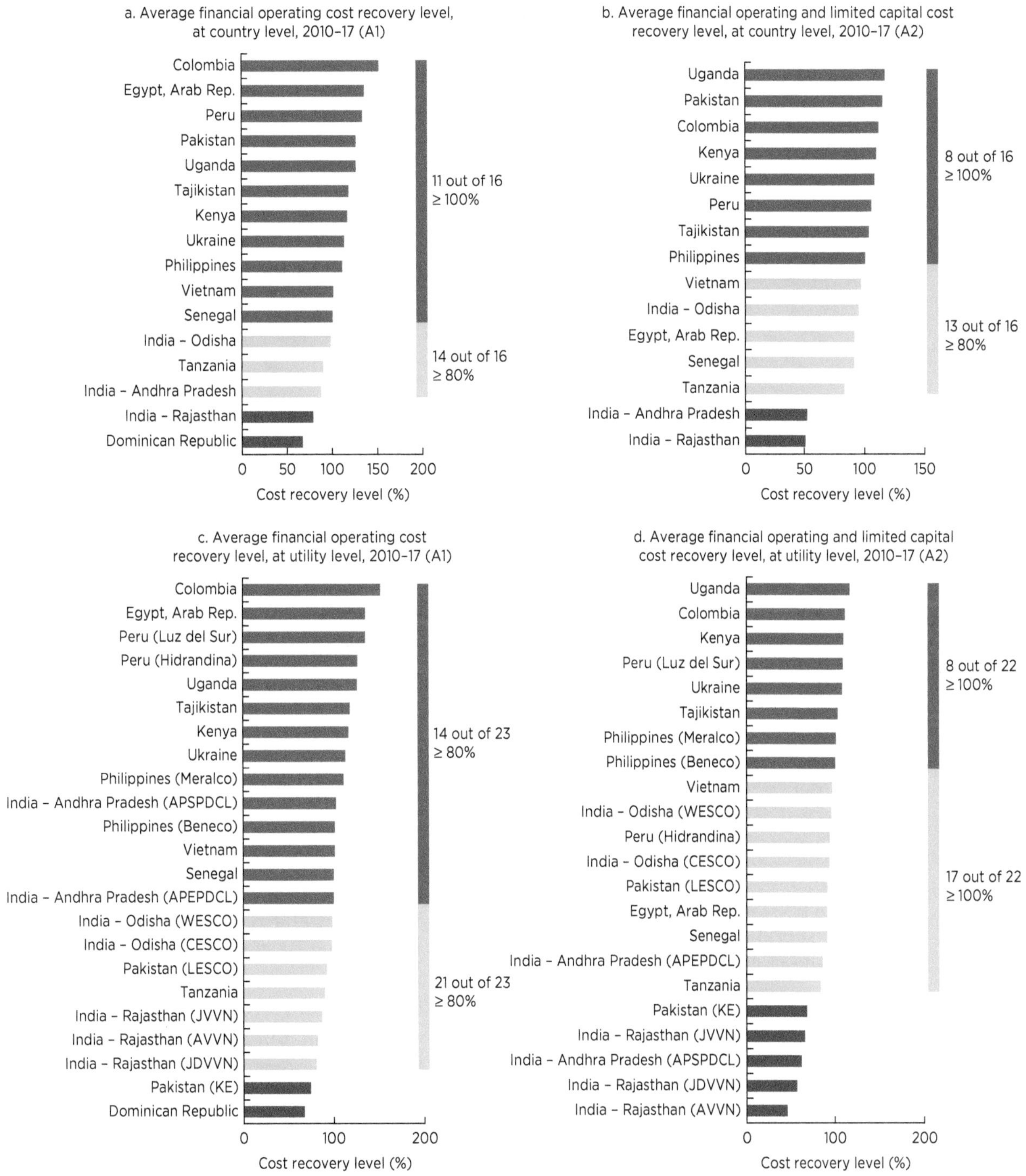

a. Average financial operating cost recovery level, at country level, 2010–17 (A1)

Colombia
Egypt, Arab Rep.
Peru
Pakistan
Uganda
Tajikistan
Kenya
Ukraine
Philippines
Vietnam
Senegal
India - Odisha
Tanzania
India - Andhra Pradesh
India - Rajasthan
Dominican Republic

11 out of 16 ≥ 100%
14 out of 16 ≥ 80%

0 50 100 150 200
Cost recovery level (%)

b. Average financial operating and limited capital cost recovery level, at country level, 2010–17 (A2)

Uganda
Pakistan
Colombia
Kenya
Ukraine
Peru
Tajikistan
Philippines
Vietnam
India - Odisha
Egypt, Arab Rep.
Senegal
Tanzania
India - Andhra Pradesh
India - Rajasthan

8 out of 16 ≥ 100%
13 out of 16 ≥ 80%

0 50 100 150
Cost recovery level (%)

c. Average financial operating cost recovery level, at utility level, 2010–17 (A1)

Colombia
Egypt, Arab Rep.
Peru (Luz del Sur)
Peru (Hidrandina)
Uganda
Tajikistan
Kenya
Ukraine
Philippines (Meralco)
India - Andhra Pradesh (APSPDCL)
Philippines (Beneco)
Vietnam
Senegal
India - Andhra Pradesh (APEPDCL)
India - Odisha (WESCO)
India - Odisha (CESCO)
Pakistan (LESCO)
Tanzania
India - Rajasthan (JVVN)
India - Rajasthan (AVVN)
India - Rajasthan (JDVVN)
Pakistan (KE)
Dominican Republic

14 out of 23 ≥ 80%
21 out of 23 ≥ 80%

0 100 200
Cost recovery level (%)

d. Average financial operating and limited capital cost recovery level, at utility level, 2010–17 (A2)

Uganda
Colombia
Kenya
Peru (Luz del Sur)
Ukraine
Tajikistan
Philippines (Meralco)
Philippines (Beneco)
Vietnam
India - Odisha (WESCO)
Peru (Hidrandina)
India - Odisha (CESCO)
Pakistan (LESCO)
Egypt, Arab Rep.
Senegal
India - Andhra Pradesh (APEPDCL)
Tanzania
Pakistan (KE)
India - Rajasthan (JVVN)
India - Andhra Pradesh (APSPDCL)
India - Rajasthan (JDVVN)
India - Rajasthan (AVVN)

8 out of 22 ≥ 100%
17 out of 22 ≥ 100%

0 100 200
Cost recovery level (%)

Source: World Bank elaboration based on Rethinking Power Sector Reform utility database 2015.

FIGURE 8.3 **Below-cost tariffs remain an issue whether underlying costs are high or low**

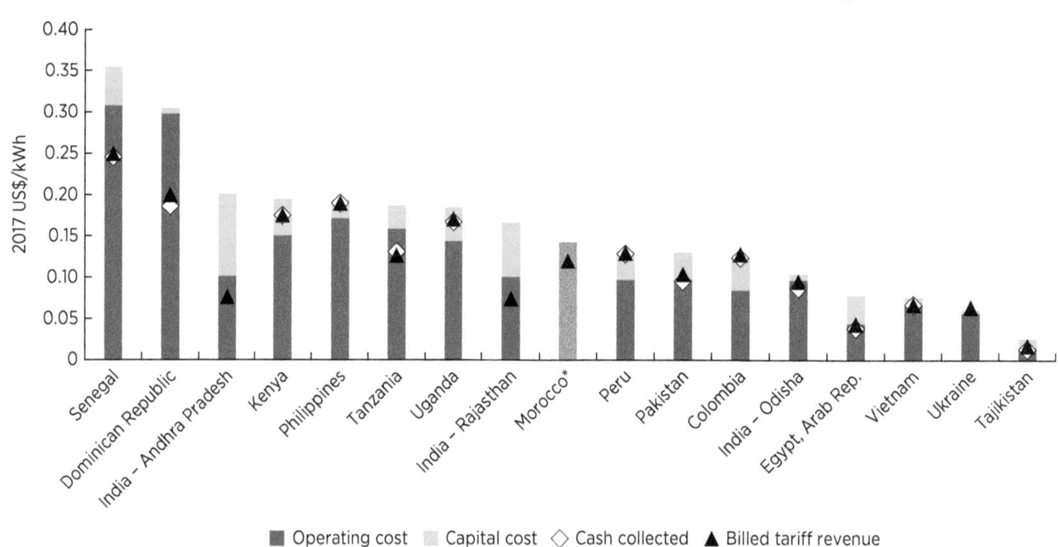

Source: World Bank elaboration based on Rethinking Power Sector Reform utility database 2015.
Note: No breakdown of operating and capital cost is available for Morocco. kWh = kilowatt-hour.

pay contracts (the Philippines, Senegal, and Uganda) or because of high system losses (Dominican Republic, the Indian states of Odisha and Rajasthan, Kenya, and Tanzania), it is difficult to raise tariffs to cover the full cost of service. In Senegal, electricity tariffs have been frozen since 2009 amid fears of profound social discontent and the eruption of protests. Nevertheless, even when cost is low, such as in Egypt and Vietnam, the extreme public sensitivity toward electricity pricing prevents the government from abolishing subsidies. In Vietnam, whereas end-user tariffs follow cost, tariff increases above 7 percent require approval from the ministry of industry and trade or above 10 percent from the prime minister. In Egypt, bold tariff adjustments of up to 40 percent were enacted in 2016 and 2017, but the low tariff base means that even heftier increases are needed to cover costs. Political pressure to provide subsidized or free power to certain groups of consumers (for example, farmers) also undercut the cost recovery effort, particularly in India.

Finding #3: Inefficiencies remain a big contributor to the underrecovery of costs

Several of the utilities under study could reach or get much closer to cost recovery by improving efficiency without altering their current level of tariffs. Good performance is considered to be when a utility loses less than 5 percent of revenues to inefficiency, because this share is equivalent to system (transmission and distribution [T&D]) losses of about 10 percent and revenue collection close to 100 percent.[6] Both India (Odisha) and Pakistan would have achieved full cost recovery in the most recent year of data had they managed full bill collection, shrinking T&D losses to 5 percent of power fed into the transmission grid. Tajikistan and Uganda would have both been 11 percentage points closer to full cost recovery with these changes, though neither would reach full cost recovery with these improvements alone. India (Rajasthan), Kenya, Senegal, Tanzania, and Vietnam would also see improvements in cost recovery (all between 1 and 9 percentage points), although full cost recovery could not be achieved without tariff adjustments.

More than half of the cases exhibits inefficiencies exceeding 5 percent of revenue. As figure 8.4 shows, nine countries lose more than 5 percent of revenue to excessive T&D losses (dark-shaded bars) and noncollection of bills (light-shaded bars) (Dominican Republic, Egypt, India (Odisha and Rajasthan), Pakistan, Senegal, Tajikistan, Tanzania, and Uganda). Some of the worst performers in the sample lose about 20 percent of utility revenues to inefficiency, which is equivalent to system losses of about 10 percent above efficient benchmarks and revenue collection of about 90 percent.

FIGURE 8.4 Inefficiencies can account for as much as 10 to 20 percent of utility revenues

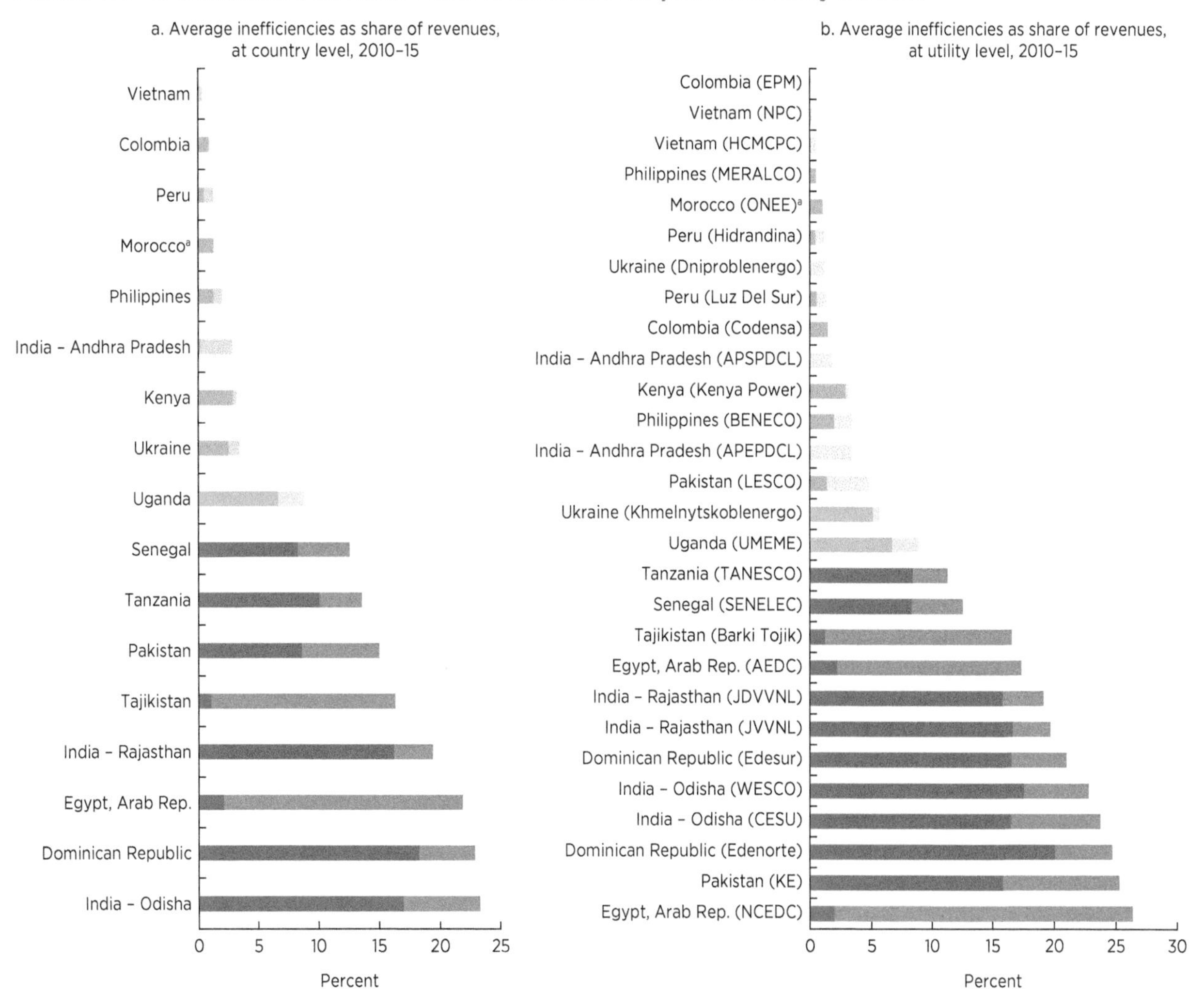

Sources: World Bank elaboration based on Rethinking Power Sector Reform utility database 2015 and World Bank World Development Indicators database 2018.

Note: Dark-shaded parts of bars in figure show inefficiencies lost because of excessive transmission and distribution losses; light-shaded parts of bars show inefficiencies lost because of noncollection of bills. Countries/utilities with inefficiencies lower than 5 percent are considered good performers and are shaded green; countries/utilities with inefficiencies between 5 and 10 percent are considered moderate and are shaded yellow; countries/utilities with inefficiencies above 10 percent are poor performers and shaded red.

a. ONEE (Morocco): collection rate data are not available.

Even within specific countries, the variation in efficiency performance can be huge, illustrated clearly by the case of India, with states such as Odisha and Rajasthan losing about 20 percent of revenues to inefficiency, and Andhra Pradesh losing less than 5 percent. By region, the cases in East Asia and Pacific and Latin America and the Caribbean perform best; the Middle East and North Africa and South Asia perform worst. The key question then is who should bear the cost of these inefficiencies—that is, to what extent should these inefficiencies be included in tariffs as a cost to consumers or absorbed in subsidies as a cost to society as a whole. In Pakistan, the tariff-setting formula includes an allowance for normal T&D losses, but the excess losses are in any case incorporated in circular debt when companies with scarce revenues fail to fully pay their suppliers. The government is responsible for this debt in the long run through its ownership of state-owned companies.

Although more difficult to measure and benchmark, inefficiencies in generation also factor into the high cost. Generation accounts for most of the total cost of electricity supply. In addition, the least-cost planning and competitive procurement of generation encourages utilities to commit to suboptimal power purchase during periods of supply crisis. Both the Philippines and Senegal experienced a power crisis (1990s and 2015, respectively). Their resulting power purchase agreements were costly and have burdened the utilities for years. Similarly, Tanzania's reliance on noncompetitive procurement for generation projects has inflated costs and financially crippled the sector.

Finding #4: The inherent volatility of costs means few countries can sustain cost recovery over time

None of the cases maintained full cost recovery throughout the observation period. Six of the cases reached full cost recovery in at least one of the observed years; however, even when tariffs attain cost recovery levels for certain periods of time, cost volatility threatens the sustainability of this outcome. This finding is true for countries at both ends of the cost recovery spectrum (for example, for both Colombia and Tanzania), as well as countries that are completely reliant on fossil-fuels (for example, Dominican Republic) and countries relying entirely on renewable energy (for example, Tajikistan). See the full data in annex 8C. Year-on-year volatility of cost recovery is high in the Indian state of Rajasthan (coefficient of variation of 16 percent), Pakistan (17 percent), Senegal (16 percent), Tajikistan (19 percent), and Tanzania (23 percent). The Indian states of Andhra Pradesh and Odisha are the two exceptions, with low volatility in cost recovery levels (table 8.2).

The sources of volatility most often observed are exchange rate, fuel cost, fuel mix, and debt service costs. Colombia, for example, was above full capital cost recovery in 2011–16, but below that level in 2010 because of the impact of drought on hydropower costs. Kenya Power and Lighting Company (KPLC) experienced worse financial performance in 2009–12 when drought reduced hydro capacity. The Tanzania Electric Supply Company (TANESCO) has had consistent losses in recent years but ran an operating profit in fiscal year 2013/14 (FY2013/14) and FY2014/15 because of favorable hydrological conditions that greatly reduced the cost of sales and reduced the need for power purchases from third parties. Fuel prices have been a major factor in cost variations seen in the Philippines, which benefited from the decline of fuel prices since 2013. Senegal was also able to improve cost recovery despite tariffs having been frozen at 2009 levels, as fuel purchase costs declined 44 percent between 2012 and 2016. Because Tajikistan imports most materials and equipment, Barki Tojik is susceptible to the devaluation of its local currency.

Despite consistent exposure to volatility, few utilities have any hedging strategy. Some utilities have opted for hydropower imports from neighbor countries (Senegal and Vietnam) to reduce

TABLE 8.2 **Full cost recovery is a moving target because of exogenous cost shocks**

Percent

Country/state	Average	Min	Max	Coeff. of var[b]	Main sources of volatility
Colombia	104	94	117	7.0	Hydro availability, debt service costs, FEX rate
Dominican Republic	66	62	73	7.0	Fuel prices
Egypt, Arab Rep.	55	49	68	**14.3**	Gas availability, FEX rate, fuel prices
India - Andhra Pradesh	38	36	44	8.9	Debt service costs
India - Odisha	91	90	94	1.7	n.a.
India - Rajasthan	45	38	57	**16.0**	Debt service costs
Kenya	90	80	101	9.0	Hydro availability, capital costs
Morocco[a]	84	n.a.	n.a.	n.a.	Fuel prices
Pakistan	80	66	97	**17.0**	Tariff increases, fuel prices
Peru	98	93	102	3.6	Fuel prices, hydro availability
Philippines	98	95	100	1.7	Fuel prices, legacy PPAs
Senegal	70	55	87	**16.0**	Fuel prices, expensive emergency power, debt service costs
Tajikistan	67	52	83	**18.7**	FEX rate
Tanzania	68	56	90	**22.6**	Hydro availability, FEX rate, fuel prices, expensive IPPs, emergency power, debt service costs
Uganda	92	81	110	**12.2**	Capital costs, high rate of return for private utility, fuel prices, FEX rate
Ukraine	106	93	115	7.1	FEX rate, fuel prices
Vietnam	89	86	91	2.1	Hydro availability

Source: World Bank elaboration based on Rethinking Power Sector Reform utility database 2015.

Note: Coefficients in bold highlight high year-on-year volatility. FEX = foreign exchange; IPP = independent power producer; PPA = power purchase agreement; n.a. = not applicable.

a. Data for Morocco are available for only one year (2013).

b. Coefficient of variation (standard deviation divided by the geometric mean).

reliance on costly oil-fired generation or during major shortages. Going forward, many countries have committed to increasing the share of renewables in their energy mix, not only for their environmental benefits but also to enhance the security of supply and diversify the fuel mix.

Finding #5: Financial losses due to the underrecovery of costs are seldom fully compensated by government subsidies but are instead absorbed through the accumulation of arrears and short-term debts

Different mechanisms exist for absorbing the underrecovery of costs, with different implications for the utilities' ability to adequately serve their customers. In principle, any shortfall in cost recovery by utilities should be covered through a compensating fiscal transfer. In practice, however, government support falls short of restoring utilities' financial viability in almost all cases, which leaves utilities adopting other coping strategies to make ends meet financially (table 8.3).

Most jurisdictions with below full cost recovery receive fiscal support in the form of operational or capital transfers. Of the 11 countries where fiscal support was registered, it was found to be more common in the jurisdictions with lower levels of cost recovery, and less common for private utilities. TANESCO, for example, relies on government grants to finance investments—totaling US$833 million in 2016 (1.72 percent of GDP) and subsidized capital from donor sources, reducing the average

TABLE 8.3 **Utilities absorb financial shortfalls in various ways**

Power utility	Type	C3 Cost recovery (%)	A1 Cost recovery (%)	Large government transfers	Subsidized fuels	Sustained net losses	Large debt service	Excessive payables	Excessive receivables	Sustained negative operating cash flow
Colombia (Codensa)	Private	104	152	No	No	No	No	No	No	Yes
Dominican Republic (Edesur)	Public	66	67	No	No	Yes	—	—	—	Yes
Egypt, Arab Rep. (EEHC)	Public	55	135	No	Yes	No	No	Yes	Yes	No
India - Andhra Pradesh (APEPDCL)	Public	36	103	—	No	Yes	No	—	—	No
India - Andhra Pradesh (APSPDCL)	Public	42	100	—	No	Yes	No	—	—	No
India - Odisha (CESU)	Public	95	99	—	No	Yes	No	—	—	No
India - Odisha (WESCO)	Public	89	98	—	No	Yes	No	—	—	No
India - Rajasthan (AVVN)	Public	40	82	—	No	Yes	Yes	No	No	No
India - Rajasthan (JDVVN)	Public	48	81	—	No	Yes	Yes	No	No	No
India - Rajasthan (JVVN)	Public	56	87	—	No	Yes	Yes	No	No	No
Kenya (KPLC)	Public	90	117	No	No	No	No	No	No	Yes
Morocco (ONEE)	Public	84	—	—	—	—	—	—	—	—
Pakistan (KE)	Public	86	92	—	No	No	No	Yes	Yes	No
Pakistan (LESCO)	Private	66	74	No	No	No	No	No	No	Yes
Peru (Hidrandina)	Private	102	135	—	No	No	Yes	No	—	No
Peru (Luz del Sur)	Public	87	127	—	No	No	No	No	—	No
Philippines (Beneco)	Private	99	111	—	No	No	No	No	No	Yes
Philippines (Meralco)	Private	89	101	—	No	No	No	No	No	No
Senegal (Senelec)	Public	70	100	No	Yes	No	No	Yes	No	No
Tajikistan (Barki Tojik)	Public	67	118	No	No	Yes	Yes	Yes	No	No
Tanzania (TANESCO)	Public	74	90	Yes	No	Yes	Yes	Yes	No	No
Uganda (UMEME)	Private	92	126	No	No	No	No	No	No	No
Ukraine (Khmelnytskoblenergo)	Mixed	109	113	—	Yes	—	—	—	—	—
Vietnam (NPC)	Public	89	101	No	No	No	No	—	—	No

Source: World Bank elaboration based on Rethinking Power Sector Reform utility database 2015.
Note: Large government transfers = If total government transfers exceed 10 percent of C3 cost recovery. *Subsidized fuels* = If utility receives fuels below market prices (qualitative information). *Sustained net losses* = net losses exceed 5 percent of revenue (average of last three years). *Large debt service* = total debt service (interest and principal payments) exceed 20 percent of C3 cost recovery. *Excessive payables* = payables exceed 50 percent of annual revenues. *Excessive receivables* = receivables exceed 50 percent of annual revenues. *Sustained negative operating cash flow* = negative net operating cash flow of more than 5 percent of revenue. Refer to table 8.1 for full definitions of A1 and C3 cost recovery. — = not available.

interest rates on TANESCO's total borrowings from 5 percent in 2012 to 3 percent in 2016. In addition, TANESCO does not always repay the ministry of finance for its loans, which then become pseudogrants. Pakistan's privatized Karachi Electric (KE) relies extensively on operational transfers. Subsidies are provided to distribution companies in the form of a tariff differential subsidy, totaling US$418 million to KE in 2015, which compensates distribution companies for the difference between the regulator-determined cost-based tariff (accounting for only efficient costs) and the uniform tariffs (based on the costs of the most efficient distribution company). The government envisioned that, as distribution companies were privatized, the efficiencies of private management would result in lower costs and lower subsidies. After almost 10 years of privatization, however, KE still receives a subsidy. Total tariff differential subsidies provided to the sector in 2016 (including KE and ex-Wapda [Water and Power Development Authority] distribution companies, XWDISCOs) represented 0.4 percent of GDP. Other governments also provide subsidies in the form of direct transfers to suppliers for fuel cost or for available capacity payment (Senegal and Uganda).

Fiscal support meant to compensate for shortfalls in tariff revenue often falls short of required levels. Although fiscal transfers are provided in most countries that fall short of cost recovery, rarely are these transfers large enough to fully compensate utilities for the financial shortfall. In Senegal, for example, tariffs were frozen at their 2009 level, with the government agreeing to make quarterly payments to compensate the national electric company, Senelec. These payments were not timely, however, causing Senelec to take on costly commercial debt. In 2011, Senelec's concession contract was updated so that, if the government is unable to make payments, making Senelec borrow from commercial banks, the government must then assume responsibility for financial fees and principal repayment.

Insufficient government transfers are aggravated when public institutions don't pay their electricity bills. Although utilities may benefit from fiscal transfers from the finance ministry, they are vulnerable as well to nonpayment of regular electricity bills by a range of public institutions, including central government departments, state-owned enterprises, and subnational jurisdictions. The government is a major contributor to receivables in Pakistan, Senegal, and Tanzania. The issue of nonpayment is especially important in Pakistan, where collection losses make up 47 percent of the QFD. Pakistan has the highest number of receivable days in the country sample (190 days). As an illustration, Pakistan's KE is contractually obligated to provide uninterrupted service to Karachi Water & Sewerage Board and City District Government Karachi, but their unpaid bills have been accumulating since before 2010. Taken together, government and autonomous bodies make up 56 percent of KE's trade receivables. In Tanzania, government nonpayment used to be a major problem, but the mainland government and TANESCO recently took steps to slash accounts payable, settling US$71 million in unpaid invoices in 2015. Collections losses make up 6 percent of Tanzania's QFD.

Utilities receiving no government support or facing large unpaid bills from public sector institutions must use other coping strategies to deal with underpricing. Often nonpayment means utilities fall into arrears with suppliers, creating contingent liabilities for the government. Typically, energy providers and goods and services providers make up the bulk of a utility's payables. Pakistan experiences this problem of "circular debt." Distribution companies often do not have the cash to pay the National Transmission and Dispatch Company (NTDC) because of low collections, low cost recovery (even with the subsidies), or lack of timely payment of subsidies. NTDC then cannot pay power producers, and power producers cannot pay fuel suppliers. Tanzania has had similar difficulty with its loan payments, with payable days of 299.

These issues of financial viability affect utilities' creditworthiness and ability to raise capital. As a result, many utilities in developing countries are overly reliant on high-cost, short-term debt, which they then can't service. Debt repayment has been particularly problematic for utilities with high investment needs owing to electrification (Kenya and Tanzania). TANESCO's inability to pay its debts is apparent in its low debt–service coverage ratio (0.16 in FY2014/15 and –0.11 in FY2015/16). In 2015, TANESCO's financial reports also show that it defaulted on government loans and World Bank loans on-lent by the finance ministry, or about 19 percent of its 2016 capital expenditure. For countries with detailed loan information (Kenya, Senegal, and Tajikistan), the average tenor of debt is between 5 and 13 years. These countries rely primarily on long-term loans but use some short-term financing, which tends to be much more expensive. For example, KPLC has one short-term loan on record with an interest rate of 16 percent, well above its average commercial loan rate of 3.6 percent (otherwise comprising medium- and long-term loans).

Overall, utilities benefit from substantial subsidies on investment financing. These subsidies lower the effective cost of capital relative to the true cost of either public or private finance. Weak financial performance persists despite the fact that utilities benefit concessional sources of finance. The average cost of debt is 6 percent across the 13 jurisdictions with available data. This cost of debt is less than half the average commercial rate for these jurisdictions (14 percent). In fact, every jurisdiction except Rajasthan in India receives lower-cost loans than the average commercial borrowing rate in each country (see figure 8.5).

FIGURE 8.5 Utilities often borrow at rates well below commercial benchmarks

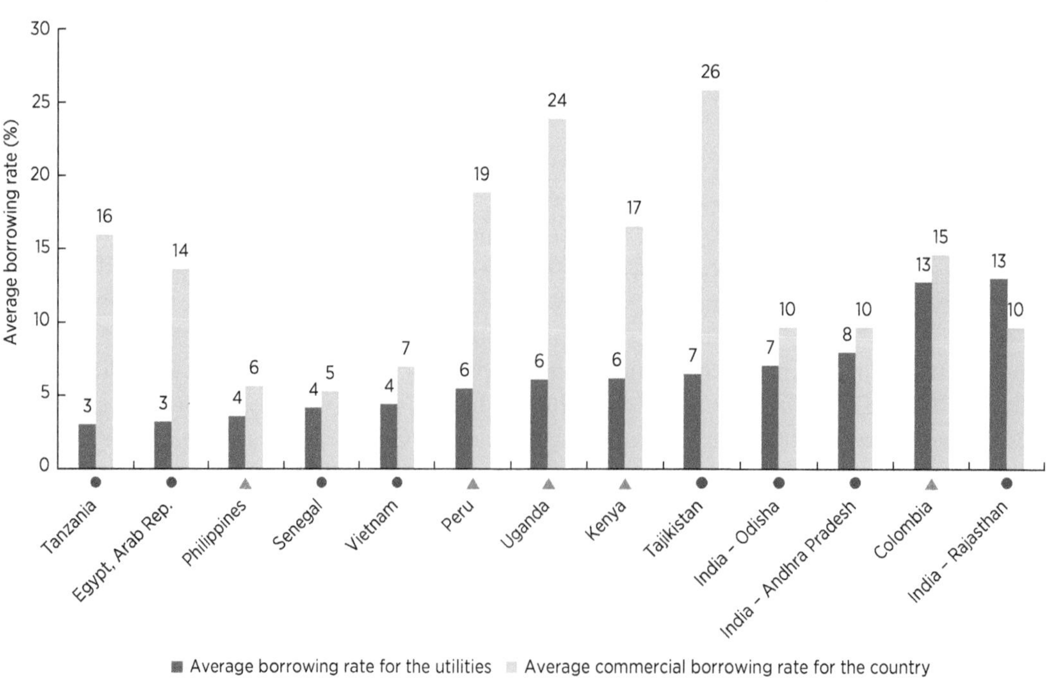

Sources: World Bank elaboration based on Rethinking Power Sector Reform utility database 2015 and World Bank World Development Indicators database 2018.

Government and international financial institution (IFI) loans have lower rates than the commercial loans, as shown in figure 8.6 (except for Senegal's IFI loans, most of which are from the West African Development Bank and have a standard 8.5 percent rate). Uganda is entirely dependent on IFI loans, while the bulk of Senegal's and Tajikistan's loans are from IFIs, and the lion's share of Kenya's and Tanzania's loans are from commercial sources. Vietnam Electricity (EVN) benefits from preferential treatment, such as privileged access to credit, land, and contracts.

Public investment subsidies are often the main driver of utilities' investment levels, as opposed to financial performance. The cases suggest no clear trend between investment levels (as a percentage of revenues) and cost recovery (see figure 8.7). The data presented in figure 8.7 show, for example, that Tanzania can invest at high levels despite low cost recovery, whereas the Indian state of Odisha is investing at relatively low levels compared with other jurisdictions at similar levels of cost recovery (Kenya, Senegal, Uganda, and Vietnam). Tanzania maintains its high levels of investment through government grants (totaling US$833 million, or 48 percent of investments in 2016), in addition to loans, which it doesn't always pay and which become pseudogrants. Odisha's low levels of investment (14 percent of revenues for CESU [Central Electricity Supply Utility] and 1 percent for WESCO [Western Electricity Supply Company] in 2015) are attributed to the utilities being annual loss-makers.

In line with the above, the qualitative evidence suggests that countries that have mobilized large investment amounts in recent years—Kenya, Tanzania, and Uganda—have done so through public investment that was mobilized despite underrecovery of costs. In line with this observation, full cost recovery levels appear largely unrelated to progress in electrification, indicating a strong role for public investment.

FIGURE 8.6 International financial institutions are typically the cheapest source of borrowing

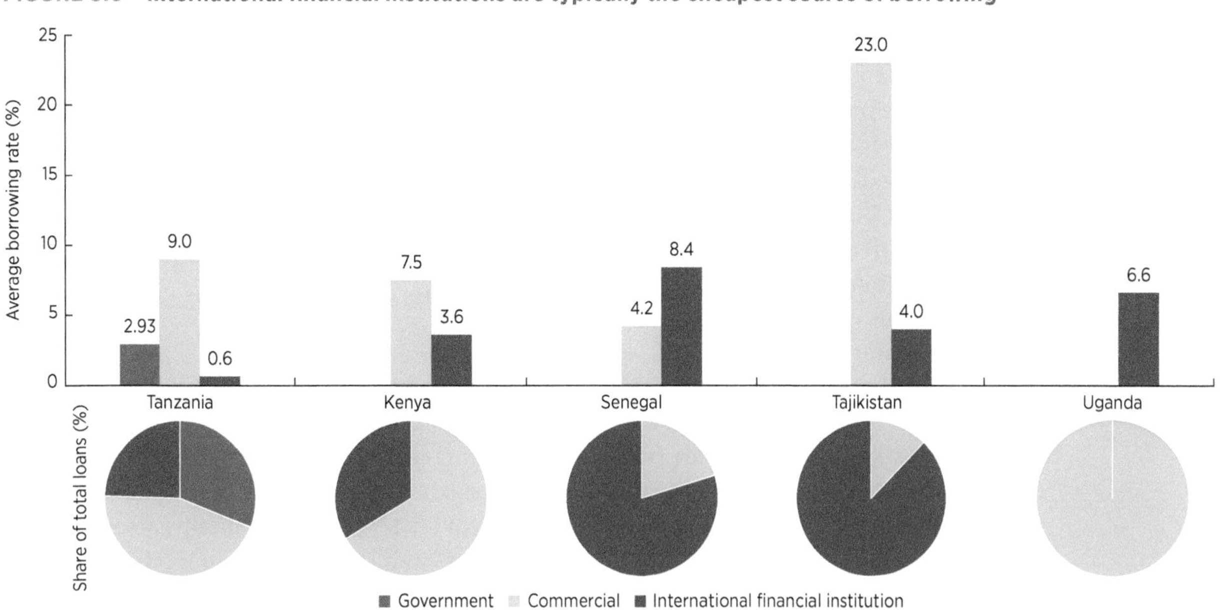

Sources: World Bank elaboration based on Rethinking Power Sector Reform utility database 2015 and World Bank World Development Indicators database 2018.

FIGURE 8.7 **Investment levels do not bear much relationship to cost recovery**

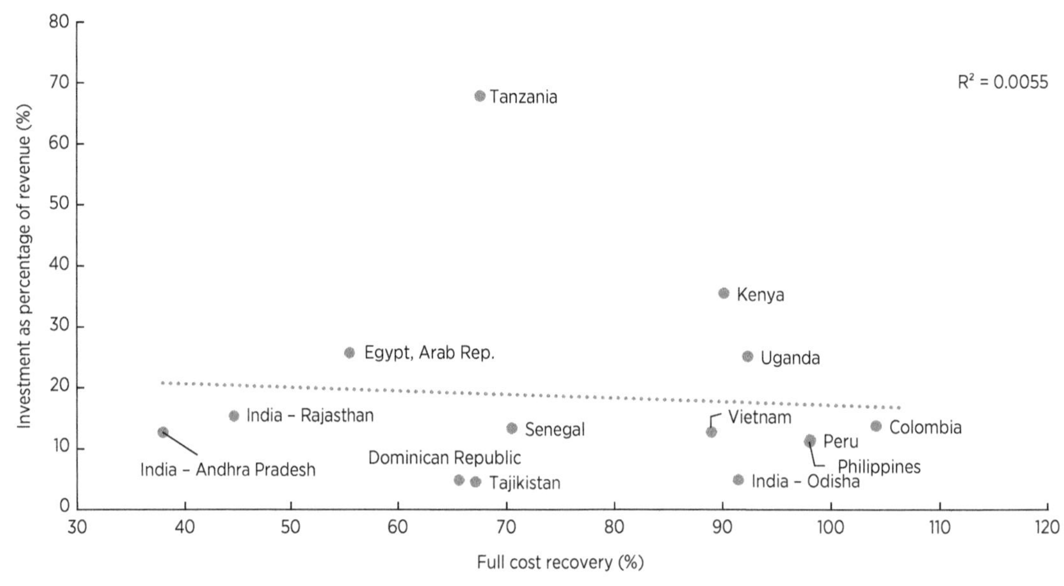

Source: World Bank elaboration based on Rethinking Power Sector Reform utility database 2015.

Finding #6: Power utilities often impose hefty fiscal burdens and contingent liabilities on government budgets

The fiscal burden associated with the power sector ranges both above and below one percentage point of GDP. Estimates of the QFD of power utilities—or hidden losses in the case of privately owned utilities—provide a strong reminder of the macroeconomic significance of the underrecovery of the cost of electricity service. The QFD of the power sector stands at 0.93 percent of GDP on average (see figure 8.8, panel a). Again, these figures are broadly in line with the literature (see annex 8A), suggesting that the sample is representative of developing countries.

Separating the analysis by ownership of the utility suggests that QFD/hidden costs are concentrated in state-owned utilities (see figure 8.8, panel b). The average QFD of sectors with publicly owned distribution is 1.63 percent of GDP, compared to 0.16 percent for mixed public/private and 0.11 percent for fully private distribution utilities.

Below-cost-recovery tariffs are the leading contributor to the QFD in 9 out of 17 cases, which suggests that tariff reforms or cost reductions would be needed to reach cost recovery. Underpricing is only one of several factors contributing to the relatively poor performance on cost recovery of many cases. The two other factors are excessive T&D losses and the noncollection of bills expressed as a percentage of total utility revenues.

Finding #7: Tariff levels are highly differentiated in many cases to make service affordable to certain consumer groups; large cross-subsidies are often associated with low levels of cost recovery

Because of the extensive practice of cross-subsidies in tariff structures,[7] industrial and commercial users are much more likely than residential and agricultural ones to pay at cost recovery levels. Systematically, across jurisdictions, industrial and commercial customers often pay a hefty tariff premium even though the costs they impose on the network are no greater (and are potentially lower) than those imposed by residential customers.

FIGURE 8.8 **Underpricing is the largest driver of quasi-fiscal deficits in the power sector**

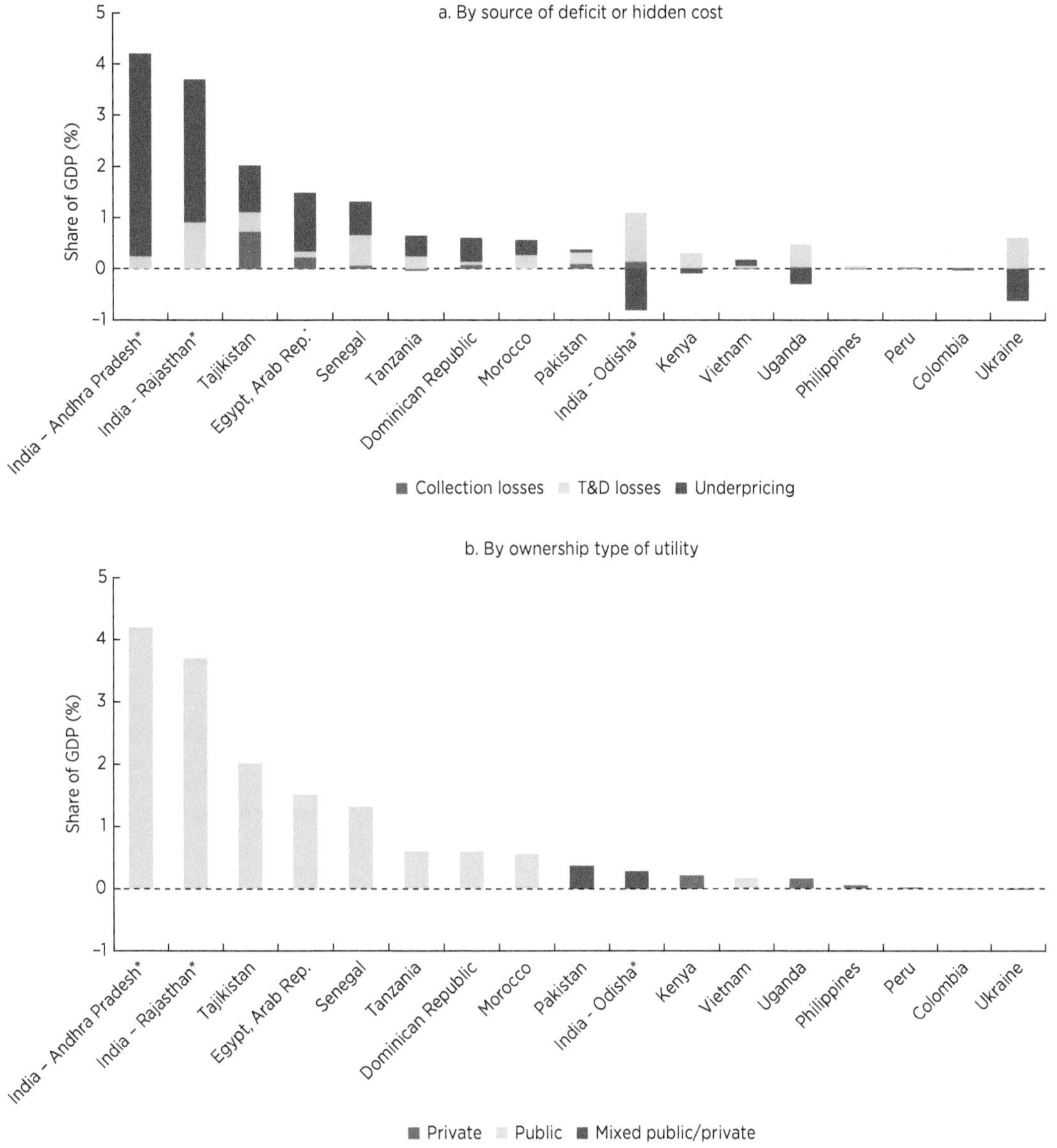

a. By source of deficit or hidden cost

Share of GDP (%)

India - Andhra Pradesh*, India - Rajasthan*, Tajikistan, Egypt, Arab Rep., Senegal, Tanzania, Dominican Republic, Morocco, Pakistan, India - Odisha*, Kenya, Vietnam, Uganda, Philippines, Peru, Colombia, Ukraine

■ Collection losses ■ T&D losses ■ Underpricing

b. By ownership type of utility

Share of GDP (%)

India - Andhra Pradesh*, India - Rajasthan*, Tajikistan, Egypt, Arab Rep., Senegal, Tanzania, Dominican Republic, Morocco, Pakistan, India - Odisha*, Kenya, Vietnam, Uganda, Philippines, Peru, Colombia, Ukraine

■ Private ■ Public ■ Mixed public/private

Source: World Bank elaboration based on Rethinking Power Sector Reform utility database 2015.
Note: T&D = transmission and distribution.

Fewer countries have cross-subsidies from industrial to commercial customers, and these cross-subsidies tend to be smaller. Several countries use such cross-subsidies to make electricity affordable to politically favored groups, typically including but not limited to the poor and vulnerable. As can be seen in figure 8.9, several countries manage to make subsistence consumption (30 kWh per month per household) affordable to the bottom 40 percent of the income spectrum. In eight cases—Egypt; the Indian states of Andhra Pradesh, Odisha, and Rajasthan; Pakistan; Senegal; Tajikistan; and Ukraine—this consumption costs less than 1 percent of gross national income of the bottom 40 percent. The Philippines and Uganda stand out as being the only countries where subsistence consumption can absorb as much

FIGURE 8.9 **Many countries seem able to reconcile cost recovery and affordability**

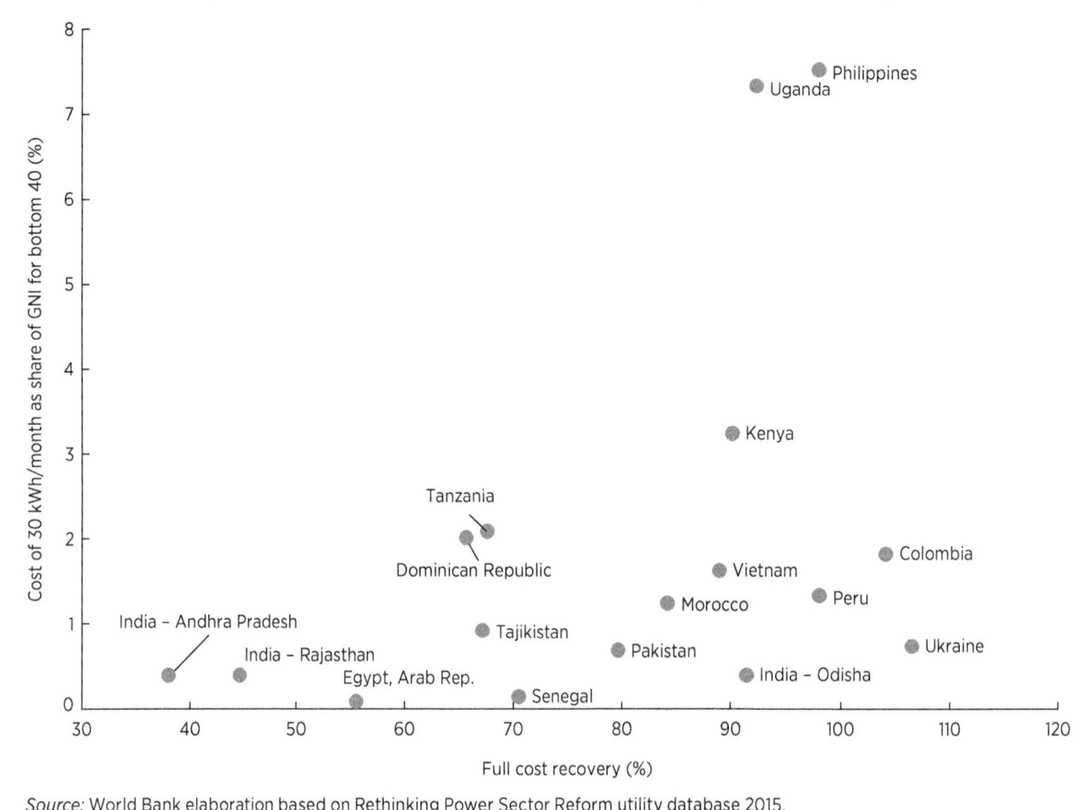

Source: World Bank elaboration based on Rethinking Power Sector Reform utility database 2015.
Note: Bottom 40 = the bottom 40 percent of the income distribution; GNI = gross national income; kWh = kilowatt-hour.

as 7–8 percent of the gross national income of the bottom 40 percent of the income distribution. Such cross-subsidies help governments maintain affordable and politically legitimate power sector reforms.

The critical question is at what point such price discrimination across customer groups starts impeding cost recovery. One way to quantify price discrimination is to use a Lorenz curve, which shows cumulative shares of consumption and revenue by customer class (see figure 8.10).[8] A straight diagonal would mean that all customers contribute shares of revenue equal to their shares of consumption, and major curvature would mean that customers' shares of revenues are not well aligned with their shares of consumption. Many utilities show a relatively even distribution of shares of revenue and consumption by customer class. This even distribution does not mean that cost recovery is being met or that there are no cross-subsidies; it only means that all customer classes are paying similar tariff rates for their consumption. Some jurisdictions with flatter distributions may not be recovering costs through tariffs on any customer class (as is the case with Rajasthan in India). In addition, it is appropriate for customers who impose lower costs on the system (such as industrial customers) to pay lower tariffs than other customer groups, so a completely flat curve is not ideal. A noticeable curvature, however, points to a misalignment of costs incurred to serve each customer group and the customers who pay those costs. The three Andhra Pradesh (Indian) distribution companies show the greatest disparity in the shares of consumption and revenue contributed by each customer class.

FIGURE 8.10 **Cross-subsidies are pronounced in some countries and states**

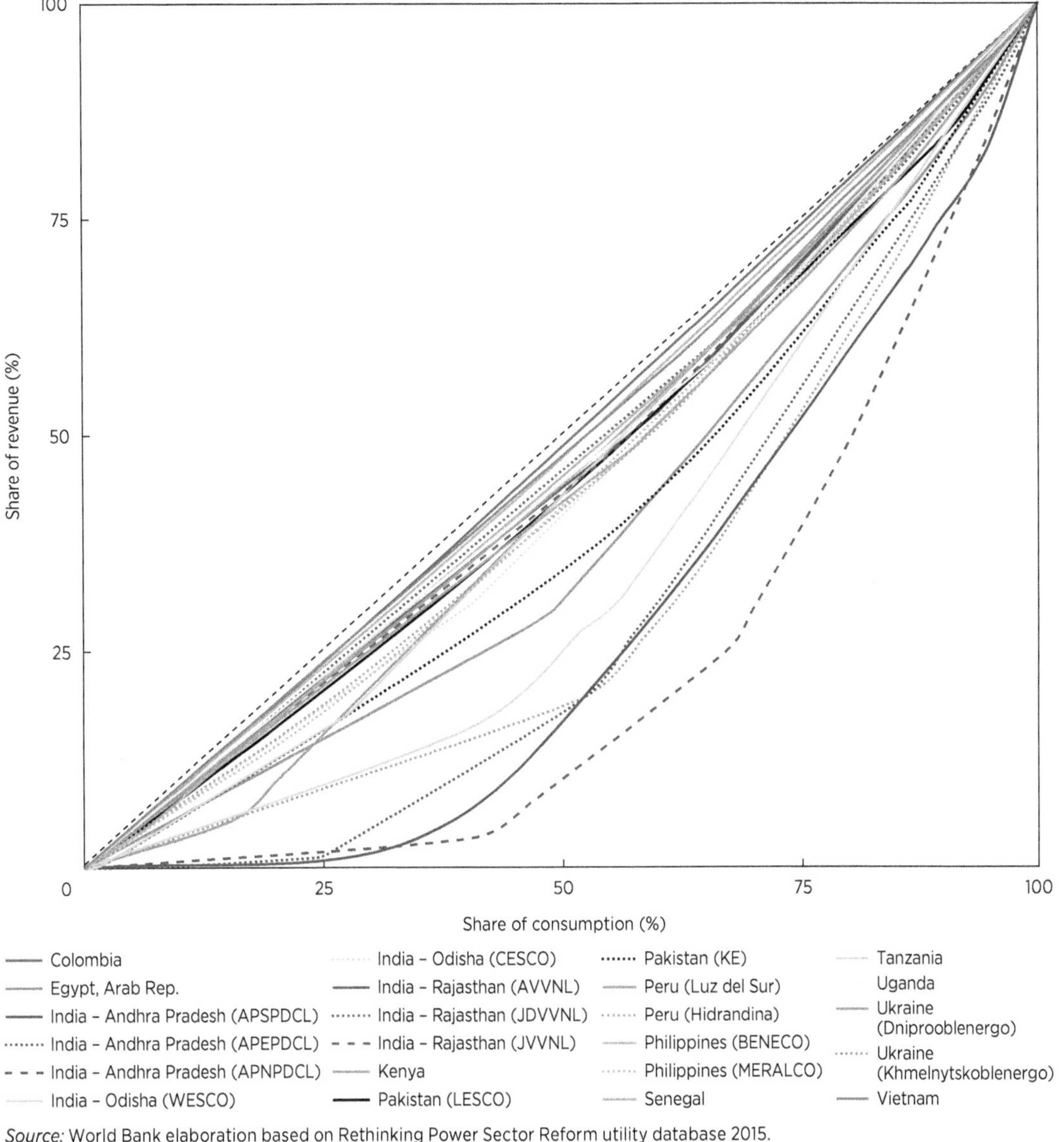

Source: World Bank elaboration based on Rethinking Power Sector Reform utility database 2015.
Note: The level of cross-subsidization is quantified here as the area between the 45-degree line and the curve defined by the cumulative shares of consumption and revenue by customer class. The formula is the same as that of a Gini-coefficient of inequality. Data for latest available year.

The distribution companies of Egypt, Tanzania, and Ukraine, and Pakistan's KE also show curves that bow out more noticeably than the rest of the sample.

There appears to be a limit of about 15 percent on the degree of price differentiation among customer categories, which is compatible with cost recovery. This limit is based on the methodology defined earlier that measures the area between the 45-degree line and the Lorenz curve representing consumption and revenue shares. Specifically, the analysis suggests that no country with a cross-subsidization indicator above 15 percent comes even close to full cost recovery (all five cases are below 80 percent; see figure 8.11). Egypt, India (Andhra Pradesh),

FIGURE 8.11 **Higher levels of cross-subsidy are associated with lower levels of cost recovery**

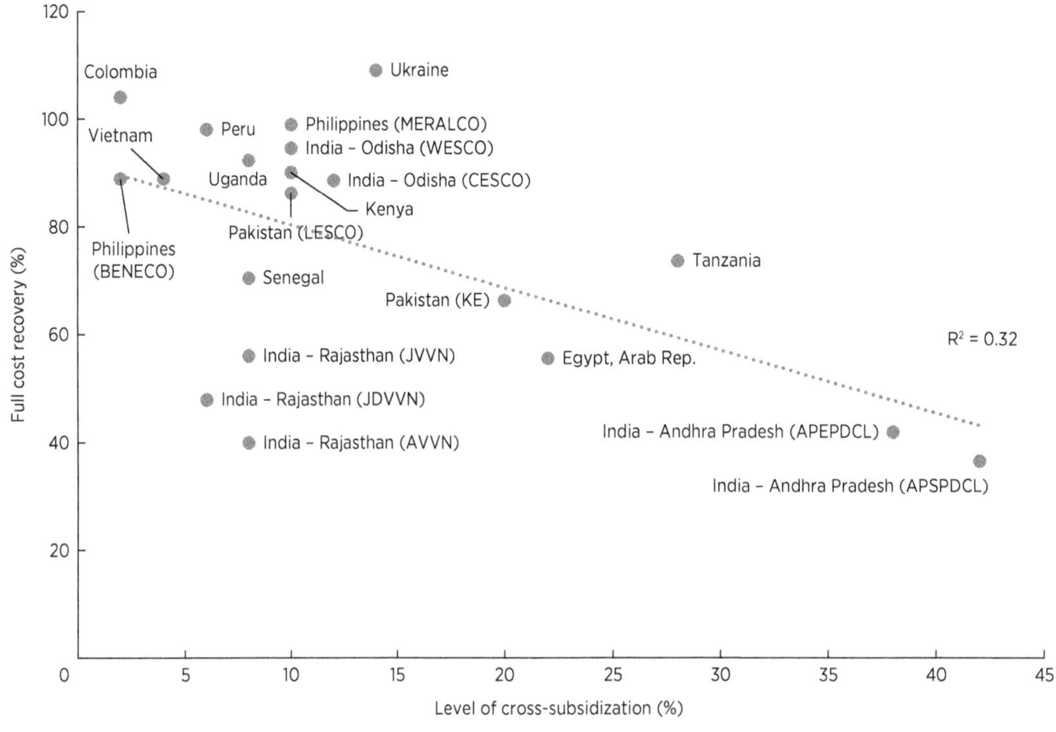

Source: World Bank elaboration based on Rethinking Power Sector Reform utility database 2015.

Pakistan, and Tanzania all feature large cross-subsidies and low cost recovery; the opposite is true for Colombia, the Philippines, and Vietnam. The Indian state of Rajasthan, with low cross-subsidization and low cost recovery, is a particular exception. This suggests that limiting cross-subsidization is a necessary but not sufficient condition for cost recovery.

Also challenging from a financial perspective are geographical cross-subsidies that preserve a uniform tariff across areas facing differential costs of supply. Another form of cross-subsidization is to maintain uniform tariffs across a country's regions and distribution companies with differences in cost of service. These forms of implicit cross-subsidization are common because differentiated tariffs, particularly across urban and rural areas, often pose challenges to rural electrification because the countryside has higher connection costs and lower income levels. KPLC has a uniform tariff for all domestic customers, despite rapid expansion of its network into lower-demand regions, with over a million new customers connected in both the 2015–16 and 2016–17 financial years.

Nevertheless, several jurisdictions that had reached full cost recovery preserved major lifelines to protect low-income consumers via an increasing block tariff. Cross-subsidization aimed at protecting the poor often takes the form of so-called lifeline tariffs up to a certain number of kWhs per month. Notably, several of the countries with the highest lifeline tariff block thresholds, such as Colombia (130–173 kWh), the Philippines (100 kWh), and Ukraine (100 kWh or higher, based on household type) are all at the high end of cost recovery. This suggests that targeted cross-subsidies to protect the poor can be part of a sustainable electricity pricing strategy.

Electricity tariff structures avoid incentivizing grid defection in view of recent developments in distributed energy. Most low-income countries rely on energy charges to recover costs. Consumers can self-generate to save almost their entire electricity bill, while benefiting from the grid's backup services. If fixed costs are recovered through kWh energy charges, rather than reflected in a separate fixed charge, each grid defection shifts these costs onto a smaller group of customers and further incentivizes grid defection. Given that the costs of distributed generation may be even lower for nonresidential customers consuming at larger scales, cross-subsidies further exacerbate incentives for grid defection. Odisha (India) illustrates the role of cross-subsidies in grid defection. After privatization, support to low-income residential customers in the form of free connections was reduced in favor of cross-subsidies from industrial customers, which led the industrial customers to seek alternative sources to avoid paying higher tariffs. This sort of grid defection, combined with residential tariffs that are kept artificially low, can put utilities in increasing financial distress because they are unable to recover their costs of service.

Finding #8: Cost recovery levels have risen on average, but progress has been uneven, with more than half of case studies showing a drop compared with the prereform period

Although cost recovery remains challenging, a key question is whether it it has improved since the 1990s reforms. To understand the impact of reforms, this section compares prereform cost recovery benchmarks from the late 1980s and early 1990s with cost recovery for 2010–17. As laid out in box 8.1, the prereform benchmarks are based on a comparison of actual prereform tariffs and prereform estimates of long-run marginal costs. The prereform cost recovery benchmarks can be understood as a *counterfactual of today's cost recovery levels if real tariffs remained the same and actual costs materialized exactly as anticipated in estimates of long-run marginal cost (LRMC)*. Comparing these benchmarks to actual cost recovery levels in 2010–17 allows us to draw conclusions about (1) the change in electricity tariffs in real terms compared to the prereform period; (2) if actual costs are now higher or lower than expected in the LRMC estimates from the late 1980s and 90s; and (3) if the combination of changes in real tariffs compared to the prereform actuals and changes in real costs compared to prereform estimates of LRMC led to a net increase or decrease in cost recovery compared to the counterfactuals.

There is evidence of some convergence in full cost recovery levels across countries (figure 8.12). As shown in figure 8.13,

FIGURE 8.12 Only a handful of countries and states had achieved full cost recovery prior to reform

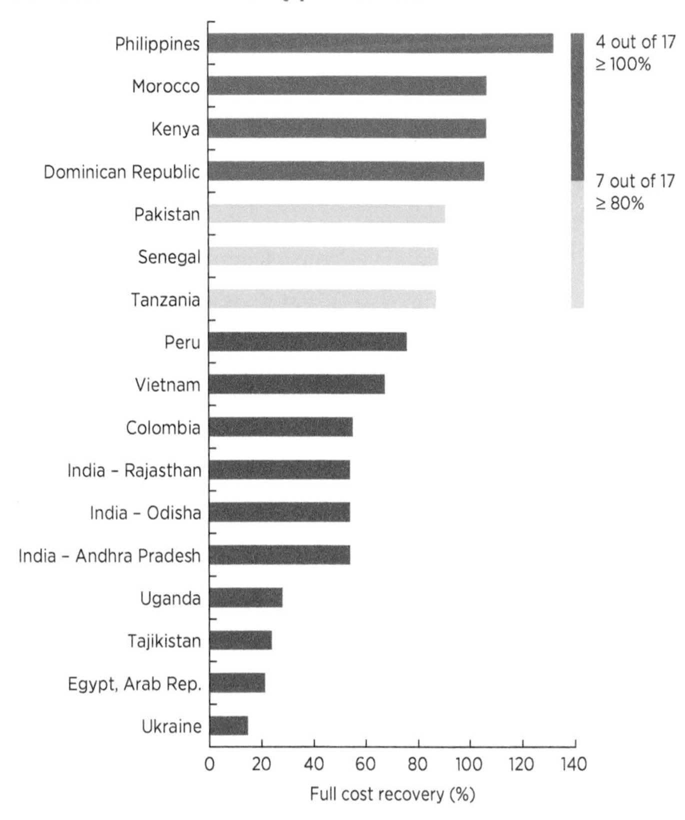

Source: World Bank elaboration based on Rethinking Power Sector Reform utility database 2015.

FIGURE 8.13 **Prereform cost recovery was not necessarily sustained over time**

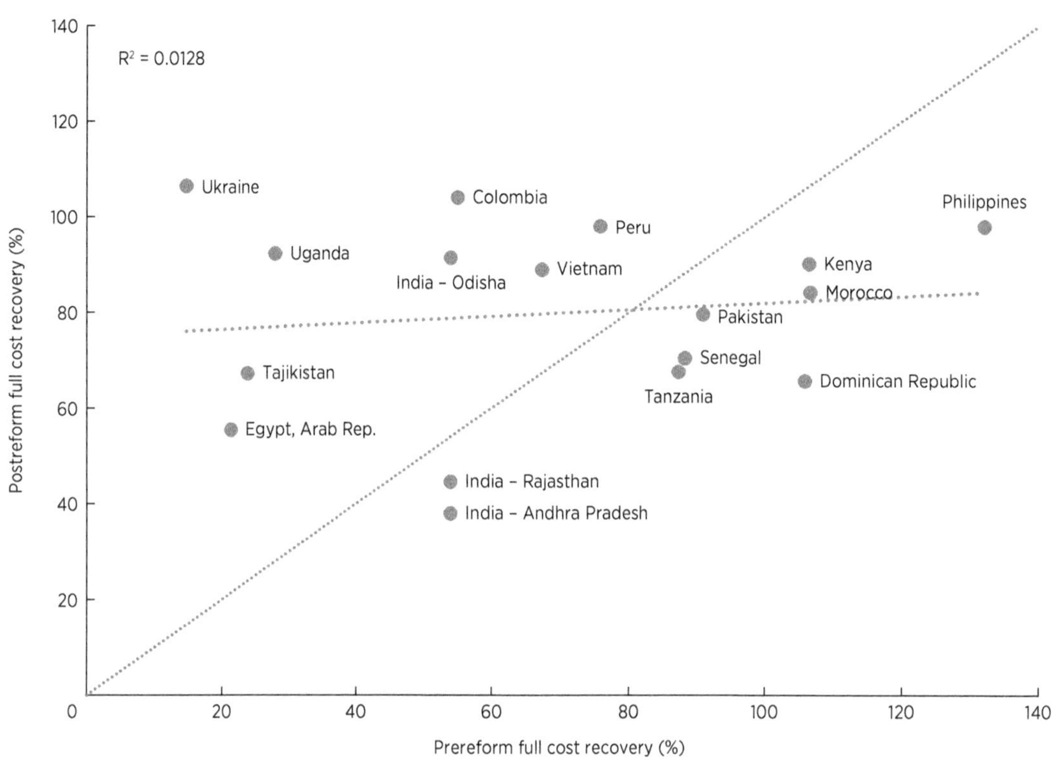

Source: World Bank elaboration based on Rethinking Power Sector Reform utility database 2015.

the largest improvements were made by countries with low cost recovery in the prereform period, and vice versa (compare to figure 8.12). The largest improvements in cost recovery were observed in Europe and Central Asia (Ukraine +92 percentage points and Tajikistan +43 percentage points) and Latin America and the Caribbean (Colombia +49; Peru +22), with South Asia and East Asia and the Pacific seeing, on average, relatively little change compared to the prereform period.

Although the average level of cost recovery improved slightly since 1990, the record was decidedly mixed, with about half of countries seeing an improvement in cost recovery and the other half a deterioration. Average full cost recovery increased from 69 percent around 1990 to 79 percent in 2010–17, but the increase was driven by a few strong performers and over half saw a decline (9 out of 17). Underlying drivers vary considerably across countries (figure 8.14). For example, Kenya saw cost recovery decline despite hefty tariff increases because costs increased even more rapidly. In Peru, by contrast, cost recovery improved despite a drop in real tariffs, because costs came down even faster. These findings are broadly in line with the literature, which suggests that, despite greater awareness about the broad adverse impacts of electricity subsidies, the aggregate level of cost recovery and financial viability in developing countries has improved only slightly between the late 1980s and the early 2010s (Huenteler and others 2017).

Notably, the average improvement in cost recovery from 69 percent around 1990 to 79 percent in 2010–17 was largely driven by cost reductions rather than tariff increases. In principle, improvements in cost recovery may result either from a reduction in costs or an increase in tariffs, and these two effects may be disaggregated (figure 8.14). Average real tariffs fell slightly in all 17 cases from US$0.122/kWh to US$0.119/kWh between the prereform (1990) and postreform (2010–17) periods—or about 2.5 percent on average. Over the same time period, costs fell slightly more on average in real terms from US$0.168/kWh to US$0.156/kWh—a drop of about 7 percent. Thus, observed improvements in cost recovery owe more to cost reductions than to tariff increases.

In countries where cost recovery deteriorated, large investment programs promoting universal electrification were a contributing factor. Tanzania's TANESCO, which saw a 20 percent decline in cost recovery over this period, has made increasingly large investments in recent years, totaling US$435.6 million in 2015–16 (69 percent of revenues) to fulfill its requirements of funding distribution expansion and some of the cost of new connections. These large investments coupled with low tariffs resulted in accounts payable exceeding revenues in 2015–16. Kenya's KPLC, which faced a 16 percent decline in cost recovery, finds itself in a similar situation, making hefty investments to expand its network despite poor cash flow. Its investments totaled US$481 million in 2016 (about 45 percent of its revenue).

Another cause of decline in cost recovery is a reversal of tariff reforms or stalled tariff increases due to sociopolitical pressure. The government of Senegal has been reluctant to authorize tariff increases and instead subsidizes the national utility, Senelec, for the difference between tariffs and the cost of service determined at quarterly tariff revisions, equivalent to 18 percent of full cost recovery for Senelec. These subsidies were not provided in 2015 and 2016, and in 2017 tariffs were lowered by 10 percent. In the Indian state of Andhra Pradesh, where cost recovery declined by 16 percentage points, the regulator was unable to increase tariffs from 2004 to 2010, and in 2004 the government also announced a policy of free power to agriculture.

The largest improvements in cost recovery compared to the prereform period were the result of both cost reductions and real tariff increases: Five out of eight countries with improved cost recovery, including three out of the four best performers, witnessed both a hike in tariffs and a fall in costs. Findings are in line

FIGURE 8.14 Postreform cost recovery improved in about half of cases and deteriorated in half

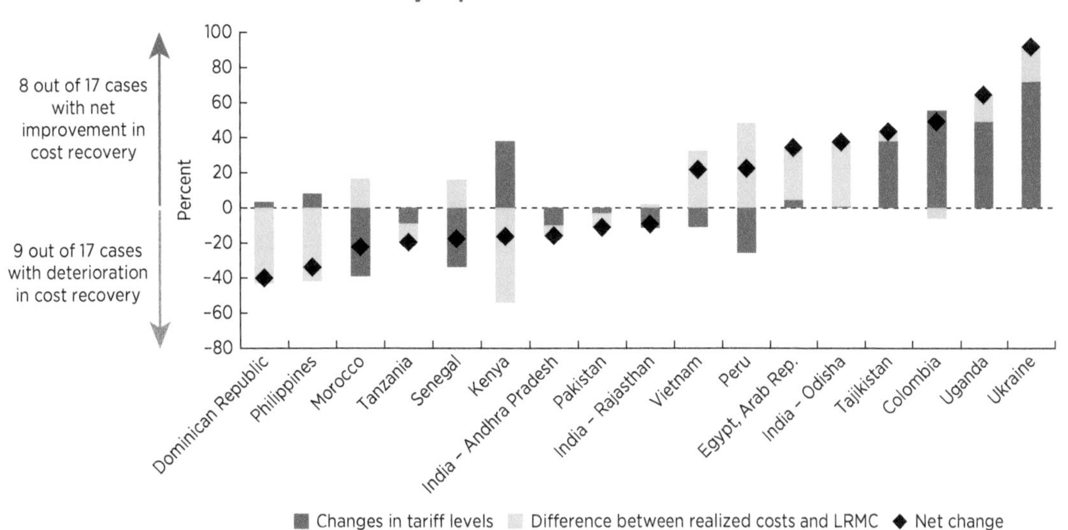

Source: World Bank elaboration based on Rethinking Power Sector Reform utility database 2015.
Note: LRMC = long-run marginal cost.

with the literature (see annex 8A). Notably, cost recovery improved in four cases—Egypt, India, Peru, and Vietnam—without large increases in real tariffs. Kenya, conversely, saw cost recovery drop despite major tariff hikes.

Reforms on cost recovery were often prompted by crisis. Countries started to adopt drastic cost reductions and tariff increases after having experienced large power deficits that require extensive investments that the government can no longer support alone. Uganda's UMEME (+64-percentage-point increase in cost recovery) has taken on aggressive investments (totaling US$93 million in 2016), allowing it to upgrade the distribution grid to keep up with growing access and demand. Uganda is also improving operating efficiency and reducing losses and operating costs. Uganda's tariffs have allowed UMEME to keep up with debt service payments, because tariffs are updated annually and subject to quarterly automatic adjustments for inflation, exchange rate, and oil price fluctuations. Investments that improve service delivery are vital to sustain cost recovery improvements, because efficiencies in service delivery will reduce costs and, if tariffs are not yet at cost recovery level, improved quality of service eases the sociopolitical pressure that might otherwise suppress increases. Similarly, the 2012 tariff increase in Tanzania was possible under emergency procedures because of drought. In Pakistan, the 2011 circular debt was crippling the energy sector as independent power producers threatened to call in sovereign guarantees because of nonpayment, pressuring the government to raise tariffs for the higher consumer blocks.

Finding #9: Tariff erosion brought on by inflation outweighed tariff increases in most cases

Real tariffs declined on average as nominal tariff increases did not keep pace with inflation in most cases. As mentioned in the previous section, average real tariffs dipped in all 17 cases from US$0.122/kWh to US$0.119/kWh between the prereform (1990) and postreform (2010–17) periods, or about 2.5 percent on average. Figure 8.15 breaks down the change

FIGURE 8.15 Nominal tariff increases were largely eroded by inflationary pressures

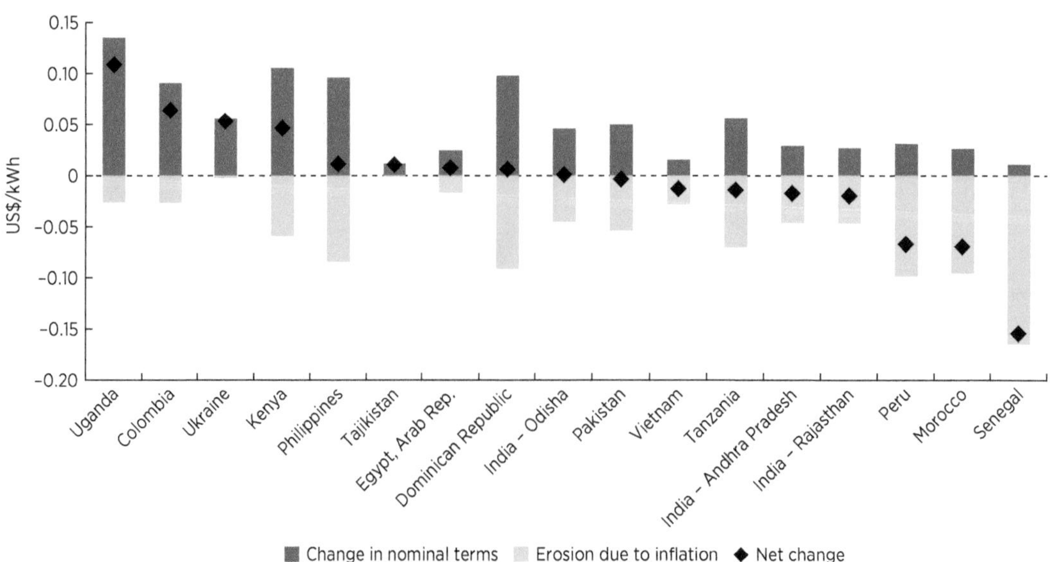

Source: World Bank elaboration based on Rethinking Power Sector Reform utility database 2015.
Note: kWh = kilowatt-hour.

in real tariffs into inflationary effects and tariff increases in nominal U.S. dollars. The graph shows that real tariffs fell in 9 cases and increased in 8—this despite a nominal tariff increase, on average, in every one of the 17 observed cases. In Egypt and Tanzania, currency devaluation reduced the tariff value by 100 and 21 percent, respectively.

Finding #10: Declining transmission and distribution losses have led to major improvements in cost recovery

Average system losses fell from 24 percent in the prereform era to 17 percent in the 2010–15 period, contributing to the decline in average costs across the case studies (figure 8.16). In contrast to the story for cost recovery, where roughly half the sample improved and half deteriorated, there is a striking consistency with 14 of the 17 jurisdictions showing reduced system losses and only 1 jurisdiction (India – Odisha) showing any sizable deterioration. Even this instance can be attributed to artificially low numbers expressed by the government in the runup to reforms. The state government estimated losses at 29 percent in 1994, although some analysis done postreform estimated them to be closer to 40 percent. The largest improvements in system losses were observed in East Asia and Pacific and Latin America and the Caribbean, followed closely by South Asia, though absolute levels are still high in South Asia. The cases from Sub-Saharan Africa showed relatively little improvement, in contrast. It is important to note that system loss reduction often requires major investments in the network; without external financing or government subsidies, loss-making utilities are then unable to maintain or improve the network, further undermining their cost recovery, as is the case for Kenya, India (Odisha), and Senegal.

The reduction in transmission and distribution losses decreased the full economic cost of service by an average of US$0.015/kWh across the sample cases. Had T&D losses remained at prereform levels for all countries, the average cost of service across the cases would be US$0.171/kWh, as compared to the average prereform cost of US$0.168 and average postreform cost of US$0.156. One of the policy tools to improve performance has been the setting of target losses under the utility's calculation of revenue requirement. Privatization, and performance-based regulation, was a contributor to the reduction of losses. In Colombia, losses fell from 25 to 19 percent in the first year of privatization. Losses in Uganda came down from about 38 percent in the mid-2000s to 17 percent in 2017. Privatization also helped keep losses in the Philippines below 10 percent. This policy has not always been successful, however: Pakistan is an example where unbundling

FIGURE 8.16 In most cases, system losses declined substantially over time

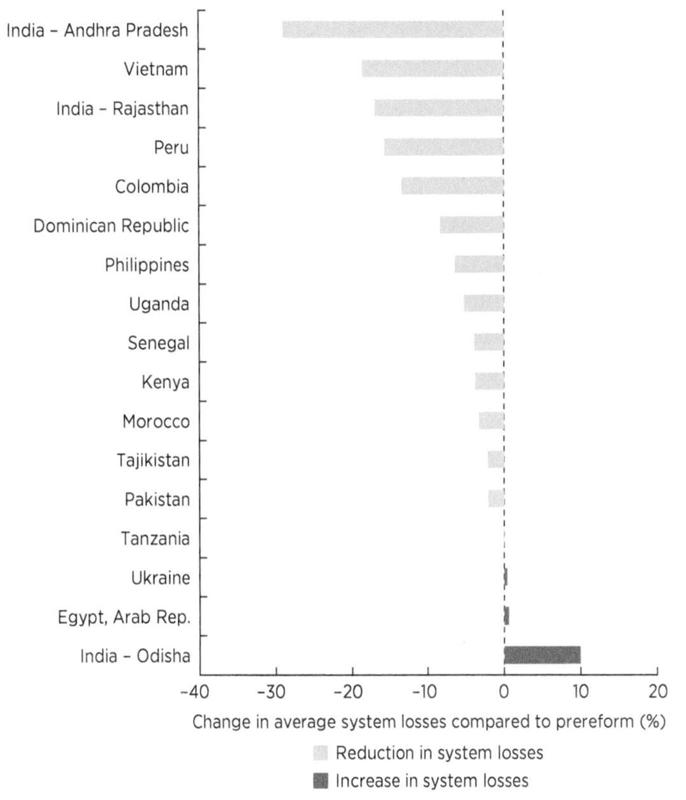

Source: World Bank elaboration based on World Bank World Development Indicators database 2018.

and privatization brought limited improvements in efficiency because the incentive for loss reduction was weak. In India (Odisha), losses were underestimated when distribution companies were transferred to private licensees; they continue at a high level (39 percent in 2015). In Senegal, confusion over ownership of T&D assets also contributed to high losses.

The effect of variability in the cost of supply is larger than the effect of T&D loss changes in most cases. Figure 8.17 breaks down the net change in average full cost of service compared to the prereform period into the effect of T&D loss reduction and changes in the cost of supply. On average, the decline due to the reduced T&D losses was outweighed by the higher average cost due to other factors in the observed cases (figure 8.17).

CONCLUSION

Strong performance on efficiency and cost recovery remains the exception, not the norm, among power utilities in developing countries. Of the 17 jurisdictions, only 2 had achieved full economic cost recovery (C3) for the power sector as of 2010–17, and as many as 7 jurisdictions had not yet achieved even operating cost recovery (C1). Even when cost recovery is attained, it remains permanently vulnerable to cost shocks arising from higher oil prices and exchange rate fluctuations, as well as supply shocks in the form of drought. Cross-subsidies among customer classes further undermine cost recovery, although the experience of some utilities suggests it is possible to attain cost recovery while offering limited discounts for so-called lifeline consumption by residential users. Although most utilities experiencing a shortfall in cost recovery receive some degree of fiscal transfer from the state, in the form of operating or capital subsidies, these are seldom high enough to fully compensate for the shortfall and are often at least partially offset by the failure of many public institutions to pay their electricity bills. As a result, utilities often find their financial viability compromised, so they

FIGURE 8.17 Improvements in system losses were often outweighed by other cost movements

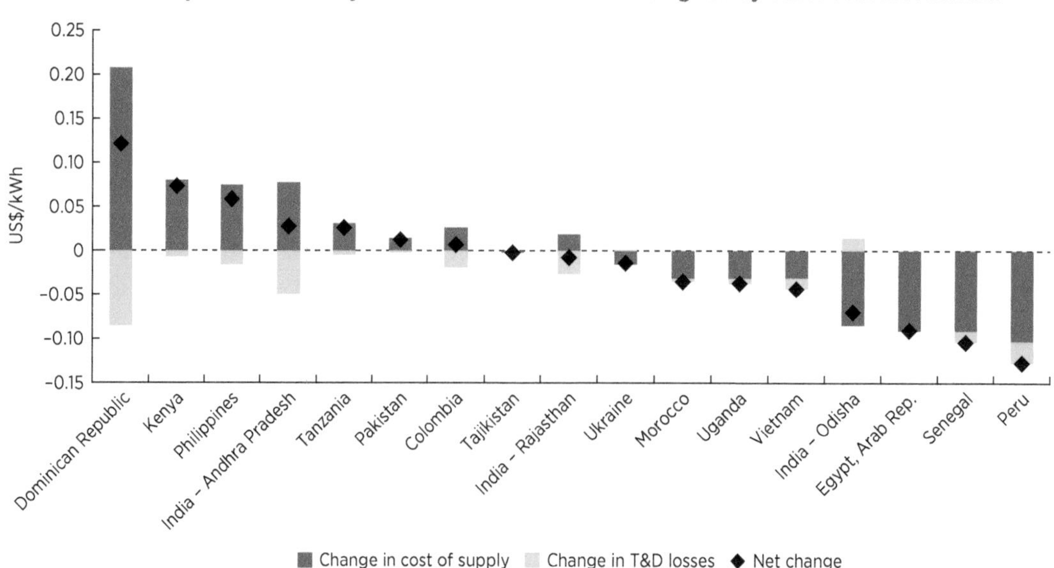

Sources: World Bank elaboration based on Rethinking Power Sector Reform utility database 2015 and World Bank World Development Indicators database 2018.
Note: kWh = kilowatt-hour; T&D = transmission and distribution.

resort to a range of coping strategies including costly short-term borrowing from commercial banks or informal borrowing in the form of payment arrears to various kinds of suppliers.

Over time, greater progress has been made on improving efficiency than on recovering costs; indeed, countries making greatest progress on cost recovery have done so at least in part by controlling costs. On aggregate, progress toward cost recovery has been relatively modest—from about 69 percent in the prereform period (1990) to 79 percent in the postreform period (2010–17). Moreover, only about half of the jurisdictions saw cost recovery improve, whereas the other half saw a deterioration. Indeed, average real tariffs *fell* slightly over this time period by about 2.5 percentage points, even if average real costs fell slightly faster at 7 percentage points. Thus, the limited progress on cost recovery over this period owed more to cost reductions than to tariff increases. More encouragingly, over the same period, system losses fell from 24 to 17 percentage points with almost all jurisdictions registering sizable improvements. Nevertheless, only about half the utilities in the sample had achieved system losses commensurate with best-practice levels.

ANNEX 8A. MAJOR STUDIES OF COST RECOVERY AND FINANCIAL VIABILITY IN THE POWER SECTOR IN DEVELOPING COUNTRIES

Serial No.	Study	Coverage	Time	KPIs	Main findings	Observed trends
1	World Bank (1972)	Argentina, Brazil, Colombia, Ethiopia, Ghana, Malaysia, Mexico, Singapore	1955–70	Rate of return on assets (based on utility financial statements)	All 10 analyzed utilities were profitable during the observation period, with return on assets mostly in the 8–9% range	Significant improvements in 1960s
2	Munasinghe, Gilling, and Mason (1989)	Recipient utilities of 123 World Bank power projects worldwide	1966–84	Four financial ratios (based on utility financial statements)	Average rate of return for the period 1966–85 was 7.9	Distinct deterioration in the trend of utilities' financial ratios for the period 1973–85
3	World Bank (1990)	60 developing countries worldwide, comparison to OECD	1979–88, with LRMC for 1990s	Comparison of existing tariffs to LRMC with shadow prices	Tariffs on (weighted) average sufficient to recover 62% of LRMC; average tariff level 55% of the average level in OECD countries	Real average tariffs constant in 1979–83, then fell sharply until 1988
4	IEA (1999)	China, India, Indonesia, Islamic Republic of Iran, Kazakhstan, Russian Federation, South Africa, Venezuela,	1998	Price gap between tariffs and reference price (LRMC based on current fuel mix)	Cost recovery ratio ranged between 37% (Venezuela) and >100% (Indonesia); average: 62.3%	n.a.
5	Foster and Yepes (2006)	83 OECD and non-OECD countries worldwide	1994–2002	Average tariff compared to global benchmark values	15% of countries did not cover O&M costs; 59% did not cover total cost; strong correlation with income per capita	Slight real increase in tariffs in some regions but no significant trend across sample
6	Ebinger (2006)	20 countries in Eastern Europe and Central Asia	2000–03	Disaggregated QFD: T&D losses, collection losses, underpricing	QFD between 0 (Belarus) and 16.53% (Tajikistan) in 2003; mostly driven by underpricing (67%)	Decline in the QFD in 17 out of 20 countries between 2000 and 2003, by 48% (from US$30 billion to US$16 billion overall)

(Annex continued next page)

Serial No.	Study	Coverage	Time	KPIs	Main findings	Observed trends
7	Saavalainen and ten Berge (2006)	8 countries in Eastern Europe and Central Asia	2002	Disaggregated QFD: T&D losses, collection losses, underpricing	Cost recovery between 11.21% and 81.6%; QFD between 1.1% and 21.4% of GDP	n.a.
8	Eberhard and others (2008)	21 Sub-Saharan African countries	2001–05	Average tariff revenues compared to average historical cost, LRMC	Despite comparatively high power prices only 57% of Sub-Saharan African countries recovered OPEXs; 36% recovered LRMC	Real tariffs almost doubled over the period 2001–05, but cost recovery ratio declined
9	Briceño-Garmendia, Smits, and Foster (2009)	20 Sub-Saharan African countries	2006	Disaggregated QFD: T&D losses, collection losses, underpricing	6 out of 20 countries recovered average historic costs; hidden costs of power mispricing amount to about 1% of GDP or 60% of total hidden costs	n.a.
10	Briceño-Garmendia and Shkaratan (2011)	27 Sub-Saharan African countries	2004–08 (latest available)	Average effective tariff and LRMC compared to OPEX (income statements) and CAPEX (LCOE benchmarks)	80% countries recovered OPEX; 30% also recovered CAPEX; 38% recovered LRMC	n.a.
11	Vagliasindi and Besant-Jones (2013)	19 developing countries worldwide + 3 Indian states	Late 1990s to late 2000s	Cost recovery index (average revenue divided by average supply cost)	Cost recovery index correlated with indexes of competition and vertical unbundling	Tariffs increased over the period, but cost recovery fluctuated
12	Alleyne and Hussain (2013)	Large sample of Sub-Saharan African countries (unspecified)	2005–09 (latest year available)	Disaggregated QFD: T&D losses, collection losses, underpricing	Average tariffs were of 70% of cost; QFD was about 1.7% of 2009, half of which from underpricing	Average QFD constant at 1.7% of GDP between 2005–06 and 2009–10
13	Mayer, Banerjee, and Trimble (2015)	Residential electricity use in 29 states in India	2005, 2010	Average effective tariff (based on household surveys)	87% of residential consumption was subsidized in 2010; average cost recovery was 68%; 2 out of 29 states had effective tariffs > average cost	In real terms, the net cost of the average household subsidy in 2010 was 70 times larger than in 2005
14	Khurana and Banerjee (2013)	29 states in India	2003–11	Comparison of average billed tariff was higher than AC	Cost recovery averaged 82% in 2003–11; 7 states had tariffs below cost in 2003, 14 in 2011	Cost recovery fluctuated within a band of 76–85%; with a low point in 2010
15	Di Bella and others (2015)	32 countries in Latin America and the Caribbean	2011–13 (average)	Price-gap approach pretax subsidies (% of GDP)	Electricity subsidies in Latin American and the Caribbean were almost as large as direct fuel subsidies, on average 0.8% of GDP in 2011–13	n.a.
16	IEA (2015)	40 non-OECD countries worldwide	2012–14	Price gap approach (based on average cost of production)	All but four countries subsidize electricity (excl. renewable energy subsidies)	Decline in total subsidies by 10.4% in 2012–14; 5 additional countries reached cost recovery

(Annex continued next page)

Serial No.	Study	Coverage	Time	KPIs	Main findings	Observed trends
17	Trimble and others (2016)	39 countries in Sub-Saharan Africa	2011–15	Disaggregated QFD: Collection losses; T&D losses; overstaffing; underpricing	Average cash collected 57% was of cost; 2 countries have a financially viable electricity sector; 19 countries cover OPEX; QFD average 1.5% of GDP	Most of the countries with low QFDs improved over past decade, while most of the countries with high QFDs remained high
18	World Bank (2016)	Utilities in 40 developing countries worldwide	2003–13	Utilities' profitability (based on utility financial statements)	10 out of 40 utilities were profitable; 2 out of 17 Sub-Saharan African utilities were profitable in 2000, 4 in 2013	Share of profitable utilities increased from 10% to 35% in 2010, then fell to 25% in 2013
19	Coady and others (2015)	153 OECD and non-OECD economies worldwide	2013, 2015	Price-gap approach (reference price including consumption taxes; excl. renewable energy subsidies)	79 out of 119 developing countries had electricity subsidies in 2015, compared to 1 out of 34 "advanced economies" (Taiwan, China)	Absolute decline of subsidies by 36.5%; numbers of countries with subsidies from 75% to 66%

Source: Huenteler and others 2017.
Note: AC = average cost; CAPEX = capital expenditure; KPI = key performance indicator; LCOE = levelized cost of energy; LRMC = long-run marginal cost; O&M = operation and maintenance; OECD = Organisation for Economic Co-operation and Development; OPEX = operating expenditure; QFD = quasi-fiscal deficit; T&D = transmission and distribution; n.a. = not applicable.

ANNEX 8B. COVERAGE OF QUANTITATIVE COST RECOVERY ANALYSIS UNDERTAKEN FOR THIS CHAPTER

Country/state	Year for prereform analysis	Years for postreform analysis	Scope of utility-level analysis (majority ownership)
Colombia	1987	2010–16	Codensa (Private)
Dominican Republic[a]	1987	2010–15	Edesur (Public)
Egypt, Arab Rep.	1987	2011–16	EEHC (Public)
India - Andhra Pradesh	1987	2011–15	APEPDCL (Public), APSPDCL (Public)
India - Odisha	1987	2011–15	CESU (Public), WESCO (Private)
India - Rajasthan	1987	2011–15	AVVN (Public), JDVVN (Public), JVVN (Public)
Kenya	1987	2010–16	KPLC (Private)
Morocco	1987	2013	ONEE (Public)[b]
Pakistan	1987	2013–16	KE (Private), LESCO (Public)
Peru	1987	2010–16	Hidrandina (Public), Luz del Sur (Private)
Philippines	1987	2010–16	Beneco (Private), Meralco (Private)
Senegal	1991	2010–16	Senelec (Public)
Tajikistan	2003	2010–16	Barki Tojik (Public)
Tanzania	1991	2012–16	TANESCO (Public)
Uganda	1987	2012–16	UMEME (Private)
Ukraine[c]	1994	2012–17	Dniprooblenergo (Private), Khmelnytskoblenergo (Public)
Vietnam	1993	2010–16	NPC (Public)

Source: World Bank elaboration based on Rethinking Power Sector Reform utility database 2015.
Note: Benchmark system loss level is 5 percent for the analysis in this table.
a. Approximated by data for Edesur because full-country data were not available.
b. Approximated by sector-wide data.
c. C1–C3 cost recovery is approximated by A1–A3 because systematic information on government support to the utilities was not available.

ANNEX 8C. INDICATORS OF COST RECOVERY AND FINANCIAL VIABILITY OF POWER SECTORS AND UTILITIES IN 17 CASE STUDIES

TABLE 8C.1 Full cost recovery for power sectors

Percent

			Full cost recovery (C3 of approximate)[a]									
Country/state	Type	Region	Prereform	2010	2011	2012	2013	2014	2015	2016	2017	Average 2010–17
Colombia	Private	LAC	55	94	109	117	103	100	105	102	—	104
Dominican Republic[b]	Public	LAC	106	71	65	62	65	62	73	—	—	66
Egypt, Arab Rep.	Public	MENA	21	—	62	62	68	50	49	50	—	55
India - Andhra Pradesh	Public	SAR	54	—	36	36	38	37	44	—	—	38
India - Odisha	Public	SAR	54	—	94	92	90	92	91	—	—	91
India - Rajasthan	Public	SAR	54	—	38	41	47	43	57	—	—	45
Kenya	Public	SSA	106	99	101	94	87	89	81	80	—	90
Morocco[c]	Public	MENA	107	—	—	—	84	—	—	—	—	84
Pakistan	Public	SAR	91	—	—	—	66	75	87	97	—	80
Peru	Public	LAC	76	102	102	101	98	95	97	93	—	98
Philippines	Public	EAP	132	98	99	100	96	98	99	95	—	98
Senegal	Public	SSA	88	77	68	55	66	72	86	87	—	70
Tajikistan	Public	ECA	24	—	—	—	52	73	83	70	—	67
Tanzania	Private	SSA	87	—	—	60	56	—	90	71	—	68
Uganda[c]	Private	SSA	28	—	—	110	96	91	85	81	—	92
Ukraine	Public	ECA	15	—	—	107	115	110	105	103	93	106
Vietnam	Private	EAP	67	91	86	91	88	89	89	88	—	89

Source: World Bank elaboration based on Rethinking Power Sector Reform utility database 2015.
Note: EAP = East Asia and Pacific; ECA = Europe and Central Asia; LAC = Latin America and the Caribbean; MENA = Middle East and North Africa; SAR = South Asia; SSA = Sub-Saharan Africa; — = not available
a. Excluding externalities.
b. Approximated by Edesur.
c. Approximated by full financial cost recovery.

TABLE 8C.2 Full financial cost recovery for power sectors

Percent

			Full financial cost recovery (A3 of approximate)[a]									
Country/state	Type	Region	Prereform	2010	2011	2012	2013	2014	2015	2016	2017	Average 2010–17
Colombia	Private	LAC	—	94	109	117	103	100	105	102	—	104
Dominican Republic[b]	Public	LAC	—	71	65	62	65	62	73	—	—	66
Egypt, Arab Rep.	Public	MENA	—	—	62	62	68	70	65	66	—	65
India - Andhra Pradesh	Public	SAR	—	—	40	41	39	39	48	—	—	41
India - Odisha	Public	SAR	—	—	94	92	90	92	91	—	—	91
India - Rajasthan	Public	SAR	—	—	40	44	51	44	59	—	—	47

(Table continued next page)

TABLE 8C.2 Full financial cost recovery for power sectors *(Continued)*

Percent

			Full financial cost recovery (A3 of approximate)[a]									
Country/state	Type	Region	Prereform	2010	2011	2012	2013	2014	2015	2016	2017	Average 2010–17
Kenya	Public	SSA	—	100	101	95	88	90	83	83	—	91
Morocco	Public	MENA	—	—	—	—	84	—	—	—	—	84
Pakistan	Public	SAR	—	—	—	—	76	89	96	104	—	90
Peru	Public	LAC	—	102	102	101	98	95	97	93	—	98
Philippines	Public	EAP	—	98	99	100	96	98	99	95	—	98
Senegal	Public	SSA	—	85	97	75	86	93	97	91	—	88
Tajikistan	Public	ECA	—	—	—	—	101	103	108	94	—	101
Tanzania	Private	SSA	—	—	—	73	68	—	107	83	—	81
Uganda	Private	SSA	—	—	—	114	105	100	94	88	—	100
Ukraine	Public	ECA	—	—	—	107	115	110	105	103	93	106
Vietnam	Private	EAP	—	92	87	93	89	91	90	89	—	90

Source: World Bank elaboration based on Rethinking Power Sector Reform utility database 2015.
Note: EAP = East Asia and Pacific; ECA = Europe and Central Asia; LAC = Latin America and the Caribbean; MENA = Middle East and North Africa; SAR = South Asia; SSA = Sub-Saharan Africa; — = not available
a. Excluding externalities.
b. Approximated by Edesur.

TABLE 8C.3 Financial operating cost recovery for power sectors, based on cash collected

Percent

			Financial operating cost recovery (A1 of approximate), adjusted for bill collection losses									
Country/state	Type	Region	Pre-Reform	2010	2011	2012	2013	2014	2015	2016	2017	Average 2010–17
Colombia	Private	LAC	—	130	162	152	149	146	145	139	—	147
Dominican Republic[a]	Public	LAC	—	67	62	58	61	59	71	—	—	62
Egypt, Arab Rep.	Public	MENA	—	—	143	125	100	103	128	119	—	120
India - Andhra Pradesh	Public	SAR	—	—	—	—	—	—	—	—	—	—
India - Odisha	Public	SAR	—	—	88	88	89	92	92	—	—	90
India - Rajasthan	Public	SAR	—	—	—	—	—	—	—	—	—	—
Kenya	Public	SSA	—	112	114	115	113	115	123	124	—	116
Morocco	Public	MENA	—	—	—	—	—	—	—	—	—	—
Pakistan	Public	SAR	—	—	—	—	90	113	124	144	—	116
Peru	Public	LAC	—	137	135	132	129	130	130	131	—	132
Philippines	Public	EAP	—	108	107	110	110	115	115	115	—	111
Senegal	Public	SSA	—	89	105	93	97	105	104	102	—	99
Tajikistan	Public	ECA	—	—	—	—	68	84	108	98	—	88
Tanzania	Private	SSA	—	—	—	81	87	—	120	88	—	93
Uganda	Private	SSA	—	—	—	117	129	124	119	129	—	123
Ukraine	Public	ECA	—	—	—	—	—	—	—	—	—	—
Vietnam	Private	EAP	—	99	98	104	102	101	102	100	—	101

Source: World Bank elaboration based on Rethinking Power Sector Reform utility database 2015.
Note: EAP = East Asia and Pacific; ECA = Europe and Central Asia; LAC = Latin America and the Caribbean; MENA = Middle East and North Africa; SAR= South Asia; SSA = Sub-Saharan Africa; — = not available
a. Approximated by Edesur.

TABLE 8C.4 Full cost recovery for power utilities

Percent

Power utility	Type	Full cost recovery (C3 of approximate)[a]								
		2010	2011	2012	2013	2014	2015	2016	2017	Average 2010–17
Colombia (Codensa)	Private	94	109	117	103	100	105	102	—	104
Dominican Republic (Edesur)	Public	71	65	62	65	62	73	—	—	66
Egypt, Arab Rep. (EEHC)	Public	—	62	62	68	50	49	50	—	55
India - Andhra Pradesh (APSPDCL)	Public	—	35	35	34	36	44	—	—	36
India - Andhra Pradesh (APEPDCL)	Public	—	42	35	48	42	44	—	—	42
India - Odisha (WESCO)	Public	—	94	97	94	94	94	—	—	95
India - Odisha (CESU)	Public	—	94	87	86	90	87	—	—	89
India - Rajasthan (AVVN)	Public	—	37	37	46	32	56	—	—	40
India - Rajasthan (JDVVN)	Public	—	34	43	49	56	64	—	—	48
India - Rajasthan (JVVN)	Public	—	57	53	57	53	61	—	—	56
Kenya (KPLC)	Public	99	101	94	87	89	81	80	—	90
Morocco (ONEE)[b]	Public	—	—	—	84	—	—	—	—	84
Pakistan (KE)	Public	—	—	—	76	86	90	94	—	86
Pakistan (LESCO)	Private	—	—	—	53	61	71	89	—	66
Peru (Luz del Sur)	Private	105	107	104	100	96	101	101	—	102
Peru (Hidrandina)	Public	91	86	91	93	90	85	74	—	87
Philippines (Meralco)	Private	99	100	101	98	98	100	96	—	99
Philippines (Beneco)	Private	96	88	88	88	88	87	88	—	89
Senegal (Senelec)	Public	77	68	55	66	72	86	87	—	70
Tajikistan (Barki Tojik)	Public	—	—	—	52	73	83	70	—	67
Tanzania (TANESCO)	Public	—	—	60	56	—	108	93	—	74
Uganda (UMEME)	Private	—	—	110	96	91	85	81	—	92
Ukraine (Khmelnytskoblenergo)	Mixed	—	—	107	115	110	105	103	110	109
Vietnam (NPC)	Public	91	86	91	88	89	89	88	—	89

Source: World Bank elaboration based on Rethinking Power Sector Reform utility database 2015.
Note: — = not available

a. Excluding externalities.

b. Approximated by sector-wide cost recovery.

ANNEX 8D. INDICATORS OF EFFICIENCY OF UTILITIES IN 17 CASE STUDIES

TABLE 8D.1 Revenue lost due to undercollection for power utilities

		Percent of revenue lost						
Power utility	Type	2010	2011	2012	2013	2014	2015	Average 2010–15
Colombia (Codensa)	Private	—	0	0	0	0	0	0
Colombia (EPM)	Public	0	0.44	0	0	0	0	0.1
Dominican Republic (Edesur)	Public	6.66	6.67	3.55	3.71	2.80	3.21	4.4
Dominic Republic (Edenorte)	Public	7.99	9.32	3.25	2.87	2.89	1.48	4.6
Egypt, Arab Rep. (Cairo North)	Public	24.97	24.33	24.38	25.52	24.47	22.41	24.3
Egypt, Arab Rep. (Alexandia)	Public	16.24	15.49	15.48	14.90	15.00	13.11	15.0
India - Andhra Pradesh (APEPDCL)	Public	0.67	7.47	3.74	3.67	2.30	2.91	3.5
India - Andhra Pradesh (APSPDCL)	Public	3.60	2.21	1.00	2.23	1.22	0.97	1.9
India - Odisha (CESU)	Public	5.40	12.60	9.40	7.61	4.68	4.00	7.3
India - Odisha (WESCO)[a]	Private	3.23	6.14	5.58	4.44	5.58	6.67	5.3
India - Rajasthan (JDVVNL)	Public	6.06	1.35	4.92	0.85	3.51	2.98	3.3
India - Rajasthan (JVVNL)	Public	3.50	2.78	4.58	1.83	3.50	1.77	3.0
Kenya (KPLC)[b]	Public	0	0	0	0	0	1.60	0.3
Morocco (ONEE)	Public	0	0	0	—	0	0	0
Pakistan (K-Electric Ltd)	Private	8.68	10.14	8.24	11.12	10.19	8.05	9.4
Pakistan (LESCO)	Public	6.24	1.84	3.67	2.08	2.07	3.95	3.3
Peru (Luz Del Sur)	Private	0.55	0.60	0.69	1.43	0.13	1.18	0.8
Peru (Hidrandina)	Public	0.11	0.53	0.92	0.40	0.55	1.59	0.7
Philippines (MERALCO)	Private	0	0	0	0	0	0.04	0
Philippines (BENECO)	Private	1.45	2.91	2.59	1.50	0	0	1.4
Senegal (Senelec)	Public	5.80	2.61	3.50	6.26	0.37	6.88	4.2
Tajikistan (Barki Tojik)	Public	13.04	0.98	26.74	23.41	14.28	12.84	15.2
Tanzania (TANESCO)	Public	26.17	-16.79	11.14	-6.39	—	3.28	3.5
Uganda (UMEME)	Private	4.35	0.92	5.20	0	0.85	1.70	2.2
Ukraine (Dniproblenergo)	Private	1.62	1.73	0	1.69	1.82	0.25	1.2
Ukraine (Khmelnytskoblenergo)	Public	0.31	1.04	0	0.79	0	1.21	0.6
Vietnam (NPC)	Public	0	0.36	0.44	0	0.44	0	0.2
Vietnam (HCMCPC)	Public	0.60	0.69	0.30	0.39	0.44	0.45	0.5

Source: World Bank elaboration based on Rethinking Power Sector Reform utility database 2015.
Note: — = not available.
a. WESCO in Odisha was under private ownership until 2015.
b. KPLC is 51 percent government owned whereas 49 percent of its equity is floated in the Nairobi stock exchange.

TABLE 8D.2 **Revenues lost due to excessive system losses for power utilities**

			Percent of revenue lost						
Power utility	**Type**	**Threshold losses[c]**	**2010**	**2011**	**2012**	**2013**	**2014**	**2015**	**Average 2010–15**
Colombia (Codensa)	Private	7	—	1.6	2.0	1.7	1.8	1.8	1.8
Colombia (EPM)	Public	10	0	0	0	0	0	0.2	0
Dominican Republic (Edesur)	Public	10	18.2	16.9	17.1	15.5	16.6	14.8	16.5
Dominic Republic (Edenorte)	Public	10	17.7	19.6	25.2	20.8	19.5	17.7	20.1
Egypt (Cairo North)	Public	7	2.4	2.2	1.7	1.6	1.4	2.9	2.0
Egypt (Alexandia)	Public	7	1.3	3.0	3.0	2.3	1.7	2.3	2.3
India- Andhra Pradesh (APEPDCL)	Public	12	0	0	0	0	0	0	0
India- Andhra Pradesh (APSPDCL)	Public	12	1.1	0.7	0	0	0	0	0.3
India- Odisha (CESU)	Public	12	15.3	16.8	17.5	17.8	15.8	15.7	16.5
India- Odisha (WESCO)[a]	Private	12	15.5	17.9	19.8	19.2	16.8	15.9	17.5
India- Rajasthan (JDVVNL)	Public	12	21.8	16.1	14.5	16.1	13.1	13.4	15.8
India- Rajasthan (JVVNL)	Public	12	16.4	16.6	13.6	16.8	18.3	18.2	16.7
Kenya (KPLC)[b]	Public	13	2.1	2.1	3.1	3.8	3.4	3.0	2.9
Morocco (ONEE)	Public	10	0.3	0.7	1.2	—	2.0	2.3	1.3
Pakistan (K-Electric Ltd)	Private	12	21.9	19.6	18.8	15.4	11.1	8.3	15.9
Pakistan (LESCO)	Public	12	1.8	1.1	1.6	1.2	1.2	1.6	1.4
Peru (Luz Del Sur)	Private	7	0.4	0.4	0.5	0.3	1.1	0.8	0.6
Peru (Hidrandina)	Public	10	0.3	0	1.0	0.6	0.6	0.3	0.5
Philippines (MERALCO)	Private	7	0.8	0.3	0.1	0.4	0.8	0.8	0.5
Philippines (BENECO)	Private	7	1.8	1.9	2.8	1.9	1.8	1.9	2.0
Senegal (Senelec)	Public	13	10.1	10.5	9.1	7.9	6.2	6.1	8.3
Tajikistan (Barki Tojik)	Public	10	0	1.1	1.0	1.1	1.7	1.6	1.3
Tanzania (TANESCO)	Public	13	7.1	13.6	10.0	16.2	—	3.5	10.1
Uganda (UMEME)	Private	13	8.6	6.7	8.1	7.6	5.1	3.8	6.7
Ukraine (Dniproblenergo)	Private	7	0	0	0	0	0	0	0
Ukraine (Khmelnytskoblenergo)	Public	10	5.1	5.1	5.5	5.2	4.8	4.9	5.1
Vietnam (NPC)	Public	10	0	0	0	0	0	0	0
Vietnam (HCMCPC)	Public	7	0	0	0	0	0	0	0

Source: World Bank elaboration based on Rethinking Power Sector Reform utility database 2015.
Note: — = not available.
a. WESCO in Odisha was under private ownership until 2015.
b. KPLC is 51 percent government owned while 49 percent of its equity is floated in the Nairobi stock exchange.
c. Threshold losses: Level of distribution losses of a comparable efficient utility in the region.

NOTES

1. This chapter draws on Huenteler and others (2017) and further original research conducted by a team led by Ani Balabanyan and comprising Arthur Kochnakyan, Joern Huenteler, Denzel Hankinson, Nicole Rosenthal, Arun Singh, and Tu Chi Nguyen. The chapter is based on a methodologically consistent set of financial models prepared for the 25 utilities across 14 countries and 3 Indian states. The financial analysis was led by Arthur Kochnakyan, supported by a team of consultants including Martin Tarzyan, Vazgen Sargsyan, Adrian Ratner, and Emiliano Lafalla. Vivien Foster and Anshul Rana coordinated the work program.
2. Conceptually, cost recovery can be viewed from the perspective of the power utility/sector, the fiscal perspective, or the overall economic perspective. In each case, the full costs would be defined differently, and which perspective is appropriate depends on the research question. Further, depending on the purpose, cost recovery may include "full costs" with any inefficiencies (including excess losses) the power company/sector has or cost recovery assuming efficient operation of the company/sector. The latter approach is ideally that taken by the regulators so as not to pass inefficiencies to consumers. Importantly, full cost recovery of tariffs for the sector does not necessarily mean that all individual parts of the supply chain (generation, transmission, and distribution) recover their costs, depending on how tariffs are set for the different services. Furthermore, some studies in the literature approximate tariffs with revenues and cost with actual cost incurred by the utilities, bringing the concept of "cost recovery" closer to the common understanding of "financial viability."
3. Financial viability is also an attribute of investment projects and in fact early World Bank studies of financial viability in the power sector were primarily interested in the ability of individual investments to make adequate returns. But this view has evolved (see Huenteler and others 2017) and now the primary unit of analysis in the literature is the utility. This is reflected in the term's usage in this chapter given the focus on leveraging private solutions and improving utility performance.
4. According to the most common definition, QFD is the difference between the actual revenue charged and collected at regulated electricity prices and the revenue required to fully cover prudently incurred operating costs of service provision and capital depreciation: QFD (as percent of GDP) = Cost of Underpricing of Electricity + Cost of Nonpayment of Bills + Cost of Excessive Technical Losses (Alleyne and Hussain 2013).
5. The literature includes relatively minor variations of this generally accepted QFD formula. For example, Briceño-Garmendia, Smits, and Foster (2009) and Kojima and Trimble (2016) introduce overstaffing as an additional "hidden cost" item.
6. Because of lack of information, the chapter does not assess labor cost inefficiencies.
7. The term *tariff structure* is used here to describe the composition of end-consumer prices (for example, one aggregate service tariff compared with separate tariffs for generation, transmission, and distribution) as well as the differentiation of end-consumer tariffs by consumer groups (do tariffs differ between groups and by how much?).
8. Specifically, the level of cross-subsidization is quantified here as the area between the 45-degree line and the curve defined by the cumulative shares of consumption and revenue by customer class. The formula is the same as that of a Gini-coefficient of inequality.

REFERENCES

Alleyne, T. S. C., and M. Hussain. 2013. *Energy Subsidy Reform in Sub-Saharan Africa: Experiences and Lessons*. Washington, DC: International Monetary Fund.

Di Bella, G., L. Norton, J. Ntamatungiro, S. Ogawa, I. Samake, and M. Santoro. 2015. "Energy Subsidies in Latin America and the Caribbean: Stocktaking and Policy Challenges." Working Paper 15, International Monetary Fund, Washington, DC.

Briceño-Garmendia, C., and M. Shkaratan. 2011. "Power Tariffs: Caught Between Cost Recovery and Affordability." Policy Research Working Paper 5904, World Bank, Washington, DC. doi:10.1596/1813-9450-5904.

Briceño-Garmendia, C., K. Smits, and V. Foster. 2009. *Financing Public Infrastructure in Sub-Saharan Africa: Patterns and Emerging Issues*. Washington, DC: World Bank Group.

Camos, D., R. Bacon, A. Estache, and M. M. Hamid. 2018. *Shedding Light on Electricity Utilities in the Middle East and North Africa: Insights from a Performance Diagnostic*. Washington, DC: World Bank.

Coady, D., I. Parry, L. Sears, and B. Shang. 2015. "How Large Are Global Energy Subsidies?" IMF Working Paper 105, International Monetary Fund, Washington, DC. doi:10.5089/9781513532196.001.

Eberhard, A., V. Foster, C. Briceño-Garmendia, F. Ouedraogo, D. Camos, and M. Shkaratan. 2008. "Underpowered: The State of the Power Sector in Sub-Saharan Africa." Background Paper 6, Africa Infrastructure Country Diagnostic, World Bank, Washington, DC.

Ebinger, J. O. 2006. *Measuring Financial Performance in Infrastructure: An Application to Europe and Central Asia*. Washington, DC: World Bank.

Foster, V., and T. Yepes. 2006. "Is Cost Recovery a Feasible Objective for Water and Electricity? The Latin American Experience." Policy Research Working Paper 3943, World Bank, Washington, DC.

Huenteler, J., I. Dobozi, A. Balabanyan, and S. G. Banerjee. 2017. "Cost Recovery and Financial Viability of the Power Sector in Developing Countries: A Literature Review." Policy Research Working Paper 8287, World Bank, Washington, DC.

IEA (International Energy Agency). 1999. "Looking at Energy Subsidies: Getting the Prices Right." World Energy Outlook 1999 Insights, IEA, Paris.

———. 2015. "Energy Subsidies: Fossil Fuel Subsidy Database." IEA, Paris.

Khurana, M., and S. G. Banerjee. 2013. *Beyond Crisis: The Financial Performance of India's Power Sector*. Washington, DC: World Bank.

Kojima, M. 2017. *"Identifying and Quantifying Energy Subsidies." Energy Subsidy Reform Assessment Framework (ESRAF) Good Practice Note 1*, World Bank, Washington, DC: World Bank Group.

Kojima, M., and C. Trimble. 2016. "Making Power Affordable for Africa and Viable for Its Utilities." World Bank, Washington, DC.

Mayer, K., S. Banerjee, and C. Trimble. 2015. *Elite Capture: Subsidizing Electricity Use by Indian Households*. Washington, DC: World Bank. doi:10.1596/978-1-4648-0412-0.

Munasinghe, M., J. Gilling, and M. Mason. 1989. *A Review of World Bank Lending for Electric Power*. Washington, DC: World Bank.

Saavalainen, T., and J. ten Berge. 2006. *Quasi-Fiscal Deficits and Energy Conditionality in Selected CIS Countries*. Washington, DC: International Monetary Fund.

Trimble, C., M. Kojima, I. Perez Arroyo, and F. Mohammadzadeh. 2016. "Financial Viability of Electricity Sectors in Sub-Saharan Africa: Quasi-Fiscal Deficits and Hidden Costs." Policy Research Working Paper 7788, World Bank, Washington, DC.

Vagliasindi, M., and J. Besant-Jones. 2013. "Power Market Structure: Revisiting Policy Options." World Bank, Washington, DC.

World Bank. 1972. "Operations Evaluation Report: Electric Power." Report 2-17, World Bank, Washington, DC.

———. 1990. "Review of Electricity Tariffs in Developing Countries during the 1980s." World Bank, Washington, DC.

———. 1993. "Tanzania: Sixth Power Project." Staff Appraisal Report, World Bank, Washington, DC.

———. 1994. "Ukraine: Electricity Market Development Project." Staff Appraisal Report, World Bank, Washington, DC.

———. 1995. "Vietnam: Power Sector Rehabilitation Project." Staff Appraisal Report, World Bank, Washington, DC.

———. 1998. "Senegal: Energy Sector Adjustment Operation." Staff Appraisal Report, World Bank, Washington, DC.

———. 2004. "Central Asia: Regional Electricity Export Potential Study." World Bank, Washington, DC.

———. 2016. "Financial Viability of the Electricity Sector in Developing Countries: Recent Trends and Effectiveness of World Bank Interventions." IEG Learning Product, World Bank, Washington, DC.

Did Power Sector Reform Deliver Better Sector Outcomes?

9

Guiding questions

- *Did the 1990s model deliver on its stated objective of improving security of supply through improved utility performance and increased investment?*
- *Did the reforms undertaken as part of the 1990s model help to support the wider policy goals of universal access and decarbonization that emerged in the 21st century?*
- *To what extent were countries that did not espouse the 1990s model able to deliver strong power sector outcomes by other means?*

Summary

- *While country-specific factors are the largest determinants of utility performance, there is evidence that regulation, restructuring, liberalization, and governance lead to significant improvements in some dimensions of efficiency for some groups of countries. When it comes to cost recovery, this is much more likely in countries that have introduced private sector participation.*
- *Although almost all private utilities attain high levels of distribution efficiency, their performance is matched by a substantial minority of public utilities.*
- *While market reforms can be helpful in improving the overall efficiency and financial viability of the power sector, and creating a better climate for investment, they cannot—in and of themselves—deliver on the social and environmental aspirations of the twenty-first century. Complementary policy measures are needed to direct and incentivize the specific investments that are needed.*
- *Progress on electrification was made by countries with widely differing approaches to power sector reform and appears to be driven primarily by targeted public investment as countries reach higher income levels. At the same time, progress (or otherwise) on decarbonization of the electricity sector has historically been an unintended by-product of measures to improve security of supply.*
- *Overall, although some of the deepest reformers delivered good sector outcomes, other countries with weaker starting conditions reformed without notable impacts on performance. Furthermore, a third group of countries showed that it was sometimes possible to achieve comparable performance by means of different institutional approaches.*

INTRODUCTION

This chapter evaluates the extent to which power sector reforms are associated with better power sector outcomes.[1] Reform is a means to an end, premised on the assumption that reform will lead to improved utility performance and better outcomes for citizens. It is therefore of fundamental importance to examine this hypothesis against the available evidence. Despite some statistical limitations in definitively determining the direction of causality, correlations between reform effort and various measures of outcomes, combined with more rigorous econometric analysis, shed some light on the underlying relationships. When this quantitative evidence is combined with the rich qualitative evidence underpinning the Power Sector Reform Observatory, some general inferences can be drawn. Thus, the guiding questions for this chapter are as follows.

Did reforms lead to improved utility performance on the key dimensions of cost recovery and operational efficiency? Did the 1990s model deliver on its stated objective of improving security of supply through increased investment? Did the reforms undertaken as part of the 1990s model help to support the wider policy goals of universal access and decarbonization that emerged in the 21st century? To what extent were countries that did not espouse the 1990s model able to deliver strong power sector outcomes by other means?

The 1990s power sector reform model was based on an implicit theory of change that linked reforms to improved utility performance and better sector outcomes (figure 9.1). The move toward commercially oriented utilities, ideally through private sector participation—or at least through restructured corporatized utilities—was expected to create incentives for more efficient decision making, which would be reinforced by competition where possible, or incentive-based regulation where necessary. At the same time, competition or regulation or both would create the required environment for cost-recovery pricing, which would itself become easier to achieve thanks to greater efficiency. The resulting improvement in financial viability would make it possible for utilities to attract the larger volume of investment needed to attain better sector outcomes.

FIGURE 9.1 The theory of change underpinning the 1990s power sector reform model

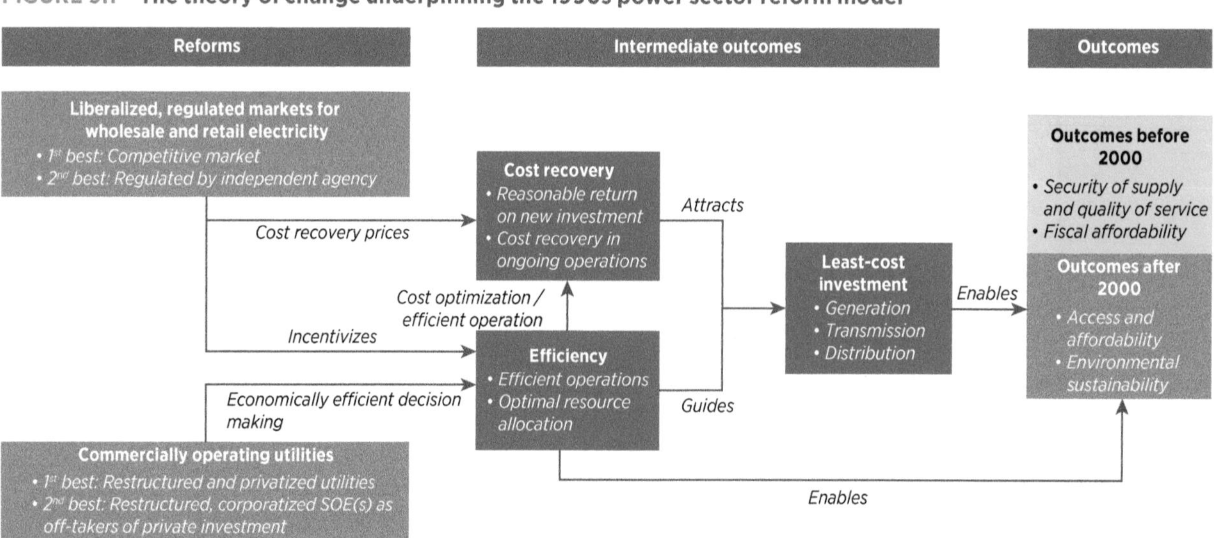

Source: World Bank elaboration.
Note: SOE = state-owned enterprise.

In the late 20th century, the main outcome of interest was security of supply; by the early 21st century, policy objectives had broadened to encompass electrification and decarbonization. Under the 1990s power sector reform model, the main purpose of attracting investment was to expand generation capacity so that supply could keep pace with rapidly growing demand in the developing world, and, to a lesser extent, also strengthen related transmission and distribution infrastructure. Although some countries (such as Morocco and Vietnam) had already chosen to pursue ambitious electrification programs in the late 20th century, this was never an explicit emphasis of the 1990s model, nor of the wider contemporaneous development agenda as articulated in the Millennium Development Goals. It was only with the adoption of the Sustainable Development Goals (SDGs) in 2015, and the associated universal access target for 2030 (SDG7.1.1), that electrification became more prominent on the political agenda and widely embraced across the developing world. Similarly, the decarbonization of the energy sector was widely adopted as a political goal only after the Paris Climate Agreement of 2015. It was also reflected in the SDGs in the form of explicit targets for renewable energy penetration and acceleration of energy efficiency (SDG7.2 and SDG7.3).

Care therefore needs to be taken when evaluating the performance of the 1990s sector reform model against electrification and decarbonization goals. It is certainly legitimate to evaluate the 1990s sector reform model against the achievement of its intended objectives of improving utility performance and security of supply. It does not seem fair, however, to judge the *past performance* of the 1990s model in terms of whether or not it supported delivery of the later policy goals of electrification and decarbonization. Nevertheless, it is relevant to consider to what extent the 1990s model was at least compatible with such goals, because this factor is key in determining its continued relevance to policy makers.

A mixture of quantitative and qualitative methods will be used to explore the effects of the 1990s model on intermediate and final sector reform outcomes. In evaluating the impact of the 1990s model, a suite of complementary approaches will be used (see box 9.1 for a methodological explanation). First, the most rigorous statistical analysis to be used is based on cross-country panel data econometrics covering a sample of 88 developing countries over the 20 years between 1995 and 2015. Because of data limitations, this analysis, however rigorous, is necessarily based on relatively simplistic measures of both reform effort and sector outcomes. Second, narrowing our focus to the 2015 cross-section of performance of the 15 observatory countries allows us to use a much richer set of variables that more effectively capture the quality of the institutional framework and the nature of the associated outcomes.[2] This comes at the cost of employing more simplistic statistical methods, including simple cross-sectional regressions and scatter plots, that cannot pretend to do more than identify patterns of correlation (as opposed to causation).[3] Third, both types of statistical analysis are complemented by qualitative information on the reform stories and associated outcomes from the 15 observatory countries. Going behind the quantitative analysis to look at the qualitative stories that underlie the observed outcomes enables us to gain some insights into patterns of causality.

KEY FINDINGS

The relationship between specific dimensions of power sector reform and resulting intermediate outcomes for utility performance in terms of efficiency and cost recovery are explored in findings #1–5 and correspond to the left-hand loop of figure 9.1. The latter part of the discussion, surrounding findings #6–9, examines the impact of reforms on final outcomes in terms

BOX 9.1 The methodological challenges of inferring the impact of power sector reforms

For purposes of the impact analysis presented in this chapter, two samples of countries are considered: a small one and a large one, the former being nested within the latter (see table 9.1 in the main text of the chapter for full details).

The *small sample* comprises 17 economies from the Power Sector Reform Observatory. The information available for this sample includes a very detailed characterization of the depth of power sector reform measures developed through a rich variety of indexes covering restructuring, privatization, utility governance, regulation, and competition that aggregate information from several hundred specific dimensions of the reform environment. These indexes were developed and presented in chapters 4–7. This detailed information is available only for 2015.

The *large sample* comprises 88 developing countries from the Regulatory Indicators for Sustainable Energy (RISE) project (Banerjee and others 2017). The information available for this sample includes a long-time series dating back to 1990, but for a much more basic set of variables than those available in the small sample. The descriptors of the reform process are significantly cruder and follow the approach described in chapter 2. For example, regulatory reform for this larger sample is represented simply as a dummy variable denoting the presence or absence of a regulatory agency, whereas for the smaller sample some 50 dimensions of the regulatory framework are quantified. The availability of data on sector outcomes is also considerably more limited.

The analysis first examines the impact of reforms on intermediate outcomes, including efficiency and cost recovery. Efficiency is captured both at the generation and distribution level and is available for both the small and the large samples.

- *Generation efficiency,* defined as the average thermal efficiency of combustion in power plants, can be expected to improve over time either through greater efficiency in plant dispatch or through shifts in the generation portfolio toward more modern plants and more efficient technologies.
- *Distribution efficiency,* defined as the percentage of the revenues of an efficient utility that are captured by the utility, expresses the shortfall attributable to both technical and commercial losses. (When this variable is not available, as in the case of the larger sample, system losses are used as a proxy for distribution efficiency.)

Cost recovery is tracked at both the operating and the full (capital) cost recovery levels on the basis of the framework presented in chapter 8. This rich information is available only for the small sample of 15 countries during the period 2010–15.

- *Operating cost recovery* (A1) captures the ratio of average billed revenue to the average operating costs borne directly by the utility (leaving aside any subsidized items).
- *Full cost recovery* (C3) captures the ratio of average billed revenue to the average total cost, including all operating costs and associated operating subsidies, as well as all capital costs and associated capital subsidies required to fund historic investments as well as the utility's forward-looking five-year business plan.

Various performance measures are used to track the extent to which final outcomes in the sector improved over time (table 9.2 in the main text of the chapter offers full details). Measurement of *security of supply* is notoriously difficult, both conceptually and empirically. The ultimate measure of interest to citizens is reliability, and this can be captured through the SAIFI (System Average Interruption Frequency Index), which is available only for 2015. Capacity-based measures suffer from fewer data limitations, and an often-used measure is the reserve margin (or ratio of peak demand to installed capacity). The use of this measure is problematic for cross-country comparisons, because the magnitude of the required reserve margin will vary across countries according to legitimate differences in their power sector structure, related particularly to renewable energy shares. Instead, several other capacity-based measures are calculated, in recognition of the fact that no single one of them is ideal.

- The *normalized capacity* measure simply facilitates meaningful comparsons of how much capacity is available across countries and over time by normalizing against population.
- *Capacity diversification* captures another aspect of security of supply by looking at the index of concentration across different types of generation technologies to examine to what extent countries are exposed to shocks affecting any particular energy source.
- The *capacity growth* indicator looks at the ratio of peak demand growth to capacity growth to give a sense of whether supply and demand are keeping pace.

As regards environmental sustainability, one popular measure used to track Sustainable Development Goal (SDG) 7.2 is the share of *modern renewables* in total final energy consumption. Nevertheless, this measure only partially captures the carbon footprint of the energy sector, which is also affected by the different types of fossil fuel generation that may be in use, as well as the presence of nuclear energy. To capture this wider perspective on decarbonization, a second indicator is used, which is the *carbon intensity* of electricity generation.

As regards social inclusion, *electrification* is comparatively easy to measure through the household access rate used to track progress on SDG7.1. For the year 2015, it can be complemented by a measure of *affordability,* measured as the percentage of the budget of those in the bottom 40 percent of the income distribution that is needed to purchase the average level of electricity consumption for the country at the prevailing tariff structure.

For each indicator, performance is benchmarked against normative bands. On the basis of benchmarking across the sample and with reference to international norms, threshold values are designated for each indicator in terms of what describes a good (green), mediocre (yellow), or poor (red) performance (see table 9.2 in the main text). In a similar way, threshold values are defined for what constitute a good (green), mediocre (yellow), or poor (red) performance improvement over time.

(*Box continued next page*)

BOX 9.1 The methodological challenges of inferring the impact of power sector reforms (*Continued*)

Two types of statistical analysis are performed. For the large sample, it is possible to perform cross-country panel data econometrics. Owing to the relatively large sample (some 1,760 observations resulting from the panel of 88 countries over the 20 years between 1995 and 2015), each outcome measure can be regressed upon indicators of the adoption of the four major reform measures (regulation, restructuring, privatization, and competition). Several different control variables are also explored, including a time trend, country fixed effects, and, as conditioning variables, power-system size and the country's income per capita. The inclusion of country fixed effects goes some way toward addressing concerns about potential reverse causality from outcomes to reforms. Full statistical results of these models are reported in annex 9B, whereas the main highlights are incorporated into the text. In general, the explanatory power of the regressions increases dramatically (with R-squared increasing from under 0.10 to over 0.50) when country fixed effects are introduced into the specification, underscoring the importance of context in determining reform outcomes.

For the small sample, simple cross-sectional regressions are performed. Because of the limited number of observations, individual regressions are performed for each outcome variable against each reform measure, always controlling for power system size and the country's income per capita. Where relevant, these regressions are further clarified through visual cross-plots and qualitative analysis, with a particular focus on explaining outliers. Outlier analysis is also performed to check upon the robustness of the visual cross-plots.

TABLE 9.1 Mapping of data availability across small and large country samples

	Small sample	Large sample
Source	Power Sector Reform Observatory	Regulatory Indicators for Sustainable Energy (RISE)
Number of countries	15	88
Basic reform indexes		Restructuring, regulation, competition, privatization, for period 1990–2015
Refined reform indexes	Utility governance, restructuring, competition, regulation, private sector participation, 2015 only	None
Intermediate outcomes		
Generation efficiency	Thermal efficiency of power generation (the percentage of energy content of fuel that is turned into electricity)	Thermal efficiency of power generation (the percentage of energy content of fuel that is turned into electricity)
Distribution efficiency	Combined measure of system losses and commercial efficiency at the utility level for 2010–17, plus prereform benchmark for system losses	Average national-level system losses since 1990
Cost recovery	Wide range of operating and capital cost recovery measures for 2010–17, plus prereform benchmark	None
Final outcomes		
Normalized capacity	1990–2015	1995–2015
Capacity diversification	1990–2015	None
Capacity growth	2010–17	None
Reliability of supply (SAIFI)	2015 only	2015 only
Fiscal sustainability (QFD)	2010–17	None
Electrification rate	1990–2015	1995–2015
Carbon intensity of electricity	1990–2015	1995–2015
Modern renewable energy share of TFEC	1990–2015	1995–2015

Source: Rethinking Power Sector Reform Observatory.
Note: QFD = quasi-fiscal deficit; SAIFI = System Average Interruption Frequency Index; TFEC = total final energy consumption.

TABLE 9.2 Framework for evaluating impact on final sector performance outcomes

	Security of supply				Social inclusion		Environmental sustainability	
	Reliability of supply	Normalized capacity	Capacity diversification	Capacity growth	Access	Affordability	Carbon intensity	Modern renewable energy share
Current performance, 2015								
Good	SAIFI up to 12	Capacity greater than 200 MW per million population	Concentration index up to 0.33	Capacity growth to peak growth >1	Electrification rate exceeds 80%	Bottom 40% spend <5% of GNI per capita on average consumption	Carbon intensity up to 250 gCO_2/kWh	Over 10% of TFEC
Moderate	SAIFI between 13 and 52	—	Concentration index between 0.34 and 0.66		Electrification rate of 60–80%	Bottom 40% spend 5%–10% of GNI per capita on average consumption	Carbon intensity of 250–500 gCO_2/kWh	5–10% of TFEC
Poor	SAIFI greater than 52	Capacity less than 200 MW per million population	Concentration index exceeds 0.66	Capacity growth to peak growth <1	Electrification rate up to 60%	Bottom 40% spend > 10% of GNI per capita on average consumption	Carbon intensity exceeds 500 gCO_2/kWh	<5% of TFEC
Performance Improvement, 1990–2015								
Good	n.a.	Capacity growth exceeds prereform capacity	Concentration index dropped by more than 0.1	n.a.	Electrification rate rises by over 25%	n.a.	Carbon intensity fell by over 100 gCO_2/kWh	Grew by >5 percentage points
Moderate	n.a.	Capacity growth is 50–100% of prereform	Concentration index dropped by less than 0.1	n.a.	Electrification rate rises by 12–25%	n.a.	Carbon intensity fell by up to 100 gCO_2/kWh	Grew by <=5 percentage points
Poor	n.a.	Capacity growth is <50% of prereform capacity	Concentration index increased	n.a.	Electrification rate rises by up to 12%	n.a.	Carbon intensity increased	Decreased

Note: gCO_2/kWh = grams of carbon dioxide per kilowatt hour; GNI = gross national income; MW = megawatt; SAIFI = System Average Interruption Frequency Index; TFEC = total final energy consumption; n.a. = not applicable; — not available.

of security of supply, social inclusion, and environmental sustainability. These correspond to the right-hand loop of figure 9.1.

Finding #1: Country context is the largest long-term driver of utility performance, but regulation seems to be a consistent contributor to improved outcomes, particularly in low-income countries

Econometric analysis underscores the major role of specific country conditions in explaining the level of utility performance. As noted above, the panel data regression provides a statistically rigorous overview of relationships between high-level reform indicators and simple measures of sector efficiency. Five different model specifications are considered, and they vary according to whether control variables are used, and whether time trends and country fixed effects are added to the regression. The statistical analysis shows that the results are quite sensitive to the model specification in many cases (see annex 9B). In general, the incorporation of country fixed effects hugely improves the overall explanatory power of the regressions, indicating that local specificities appear to have a much larger impact on power sector outcomes than the extent of power sector reform. For this reason, attention in table 9.3 will focus on the specification incorporating both time trend and country fixed effects. In addition to estimating the model on the full sample of 88 countries, two separate nested models are estimated for the low-income and middle-income countries of the sample, respectively. Statistical tests reject the hypothesis of no structural differences between these subcategories of countries, indicating that coefficients are significantly different between the two and hence that there is value in estimating separate models.

When it comes to system losses, regulation and restructuring reforms seem to have the strongest impact. They are the only two reform actions that remain statistically significant when country fixed effects are introduced into the model and display the correct negative sign. (Although only reported in the annex 9B, table 9B.1, private sector participation appears to have a large impact on system losses under some specifications; however, these losses disappear once country fixed effects are added, suggesting that the impact of private sector participation is subsumed within the country context.) In terms of magnitude, the impact of regulation and restructuring on system losses is only moderate, because introducing a regulatory agency or undertaking full vertical and horizontal unbundling of the sector is associated with just a two-percentage-point reduction in system losses. Disaggregation of the sample between low- and middle-income countries indicates that the impact of regulation and restructuring on system losses is primarily a low-income country phenomenon, with statistical significance disappearing in the middle-income sample. Particularly striking is the larger magnitude of the coefficient for restructuring in

TABLE 9.3 Impact of reform measures on intermediate outcomes for the large sample

Regression coefficients	Full sample	Low-income countries	Middle-income countries
Dependent variable: system losses			
Regulation	-2.167*	-2.652*	~
Restructuring	-2.024**	-7.362***	~
Competition	~	~	~
Private sector participation	~	~	~
Observations	1,358	635	720
R-squared	0.557	0.502	0.593
Dependent variable: generation efficiency			
Regulation	1.367***	1.423**	1.890***
Restructuring	~	~	~
Competition	~	4.201*	~
Private sector participation	~	~	~
Observations	1348	622	722
R-squared	0.678	0.613	0.764

Note: Ordinary least squares regressions on panel data set of 88 countries for the period 1995–2015. A separate regression is estimated for each of the intermediate outcomes on the full set of reform variables capturing regulation, restructuring, competition, and privatization. The analysis considers a number of different ways of capturing control variables in terms of time trend, country-specific fixed effects, and control variables capturing system size and GDP per capita. The preferred specification, incorporating time trend and country-specific fixed effects (but no control variables), is reported here. All the other specifications can be viewed in the annex (for complete econometric results refer to annex 9B, tables 9B.1 and 9B.2).
Significance level: *** = 1 percent, ** = 5 percent, * = 10 percent, ~ = not significant.

the low-income country sample, suggesting that full vertical and horizontal unbundling is associated with system losses that are seven percentage points lower than full vertical integration.

When it comes to generation efficiency, regulation and private sector participation seem to have the strongest impact. Although all reform measures are statistically significant in the simpler model specifications, once again most of these effects disappear once country fixed effects are introduced. Regulation is the one reform that seems to have a consistently positive impact on generation efficiency, with the introduction of a regulatory agency associated with an increase of about two percentage points in the thermal efficiency of generation plants. (In some specifications, reported only in annex 9B, tables 9B.1 and 9B.2, private sector participation is also significantly related to generation efficiency, with a similar order of magnitude in terms of the size of the effect.) The impact of regulation on generation efficiency is consistent in significance and in size between the low-income and middle-income subsamples. Additionally, the introduction of competition is found to have a large impact on generation efficiency for the specific case of low-income countries, where the effect can be up to four percentage points.

Finding #2: Certain elements of utility governance—associated with financial discipline and human resource management—are associated with improved cost recovery and efficiency

The association between utility governance and utility performance is particularly strong for certain aspects of governance (table 9.4). The panel data analysis reported in table 9.3 was not able to capture the issue of utility corporate governance, because of a lack of time series variables on this issue. Focusing attention on the small sample makes it possible to explore how utility governance interacts with utility performance. The Utility Governance Index is built up from a series of subindexes that capture different components of governance, including corporate governance of the board and quality of utility management in terms of financial discipline, human resources, and adoption of information technology (recall chapter 4). Statistical analysis at the subindex level helps to pinpoint which might be the most critical aspects of utility corporate governance from a performance perspective. The analysis finds statistically significant relationships between managerial practices relating to financial discipline and human resources, as well as various measures of

TABLE 9.4 Impact of utility governance and intermediate outcomes for the 15-country sample

Regression coefficients	Corporate governance		Managerial practices		
Reforms	Autonomy	Accountability	Financial discipline	Human resources	Information and technology
Operating cost recovery	-	-	0.995**	1.182**	-
Full capital cost recovery	-	-	-	0.857*	-
Distribution efficiency	-	-	0.352**	-	-
Generation efficiency	-	-	-	-	-

Note: Ordinary least squares multivariate regression analysis for the Rethinking Power Sector Reform country sample (17 economies). The table reports the results of a series of cross-sectional regressions modeling the impact of each individual utility level governance reform measure (board autonomy, board accountability, financial discipline, human resource management, and information technology) on each individual intermediate outcome (cost recovery, distribution, and thermal efficiency). GDP per capita and system size have been used as control variables. For complete results refer to annex 9C, table 9C.3. Significance level: *** = 1 percent, ** = 5 percent, * = 10 percent, - = not significant.

efficiency and cost recovery (table 9.4), but not for measures relating to the autonomy and accountability of the utility's board, or for the adoption of information technology.

Stronger financial discipline is associated with improvements in operating cost recovery and distribution efficiency. It makes sense that tighter financial discipline would lead to improved operating cost recovery and distribution efficiency. The analysis shows relatively high correlations between financial discipline and operating cost recovery (R-squared of 0.40) and between financial discipline and distribution efficiency (R-squared of 0.30) (figure 9.2). The effects of governance on cost recovery are relatively material in size, with a one-percentage-point increase in the respective governance indexes associated with roughly one additional percentage point in operating cost recovery. Drilling deeper and looking at the practices of financial discipline that underpin the index sheds further light on the drivers of improved performance (figure 9.3). Comparing countries with good and poor performance on operating cost recovery reveals major differences in terms of whether they prepare accounts in compliance with international financial reporting standards (100 percent of countries with higher operating cost recovery versus 21 percent of the others), whether they have the liberty to issue equity (45 versus 14 percent), whether they are required to pay dividends to shareholders (67 versus 0 percent), and whether they publicly disclose their accounts (100 versus 57 percent). Similarly, comparing countries with good and poor performance on distribution efficiency reveals that the former are much more likely to have clearly defined public service obligations (69 versus 17 percent), financial accounts produced in compliance with international financial reporting standards (81 versus 25 percent), and the requirement to pay dividends to shareholders (50 versus 0 percent).

Examples from Sub-Saharan Africa serve to illustrate this point. Kenya and Uganda are two countries whose utilities score relatively highly in terms of financial discipline and that are also doing well on distribution efficiency and

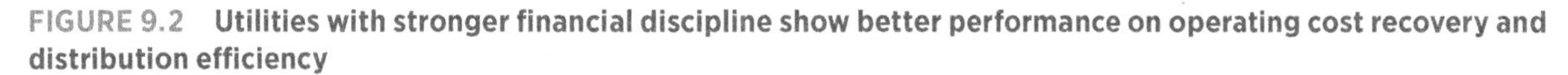

FIGURE 9.2 Utilities with stronger financial discipline show better performance on operating cost recovery and distribution efficiency

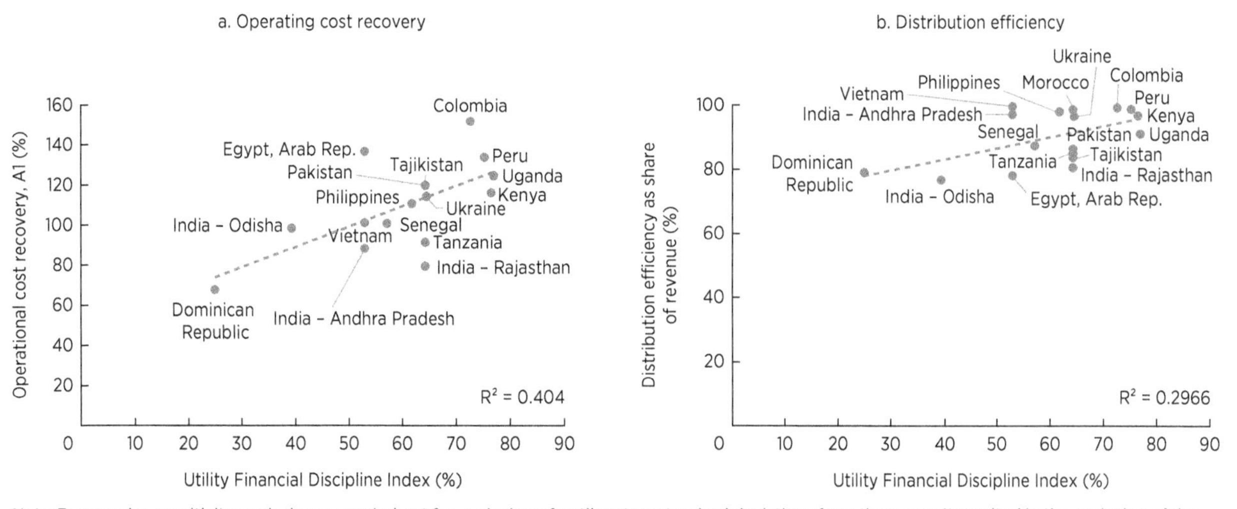

Note: For panel a, sensitivity analysis was carried out for exclusion of outliers two standard deviations from the mean. It resulted in the exclusion of the Dominican Republic and the R^2 of 0.2116, and does not materially affect the analysis. For panel b, a sensitivity analysis was carried out for exclusion of outliers two standard deviations from the mean. It resulted in the exclusion of the Dominican Republic and the R^2 of 0.2069 and does not materially affect the analysis. A1 = operating cost recovery.

FIGURE 9.3 **Certain financial discipline practices are much more widely practiced among utilities that perform well in terms of operating cost recovery and distribution efficiency**

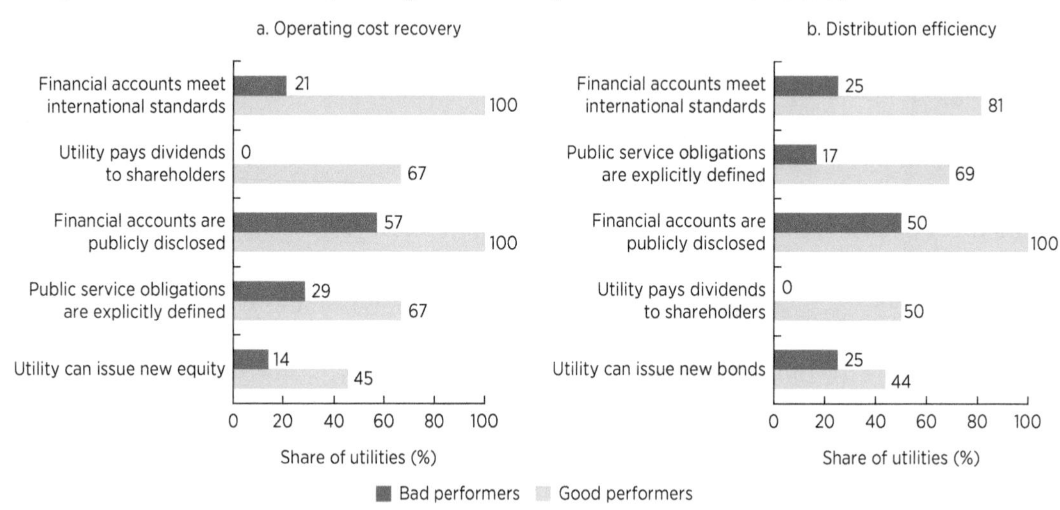

operating cost recovery. Although these two countries adopted different reform paths, each of them incorporated comparable measures to boost the financial discipline of the utility. In the case of Uganda, the national distribution company was awarded as a concession to a private operator, introducing the stronger commercial orientation of a private sector board and management. Additionally, the listing of the utility on the stock exchanges in Kampala and Nairobi makes the financial reporting more transparent. In the case of Kenya, the utility remains under majority public ownership, with a 49 percent stake floated on the Nairobi Stock Exchange. This situation introduces a series of requirements in terms of the quality of financial reporting and the degree of transparency that serve to enhance financial discipline.

Stronger human resource management is associated with improvements in operating cost recovery and capital cost recovery. The analysis shows relatively high correlations between human resource management and operating cost recovery (R-squared of 0.50), and between human resource management and full capital cost recovery (R-squared of 0.42) (figure 9.4). Drilling deeper into the practices of human resource management underpinning the index puts the drivers of improved performance into greater relief (figure 9.5). Comparing countries with good and poor performance on operating cost recovery, one sees major differences in terms of the following human resource practices: good employee performance awarded with bonuses (91 versus 50 percent); transparency in hiring of staff (92 versus 57 percent); and freedom to fire employees for bad performance (100 versus 57 percent). Similarly, comparing countries with good and poor performance on full capital cost recovery reveals major differences in freedom to fire (75 versus 14 percent) or hire employees (50 versus 5 percent) and having no link to government pay scales (75 versus 43 percent) or public employment regulations (50 versus 22 percent).

Finding #3: The quality of the legislated regulatory framework for tariffs has no real impact on cost recovery, but the quality of implementation of that regulatory framework does seem to have an effect

A closer look at the components of regulation suggests that it is the way that tariff and quality

FIGURE 9.4 Utilities with better human resource management show better performance on operating cost recovery and full capital cost recovery

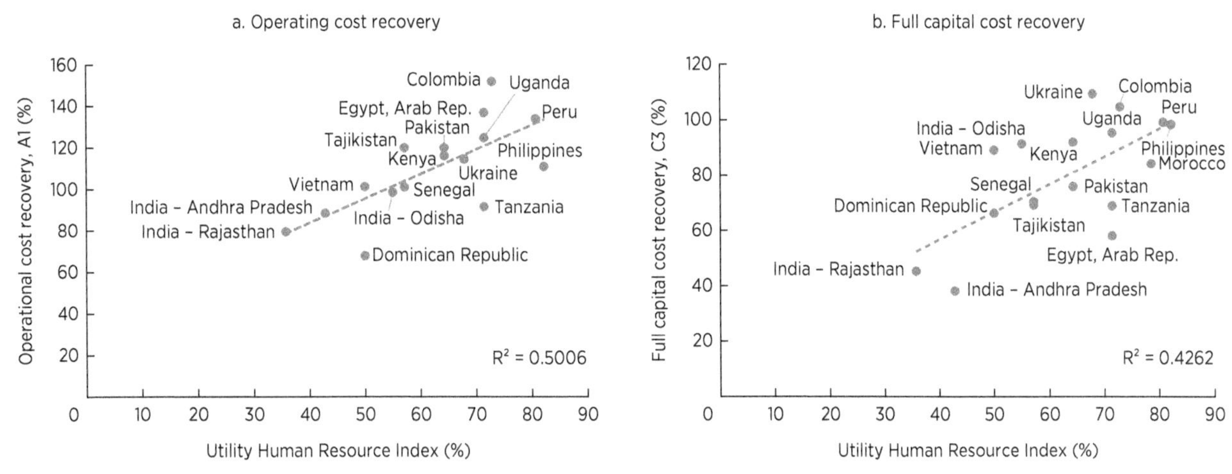

Note: For panel a, a sensitivity analysis was carried out for exclusion of outliers two standard deviations from the mean. It resulted in the exclusion of the Indian state of Rajasthan and the R^2 of 0.4236 and does not materially affect the analysis. For panel b, a sensitivity analysis was carried out for exclusion of outliers two standard deviations from the mean. It resulted in the exclusion of the Indian states of Andhra Pradesh and Rajasthan and the R^2 of 0.1372 and does not materially affect the analysis. A1 = operating cost recovery; C3 = full cost recovery.

FIGURE 9.5 Certain financial human resource practices are much more widely practiced among utilities that perform well in terms of operating cost recovery and full capital cost recovery

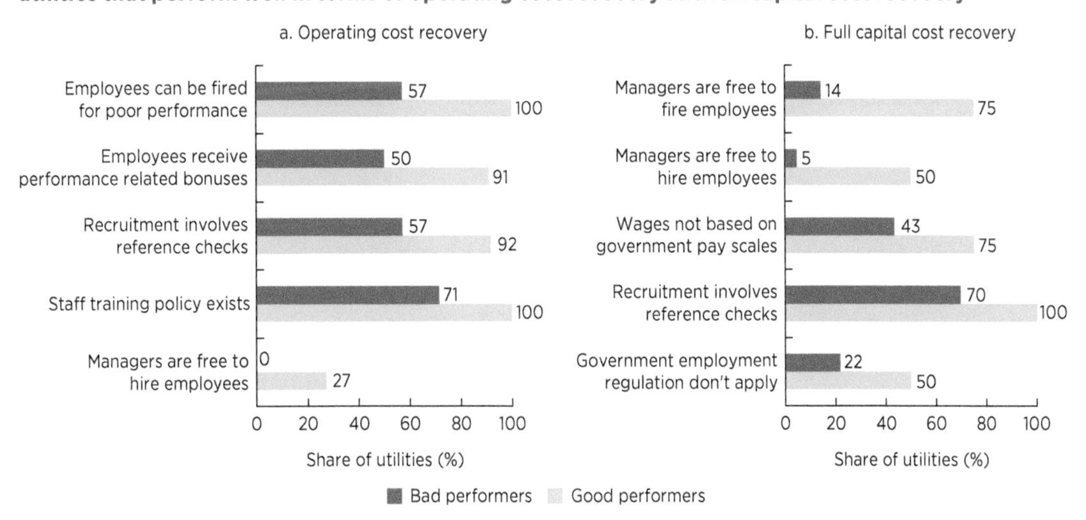

regulation are actually practiced that matters the most. The panel data regression analysis presented in table 9.3 was able to capture regulation only in terms of a single dummy variable denoting the date when a regulatory agency was created. Confining attention to the small sample, there is a much richer characterization of the regulatory environment, which serves to explore how the quality of regulatory framework affects utility performance. This regulatory index is built up from a series of subindexes that capture different components of regulation, including regulatory governance and the substantive aspects of regulating tariffs, quality, and market entry (recall chapter 6). Statistical analysis at the subindex level helps to pinpoint which might be the most critical aspects of regulation from a performance perspective. This exercise reveals that it is the substantive aspects of tariff and quality regulation—rather than overall regulatory governance considerations—that are most strongly associated with outcomes for efficiency and cost recovery (table 9.5). It also shows little association between de jure regulation and utility performance; rather it is perceived regulatory practice that is associated with better outcomes.

Cost recovery is strongly associated with the actual practice of tariff regulation. Because achieving cost recovery is the purpose of tariff regulation, it makes sense that the two would be associated. The visual cross-plots serve to contrast the impact of regulation on paper versus regulation in practice. The relationship between tariff regulation on paper and cost recovery is very weak, with an R-squared of just 0.03 (figure 9.6). The reason is that almost every

TABLE 9.5 Impact of regulatory reforms on intermediate outcomes for small sample

	De jure			Perceived		
Regression coefficients	**Regulatory governance**	**Quality regulation**	**Tariff regulation**	**Regulatory governance**	**Quality regulation**	**Tariff regulation**
Operating cost recovery	~	~	~	~	~	0.466**
Full capital cost recovery	~	~	~	~	0.389*	~
Distribution efficiency	~	~	~	~	0.181*	~
Generation efficiency	~	~	~	~	~	~

Note: Ordinary least squares multivariate regression analysis for the Rethinking Power Sector Reform country sample (17 economies). The table reports the results of a series of cross-sectional regressions modeling the impact of each regulatory reform measure (regulatory governance, tariff regulation, and quality regulation) on intermediate outcomes (various measures of efficiency and cost recovery). GDP per capita and system size have been used as control variables. For complete results refer to annex 9C, table 9C.4.
Significance level: *** = 1 percent, ** = 5 percent, * = 10 percent, ~ = not significant.

FIGURE 9.6 Operating cost recovery bears little relation to the quality of tariff regulations on paper, but it is closely associated with the quality of tariff regulations in practice

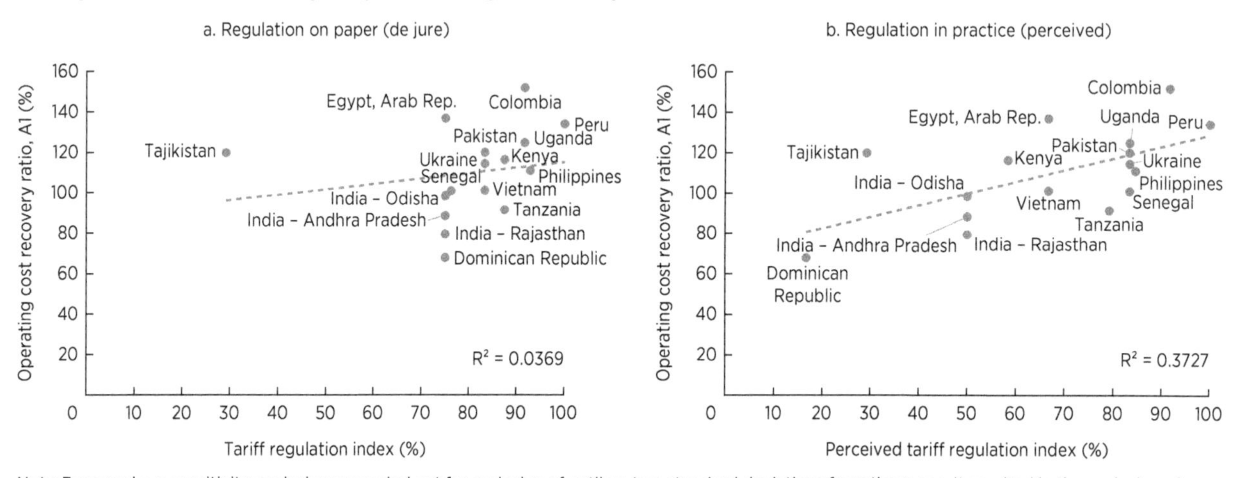

Note: For panel a, a sensitivity analysis was carried out for exclusion of outliers two standard deviations from the mean. It resulted in the exclusion of Tajikistan and the R^2 of 0.3554 and does not materially affect the analysis. For panel b, a sensitivity analysis was carried out for exclusion of outliers two standard deviations from the mean. It resulted in the exclusion of Tajikistan and the R^2 of 0.5529 and does not materially affect the analysis. A1 = operating cost recovery.

country has a well-written regulatory framework for tariffs, as can be visualized in the cross-plot. This makes sense because sound regulations can simply be drafted at the outset of a reform by lifting existing good practices from elsewhere, but their existence does not translate into any guarantee that such good principles will be followed in practice. By contrast, the relationship between tariff regulation as practiced and cost recovery is much stronger with an R-squared of 0.37 (figure 9.6). To put this result in perspective, a one-percentage-point increase in the index for the quality of perceived tariff regulation is associated with a half-percentage-point increase in operating cost recovery. Digging a little deeper, countries that achieve operational cost recovery have particular attributes of the regulatory framework that distinguish them from the rest, including the adoption of regulatory accounting standards (91 percent of cost-recovering countries versus 0 percent of those that do not achieve operational cost recovery), a clear definition of cost recovery in the regulatory framework (100 percent versus 17 percent), and clearly articulated principles for the setting of end-user tariffs (91 percent versus 17 percent).

These results are also consistent with what is known about the actual practice of tariff regulation in the countries concerned. For example, the Dominican Republic was a relatively late reformer and adopted a sophisticated regulatory framework for tariff setting that was able to incorporate good practice lessons from earlier Latin American reformers. Shortly after the adoption of the tariff framework came an oil price shock that had a major impact on the costs of the utilities that were heavily reliant on oil-fired power generation, costs that—according to the regulatory framework—should have been passed on in consumer prices. Nevertheless, because of political concerns, tariffs remained frozen for several years. The resulting financial distress to the privatized utilities eventually led to their renationalization. A similar story can be told for the Indian state of Rajasthan, where a modern regulatory framework was introduced as required by federal regulation. Although the framework was respected for several years, a change of government eventually led to tariffs being frozen for as long as 10 years in contradiction of the regulatory framework. This weaker perceived performance of regulation for the Dominican Republic and India (Rajasthan) is shown by the fact that both jurisdictions appear much farther to the left in figure 9.6, panel b (perceived regulation) than in panel a (de jure regulation).

Distribution efficiency is much more closely associated with the practice of quality regulation than the practice of price regulation. In principle, tariff regulation—particularly when it is incentive-based—is expected to lead to greater operational efficiency among utilities. Despite some association between the practice of tariff regulation and distribution efficiency of the utilities (with an R-squared of 0.13) (figure 9.7, panel a), the relationship with the practice of quality regulation is much stronger (with an R-squared of 0.33) (figure 9.7, panel b). Once again, what might be at play is a reverse causality with utilities performing at high levels of operational efficiency being better placed to respond to quality of service regulation. Digging deeper, we find that the countries with more efficient utilities tend to share certain aspects of quality regulation that separate them from the underperformers. These aspects include the following: financial incentives exist for utilities to meet customer service standards or increase customer satisfaction (75 percent of the more efficient utilities versus 22 percent of the others); utilities are required to use an automated information management system to measure the quality or reliability of the power supply (63 percent versus 11 percent); and utilities are fined for failing to meet quality of service standards (50 percent versus 0 percent).

FIGURE 9.7 Distribution efficiency follows the practice of quality regulation more than the practice of tariff regulation

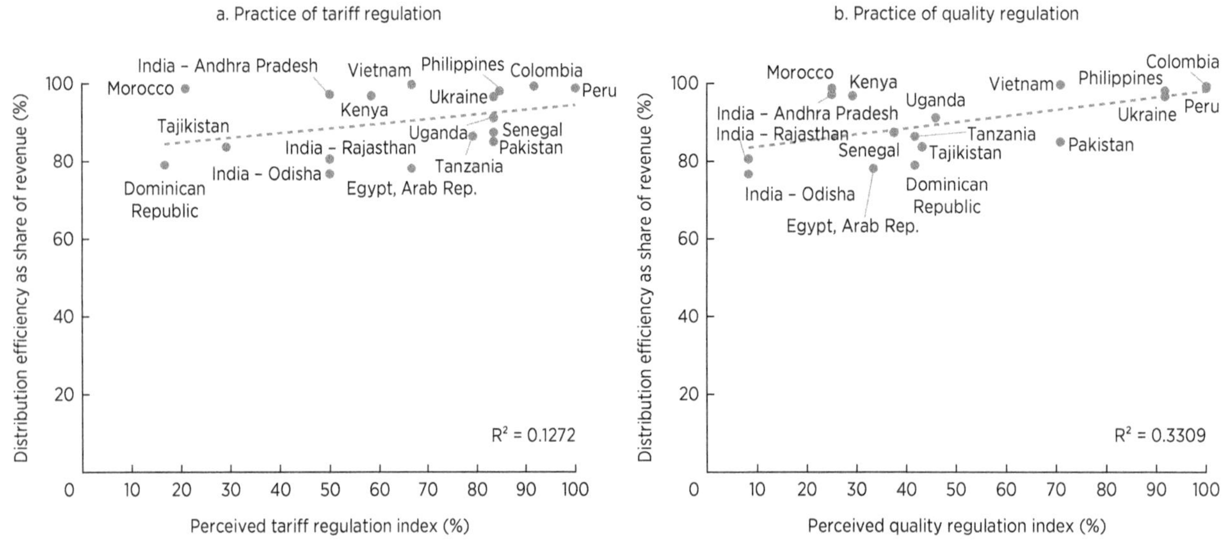

Note: A sensitivity analysis was carried out for exclusion of outliers two standard deviations from the mean, but it found no outliers.

Finding #4: Private sector participation is much more strongly associated with full capital cost recovery than any other reform, but this finding may simply reflect that one is a precondition for the other

The relationship between private sector participation and full cost recovery potentially runs in both directions. The lack of availability of time series data for cost recovery did not make it possible for the panel data regression analysis reported in table 9.3 to examine the relationship between private sector participation and cost recovery. This relationship could potentially run in both directions: from higher private sector participation to greater cost recovery or from greater cost recovery to higher private sector participation. On the one hand, countries that have private sector participation find it easier to commit politically to tariffs that recover full capital costs, because of the presence of a nongovernment entity in the sector. On the other hand, the private sector is unlikely to take much interest in entering a power sector where full capital cost recovery has not been achieved; however, the story may be different for generation and distribution. When the private sector comes into generation, full cost recovery—though desirable—is not essential. Private generators do not typically sell to end-consumers but to distribution companies and are willing to do so even if these off-takers have not achieved cost recovery—as long as adequate contractual protections and credit enhancements can be put in place. In the distribution segment, by contrast, private operators will depend entirely on consumer revenues; therefore, privatization is unlikely to go ahead unless tariffs are relatively close to full capital cost recovery levels.

Almost all countries with significant private sector participation in the distribution sector have reached close to full capital cost recovery (table 9.6). There is a strong positive association between private sector participation and full capital cost recovery, with an R-squared coefficient of 0.49. Countries scoring at least 40 percent on the private sector participation index, typically have the private

TABLE 9.6 **Impact of private sector participation on intermediate outcomes for small sample**

	Private sector participation			
Regression coefficients	**Overall**	**Generation**	**Distribution**	**Transmission**
Operating cost recovery	~	~	~	~
Economic cost recovery	0.77***	~	0.63***	0.38**
Distribution efficiency	0.29**	~	0.20**	0.17**
Thermal efficiency	~	~	~	~

Note: Ordinary least squares multivariate regression analysis for the Rethinking Power Sector Reform country sample (17 economies). The table reports the results of a series of cross-sectional regressions modeling the impact of private sector participation (overall, in generation, in distribution, and in transmission) on intermediate outcomes (various measures of efficiency and cost recovery). GDP per capita and system size have been used as control variables. Significance level: *** = 1 percent, ** = 5 percent, * = 10 percent, ~ = not significant.

sector active not only in generation but also in distribution. These countries include Colombia, Peru, Philippines, Uganda, and Ukraine, all of which are close to full capital cost recovery (figure 9.8). This pattern becomes even clearer in figure 9.9, which presents results at the utility level as opposed to the country level.

Unlike their publicly owned counterparts, private utilities almost universally operate in jurisdictions with full capital cost recovery—except for those where privatization has been reversed. The cost recovery ratio for the publicly owned utilities appears on the left-hand axis, whereas that for the private utilities appears on the right-hand axis. Several countries have utilities that appear on both the public and private sides of the ledger (such as Pakistan and Peru). The public utilities exhibit a wide range of performance on cost recovery, ranging from 30 percent to 90 percent. Some of the public utilities scoring highly on cost recovery hail from jurisdictions where public and private utilities coexist under a common framework for tariff regulation (India [Odisha], Pakistan, and Peru) although this is not the case for Morocco's Office National de l'Electricité et de l'Eau Potable (ONEE) and Vietnam Electricity (EVN), which both do relatively well on cost recovery. By contrast, the private utilities are all tightly clustered around the 90 to 110 percent range for capital cost recovery. A notable exception is Karachi Electric from Pakistan, which although private remains distant from full capital cost recovery tariffs; however, under the Pakistani regulatory framework, Karachi Electric receives a compensating subsidy that (at least partially) compensates for this shortfall. Another striking result is that the utilities involved in the failed privatizations of the Dominican Republic and Senegal are only about 70 percent of the way toward full capital cost recovery—a factor which contributed to the failure of these privatizations.

FIGURE 9.8 **Private sector participation is strongly associated with full capital cost recovery**

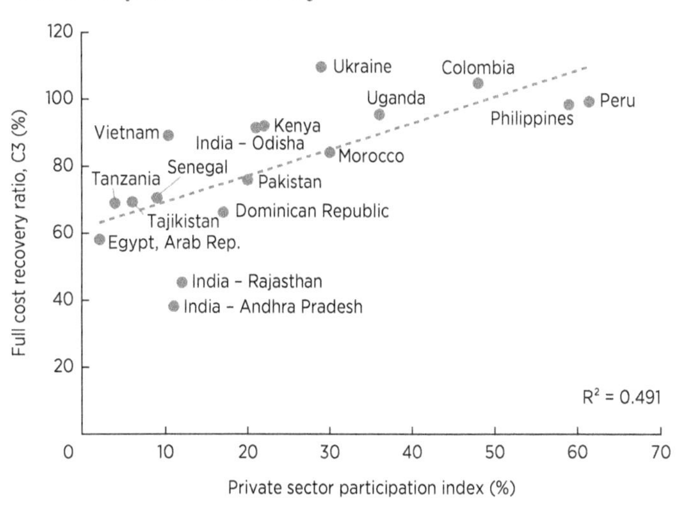

Note: A sensitivity analysis was carried out for exclusion of outliers two standard deviations from the mean, which resulted in the exclusion of the Indian state of Andhra Pradesh and of Peru, and the R^2 of 0.3381. It does not materially affect the analysis. C3 = full cost recovery.

FIGURE 9.9 Only a few publicly owned utilities are allowed to charge cost recovery tariffs

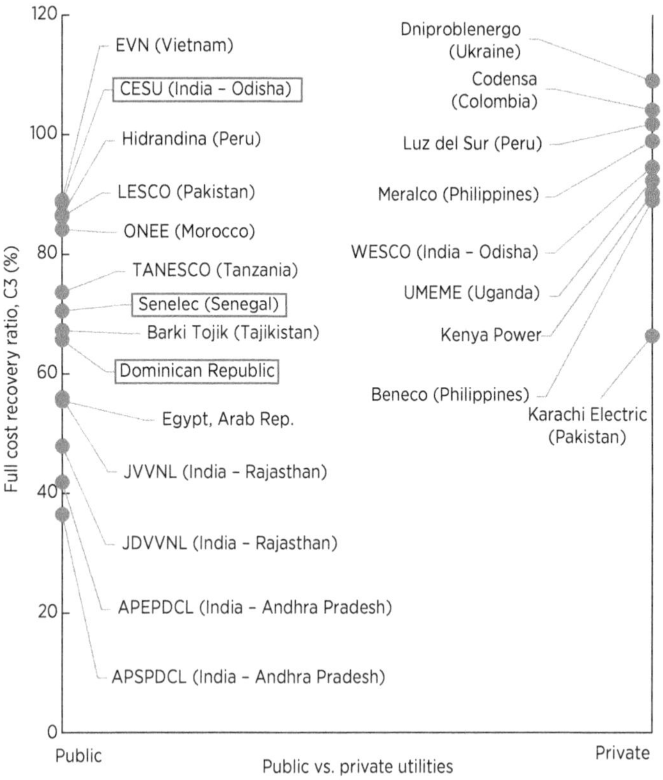

Note: Red boxes indicate utilities that have seen privatization rollback. The cost recovery ratio for the publicly owned utilities appears on the left-hand axis, whereas that for the private utilities appears on the right-hand axis. C3 = full cost recovery.

Finding #5: Although private sector participation is also strongly associated with distribution efficiency, a significant subset of publicly owned utilities is able to match this performance

A moderately strong association exists between private sector participation and distribution efficiency. The panel data regression analysis reported in table 9.3 was able to capture only the relationship between private sector participation and system losses. The small sample permits a deeper analysis of efficiency, which incorporates both system losses and collection losses into a single aggregate measure of distribution efficiency. For every percentage-point increase in private sector participation in the power sector, the utility efficiency improves by 0.28 percentage points (table 9.6). The relationship between private sector participation and distribution efficiency shows a relatively strong correlation, with an R-squared coefficient of 0.33 (figure 9.10). Nevertheless, it is clear from the scatter plot that the countries with publicly owned distribution utilities fall into two distinct groups. Publicly owned utilities in jurisdictions such as the Dominican Republic, the Arab Republic of Egypt, India (Odisha and Rajasthan), Pakistan, Senegal, and Tanzania display exceptionally low levels of efficiency, with utilities losing 15 to 25 percent of revenues. At the same time, publicly owned utilities in jurisdictions such as India (Andhra Pradesh), Morocco, and Vietnam display exceptionally high levels of efficiency, losing no more than 5 percent of revenues. Strikingly, the performance of this second group is as good as that of the handful of countries that have largely privatized their power sectors: Colombia, Peru, and the Philippines.

A strong overlap exists, however, between the best-performing public utilities and the worst-performing private utilities. The picture becomes clearer by disaggregating results from the country level to the utility level (figure 9.11). As before, the efficiency performance of public utilities is plotted on the left axis and that of private utilities is plotted on the right axis. In contrast to the analysis of cost recovery (see figure 9.9) where public and private utilities were clearly segmented, in the case of efficiency there is substantial overlap in performance between the two groups, with the overall range running from 5 percent to 25 percent of revenues lost to inefficiency in both cases. As noted in figure 9.10, a significant cluster of public utilities performs as efficiently as some of the private ones, including ONEE (Morocco), Hidrandina (Peru), EVN (Vietnam), and the utilities from the Indian state of Andhra Pradesh. It is also striking to see privatized utilities such as Karachi Electric (Pakistan) and WESCO (Indian state of Odisha) whose performance is comparable to the worst of the publicly owned utilities. At the same time, some of the worst-performing publicly owned utilities include cases of failed privatization such as the Dominican Republic, CESU in the Indian state of Odisha, and Senegal.

These findings point to potential selectivity and survivor effects with the privatized utilities. The comparison between public and private utilities is not entirely even-handed for two reasons, as the results for cost recovery (figure 9.9) and efficiency (figure 9.11) appear to show. First, there is a selectivity effect: utilities that already perform relatively well (such as those close to cost recovery) tend to be those most likely to be selected for privatization. Second, there is a survivor effect: privatized utilities that experience poor performance on efficiency or cost recovery over time are likely to revert to state control. These two considerations mean that the causality between private sector participation and utility performance is likely to run in both directions.

Well-performing public utilities show governance practices that are much closer to those of private utilities than their underperforming state-owned counterparts. Comparing well-performing public utilities with their under-performing state-owned counterparts serves to shed light on the governance practices that seem to be most critical to enhancing the performance of public utilities. Interestingly, the governance practices of this well-performing public sector group have more in common with well-performing private utilities than with their state-owned counterparts. In terms of board governance, the largest differences across groups can be found in their greater freedom to appoint a chief executive officer, higher exposure to audits, and increased tendency to publish annual reports (figure 9.12, panel a). In terms of financial discipline, the largest differences across groups can be found in their greater propensity to publish accounts that conform with international financial reporting standards and their explicit costing of their public service obligations (figure 9.12, panel b). Human resource management is where the largest differences across groups appear, with greater ability to fire underperforming employees and increased tendency to conduct transparent and objective hiring processes (figure 9.12, panel c). Finally, well-performing public and private utilities are much more likely to make use of modern information and technology systems (figure 9.12, panel d).

FIGURE 9.10 A substantial minority of publicly owned utilities performs as efficiently as private ones

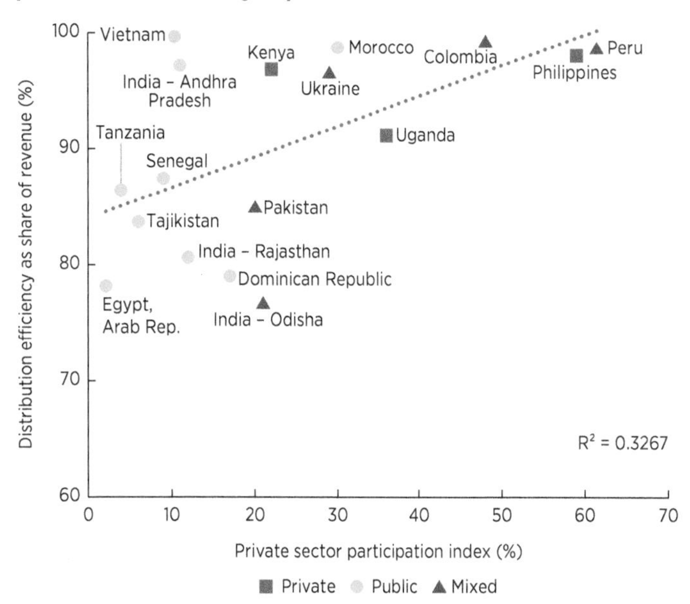

Note: A sensitivity analysis was carried out for exclusion of outliers two standard deviations from the mean. It resulted in the exclusion of Peru and the R^2 of 0.2803, and does not materially affect the analysis. KPLC in Kenya is a majority government-owned utility (51 percent of the shares owned by the government); however, it is listed on the Nairobi stock exchange, and its day-to-day functioning is more in line with a private utility. Thus, for the purpose of the Rethinking project we classify KPLC as private.

Finding #6: Country context is again a strong determinant of sector outcomes; however, reform measures are also found to contribute significantly, particularly in low-income countries

Econometric analysis shows that certain sector reform measures are significantly associated with improved sector outcomes. As noted earlier in the chapter, the panel data regression provides a statistically rigorous overview of

FIGURE 9.11 Among the least efficient utilities are those that experienced privatization reversals

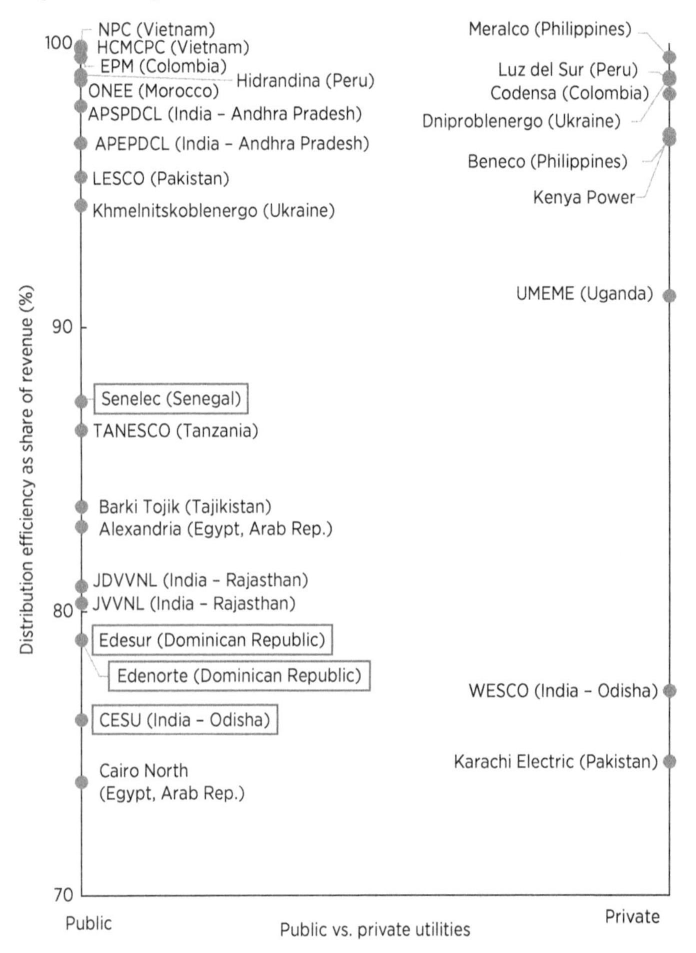

Source: World Bank elaboration based on Rethinking Power Sector Reform utility database 2015.
Note: Red boxes indicate utilities that have seen privatization rollback. The efficiency performance of the publicly owned utilities appears on the left-hand axis, whereas that for the private utilities appears on the right-hand axis.

relationships between high-level reform indicators and different dimensions of sector outcome. Five different model specifications are considered, which vary according to whether control variables are used, and whether time trends and country fixed effects are added to the regression. The statistical analysis shows that the results are quite sensitive to the model specification in many cases (annex 9B). In general, however, the incorporation of country fixed effects greatly improves the overall explanatory power of the regressions, indicating that local specificities appear to have a much larger impact on power sector outcomes than the extent of power sector reform. For this reason, attention in table 9.7 focuses on the specification incorporating both time trend and country fixed effects. In addition to estimating the model on the full sample of 88 countries, two separate nested models are estimated for the low- and middle-income countries of the sample. Statistical tests reject the hypothesis of no structural differences between these subsets, indicating that coefficients are significantly different between the two.

When it comes to security of supply, private sector participation seems to be the only reform with a significantly positive effect, particularly in low-income countries. Normalized capacity per million population is the only measure of security of supply that is available as a long-term time series. None of the reform measures except private sector participation shows any statistically significant relationship with normalized capacity (table 9.7). This effect is both larger and more significant for the low-income country group, where one megawatt per million population is added for every percentage-point increase in private sector participation; however, there is no statistical significance for the middle-income country group.

When it comes to electrification, private sector participation again emerges as significant, particularly in low-income countries. For the electrification regression, the sample is curtailed to those countries that had not yet reached 90 percent at the start of the study period in 1995, because otherwise the dependent variable becomes truncated as countries reach universal access. There are some puzzling results with both regulation and restructuring appearing with significantly negative relationships. Private sector participation is the only reform that emerges with a significant positive effect on electrification. As before, this effect is larger for low-income countries, where every percentage-point increase in private sector participation is associated with a

FIGURE 9.12 **Efficient public and private utilities score relatively higher on management and governance good practices**

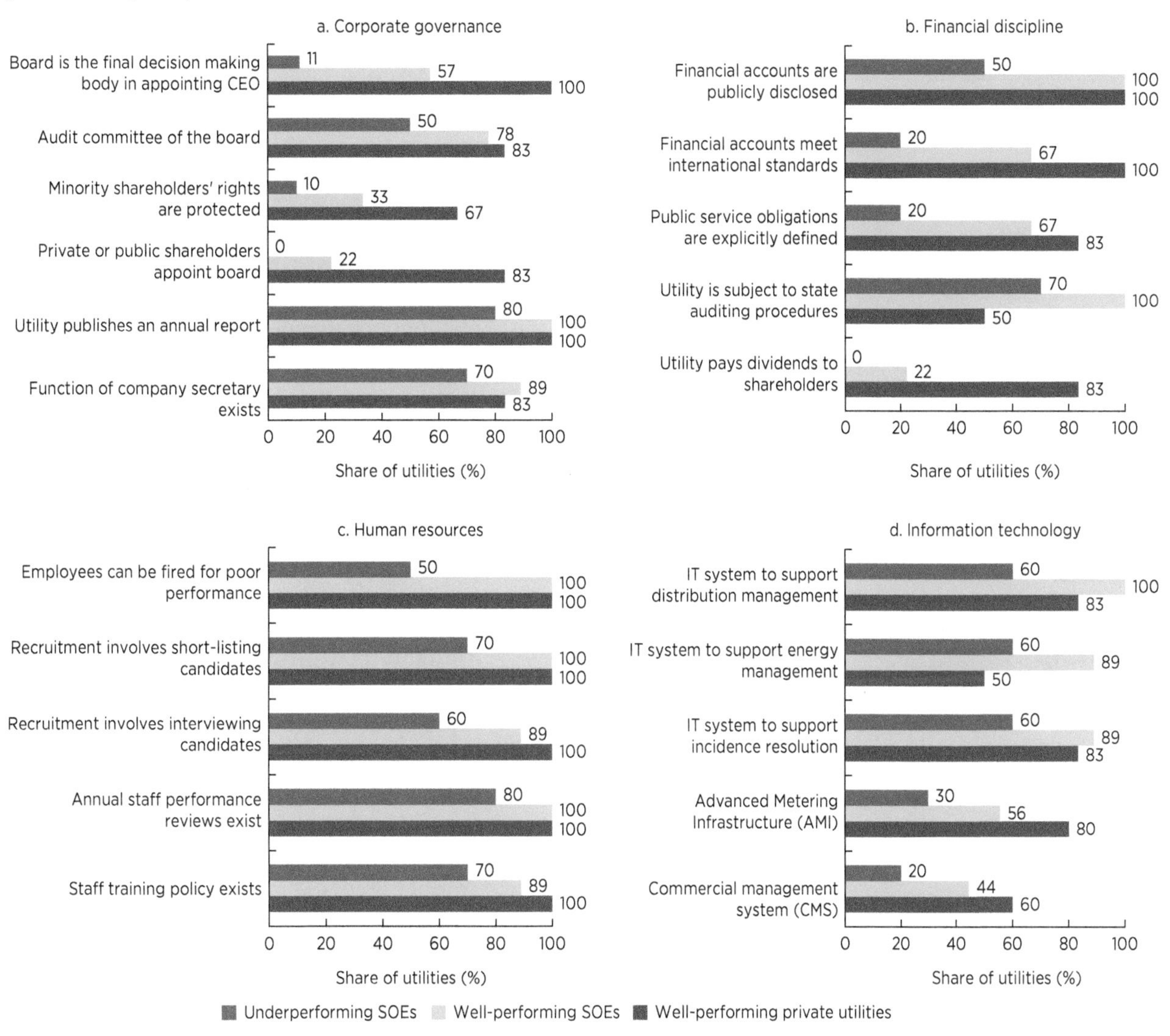

Note: CEO = chief executive officer; IT = information and technology; SOE = state-own enterprise.

0.3-percentage-point increase in electricity access, but disappears altogether in middle-income countries.

When it comes to renewable energy, there is a consistent positive impact from regulation and private sector participation, whereas other effects differ by income grouping. The presence of a regulator has a significant positive effect on a country's renewable energy share across income levels, although the effect is rather small, with an increase of just 0.4 percentage points. In the case of middle-income countries, competition and private sector participation also have a significant positive effect, although again the effect is not all that large. For low-income countries, the coefficient on competition, though significant, is surprisingly negative.

These broad statistical results are complemented by a more detailed analysis of reform

TABLE 9.7 Impact of reform measures on final outcomes for large sample

Regression coefficients	Full sample	Low-income countries	Middle-income countries
Dependent variable: normalized capacity			
Regulation	~	~	~
Restructuring	~	~	~
Competition	~	~	~
Private sector participation	61.14**	134.0***	~
Observations	1831	1008	819
R-squared	0.962	0.928	0.958
Dependent variable: access to electricity[a]			
Regulation	-1.775***	-2.369***	~
Restructuring	~	~	-6.127***
Competition	~	~	~
Private sector participation	19.25***	29.66***	~
Observations	1314	924	378
R-squared	0.973	0.958	0.978
Dependent variable: renewable energy share			
Regulation	0.461***	0.417***	0.437***
Restructuring	~	~	~
Competition	~	-2.012***	0.613*
Private sector participation	1.487***	~	1.733***
Observations	1760	960	780
R-squared	0.974	0.975	0.973

Note: Ordinary least squares regressions on panel data set of 88 countries for the period 1995–2015. A separate regression is estimated for each of the intermediate outcomes on the full set of reform variables capturing regulation, restructuring, competition, and privatization. The analysis considers a number of different ways of capturing control variables in terms of time trend, country-specific fixed effects, and control variables capturing system size and GDP per capita. The preferred specification, incorporating time trend and country-specific fixed effects (but no control variables) is reported here. All the other specifications can be viewed in the annex (for complete results refer to annex 9B, tables 9B.3, 9B. 5, and 9B.6). Coefficients are scaled in units of the intermediate outcome variable.

a. For the specific case of the regression for access to electricity, observations are dropped for any countries that had already reached electrification rates above 90 percent in 1995.

Significance level: *** = 1 percent, ** = 5 percent, * = 10 percent, ~ = not significant.

measures and associated sector outcomes based on the smaller sample. This additional analysis makes it possible to examine a wider range of sector outcome indicators than those that were available for the long-term panel data analysis. It also allows for a deeper qualitative discussion of the actual reform path of each country and its relationship to sector outcomes.

Finding #7: Although many countries made some progress toward security of supply with the 1990s reform model, few were able to achieve that goal completely, despite harnessing substantial private investment

Almost all developing countries face security of supply challenges; addressing these challenges was one of the central concerns of the 1990s power sector reform model. With electricity demand typically growing above five percentage points per annum, maintaining supply–demand balance in the developing world calls for substantial and timely investments in new capacity. Furthermore, many countries find their power supply overly exposed to exogenous factors, in particular droughts and oil price shocks, calling for greater diversification of power generation capacity. Even when these issues are overcome, weaknesses in the transmission and distribution grids may continue to undermine the reliability of supply to consumers. According to the 1990s power sector reform model, security of supply could best be addressed by opening the generation segment to entry and investment by the private sector, ideally in the context of a competitive wholesale market, backstopped by a financially robust distribution sector.

Although such reforms were among the most widely adopted, achieving full security of supply has proved quite challenging. Reflecting the multidimensional nature of this concept, several different measures of security of supply, which do not necessarily correlate closely with each other, are used here. They are normalized capacity, capacity diversification, supply reliability, and capacity growth. Overall, it is striking that, although countries embracing extensive power sector reforms tend to do quite well overall on these various measures of security of supply, the converse is not necessarily the case (figure 9.13). Moreover, some of the worst performers (notably, India and Pakistan) are among those that have pursued a significant amount of reform.

Normalized capacity metrics reached levels typical of middle-income countries in most parts of the world, except for Sub-Saharan Africa. The first measure relates to capacity normalized by population (figure 9.13, panel

FIGURE 9.13 Performance on various measures of security of supply with countries ranked by extent of power sector reforms from most reformed (Philippines) to least reformed (Tajikistan)

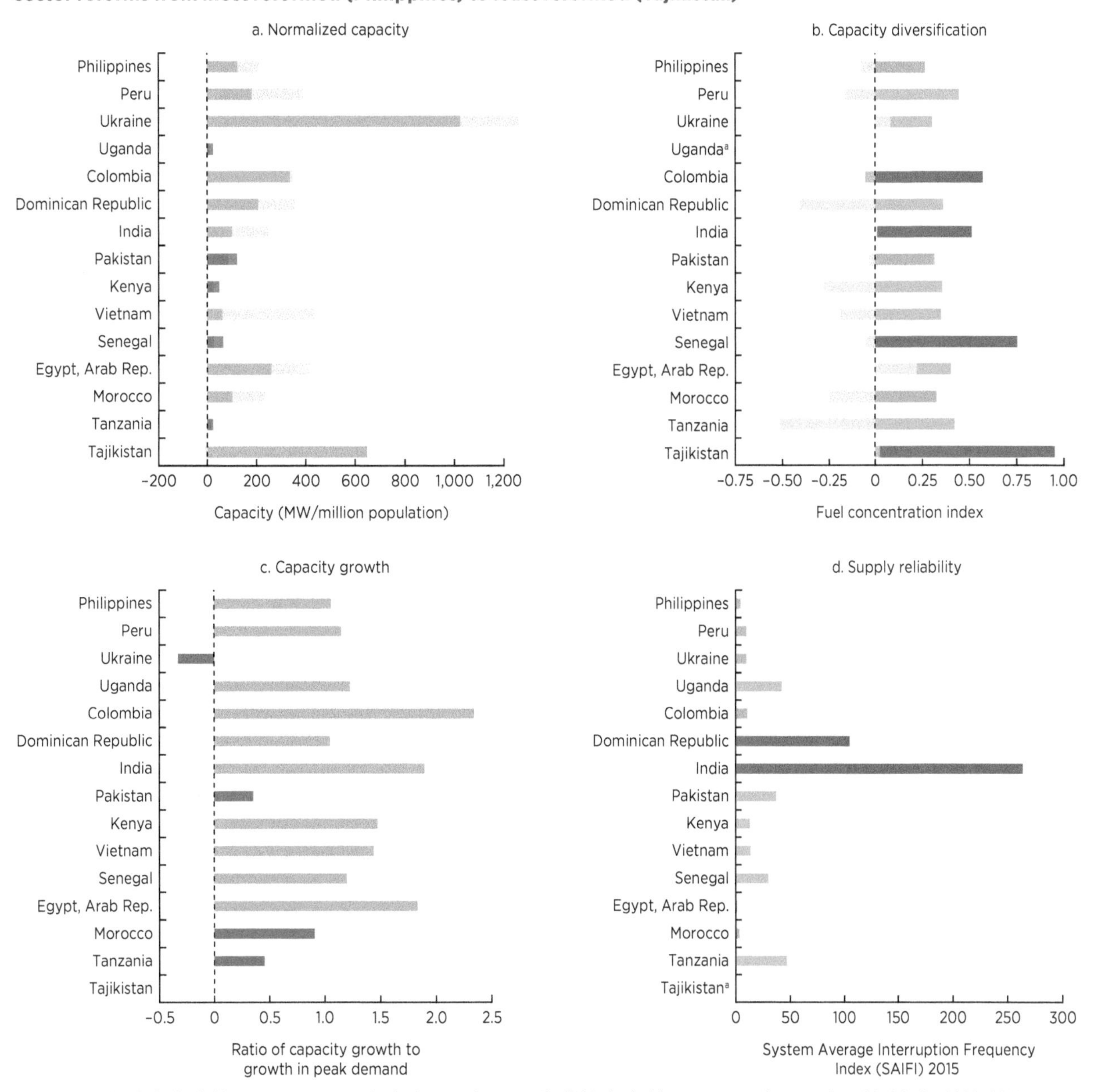

Note: In panel a, dark-shaded bars represent capacity in the prereform era; the light-shaded bars represent the capacity added during 1990–2015. In panel b, dark-shaded bars represent average value in 2010–15; light-shaded bars represent the change in values from prereform era. For all panels the color of the bar follows the evaluation framework set up in table 9.2. MW = megawatt.

a. Data not available.

a). For the Eastern European countries, available capacity was already unusually high prior to the reform, reflecting the Soviet legacy. Most of the middle-income countries in the sample appear to operate adequately in the range of 200–400 megawatts per million population. Many low-income countries started out in 1990 with capacity well below these levels; however, there is a marked contrast between Sub-Saharan African countries that despite some progress remain below this threshold, and countries elsewhere that succeeded in growing their capacity into this range (such as India, Morocco, and the Philippines). The most dramatic example is Vietnam, whose normalized capacity grew sixfold over the period 1990–2015. Overall, normalized capacity is most strongly associated with income per capita, which alone explains about 67 percent of the cross-country variation in this indicator (figure 9.14).

Few countries have achieved a fully diversified power mix. The second measure relates to capacity diversification and indicates the extent to which supply is concentrated rather than spread across a variety of different energy sources. Almost all countries succeeded in improving capacity diversification over the period 1990–2015, with Egypt being the main exception. Nevertheless, only a handful of countries has actually achieved a well-balanced portfolio of energy sources, with Morocco, Pakistan, and the Philippines being the best examples (figure 9.13, panel b). Looking across countries in the small sample, diversification shows a significant relationship with reforms such as restructuring, regulation, and competition, with competition having by far the largest effect (figure 9.15). A one-percentage-point increase in competition leads to improvement of two percentage points on diversification.[4]

Almost all countries are managing to grow generation capacity at least as rapidly as peak demand. The third measure relates to the extent to which capacity expansions are keeping pace with the growth of peak demand in recent years. On this measure, almost all countries do well, managing to grow capacity more rapidly than demand. Notable exceptions, however, are Pakistan and Tanzania.

FIGURE 9.14 Normalized capacity is strongly associated with income per capita

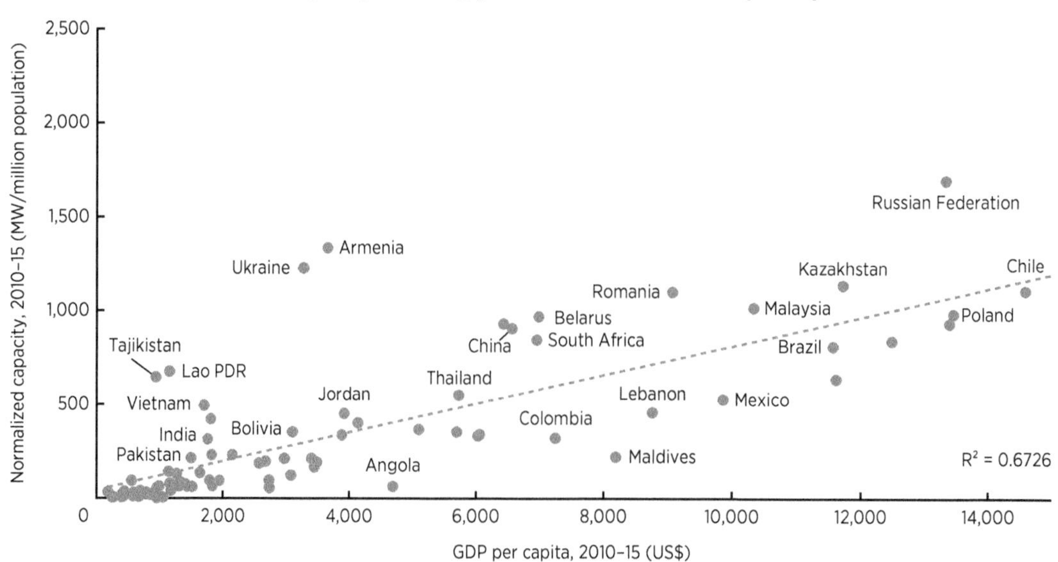

Note: A sensitivity analysis was carried out for exclusion of outliers two standard deviations from the mean. It resulted in the exclusion of 11 countries and the R^2 of 0.6531, and does not materially affect the analysis. MW = megawatt.

Performance on supply reliability varies across different geographical regions. The fourth measure relates to supply reliability from the consumer perspective as measured by the standard System Average Interruption Frequency Index (SAIFI). Generally, performance on this measure is good across Latin America, Eastern Europe, and North Africa and moderate across Asia and Sub-Saharan Africa. Nonetheless, the Dominican Republic and especially India stand out as having exceptionally high levels of service interruptions (figure 9.13, panel d). Perhaps not surprisingly, SAIFI also seems to improve with more efficient utilities (figure 9.16)

Overall, the qualitative analysis shows that the implementation of the 1990s power sector reform measures did not of themselves fully guarantee the achievement of security of supply (table 9.8). By looking at the underlying narratives for security of supply in the 15 countries of the Power Sector Reform Observatory, it becomes possible to gain some insights into the channels of causality. Peru and the Philippines stand out as the countries where the model was most successful. These countries went as far as introducing wholesale power markets and, as a result, were able to finance generation capacity expansions almost entirely through private investment, while maintaining capacity adequacy and diversification and providing quite reliable supply to consumers. Another group of countries—comprising Colombia, Morocco, and Vietnam—also did relatively well, although their private financing shares were substantially lower than the first group. A third group of countries—including the Dominican Republic, India, Kenya, and Uganda—made some progress, but the countries continue to face challenges with either reliability or diversification. A last group of countries—despite adopting significant reforms—cannot be considered to have achieved security of supply. They are Pakistan, Senegal, and Tanzania. Finally, Egypt has (after a period of supply crisis) achieved security of supply, but—following an early unsuccessful independent power producer program—did so through a completely different approach based on intergovernmental bilateral deals (primarily with Germany) to support new investments in power generation. Since 2015, a similar approach has been adopted in Pakistan with the China–Pakistan Economic Corridor, which comprises a support package including 17 gigawatts of new power generation capacity.

FIGURE 9.15 Power sector diversification is significantly related to various reform steps

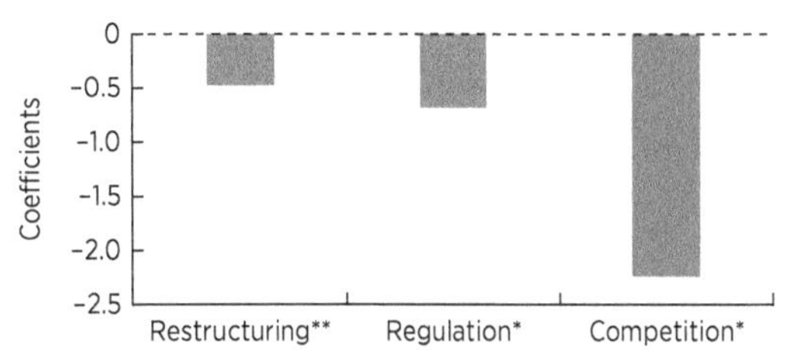

Note: Ordinary least squares multivariate regression analysis for the Rethinking Power Sector Reform country sample (17 economies). The chart reports the results of a series of cross-sectional regressions modeling the impact of various reform steps on the concentration index for power generation, with lower scores representing greater diversification. GDP per capita and system size have been used as control variables. For complete results, refer to annex 9C, table 9C.5.
Significance level: *** = 1 percent, ** = 5 percent, * = 10 percent.

FIGURE 9.16 Reliability of supply is associated with greater distribution efficiency

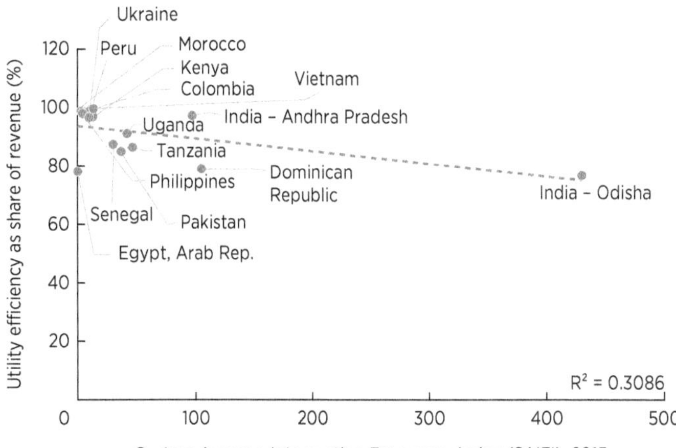

Note: A sensitivity analysis was carried out for exclusion of outliers two standard deviations from the mean. It resulted in the exclusion of the Indian state of Odisha and Egypt, Arab Rep. and the R^2 of 0.4002 and does not materially affect the analysis.

TABLE 9.8 Qualitative analysis of underlying causality for security of supply trends

Hypothesis: Security of supply would be achieved by opening up generation to the private sector

	Performance			
	Security of supply 2015	PSP share in generation 1990–2015 (%)	Policy steps	Supports hypothesis
Colombia	Medium	70	Created successful wholesale market and attracted private investment, but hydro exposure to drought risk still high.	Largely
Dominican Republic	Medium	65	Created simple power market and attractive private sector investment, but power supply still unreliable (SAIFI).	Partially
Egypt, Arab Rep.	High	8	Initial IPP program stalled and capacity expansion achieved through state-driven bilateral deals.	No
India	Medium	46	Created wholesale market for electricity attracting private investment and growing reserve margin, but SAIFI an issue.	Partially
Kenya	Medium	40	IPP program attracted significant private investment; still scope to improve SAIFI and capacity diversification.	Partially
Morocco	High	50	IPP program attracted significant private investment; adequate capacity and reliability of service.	Largely
Pakistan	Low	57	IPP program attracted private investment, but not enough to meet fast-growing demand because of sector financial crisis.	No
Peru	High	92	Created successful wholesale market largely reliant on private investment; supply reliable, diversification ongoing.	Yes
Philippines	High	96	Created successful wholesale market largely reliant on private investment; supply both reliable and diversified.	Yes
Senegal	Low	50	IPP program attracted private investment, but capacity remains low, reliability poor, and diversification weak.	No
Tanzania	Low	32	IPP program had limited success, but capacity remains low, reliability poor, and drought exposure substantial.	No
Uganda	Medium	55	IPP program eventually delivered investment, but capacity remains low, reliability mediocre, and hydro dominant.	Partially
Vietnam	High	38	IPP program attracted significant private investment; capacity grew rapidly and diversified; few reliability issues.	Largely

Note: Ukraine is excluded because it has excess capacity, demand is falling, and security of supply is therefore not an issue; the same can be said of Tajikistan. IPP = independent power producer; PSP = private sector participation; SAIFI = System Average Interruption Frequency Index.

Finding #8: Progress on electrification was made by countries with widely differing approaches to power sector reform and appears to be driven primarily by targeted public investment as countries reach higher income levels

The social inclusion agenda that became ascendant after 2010 underscored the importance of universal access to affordable energy services. SDG7.1.1 sets the objective for universal access to affordable, reliable, modern, and sustainable electricity for all by 2030. Universal access is usually measured in terms of the percentage of households with access to some form of electricity service, whether on- or off-grid. Acknowledging that access without affordability would be meaningless, the 2030 goals also underscore the importance of affordability. Although the United Nations has no official indicator for affordability, in the literature it is often defined as the ability to avail an average volume of electricity consumption for no more than 5 percent of the budget of the bottom 40 percent of the income distribution. As noted earlier in the chapter, neither access nor affordability was a central focus of the 1990s power sector reform model. Although some middle-income countries had already very much engaged in this agenda by the 1990s, that aspiration only became mainstream across the developing world following the promulgation of SDG7.

Performance on electrification does not seem to reflect the overall extent of power sector reform in a country but is closely associated with per capita income. During the prereform period around 1990, countries already had a wide disparity of electrification rates, ranging from those in Latin America and Eastern Europe that were already close to universal access all the way to countries in Sub-Saharan Africa with electrification rates no higher than 30 percent. The amount of progress made on electrification over the 25 years between 1990 and 2015 also varies greatly. At one extreme, Morocco electrified more than 50 percent of its population during this period and came close to achieving universal access. At the other extreme, Tanzania barely electrified 8 percent of its population and had below 20 percent coverage even in 2015. Strong performances on electrification can be found both among countries that embraced power sector reform (Peru and the Philippines) and among countries that took a much more cautious approach to power sector reform (Morocco and Vietnam). At the same time, relatively weak performances on electrification can be found among countries that embraced reform (Uganda) as well as those that eschewed it (Tanzania) (figure 9.17, panel a). By far the strongest and most consistent driver of electrification is gross domestic product per capita with a correlation of about 0.35 in the cross-sectional analysis, dropping to 0.17 for the difference in differences.

FIGURE 9.17 Performance on social inclusion, countries ranked by extent of power sector reforms from most reformed (Philippines) to least reformed (Tajikistan)

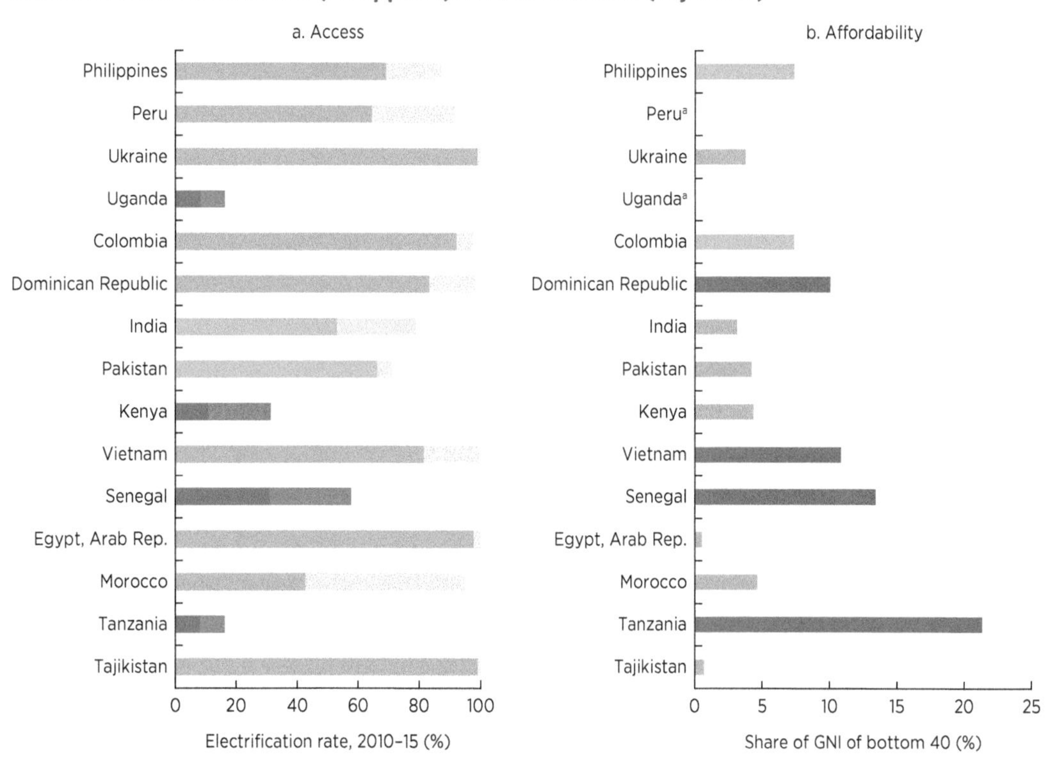

Note: Dark-shaded bars represent electrification in the prereform era, and the light-shaded bars represent electrification during 1990–2015. For all panels the color of the bar follows the evaluation framework set up in table 9.2. Bottom 40 = population in the bottom 40 percent of the income distribution; GNI = gross national income.

a. Affordability data for average consumption in Peru and Uganda not available.

FIGURE 9.18 **The strongest driver of electrification has been GDP per capita**

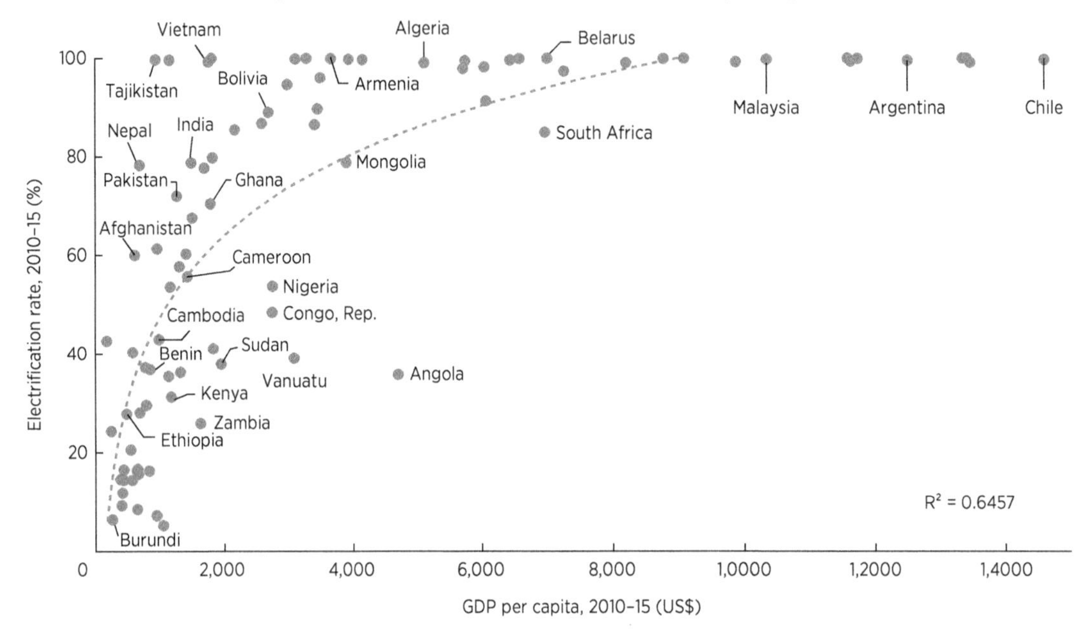

A graphical representation reveals a nonlinear relationship between gross domestic product per capita and electrification with a particularly steep ascent in the US$1,000–3,000 per capita window, and the resulting R-squared rising to over 0.60 (figure 9.18).

Regarding affordability, this factor largely affects the local economic context and little relation is found to reform measures. Most countries meet the basic affordability criterion of keeping the cost of average consumption below 5 percent of the budget of the bottom 40 percent of the income distribution (figure 9.17, panel b). The few exceptions usually reflect at least one of the following three factors: high price, low income, and high consumption. For instance, the Dominican Republic, the Philippines, and Senegal all charge average residential tariffs in excess of US$0.20 per kilowatt-hour. Because these factors are unrelated to reform, it is perhaps not surprising that little association is found between affordability and power sector reform measures, and even with respect to the intermediate outcomes of efficiency and cost recovery.

Overall, the qualitative analysis confirms a consistent pattern of progress on electrification in countries with strong government-led targets, institutions, and investment subsidies. For the small sample, it is possible to go behind the quantitative analysis to examine the underlying electrification narratives, which shed light on the process through which progress toward universal access is achieved. These qualitative stories, summarized in table 9.9, show a remarkably consistent pattern across countries. Specifically, in country after country, the drive for electrification entailed a strong political commitment backed up by sustained public investment typically channeled through a reasonably competent utility. Moreover, for most countries, the main period of acceleration in the electrification process is unrelated to the period when the 1990s reform model was being implemented. In several of the middle-income countries—such as India, Morocco, and Vietnam—the electrification drive predated the implementation of the 1990s reforms and simply continued in the background through its own established channels even as

TABLE 9.9 Qualitative analysis of underlying causality for electrification trends

Hypothesis: Progress on electrification had little to do with the 1990s reforms, but was driven by government-funded electrification programs led by public utilities and other state institutions

	Performance			
	Average 2010–15 (%)	Change 1990–2015 (%)	Policy steps	Supports hypothesis
Colombia	97.4	+5.5	Areas with low access were identified as special zones, and targeted electrification programs were backed by resources from various special funds.	Yes
Dominican Republic	98.2	+15.0	Rural and suburban electrification unit in the CDE is responsible for access and made efforts to encourage development of isolated grids to improve access.	Yes
Egypt, Arab Rep.	99.8	+2.1	Rural electrification authority established in 1971 had completed its mandate by 2007 and was disbanded.	Yes
India	78.8	+25.7	Multiple policies at national and state levels targeted resources toward utility-led electrification programs.	Yes
Kenya	31.3	+20.2	Since 2010, government sets electrification targets and directs special funding to national utility for grid rollout; measures also taken to encourage off-grid solar.	Yes
Morocco	94.6	+51.9	Rural electrification program targeted government resources to national utility for electrification; off-grid solar systems also encouraged.	Yes
Pakistan	71.0	+4.8	No national-level agency for rural electrification, and no special incentives for utilities to expand access.	Yes
Peru	91.3	+26.9	Rural electrification law in 2006–07, long after wider power sector reform law in 1992; substantial government funds targeted to rural electrification.	Yes
Philippines	86.8	+17.7	National Electrification Administration drives rural electrification program implemented by cooperatives; utilities responsible for urban grid rollout.	Yes
Senegal	57.7	+26.5	Attempt to use private rural concessions to expand access only partially successful; current focus is on public funding for utility-based electrification.	Largely
Tanzania	16.4	+8.0	Not until 2005–07 were the Rural Energy Fund and Rural Energy Agency created to channel donor resources to on- and off-grid electrification programs.	Yes
Uganda	16.4	+8.0	Rural private concessions were unsuccessful in expanding access; privatized national utility made limited contribution to access, leaving government to drive electrification.	Largely
Vietnam	99.3	+17.8	Access drive predates power sector reform, based on channeling public resources to electrification programs led by the national utility.	Yes.

Note: Tajikistan and Ukraine are excluded from this table since both countries had already achieved universal access to electricity at the time of independence. CDE = Corporación *Dominicana de* Electricidad.

other reforms took place. In many low-income countries—such as Kenya, Tanzania, and Uganda—progress in electrification came many years after the 1990s reform process and entailed the creation of additional institutions (Rural Electrification Agencies) and funds (Rural Electrification Funds). Only two countries (Senegal and Uganda) attempted to apply the principles of the 1990s reform model to the electrification process itself, by tendering rural concessions to the private sector. In both cases, the approach encountered challenges related to the balance between financial viability and tariff affordability and has largely been dropped.

Finding #9: Progress on decarbonization of the electricity sector has historically been an unintended by-product of measures to improve security of supply

The 2030 Agenda for Sustainable Development calls for rapid decarbonization of the electricity sector, with a particular emphasis on renewable energy. The 2015 Paris Climate Agreement holds the world accountable for the implementation of rapid decarbonization action across sectors aimed at limiting climate change to 1.5 degrees Celsius. As the largest carbon-emitting sector, energy has featured among the Nationally Determined Contributions of many countries. Decarbonization of the sector can be achieved by improving energy efficiency, substituting renewables for fossil fuel generation, and shifting toward fossil fuels with lower carbon content (such as natural gas). Further specificity is provided by SDG7.2, which calls for a doubling of the share of renewable energy in total final energy consumption globally by 2030. These new political agreements came right at the end of the historical period considered here, but some pioneering countries (such as Morocco) had already begun to espouse the decarbonization agenda.

In many countries, however, electricity remains carbon intensive and this situation has been just as likely to improve as to deteriorate over time historically. The only countries with low-carbon power sectors are Colombia and Tajikistan, which in 1990 already enjoyed strong reliance on hydroelectric power, a reflection of the availability and cost-effectiveness of this resource rather than a particular focus on decarbonization (figure 9.19). As of 2015, most countries in the Power Sector Reform Observatory still have highly carbon-intensive electricity sectors (more than 500 grams of carbon dioxide per kilowatt-hour). Moreover, a broadly equal number of countries saw their carbon intensity rise and fall during the years 1990–2015, and the rise and fall happened across the power sector reform spectrum. Programs to explicitly promote the uptake of modern renewable energy started to pick up only toward the end of this period, and the programs for most countries have been small relative to the overall scale of electricity consumption. Nevertheless, Colombia, Peru, the Philippines, and Vietnam all saw significant expansions in their shares of modern renewable energy.

FIGURE 9.19 Performance on environmental sustainability, with countries ranked by extent of power sector reforms from most reformed (Philippines) to least reformed (Tajikistan)

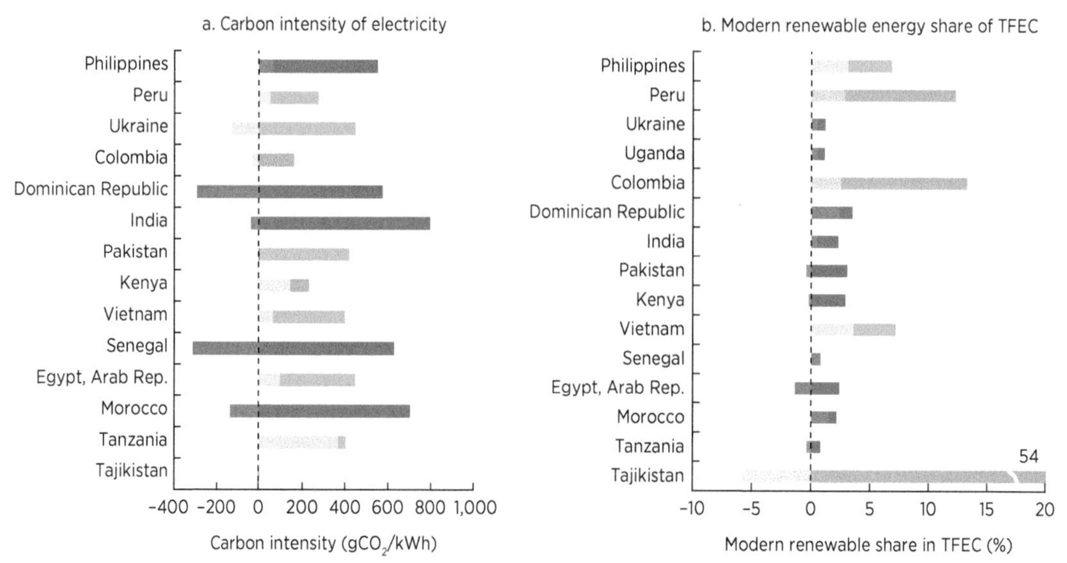

Note: Dark-shaded bars represent average value in 2010–15; light-shaded bars represent the change in values from prereform era. For all panels the color of the bar follows the evaluation framework set up in table 9.2. gCO_2/kWh = grams of carbon dioxide per kilowatt-hour; TFEC = total final energy consumption.

Countries with relatively low carbon intensity tended to do better in terms of operating cost recovery (figure 9.20). This finding is hardly surprising because fuel purchase expense tends to be one of the largest components of operating costs for the power sector in many countries. By contrast, renewable sources of energy have very low operating costs because of the fact that they do not run on fossil fuels.

Overall, the qualitative analysis shows that for most countries, changes in decarbonization have historically been an unintended by-product of measures to promote security of supply (table 9.10). In order to understand what lies behind the numbers, it is helpful to zoom in on the narrative for each of the 15 countries of the Power Sector Reform Observatory. These stories provide strong support for the view that in most countries, power generation decisions have been driven by security of supply considerations, with unintended consequences—both positive and negative—for decarbonization. For instance, several countries started out as overly dependent on hydropower and consciously diversified into fossil fuels (mainly natural gas) as a way of reducing exposure to drought risk (as in Colombia, Kenya, Peru, and Tanzania), at the same time increasing the carbon intensity of their electricity sectors. In other cases, overreliance on oil-fired power generation also created the need to diversify to reduce exposure to oil price shocks, which led to falling carbon intensity as a result of displacing oil with either hydro (as when Senegal began to import power from Manantali) or less-carbon-intensive fossil fuels (as with gas in the Dominican Republic, the Philippines, and Ukraine). The only clear exception to this pattern is Morocco, where a strategic decision was taken to pursue an aggressive renewable energy program with a view to positioning as a first mover, particularly with regard to concentrated solar power. This decision led to a sizable reduction in carbon intensity of electricity as renewables came in to displace fossil fuels. Nevertheless, as of 2015, the carbon intensity of electricity remained high in absolute terms: coal- and oil-fired plants still account for more than half of generation capacity compared to 12 percent for wind and solar.

FIGURE 9.20 Carbon intensity shows strong inverse correlation with operating cost recovery

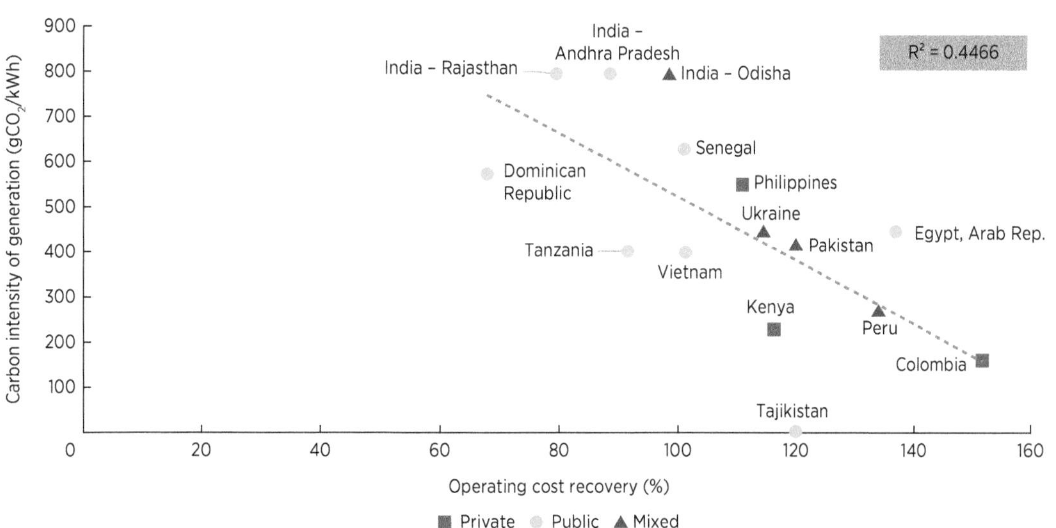

Note: A sensitivity analysis was carried out for exclusion of outliers two standard deviations from the mean. It resulted in the exclusion of Tajikistan and the R^2 of 0.5058, and does not materially affect the analysis. gCO_2/kWh = grams of carbon dioxide per kilowatt-hour. The type of ownership of utilities is represented by the color of dots.

TABLE 9.10 Qualitative analysis of underlying causality for decarbonization trends

Hypothesis: Decarbonization of the electricity sector has sometimes been an unintended consequence of policies to improve security of supply, rather than a deliberate policy effort

	Performance			
	Average 2010–15 (gCO_2/kWh)	Change 1990–2015 (gCO_2/kWh)	Policy steps	Supports hypothesis
Colombia	160.6	-27.9	Doubled natural gas generation to diversify from hydro drought risk, but renewables also increased.	Yes
Dominican Republic	572.7	-294.7	Massive financial crisis due to oil price shocks prompted diversification toward natural gas and coal.	Yes
Egypt, Arab Rep.	445.9	+99.4	Until recent renewable program, capacity expansions have been entirely done using natural gas.	Yes
India	794.2	-37.7	Generation coal dominated because of security of supply concerns; recent shift toward scaling renewables.	Yes
Kenya	228.6	+146.0	Diversification away from hydropower toward rapid expansion of oil and geothermal capacity.	Yes
Morocco	700.7	-136.5	Explicit decarbonization policy objective and creation of MASEN as specialized entity to promote solar energy.	No
Pakistan	416.9	+5.1	Most new generation added is from coal and natural gas driven by security of supply concerns.	Yes
Peru	271.0	+116.1	Shift toward natural gas generation to diversify from hydro drought risk for security of supply.	Yes
Philippines	549.0	+65.4	Almost all new generation added since EPIRA in 2001 is natural gas and coal, reducing share of oil-fired plants.	Yes
Senegal	627.1	-313.6	Almost all reduction is because of hydro capacity coming online; oil remains the mainstay.	Yes
Tanzania	401.5	+366.1	Expanded oil and gas generation since 2004 to diversify from hydro drought risk.	Yes
Ukraine	446.4	-129.6	Because of declining demand, costly oil-fired generation could be removed to result in lower carbon footprint.	Yes
Vietnam	398.6	+66.5	Generation expansion for security of supply based on oil and coal; explicit renewables program more recent.	Yes

Note: Tajikistan is excluded from the table because it was already almost entirely hydropower based in 1990 and has continued to be so. Uganda is excluded because of data limitations. EPIRA = Electric Power Industry Reform Act; gCO_2/kWh = grams of carbon dioxide per kilowatt-hour; MASEN = Moroccan Agency for Sustainable Energy.

Finding #10: Overall, although some of the deepest reformers delivered good sector outcomes, other countries showed that it was possible to achieve comparable performance through different approaches

The multidimensionality of power sector performance makes it difficult to reach very crisp conclusions regarding the relationship between reforms and outcomes. Tables 9.11 and 9.12 bring together the eight different dimensions of power sector reform outcomes that have been discussed throughout this chapter. Countries are shaded in green, yellow, and red according to whether their performance on any particular matrix is good, mediocre, or poor, as laid out in the rubric in table 9.2. Table 9.11 evaluates countries on the basis of their absolute performance in 2015, without taking starting conditions into account. Table 9.12 looks at the changes in performance from 1990 to 2015—at least for those variables for which a significantly long time series is available. This exercise serves to reveal that few countries are able to perform consistently well on all eight measures of power sector outcomes. It also illustrates

TABLE 9.11 Overview of cross-sectional performance on final outcomes as of 2015 with countries ranked by extent of power sector reforms from most reformed (Philippines) to least reformed (Tajikistan)

		Security of supply			Social inclusion		Environmental sustainability	
2015	Reliability (SAIFI)	Normalized capacity (MW/million population)	Capacity diversification (Fuel Concentration index)	Capacity growth (Capacity growth/peak dd growth)	Electrification (% population)	Affordability (% of GNI)	Carbon intensity (gCO_2/kWh)	Modern RE share in TFEC (%)
Philippines	4	209	0.26	1.05	87	7.35	549	6.8
Peru	10	391	0.44	1.14	91	—	271	12.24
Ukraine	10	1,260	0.38	—	100	3.71	446	1.15
Uganda	42	23	—	1.22	16	—	—	1.09
Colombia	11	345	0.57	2.34	97	7.32	161	13.21
Dominican Republic	105	354	0.36	1.04	98	10.01	573	3.46
India		248	0.52	1.89	79	3.09	794	2.28
Pakistan	37	121	0.31	0.35	71	4.18	417	3.05
Kenya	13	49	0.35	1.47	31	4.32	229	2.89
Vietnam	14	433	0.35	1.43	99	10.84	399	7.14
Senegal	31	64	0.75	1.19	58	13.45	627	0.79
Egypt, Arab Rep.	0	415	0.62	1.83	100	0.49	446	2.37
Morocco	3	231	0.32	0.9	95	4.59	701	2.15
Tanzania	47	22	0.42	0.45	16	21.32	401	0.78
Tajikistan	—	637	0.98	—	100	0.67	3	54.79

Note: Country results are shaded in green, yellow, and red according to whether their performance on any particular matrix is good, mediocre, or poor. dd = demand; gCO_2/kWh = grams of carbon dioxide per kilowatt-hour; GNI = gross national income; MW = megawatt; RE = renewable energy; SAIFI = System Average Interruption Frequency Index; TFEC = total final energy consumption; — = not available.

that, overall, there are some dimensions of performance where many countries are doing relatively well—notably electrification and capacity expansion—compared to others—such as supply reliability, capacity diversification, and carbon intensity—where many countries are struggling.

Although top reformers show good performance, not all reformers perform well and several cautious reformers achieve strong outcomes. Tables 9.11 and 9.12 list countries in order of their extent of adoption of the 1990s power sector reform model, from the most reformed country (the Philippines) to the least (Tajikistan). This treatment provides a straightforward visual check for the relationship between reform and sector outcomes. If such reforms are systematically associated with better performance, we would expect to see the countries toward the top of the table shaded mainly in green, and those toward the bottom of the table shaded mainly in red. Reality proves to be more complex. It is certainly true that among the deepest reformers —Colombia, Peru, the Philippines, and Ukraine—good performance is in evidence. For a number of countries engaging in substantial reforms, however, performance is more checkered—the Dominican Republic, India, Pakistan, and Uganda. Furthermore, several countries that did not reform so extensively—such as Egypt, Morocco, and Vietnam—exhibit positive performance. And Kenya, the best performer in

TABLE 9.12 Overview of performance on differences in final outcomes from 1990 to 2015 with countries ranked by extent of power sector reforms from most reformed (Philippines) to least reformed (Tajikistan)

	Change in security of supply				Social inclusion		Environmental sustainability	
1990–2015	Reliability (SAIFI)	Normalized capacity (% change in MW/million)	Capacity diversification (fuel concentration index)	Capacity growth (capacity growth/peak dd growth)	Electrification (% population)	Affordability (% of GNI)	Carbon intensity (gCO_2/kWh)	Modern RE share in TFEC (%)
Philippines	—	68	-0.07	—	18	—	65	3.10
Peru	—	114	-0.16	—	27	—	53	2.82
Ukraine	—	23[a]	0.08	—	1[a]	—	-130	0.45
Uganda	—	95	—	—	8	—	—	0.49
Colombia	—	3	-0.05	—	5[a]	—	-28[a]	2.51
Dominican Republic	—	70	-0.41	—	15[a]	—	-295	2.32
India	—	149	-0.01	—	26	—	-38	0.47
Pakistan	—	37	-0.03	—	5	—	5	-0.41
Kenya	—	79	-0.28	—	20	—	146	-0.19
Vietnam	—	611	-0.19	—	18	—	67	3.61
Senegal	—	137	-0.05	—	26	—	-314	0.79
Egypt, Arab Rep.	—	59	0.22	—	2[a]	—	99	-1.36
Morocco	—	127	-0.25	—	52	—	-136	1.47
Tanzania	—	20	-0.52	—	8	—	366	-0.36
Tajikistan	—	-1[a]	0.03	—	1[a]	—	-4[a]	-5.79

Note: Country results are shaded in green, yellow, and red according to whether their performance on any particular matrix is good, mediocre, or poor. dd = demand; gCO_2/kWh = grams of carbon dioxide per kilowatt-hour; GNI = gross national income; MW = megawatt; RE = renewable energy; SAIFI = System Average Interruption Frequency Index; TFEC = total final energy consumption; — = not available.
a. Instances where countries were already performing well and therefore the change during study period is marginal.

Sub-Saharan Africa, adopted a middle road with regard to reforms. This pattern becomes further marked when one considers performance improvements (table 9.12) as opposed to performance alone (table 9.11).

CONCLUSIONS

Certain aspects of power sector reform are strongly associated with utility performance in distribution efficiency and cost recovery. The panel data econometrics for the large sample confirm that the efficiency of the generation and distribution segments is significantly associated with regulatory reform, that restructuring has a positive influence on system losses in distribution, and that competition is significantly associated with higher thermal efficiency of generation plants. Further insights are gleaned from the richer set of variables available for the small sample. Certain aspects of utility governance—notably financial discipline and human resource management—are found to have a significant association with cost recovery and distribution efficiency. Cost recovery is also found to be significantly associated with the real functioning of the regulatory system—as opposed to the way the system is described on paper.

On private sector participation, the effects are particularly difficult to disentangle. Whereas the panel data analysis does not find any significant impact of private sector participation on utility performance once country fixed effects are taken into account, the cross-sectional analysis finds a strong link between private sector participation and cost recovery and, to some extent, distribution efficiency. This finding may reflect a certain circularity, with cost recovery being both a precondition for private sector participation and a means of committing government to ongoing cost recovery. When the detailed performance of individual public and private utilities in the sample is compared, it is striking that all of the private sector arrangements that have been sustained over time are in environments where full capital cost recovery has been attained—something that has not been fully attained anywhere with state-owned utilities. At the same time, the relatively good levels of distribution efficiency reached by most (but not all) privatized utilities are matched by a substantial minority of the state-owned utilities in the sample.

Overall, the link between power sector reforms and final sector outcomes is much weaker, despite some evidence that private sector participation has made a positive contribution. Some of the countries that carried out the deepest reforms (Colombia, Peru, and the Philippines) have also done well in terms of final outcomes. Several other countries implementing major reforms (the Dominican Republic, Pakistan, and Uganda) did not achieve anywhere near the same amount of progress. At the same time, a third group of countries that took a much more cautious approach to reform (Morocco, Vietnam) made remarkable gains. Nevertheless, econometric analysis supports the finding that, among the elements of power sector reform, private sector participation is the one that seems to be most strongly associated with improved sector outcomes in terms of security of supply, electrification, and renewable energy share.

With regard to security of supply, the picture is complex. Two countries (Peru and the Philippines) succeeded in achieving a broad-based security of supply based on almost exclusive private investment in generation capacity, as envisaged in the 1990s model. In other countries—despite significant reforms—challenges remain on one or more of the dimensions of security of supply, and private sector contributions to investment have been closer to 50 percent in most cases. In some cases, serious security-of-supply issues remain despite significant reform, with Pakistan one of the most striking examples. The panel data econometrics suggest that private sector participation may have been the only reform to contribute significantly to the development of normalized capacity, particularly in low-income countries, whereas the cross-sectional analysis suggests that regulation, restructuring, and competition may contribute to improved diversification.

With regard to electrification, countries progressed in ways that were unrelated to the 1990s reforms. Efforts proceeded in parallel to the 1990s reforms in almost every country, occasionally predating them but more often coming later, as access objectives surfaced on the political agenda in the 21st century. Just about every country that made progress on electrification did so through state-led efforts to plan and finance the rollout of grid technologies (and increasingly to catalyze the uptake of off-grid technologies) through competent national utilities sometimes supported by specialized rural electrification entities. Isolated attempts to apply the 1990s approach to rural electrification were largely unsuccessful. Nevertheless, the panel data econometrics suggest that private sector participation has had a positive influence on electrification, particularly in low-income countries.

With regard to decarbonization, progress has historically been an unintended by-product of measures to improve security of supply, rather than a conscious effort. Generally, countries needing to diversify away from hydropower increased their carbon footprint, whereas those needing to diversify away from oil-fired power generation decreased their footprint. Nevertheless, competition reforms do seem to have been associated with greater progress by encouraging entry of the private sector, which is more inclined to invest in cheaper forms of energy, such as natural gas or solar power. The panel data econometrics found that regulatory reforms provided a helpful impetus for the uptake of renewable energy, whereas competition and private sector participation seem to have played a hand in middle-income countries.

ANNEX 9A. ECONOMETRIC ANALYSIS OF POWER SECTOR REFORM IMPACTS BASED ON LARGE SAMPLE (88 COUNTRIES)

Specification

We assess the impact of reforms on intermediate output over time:

$$\textit{Int Outcome}_{c,t} = \alpha_0 + \alpha_1 \textit{Reform}_{c,t} + \alpha_2 X_{c,t} + \mu_c + \tau_t + \epsilon_{c,t}$$

where the dependent variable is the Intermediate Outcome (*Int Outcome*) in country *c*, year *t*. For the large sample (N=88), *Int Outcome* data include generation efficiency and system losses between 1995 and 2015. *Reform* is the main variable of interest and includes regulations, restructuring, competition, and privatization for each country starting from 1995. *X*s are country specific control variables that may have an impact on the *Int Outcome* such as gross domestic product per capita and system size. The variables μ_c, τ_t, and $\epsilon_{c,t}$ are country and year fixed effects and the error term, respectively.

We assess the impact of reforms on final outcomes over time:

$$\textit{Final Outcome}_{c,t} = \alpha_0 + \alpha_1 \textit{Reform}_{c,t} + \alpha_2 X_{c,t} + \mu_c + \tau_t + \epsilon_{c,t}$$

where the dependent variable is the *Final Outcome* in country *c*, year *t*. *Final Outcome* includes access, affordability, modern renewable energy share, carbon intensity, reliability of supply, and normalized capacity. Because of data limitation, however, the estimation results are presented for access, modern renewable energy share, carbon intensity, and normalized capacity.

ANNEX 9B. RESULTS OF ECONOMETRIC ANALYSIS OF POWER SECTOR REFORM IMPACTS BASED ON LARGE SAMPLE (88 COUNTRIES)

TABLE 9B.1 Impact of reforms on system losses (intermediate outcome), 1995–2015

Dep. var.: system losses	Complete sample (87)					Low-income countries (48)					Middle-income countries (39)				
	(1)	(2)	(3)	(4)	(5)	(1)	(2)	(3)	(4)	(5)	(1)	(2)	(3)	(4)	(5)
Regulation	0.438	0.571	-2.167*	-2.472**	-2.528**	-2.336*	-1.463	-2.652*	-2.218	-3.205**	0.995	1.130	-1.931	-2.223	-1.795
	(0.739)	(0.723)	(1.130)	(1.187)	(1.145)	(1.288)	(1.358)	(1.595)	(1.601)	(1.617)	(0.897)	(0.895)	(1.729)	(1.735)	(1.721)
Restructuring	-0.172	0.785	-2.024**	-1.781**	-1.815**	-4.319***	-3.662***	-7.362***	-6.930***	-6.614***	1.997***	2.509***	-0.177	-0.140	-0.173
	(0.643)	(0.642)	(0.944)	(0.903)	(0.924)	(1.313)	(1.305)	(2.674)	(2.382)	(2.415)	(0.745)	(0.748)	(0.919)	(0.872)	(0.924)
Competition	-2.117**	-0.991	1.853	2.541**	2.279*	4.134*	5.421**	5.977	9.049**	7.970*	-0.791	-0.851	0.981	1.094	0.756
	(0.866)	(0.824)	(1.325)	(1.268)	(1.350)	(2.179)	(2.125)	(4.311)	(3.867)	(4.220)	(0.960)	(0.935)	(0.915)	(0.892)	(0.912)
Privatization	-6.106***	-4.850***	1.699	0.911	0.925	-2.764	-0.735	10.69	13.29***	7.471	-7.283***	-6.929***	-2.885	-4.197*	-2.600
	(0.976)	(0.935)	(2.881)	(2.400)	(2.925)	(1.903)	(1.895)	(6.861)	(4.975)	(6.552)	(1.119)	(1.055)	(2.386)	(2.292)	(2.423)
Observations	1,358	1,350	1,358	1,350	1,350	635	627	635	627	627	720	720	720	720	720
R-squared	0.030	0.094	0.557	0.551	0.555	0.018	0.046	0.502	0.504	0.516	0.043	0.110	0.593	0.581	0.593
Year fixed effect	No	No	Yes	No	Yes	No	No	Yes	No	Yes	No	No	Yes	No	Yes
Country fixed effect	No	No	Yes	Yes	Yes	No	No	Yes	Yes	Yes	No	No	Yes	Yes	Yes
Controls	No	Yes	No	Yes	Yes	No	Yes	No	Yes	Yes	No	Yes	No	Yes	Yes

Note: Significance level: *** = 1 percent, ** = 5 percent, * = 10 percent, Dep. var. = dependent variable.

TABLE 9B.2 **Impact of reforms on generation efficiency (intermediate outcome), 1995–2015**

Dep. var.: generation efficiency	Complete sample (87)					Low-income countries (48)					Middle-income countries (39)				
	(1)	(2)	(3)	(4)	(5)	(1)	(2)	(3)	(4)	(5)	(1)	(2)	(3)	(4)	(5)
Regulation	1.440***	1.462***	1.367***	2.259***	1.559***	1.007	0.164	1.423**	1.846***	1.736**	2.330***	2.042***	1.890***	2.966***	1.857***
	(0.381)	(0.378)	(0.455)	(0.426)	(0.473)	(0.700)	(0.696)	(0.711)	(0.580)	(0.732)	(0.389)	(0.390)	(0.678)	(0.706)	(0.676)
Restructuring	-1.035**	-1.820***	0.336	0.343	0.249	-2.162**	-2.396**	-0.133	-0.308	-0.231	-1.583***	-2.247***	0.163	0.520	0.271
	(0.405)	(0.400)	(0.629)	(0.624)	(0.630)	(1.024)	(1.062)	(1.218)	(1.197)	(1.216)	(0.407)	(0.400)	(0.854)	(0.858)	(0.863)
Competition	1.930***	1.318**	1.208	1.572	1.180	3.870**	1.977	4.201*	4.443**	4.109*	0.147	0.487	-0.355	0.173	-0.357
	(0.576)	(0.530)	(0.986)	(1.000)	(1.013)	(1.605)	(1.742)	(2.273)	(2.217)	(2.330)	(0.542)	(0.497)	(1.055)	(1.087)	(1.030)
Privatization	2.001***	1.424**	0.0372	2.592**	0.639	-0.0222	-0.545	-2.142	-0.767	-0.787	2.393***	2.369***	1.375	5.231***	1.244
	(0.640)	(0.590)	(1.471)	(1.269)	(1.502)	(1.094)	(1.097)	(2.856)	(2.129)	(2.920)	(0.683)	(0.628)	(1.724)	(1.647)	(1.716)
Observations	1,348	1,339	1,348	1,339	1,339	622	613	622	613	613	722	722	722	722	722
R-squared	0.038	0.128	0.678	0.672	0.679	0.021	0.057	0.613	0.608	0.619	0.059	0.199	0.764	0.741	0.765
Year fixed effect	No	No	Yes	No	Yes	No	No	Yes	No	Yes	No	No	Yes	No	Yes
Country fixed effect	No	No	Yes	Yes	Yes	No	No	Yes	Yes	Yes	No	No	Yes	Yes	Yes
Controls	No	Yes	No	Yes	Yes	No	Yes	No	Yes	Yes	No	Yes	No	Yes	Yes

Note: Significance level: *** = 1 percent, ** = 5 percent, * = 10 percent, Dep. var. = dependent variable.

TABLE 9B.3 Impact of reforms on modern renewable share in TFEC (final outcome), 1995–2015

Dep. var.: renewables in TFEC	Complete sample (87)					Low-income countries (48)					Middle-income countries (39)				
	(1)	(2)	(3)	(4)	(5)	(1)	(2)	(3)	(4)	(5)	(1)	(2)	(3)	(4)	(5)
Regulation	−0.854***	−0.976***	0.461***	0.570***	0.424***	-1.106**	-1.520***	0.417***	0.618***	0.345**	−0.731*	−0.599	0.437***	0.459***	0.457***
	(0.330)	(0.341)	(0.109)	(0.101)	(0.117)	(0.543)	(0.577)	(0.151)	(0.150)	(0.154)	(0.422)	(0.426)	(0.160)	(0.143)	(0.171)
Restructuring	2.798***	2.853***	−0.289	−0.303	−0.290	-1.597**	-1.631**	0.473	0.353	0.475	5.051***	5.143***	−0.556	−0.545	−0.575
	(0.501)	(0.508)	(0.264)	(0.266)	(0.265)	(0.786)	(0.773)	(0.371)	(0.380)	(0.364)	(0.643)	(0.662)	(0.369)	(0.382)	(0.377)
Competition	1.281**	1.181**	−0.194	−0.185	−0.222	1.518	0.811	-2.012***	-1.863***	-2.140***	1.903***	1.800***	0.613*	0.539*	0.582*
	(0.508)	(0.506)	(0.312)	(0.302)	(0.316)	(1.135)	(1.078)	(0.640)	(0.664)	(0.665)	(0.545)	(0.535)	(0.336)	(0.305)	(0.321)
Privatization	0.0605	0.313	1.487***	1.972***	1.390	1.851**	2.678***	0.799	1.650	0.518	-1.287*	-1.306*	1.733***	1.899***	1.770***
	(0.484)	(0.506)	(0.557)	(0.599)	(0.553)	(0.748)	(0.855)	(1.147)	(1.163)	(1.140)	(0.668)	(0.669)	(0.421)	(0.350)	(0.433)
Observations	1760	1339	1760	1705	1705	960	922	960	922	922	780	780	780	780	780
R-squared	0.035	0.033	0.974	0.974	0.974	0.006	0.010	0.975	0.975	0.975	0.195	0.199	0.973	0.973	0.973
Year fixed effect	No	No	Yes	No	Yes	No	No	Yes	No	Yes	No	No	Yes	No	Yes
Country fixed effect	No	No	Yes	Yes	Yes	No	No	Yes	Yes	Yes	No	No	Yes	Yes	Yes
Controls	No	Yes	No	Yes	Yes	No	Yes	No	Yes	Yes	No	Yes	No	Yes	Yes

Note: TFEC = total final energy consumption.
Significance level: *** = 1 percent, ** = 5 percent, * = 10 percent, Dep. var. = dependent variable.

TABLE 9B.4 **Impact of reforms on carbon intensity (final outcome), 1995–2015**

Dep. var.: carbon intensity	Complete sample (87)					Low-income countries (48)					Middle-income countries (39)				
	(1)	(2)	(3)	(4)	(5)	(1)	(2)	(3)	(4)	(5)	(1)	(2)	(3)	(4)	(5)
Regulation	29.21*	34.90**	-2.639	-17.88*	-7.985	61.18*	55.09*	12.96	-25.95*	0.497	24.34	18.78	-10.76	-12.40	-11.17
	(17.32)	(17.46)	(11.12)	(10.14)	(11.29)	(32.00)	(32.49)	(19.89)	(15.71)	(19.69)	(18.22)	(19.09)	(11.16)	(11.13)	(11.23)
Restructuring	-126.7***	-133.0***	2.849	3.164	2.023	-145.9***	-139.8***	3.545	17.78	1.628	-147.8***	-150.6***	-5.099	-4.772	-2.675
	(18.23)	(18.45)	(10.45)	(10.37)	(10.35)	(38.64)	(40.31)	(25.35)	(24.24)	(23.77)	(22.18)	(22.36)	(10.88)	(10.46)	(10.82)
Competition	57.58**	40.71*	-25.15	-25.33	-22.34	126.6*	-3.614	69.11	36.28	69.56	-6.777	-2.586	-59.15***	-54.66***	-58.99***
	(23.84)	(23.48)	(18.74)	(18.64)	(19.07)	(65.94)	(67.76)	(51.42)	(49.99)	(51.28)	(23.96)	(24.14)	(15.79)	(16.58)	(16.77)
Privatization	-145.2***	-151.1***	44.61	-16.15	28.29	-277.9***	-285.3***	120.3*	-54.62	90.46	-89.30***	-87.72**	1.479	-9.267	-2.276
	(27.36)	(27.72)	(35.06)	(32.09)	(35.08)	(53.50)	(54.93)	(63.92)	(54.80)	(61.08)	(34.31)	(34.24)	(42.97)	(38.95)	(42.20)
Observations	1364	1355	1364	1355	1355	625	616	625	616	616	735	735	735	735	735
R-squared	0.054	0.066	0.897	0.897	0.899	0.047	0.090	0.885	0.884	0.891	0.135	0.139	0.922	0.921	0.923
Year fixed effect	No	No	Yes	No	Yes	No	No	Yes	No	Yes	No	No	Yes	No	Yes
Country fixed effect	No	No	Yes	Yes	Yes	No	No	Yes	Yes	Yes	No	No	Yes	Yes	Yes
Controls	No	Yes	No	Yes	Yes	No	Yes	No	Yes	Yes	No	Yes	No	Yes	Yes

Note: Significance level: *** = 1 percent, ** = 5 percent, * = 10 percent, Dep. var. = dependent variable.

TABLE 9B.5 Impact of reforms on normalized capacity (final outcome), 1995–2015

Dep. var.: normalized capacity	Complete sample (87)					Low-income countries (48)					Middle-income countries (39)				
	(1)	(2)	(3)	(4)	(5)	(1)	(2)	(3)	(4)	(5)	(1)	(2)	(3)	(4)	(5)
Regulation	-66.39***	-76.24***	-10.30	7.061**	-1.038	17.26	-29.66***	4.167	-9.298***	-8.028**	-93.34***	-76.39***	3.421	47.30***	41.43***
	(17.57)	(12.87)	(6.853)	(3.567)	(4.839)	(10.57)	(10.30)	(6.321)	(2.994)	(3.831)	(32.06)	(22.40)	(12.14)	(7.742)	(8.877)
Restructuring	262.1***	171.0***	10.14	-4.942	-7.099	112.9**	112.1**	7.757	18.11	18.55*	287.9***	221.7***	-20.82	-27.34**	-37.06***
	(25.91)	(21.09)	(11.43)	(8.770)	(8.965)	(50.28)	(43.97)	(20.92)	(11.13)	(11.02)	(30.63)	(21.42)	(15.73)	(11.46)	(11.51)
Competition	180.0***	53.98**	-6.192	-26.78*	-38.58***	103.9***	-7.542	-18.01	-57.58***	-59.37***	-2.383	-42.34	-8.594	-61.58***	-72.03***
	(32.83)	(23.88)	(19.55)	(13.68)	(13.87)	(40.17)	(35.48)	(45.80)	(21.32)	(21.40)	(42.96)	(26.97)	(20.74)	(17.73)	(17.40)
Privatization	43.53	-15.50	61.14**	104.5***	73.89***	128.4***	80.11***	134.0***	83.59**	87.59**	-92.78**	-100.3***	-8.357	86.70***	50.50
	(30.24)	(22.44)	(30.43)	(22.97)	(24.43)	(25.13)	(22.57)	(46.49)	(32.47)	(34.90)	(43.45)	(26.96)	(42.47)	(28.91)	(31.49)
Observations	1831	1792	1831	1792	1792	1008	969	1008	969	969	819	819	819	819	819
R-squared	0.176	0.548	0.962	0.976	0.977	0.145	0.344	0.928	0.970	0.971	0.090	0.581	0.958	0.975	0.977
Year fixed effect	No	No	Yes	No	Yes	No	No	Yes	No	Yes	No	No	Yes	No	Yes
Country fixed effect	No	No	Yes	Yes	Yes	No	No	Yes	Yes	Yes	No	No	Yes	Yes	Yes
Controls	No	Yes	No	Yes	Yes	No	Yes	No	Yes	Yes	No	Yes	No	Yes	Yes

Note: Significance level: *** = 1 percent, ** = 5 percent, * = 10 percent, Dep. var. = dependent variable.

TABLE 9B.6 **Impact of reforms on access (final outcome), 1995–2015**

Dep. var.: access	Complete sample (87)					Low-income countries (48)					Middle-income countries (39)				
	(1)	(2)	(3)	(4)	(5)	(1)	(2)	(3)	(4)	(5)	(1)	(2)	(3)	(4)	(5)
Regulation	1.491	-3.871***	-1.775***	5.268***	-1.313**	1.016	-5.077***	-2.369***	4.384***	-1.569***	-2.099	-13.67***	1.349	6.624***	1.286
	(1.430)	(1.153)	(0.529)	(0.469)	(0.518)	(1.393)	(1.183)	(0.642)	(0.521)	(0.608)	(3.231)	(2.227)	(0.897)	(1.191)	(0.851)
Restructuring	-12.32***	-11.69***	-1.698	-5.413***	-2.198**	-22.09***	-19.30***	-1.891	-5.817**	-2.614	7.376*	11.40***	-6.127***	-8.169***	-5.486***
	(2.696)	(2.167)	(1.063)	(1.413)	(1.081)	(3.097)	(2.875)	(1.714)	(2.253)	(1.772)	(3.842)	(2.868)	(1.452)	(1.420)	(1.481)
Competition	68.46***	48.58***	-2.650	6.289***	-1.924	72.90***	54.98***	-2.984	4.126	-1.846	37.51***	32.26***	1.110	11.76***	-0.187
	(2.744)	(2.308)	(1.692)	(1.909)	(1.663)	(3.464)	(3.557)	(2.509)	(3.015)	(2.499)	(4.361)	(3.270)	(2.043)	(1.950)	(2.146)
Privatization	-5.724**	-2.868	19.25***	41.94***	19.74***	6.107**	1.626	29.66***	50.58***	31.05***	-27.48***	-12.87***	0.611	20.79***	-2.384
	(2.658)	(2.718)	(3.263)	(4.383)	(3.306)	(2.668)	(2.641)	(5.152)	(6.694)	(5.377)	(3.968)	(4.262)	(3.138)	(3.291)	(3.421)
Observations	1314	1267	1314	1267	1267	924	885	924	885	885	378	378	378	378	378
R-squared	0.361	0.599	0.973	0.961	0.975	0.358	0.558	0.958	0.948	0.963	0.277	0.549	0.978	0.955	0.980
Year fixed effect	No	No	Yes	No	Yes	No	No	Yes	No	Yes	No	No	Yes	No	Yes
Country fixed effect	No	No	Yes	Yes	Yes	No	No	Yes	Yes	Yes	No	No	Yes	Yes	Yes
Controls	No	Yes	No	Yes	Yes	No	Yes	No	Yes	Yes	No	Yes	No	Yes	Yes

Note: Significance level: *** = 1 percent, ** = 5 percent, * = 10 percent, Dep. var. = dependent variable.

TABLE 9B.7 Correlation between reform steps and final outcomes

	Intermediate outcomes				Final outcomes							
					Security of supply				Social inclusion		Environmental sustainability	
Reforms	Operating cost recovery	Full capital cost recovery	System losses	Generation efficiency	Reliability	Normalized capacity	Capacity diversification	Capacity growth	Electrification	Affordability	Carbon intensity	Modern RE in TFEC
Total reform	n.a.	n.a.	−0.19	0.16	−0.33	0.31	n.a.	n.a.	0.41	−0.13	−0.15	0.13
Regulation	n.a.	n.a.	0	0.31	−0.19	−0.05	n.a.	n.a.	−0.05	0	0	−0.06
Restructuring	n.a.	n.a.	−0.20	−0.05	−0.25	0.41	n.a.	n.a.	0.46	−0.14	−0.21	0.18
Competition	n.a.	n.a.	−0.20	0.05	−0.32	0.38	n.a.	n.a.	0.58	−0.16	−0.09	0.18
Privatization	n.a.	n.a.	−0.18	0.13	−0.25	0.20	n.a.	n.a.	0.25	−0.10	−0.16	0.11

Note: Correlations with R squared higher than 0.1 are highlighted. RE = renewable energy; TFEC = total final energy consumption; n.a. = not applicable.

ANNEX 9C. CROSS-SECTIONAL REGRESSION ANALYSIS ON THE IMPACT OF POWER SECTOR REFORM BASED ON SMALL SAMPLE (17 ECONOMIES)

TABLE 9C.1 **Correlation between reform steps and outcomes**

	Intermediate outcomes				Final outcomes							
					Security of supply				Social inclusion		Environmental sustainability	
Reforms	Operating cost recovery	Full capital cost recovery	Distribution efficiency	Generation efficiency	Reliability	Normalized capacity	Capacity diversification	Capacity growth	Electrification	Affordability	Carbon intensity	Modern RE in TFEC
Restructuring index	−0.10	0.18	−0.04	0.21	0.26	0.11	−0.53	0.04	0.29	−0.31	0.24	−0.27
Regulation index	−0.12	−0.18	0.23	0.57	0.16	−0.25	−0.49	−0.11	−0.14	0.21	0.42	−0.64
Perceived regulation index	0.42	0.35	−0.29	0.32	−0.27	−0.06	−0.16	−0.33	−0.29	0.47	0.01	−0.35
Competition index	−0.17	0.14	−0.20	0.43	0.27	0.20	−0.39	0.18	0.50	−0.20	0.51	−0.36
Private sector participation index	0.47	0.70	−0.57	0.05	−0.17	0.04	−0.40	−0.03	0.12	−0.01	−0.17	−0.04
Utility governance index	0.40	0.45	−0.33	−0.23	−0.39	−0.11	−0.27	−0.21	−0.45	0.23	−0.30	−0.17
Regulatory framework												
De jure governance	−0.11	−0.24	0.30	0.51	0.23	−0.20	−0.37	−0.07	−0.13	0.15	0.45	−0.60
De jure accountability	−0.19	−0.15	0.07	−0.04	0.22	−0.18	−0.40	0.00	−0.26	0.19	0.37	−0.40
De jure autonomy	−0.04	−0.18	0.30	0.64	0.13	−0.18	−0.30	−0.11	−0.08	0.13	0.39	−0.57
De jure substance	−0.12	0.12	−0.02	0.39	0.01	−0.22	−0.47	0.06	−0.20	0.24	0.07	−0.47
De jure tariff regulation	0.19	0.32	−0.21	0.40	−0.09	−0.19	−0.44	0.01	−0.33	0.29	−0.04	−0.51
De jure quality regulation	0.17	0.11	0.01	0.39	0.10	0.22	−0.34	0.09	0.34	−0.18	0.05	−0.22
De jure market entry regulation	−0.51	−0.19	0.16	−0.02	−0.02	−0.52	−0.16	0.00	−0.46	0.44	0.12	−0.23
De facto governance	0.10	−0.04	0.06	0.35	0.17	−0.21	−0.16	−0.09	−0.34	0.27	0.48	−0.55
De facto accountability	0.07	0.01	−0.03	0.30	0.13	−0.28	−0.32	−0.14	−0.47	0.34	0.48	−0.64
De facto autonomy	0.05	−0.09	0.21	0.35	0.15	−0.13	−0.04	−0.04	−0.16	0.27	0.39	−0.43
De facto substance	0.59	0.65	−0.55	0.16	−0.53	0.10	−0.16	−0.29	−0.12	0.32	−0.43	−0.06
De facto tariff regulation	0.60	0.47	−0.37	0.25	−0.33	−0.04	−0.16	−0.16	−0.26	0.26	−0.26	−0.24
De facto quality regulation	0.56	0.66	−0.58	0.14	−0.51	0.38	−0.25	−0.32	0.29	0.05	−0.54	0.14
De facto market entry regulation	0.04	0.18	−0.19	−0.12	−0.28	−0.31	0.20	−0.10	−0.51	0.54	−0.03	−0.10
Utility governance reforms												
Utility corporate governance	0.29	0.34	−0.13	−0.20	−0.19	−0.13	−0.16	−0.10	−0.43	0.29	−0.17	−0.20
Accountability	0.29	0.36	−0.19	−0.02	−0.11	−0.30	−0.37	−0.03	−0.53	0.27	−0.09	−0.37
Autonomy	0.24	0.26	−0.04	−0.34	−0.23	0.06	0.06	−0.15	−0.27	0.28	−0.21	0.00
Utility management	0.49	0.50	−0.66	−0.19	−0.70	0.00	−0.42	−0.36	−0.27	−0.04	−0.47	−0.01
Financial discipline	0.64	0.37	−0.54	−0.32	−0.55	−0.06	0.02	−0.10	−0.38	−0.01	−0.50	0.14
Human resources	0.71	0.65	−0.40	0.07	−0.43	0.01	−0.26	−0.32	−0.09	0.24	−0.40	−0.02
Information and technology	−0.23	0.05	−0.39	−0.14	−0.48	0.05	−0.51	−0.29	−0.10	−0.25	−0.08	−0.12

Note: Correlations with R squared higher than 0.1 are highlighted. RE = renewable energy; TFEC = total final energy consumption.

TABLE 9C.2 Regression analysis: reforms and intermediate outcomes

	Operating cost recovery	Economic cost recovery	Distribution efficiency	Thermal efficiency
A				
Restructuring	-0.055	0.282	0.036	1.879
GDP	0	0	0	0.002
System size	0	0	0	0.004
R squared	0.232	0.369	0.119	0.182
B				
Private sector participation	0.450	0.77***	0.282**	-17.198
GDP	0	0	0	0.003*
System size	0	0	0	0.004
R squared	0.332	0.608	0.370	0.234
C				
Perceived regulation	0.413	0.290	0.110	15.985
GDP	0	0	0	0.002
System size	0	0	0	0.005
R squared	0.323	0.324	0.144	0.236
D				
Utility governance	0.417	0.468	0.154	-22.242
GDP	0	0	0	0.002
System size	0	0	0	0.002
R squared	0.293	0.368	0.170	0.253
E				
Competition	-0.911	1.703	0.750	56.395
GDP	0	0	0	0.001
System size	0	-0.0004**	-0.0001*	-0.001
R squared	0.261	0.395	0.248	0.230

Note: Significance level: *** = 1 percent, ** = 5 percent, * = 10 percent.

TABLE 9C.3 Regression analysis: utility governance reforms and intermediate outcomes

		Operating Cost recovery	Economic cost recovery	Distribution efficiency	Thermal efficiency
A					
Corporate governance	Autonomy	0.098	0.113	-0.016	-16.252
	GDP	0	0	0	0.002*
	System size	0	0	0	0.002
	R squared	0.241	0.293	0.113	0.319
B					
Corporate governance	Accountability	0.207	0.256	0.056	-4.155
	GDP	0	0	0	0.002
	System size	0	-0.0002*	0	0.004
	R squared	0.281	0.375	0.138	0.189
C					
Managerial practices	Financial discipline	0.995**	0.473	0.352**	-23.366
	GDP	0	0	0	0.002
	System size	0	0	0	0.002
	R squared	0.566	0.358	0.381	0.252

(*Table continued next page*)

TABLE 9C.3 Regression analysis: utility governance reforms and intermediate outcomes (*Continued*)

		Operating Cost recovery	Economic cost recovery	Distribution efficiency	Thermal efficiency
D					
Managerial practices	Human Resources	1.182**	0.857*	0.217	2.891
	GDP	0	0	0	0.002
	System size	0	0	0	0.005
	R squared	0.509	0.448	0.176	0.181
E					
Managerial practices	Information and Technology	-0.398	-0.012	0.206	-4.291
	GDP	0	0	0	0.002
	System size	-0.0002*	0	0	0.004
	R squared	0.313	0.276	0.254	0.184

Note: Significance level: *** = 1 percent, ** = 5 percent, * = 10 percent.

TABLE 9C.4 Regression analysis: regulatory reforms and intermediate outcomes

		Operating cost recovery	Economic cost recovery	Distribution efficiency	Thermal efficiency
A					
De jure regulation	Governance	-0.045	-0.261	-0.172	29.338
	GDP	0	0	0	0.002
	System size	0	0	0	0
	R squared	0.230	0.308	0.192	0.347
B					
De jure regulation	Quality regulation	0.384	0.224	0	11.832
	GDP	0	0	0	0.002
	System size	-0.0002*	-0.0002*	0	0.001
	R squared	0.288	0.309	0.111	0.216
C					
De jure regulation	Tariff regulation	0.109	0.286	0.074	16.687
	GDP	0	0	0	0.002
	System size	0	-0.0002*	0	0.003
	R squared	0.234	0.347	0.139	0.265
D					
Perceived regulation	Governance	0.265	0.095	-0.001	17.205
	GDP	0	0	0	0.002
	System size	-0.0002*	0	0	0.001
	R squared	0.288	0.282	0.098	0.276
E					
Perceived regulation	Quality regulation	0.334	0.389*	0.181*	-1.003
	GDP	0	0	0	0.002
	System size	0	0	0	0.004
	R squared	0.333	0.458	0.343	0.181
F					
Perceived regulation	Tariff regulation	0.466**	0.296	0.102	9.815
	GDP	0	0	0	0.002
	System size	0	0	0	0.005
	R squared	0.449	0.403	0.200	0.232

Note: Significance level: *** = 1 percent, ** = 5 percent, * = 10 percent.

TABLE 9C.5 **Regression analysis: reform steps and final outcomes**

	Security of supply				Social inclusion		Environmental sustainability	
	SAIFI	Normalized capacity	Capacity diversification	Capacity growth	Access	Affordability	Carbon intensity	Modern RE share in TFEC
A								
Utility restructuring	16.00	29.10	−0.4**	0.02	3.97	−0.05	37.47	−13.14
GDP per capita	0	0.04	0	0	0.01**	0	−0.01	0
System size (GWh)	0.22***	0.01	0	0	0.02	0	0.36**	0
B								
Private sector participation	−7.04	−238.93	−0.48	0	−32.73	−0.02	−5.34	−6.70
GDP per capita	0	0.05	0	0	0.01**	0	0	0
System size (GWh)	0.22***	0.01	0	0	0.02	0	0.37**	−0.01
C								
Regulation	78.83	−792.00	−0.68*	−0.01	−85.63*	0.15	499.73	−62.61**
GDP per capita	0	0.06	0	0	0.01***	0	−0.02	0
System size (GWh)	0.21***	0.09	0	0	0.028*	0	0.34***	0
D								
Perceived regulation	−58.29	−211.24	−0.14	−0.04	−66.34*	0.19	228.73	−32.52
GDP per capita	0	0.04	0	0	0.01***	0	−0.01	0
System size (GWh)	0.21***	0	0	0	0.01	0	0.41***	−0.01
E								
Utility governance	−54.671	−275.838	−0.358	−0.074	−93.31**	0.061	−159.861	−23.908
GDP per capita	0	0.040	0	0	0.009***	0	−0.004	0
System size (GWh)	0.21***	−0.013	0	0	0.007	0	0.35**	−0.009
F								
Competition	−83.68	439.86	−2.24*	0.05	82.05	−0.05	2108.25*	−125.51
GDP per capita	0	0.03	0	0	0.01	0	−0.05	0
System size (GWh)	0.23**	−0.02	0	0	0.01	0	0.16	0.01

Note: GWh = gigawatt-hour; RE = renewable energy; SAIFI = System Average Interruption Frequency Index; TFEC = total final energy consumption.
Significance level: *** = 1 percent, ** = 5 percent, * = 10 percent.

TABLE 9C.6 Regression analysis: regulatory reforms and final outcomes

		Security of supply				Social inclusion		Environmental sustainability	
		SAIFI	Normalized capacity	Capacity diversification	Capacity growth	Access	Affordability	Carbon intensity	Modern RE share in TFEC
De jure regulation	Regulatory governance	56.22	-527.73	-0.40	-0.02	-72.11*	0.12	366.08	-48.8**
	GDP	0	0.05	0	0	0.01***	0	-0.01	0
	System size	0.21***	0.09	0	0	0.03*	0	0.34**	0
De jure regulation	Quality regulation	-52.352	178.239	-0.388	-0.023	-5.547	-0.019	-221.426	-13.392
	GDP	0.003	0.029	0	0	0.009**	0	0.009	0.001
	System size	0.23***	-0.024	0	0	0.019	0	0.42**	-0.003
De jure regulation	Tariff regulation	1.033	-385.430	-0.410	0.049	-69.89**	0.093	-38.471	-34.87**
	GDP	0	0.049	0	0	0.01***	0	-0.004	0.001
	System size	0.22***	0.023	0	0	0.018	0	0.37**	-0.006
Perceived regulation	Regulatory governance	-5.82	-316.64	-0.13	-0.04	-65.64**	0.16*	245.24	-32.90**
	GDP	0	0.04	0	0	0.008***	0	-0.01	0
	System size	0.22***	0.07	0	0	0.03**	-0.00*	0.35**	0
Perceived regulation	Quality regulation	-100.913	413.576	-0.165	-0.016	15.531	-0.032	-222.007	4.255
	GDP	0.007	0.012	0	0	0.008*	0	0.010	-0.001
	System size	0.19***	0.148	0	0	0.023	0	0.30**	-0.005
Perceived regulation	Tariff regulation	-60.233	-76.345	-0.099	0.004	-30.628	0.048	-105.131	-15.044
	GDP	-0.001	0.041	0	0	0.009**	0	-0.004	0
	System size	0.21***	0.008	0	0	0.013	0	0.36**	-0.008

Note: RE = renewable energy; SAIFI = System Average Interruption Frequency Index; TFEC = total final energy consumption.
Significance level: *** = 1 percent, ** = 5 percent, * = 10 percent.

TABLE 9C.7 **Regression analysis: utility governance reforms and final outcomes**

		Security of supply				Social inclusion		Environmental sustainability	
		SAIFI	Normalized capacity	Capacity diversification	Capacity growth	Access	Affordability	Carbon intensity	Modern RE share in TFEC
Corporate governance	Autonomy	5.656	46.427	0.070	−0.083**	−31.263	0.046	−39.424	−2.797
	GDP	0.000	0.040	0.000	0.000	0.009**	0.000	−0.005	0.000
	System size	0.22***	0.026	0	0	0.013	0	0.36**	−0.007
Corporate governance	Accountability	19.533	−364.391	−0.266	−0.025	−63.67***	0.062	−70.035	−19.347
	GDP	0	0.043	0	0	0.009***	0	−0.003	0
	System size	0.22***	0.008	0	0	0.016	0	0.37**	−0.007
Managerial practices	Financial discipline	−231.601	−49.851	0.034	−0.011	−57.496	−0.049	−634.023	8.201
	GDP	−0.003	0.040	0	0	0.008**	0	−0.010	0
	System size	0.18**	0.013	0	0	0.011	0	0.30**	−0.005
Managerial practices	Human Resources	−23.392	−224.647	−0.457	−0.065	−41.355	0.033	126.068	−23.719
	GDP	0	0.043	0	0	0.009**	0	−0.007	0
	System size	0.22**	−0.026	0	0	0.009	0	0.39**	−0.011
Managerial practices	Information and Technology	−168.770	175.934	−0.63**	0.095	7.292	−0.111	67.803	−14.652
	GDP	−0.003	0.043	0	0	0.009**	0	−0.004	0
	System size	0.19***	0.036	0	0.00*	0.018	0	0.38**	−0.008

Note: RE = renewable energy; SAIFI = System Average Interruption Frequency Index; TFEC = total final energy consumption.
Significance level: *** = 1 percent, ** = 5 percent, * = 10 percent.

NOTES

1. This chapter is based on original research by Vivien Foster, Joern Huenteler, Anshul Rana, and Tu Chi Nguyen.
2. The methodological approach draws from recent statistical literature underpinning machine learning (Dean 2014; Friedman, Hastie, and Tibshirani 2000; Witten and others 2017). It proceeds by examining all possible bilateral correlations between variables that arise from the theoretical framework described in figure 9.1, reporting those for which the coefficient of determination (R-squared) is above 0.06 (equivalent to a correlation coefficient of over 0.25) with the expected sign.
3. In particular, where no association of any kind is found between reform and outcome variables, it is unlikely that this association would increase as a result of the adoption of more sophisticated statistical methods. Conversely, where a strong association is found, it is possible that the size of this effect could weaken if more sophisticated statistical tools were employed. In that sense, negative results could be regarded as more powerful than positive ones.
4. Diversification index is calculated such that results closer to 1 show increasing dependence on a single fuel whereas results closer to 0 show more diversification.

REFERENCES

Banerjee, S. G., F. A. Moreno, J. E. Sinton, T. Primiani, and J. Seong. 2017. "Regulatory Indicators for Sustainable Energy: A Global Scorecard for Policy Makers." World Bank, Washington, DC.

Dean, J. 2014. *Big Data, Data Mining, and Machine Learning: Value Creation for Business Leaders and Practitioners.* Hoboken, NJ: John Wiley & Sons.

Friedman, J., T. Hastie, and R. Tibshirani. 2000. "Additive Logistic Regression: A Statistical View of Boosting." *The Annals of Statistics* 28 (2): 337–407.

Witten, I., E. Frank, M. Hall, and C. Pal. 2017. *Data Mining: Practical Machine Learning Tools and Techniques* (Fourth Edition). Cambridge, MA: Morgan Kaufmann.

ECO-AUDIT

Environmental Benefits Statement

The World Bank Group is committed to reducing its environmental footprint. In support of this commitment, we leverage electronic publishing options and print-on-demand technology, which is located in regional hubs worldwide. Together, these initiatives enable print runs to be lowered and shipping distances decreased, resulting in reduced paper consumption, chemical use, greenhouse gas emissions, and waste.

We follow the recommended standards for paper use set by the Green Press Initiative. The majority of our books are printed on Forest Stewardship Council (FSC)–certified paper, with nearly all containing 50–100 percent recycled content. The recycled fiber in our book paper is either unbleached or bleached using totally chlorine-free (TCF), processed chlorine–free (PCF), or enhanced elemental chlorine–free (EECF) processes.

More information about the Bank's environmental philosophy can be found at http://www.worldbank.org/corporateresponsibility.